FINITE ELASTICITY AND VISCOELASTICITY

A Course in the Nonlinear Mechanics of Solids

FINITE ELASTICITY AND VISCOELASTICITY

A Course in the Nonlinear Mechanics of Solids

Aleksey D. Drozdov

Institute for Industrial Mathematics
Ben-Gurion University of the Negev
Israel

World Scientific
Singapore • New Jersey • London • Hong Kong

Published by

World Scientific Publishing Co. Pte. Ltd.

P O Box 128, Farrer Road, Singapore 912805

USA office: Suite 1B, 1060 Main Street, River Edge, NJ 07661

UK office: 57 Shelton Street, Covent Garden, London WC2H 9HE

Library of Congress Cataloging-in-Publication Data
Drozdov, Aleksey D.
    Finite elasticity and viscoelasticity : a course in the nonlinear
mechanics of solids / Aleksey D. Drozdov.
        p.   cm.
    Includes bibliographical references and index.
    ISBN 9810224338
    1. Elasticity.   2. Viscoelasticity.   I. Title.
QA931.D76   1996
531'.382--dc20   95-42801
                                                                      CIP

British Library Cataloguing-in-Publication Data
A catalogue record for this book is available from the British Library.

This book is printed on acid-free paper.

Printed in Singapore by Uto-Print.

To my wife, Lena,
and to my children, Marina and Dmitry

To my wife, Lena,
and to my children, Marina and Dmitry

Preface

The textbook presents extended lecture notes on the course "Mathematical Models in Finite Elasticity and Viscoelasticity" which was taught in 1988–1990 at the Mechanical Engineering Department of the Moscow Automechanical Institute (Moscow, USSR) and in 1993–1995 at the Department of Mathematics and Computer Science of the Ben–Gurion University (Be'ersheva, Israel). It corresponds to the syllabuses of lecture courses on the mechanics of continua for graduate students in Applied Mathematics and Mechanical Engineering.

The book is intended to be self-contained. The exposition is presented in such a way that students are not assumed to have any preliminary knowledge in the mechanics of continua and theoretical physics, e.g. in linear elasticity and viscoelasticity. All definitions and theoretical constructions are given independently of other sources. Nevertheless, some basic knowledge in mathematics is supposed, e.g. at the level of the first two semesters of university studies. It consists of elementary theorems in linear algebra, basic assertions in calculus for functions of several variables, and an elementary introduction to the optimization theory.

The mechanical behavior of solids is described by various models in continuum mechanics. In this book, we concentrate on two basic models: elasticity and viscoelasticity. Plasticity and failure will be the subject of the second volume of the textbook.

Elasticity is applicable whenever an instantaneous response of material is treated as an adequate approximation. Finite elasticity is a part of the elasticity theory which does not impose any *a priori* limitations on displacements and strains. Typical examples of elastic media with large deformations are natural and synthetic rubber, rubber-like polymers and elastomers, biological tissues, as well as metals and alloys under high pressure.

Viscoelasticity may be defined as a theory which takes into account the effect of the entire deformation history on stresses at current instant. Finite viscoelasticity concentrates on phenomena which are typical for the nonlinear behavior at large deformations (up to several hundred per cent). The results derived in finite viscoelasticity are applied to rubber-like polymers and plastics, metals at elevated temperatures, soils, road construction materials, biological tissues, foodstuffs, as well as to polymeric melts and solutions. Viscoelastic media demonstrate two different kinds of behaviors: liquid-like and solid-like. The present book is confined to the study of solid-like media.

The textbook deals with the equilibrium and motion of elastic and viscoelastic media with finite deformations. Our objective is to derive adequate mathematical techniques for the analysis of motions, as well as to solve several applied problems where important physical phenomena are demonstrated.

In the past decades a significant development has occurred in the nonlinear mechanics of solids. It has led to the creation of a specific language based on the presentation of constitutive and governing equations in the tensor form. On one hand, this language simplifies the analysis essentially. On the other hand, it caused a split between engineering mechanics (laying special stress on the structural mechanics with infinitesimal strains) and mathematical theory (with a bias to the analysis of the existence and uniqueness of solutions).

The textbook is suggested as an attempt to fill this gap, namely, to teach a relatively new language of finite strains to physicists, engineers, and other scientists who investigate and apply a wide variety of methods of continuum mechanics. The readership determines requirements to the book. It should be as brief as possible, self-contained, and completely independent of other bibliographical sources. The book should contain a number of exercises for self-education and for testing comprehension. Finally, it should be interesting for graduate students specializing in applied mathematics and mechanical engineering, as well as for physicists and engineers having experience in the linear mechanics of continua and its applied versions, e.g. strength of materials. To fulfill the latter condition, I provide a number of notes which, sometimes, are optional and may be omitted, but which allow physicists and engineers to visualize some known questions from a new point of view.

In the past two decades a number of high-quality textbooks have appeared which provide an introduction to continuum mechanics and to finite elasticity as its part. I note only a few of them: Atkin & Fox (1980), Ciarlet (1988), Eringen (1967), Green & Adkins (1970), Gurtin (1981), Lai, Rubin & Krempl (1993), Lurie (1990), Ogden (1984), Spencer (1980), and Truesdell (1977). As common practice, viscoelasticity is treated as an independent subject, and only some preliminary information regarding the linear viscoelastic behavior is included in textbooks on the mechanics of continua. Nevertheless, I should mention here several brilliant textbooks concerned with the behavior of viscoelastic media: Aklonis et al. (1972), Christensen (1982), Ferry (1980), Findley et al. (1989), Pipkin (1972), Skrzypek (1993), and Tschoegl (1989), as well as some books on the mathematical theory of viscoelasticity: Coleman et al. (1966), Fabrizio & Morro (1992), and Renardy et al. (1987).

The present textbook is the first attempt to provide an introduction to finite elasticity and viscoelasticity of solids in the course of continuum mechanics. The aim of the book is to give a self-contained exposition of the main models used in the nonlinear elasticity and viscoelasticity, as well as of the basic boundary problems where similarities and differences are demonstrated between phenomena which occur at infinitesimal and finite strains. For this purpose, I employ consistently the so-called direct tensor notation, and compare (where possible) theoretical results with experimental data.

Chapter 1 is concerned with tensor calculus. The exposition of the tensor algebra and analysis employs the direct tensor notation. In addition to the "classical" topics, corotational derivatives are studied for objective tensors, and basic assertions are formulated and proved in the theory of tensor functions. The purpose of this chapter is to provide the entire set of definitions and statements which are used in the textbook. I hope that this chapter will be helpful for appropriate references even for the reader acquainted with tensor analysis.

Chapter 2 deals with basic concepts in the kinematics of continua, with equations of motion and equilibrium for deformable media, as well as with the main axioms of the constitutive theory.

In Chapter 3, basic constitutive models are introduced for the Cauchy and Green elasticity. I discuss the main properties of strain energy densities and present several models in hyperelasticity both for compressible and incompressible media. The main attention is focused on the correspondence between axioms of the constitutive theory, concrete examples of elastic potentials and their agreement with experimental data. Finally, the governing equations in finite elasticity are developed and the existence and uniqueness of their solutions are discussed.

Chapter 4 is concerned with several "classical" boundary-value problems in finite elasticity. Its aim is to demonstrate mathematical techniques used to solve nonlinear boundary problems, and to discuss physical phenomena accompanying large deformations. I consider the so-called universal solutions in finite elastostatics, and concentrate on the correspondence between experimental data and their theoretical prediction. By discussing simple shear of an elastic medium with large deformation, the Weissenberg effect is derived. For torsion of a circular cylinder made of an incompressible elastic material, I derive the Rivlin solution and compare it with experimental data to demonstrate the Poynting effect. The Saint-Venant principle is proved which demonstrates that the effect of surface loads has a local character. By studying one-dimensional motions of an elastic medium and using the Riemann invariants, a blow-up result is demonstrated in elastodynamics. Finally, the modified Korteweg–de Vries equation is derived for transversal oscillations of a nonlinearly elastic rod, particular solutions of this equation are constructed in the form of solitary waves, and some physical properties of soliton solutions are derived.

In Chapter 5, the Lagrange principle in finite elastostatics is formulated and its connections with the stability conditions are discussed. Governing equations describing the first order phase transitions in elastic media are derived. A specific condition on the interface is developed by employing the generalized Lagrange variational principle.

Chapter 6 deals with basic physical phenomena in viscoelasticity and rheological models for their description. Both differential and integral constitutive relations are studied. First, I introduce constitutive models at infinitesimal strains. Afterward, these models are extended to large deformations. A special section is devoted to the discussion of creep and relaxation kernels used in applications for aging and non-aging viscoelastic media.

Chapter 7 deals with the study of several boundary problems in finite viscoelasticity. In the Rayleigh problem, the influence of singularities in relaxation kernels on wave propagation is demonstrated. It is shown that the material behavior changes from "elastic" (hyperbolic) to "viscous" (parabolic) with the growth of the strength of singularity. By analyzing one-dimensional motions, I show that under some conditions acceleration waves with small amplitudes decay exponentially, while waves with large amplitudes grow. Nonlinear radial oscillations are studied for a thick-walled spherical pressure vessel. In the classical problem of torsion for a circular cylinder, theoretical results are compared with experimental data, and the separability hypothesis is validated for elastic and viscous effects. For recovery of a viscoelastic medium that underwent a homogeneous deformation an interesting phenomenon is demonstrated: during stress-free recoil the medium may go through such states in which it has never previously been at loading. Finally, a model for the continuous accretion is developed. This model is employed to study the winding process for a viscoelastic cylindrical pressure vessel and to analyze the effect of mass influx on stresses and displacements in a growing solid.

Exercises are one of the most important parts of the textbook. They are divided into two groups. The first consists of calculations similar to the calculations carried out in the main text. These problems are of different levels of complexity, but, as a rule, they do not require special "tricks", and their results are precisely formulated. The reader is recommended to carry out all the calculations, (i) to get accustomed to the language used, and (ii) to test his/her understanding of the material. The other group consist of more sophisticated problems. To solve them, some remarks and additional references are provided.

Displayed equations are numerated consecutively by two figures separated by a period, the first indicating the section and the other designating the ordinal number of the equation in the section. Within a chapter, an equation is referred to by this pair of numbers. To refer to an equation in another chapter, I put the chapter number before the pair. The same rule is used to numerate exercises and propositions.

Calculations carried out throughout the textbook are purely formal: the appearing quantities are assumed to have sufficient smoothness to justify any operations required, e.g. differentiation or integration. In several places the imposed requirements may be essentially weakened. I do not dwell on these details remembering Birkhoff's suggestion that the development of applied mathematics would be significantly slower if the strong mathematics were not supplemented by such natural and probable hypotheses.

I am grateful to the students who have suffered my attempts to present the material of this textbook at various stages of its development. Many of my colleagues have given considerable assistance by reading and commenting on the draft manuscript. I wish to thank Prof. V. Plotkin and Dr. Y. Shtemler, whose criticisms and suggestions have led to significant improvements. Finally, I am especially grateful to my wife, without whose encouragement this book would not have been written.

Aleksey Drozdov

Contents

FINITE ELASTICITY AND VISCOELASTICITY

A Course in the Nonlinear Mechanics of Solids

Chapter 1

Tensor calculus

In this Chapter we introduce basic concepts and establish the main assertions in the tensor theory which are employed in the mechanics of continua. Our exposition of the tensor algebra and analysis is based on the so-called direct tensor notation which allows arbitrary curvilinear coordinate frames to be used. Section 1 is concerned with Eulerian and Lagrangian coordinate frames. In Section 2, basic concepts of the tensor algebra are provided. Section 3 deals with the tensor calculus. We introduce the nabla-operator and covariant derivatives of tensors and prove the Stokes theorem. In Section 4, we discuss corotational derivatives, i.e. time derivatives of objective tensors which are indifferent with respect to superimposed rigid motions. Finally, Section 5 is concerned with scalar and tensor functions of tensors.

1. Geometry of Motion

This Section is concerned with the geometrical description of motion in the mechanics of continua. We introduce Eulerian and Lagrangian coordinates and establish a correspondence between setting a motion with respect to these coordinate frames. We derive expressions for tangent vectors of the main and dual bases of Lagrangian coordinate frames, develop formulas for the volume element, and establish several properties of covariant and contravariant objects.

1.1. Description of Motion

The main problem in the mechanics of continua is to describe motion of a medium under the action of external loads which are assumed to be given. Our objective is to find what kind of motion the medium performs (i.e. to derive a law of motion and distributions of velocities and accelerations at different points), as well as to determine the stress distribution. To answer these questions various coordinate frames are introduced, where the motion is described.

1

Two different kinds of coordinate frames are employed. The first is *Eulerian (spatial) coordinate frame*, which is fixed and immobile in space. During the motion, points of the medium change their positions in space with respect to an Eulerian frame. To set a motion with respect to an Eulerian frame means to set all the kinematic characteristics of the motion (i.e. velocities and accelerations) at any point.

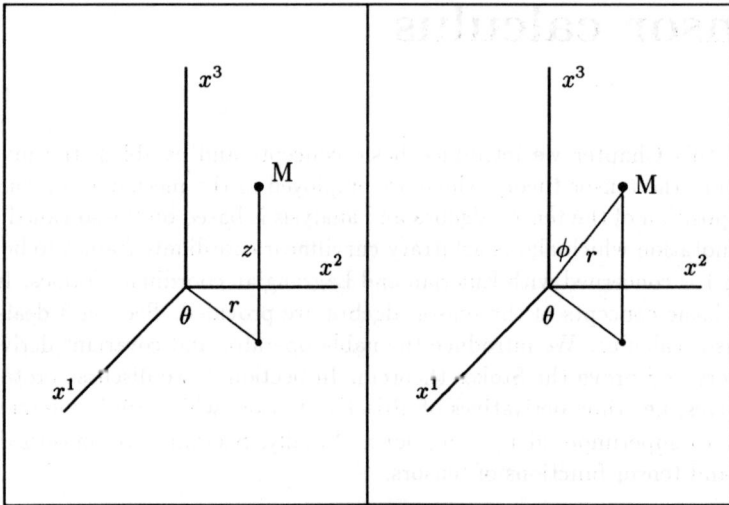

Fig. 1.1. Cylindrical and spherical coordinate frames.

As common practice, *Cartesian coordinates* $\{x^1, x^2, x^3\}$, *cylindrical coordinates* $\{r, \theta, z\}$, and *spherical coordinates* $\{r, \theta, \phi\}$ are employed as Eulerian coordinates. We will denote *the unit vectors* of Cartesian coordinates as $\bar{e}_1$, $\bar{e}_2$, and $\bar{e}_3$, the unit vectors of cylindrical coordinates as $\bar{e}_r$, $\bar{e}_\theta$, and $\bar{e}_z$, and the unit vectors of spherical coordinates as $\bar{e}_r$, $\bar{e}_\theta$ and $\bar{e}_\phi$.

Exercise 1.1. Establish connections between Cartesian, cylindrical and spherical coordinates. In particular, show that for cylindrical coordinates

$$r = \sqrt{(x^1)^2 + (x^2)^2}, \qquad \theta = \tan^{-1}\frac{x^2}{x^1}, \qquad z = x^3,$$

and for spherical coordinates

$$r = \sqrt{(x^1)^2 + (x^2)^2 + (x^3)^2}, \qquad \theta = \tan^{-1}\frac{\sqrt{(x^1)^2 + (x^2)^2}}{x^3},$$

$$\phi = \tan^{-1} \frac{x^2}{x^1}. \qquad \square$$

Exercise 1.2. Derive the following formulas for the derivatives of the unit vectors of cylindrical coordinate frames:

$$\frac{\partial \bar{e}_r}{\partial r} = 0, \qquad \frac{\partial \bar{e}_\phi}{\partial r} = 0, \qquad \frac{\partial \bar{e}_z}{\partial r} = 0,$$

$$\frac{\partial \bar{e}_r}{\partial \phi} = \bar{e}_\phi, \qquad \frac{\partial \bar{e}_\phi}{\partial \phi} = -\bar{e}_r, \qquad \frac{\partial \bar{e}_z}{\partial \phi} = 0,$$

$$\frac{\partial \bar{e}_r}{\partial z} = 0, \qquad \frac{\partial \bar{e}_\phi}{\partial z} = 0, \qquad \frac{\partial \bar{e}_z}{\partial z} = 0. \qquad \square \qquad (1.1)$$

Exercise 1.3. Derive the following formulas for the derivatives of the unit vectors of spherical coordinate frames:

$$\frac{\partial \bar{e}_r}{\partial r} = 0, \qquad \frac{\partial \bar{e}_\theta}{\partial r} = 0, \qquad \frac{\partial \bar{e}_\phi}{\partial r} = 0,$$

$$\frac{\partial \bar{e}_r}{\partial \theta} = \bar{e}_\theta, \qquad \frac{\partial \bar{e}_\theta}{\partial \theta} = -\bar{e}_r, \qquad \frac{\partial \bar{e}_\phi}{\partial \theta} = 0,$$

$$\frac{\partial \bar{e}_r}{\partial \phi} = \bar{e}_\phi \sin \theta, \qquad \frac{\partial \bar{e}_\theta}{\partial \phi} = \bar{e}_\phi \cos \theta, \qquad \frac{\partial \bar{e}_\phi}{\partial \phi} = -(\bar{e}_r \sin \theta + \bar{e}_\theta \cos \theta). \qquad \square \quad (1.2)$$

The other kind of coordinate frames are *Lagrangian (material) coordinates*. These coordinates are assumed to be frozen into a moving medium and to deform together with it. The position of points with respect to Lagrangian coordinates remains unchanged in time, while the frame is displaced together with the medium. As a rule, Lagrangian coordinates are assumed to coincide with Eulerian coordinates at the initial instant, when the motion starts. Lagrangian coordinates are denoted as $\xi = \{\xi^i\}$ $(i = 1, 2, 3)$.

The position of any point with respect to an immobile spatial coordinate frame is determined by its *radius-vector* $\bar{r}$ which changes in time. Two radius-vectors should be distinguished: the initial $\bar{r}_0(\xi)$ and the current $\bar{r}(t, \xi)$, where t stands for time. To set a motion with respect to a Lagrangian frame means to formulate the law

$$\bar{r} = \bar{r}(t, \xi) \qquad (1.3)$$

for any point ξ and for any instant t.

The main characteristics of the motion are *velocity*

$$\bar{v} = \frac{\partial \bar{r}}{\partial t} \qquad (1.4)$$

and *acceleration*

$$\bar{a} = \frac{\partial \bar{v}}{\partial t} = \frac{\partial^2 \bar{r}}{\partial t^2}. \qquad (1.5)$$

3

Proposition 1.1. Setting a motion with respect to Lagrangian coordinates leads to setting this motion with respect to Eulerian coordinates and vice versa.
Proof. First, we assume that a motion is set with respect to Lagrangian coordinates, i.e. that the function $\bar{r}(t, \xi)$ is given. Then Eqs. (1.4) and (1.5) determine vectors $\bar{v}$ and $\bar{a}$ as functions of time and Lagrangian coordinates ξ. By solving Eq. (1.3) with respect to ξ, we find ξ as a function of time and Eulerian coordinates (components of the radius-vector in the immobile coordinate frame). Substitution of this expression into Eqs. (1.4) and (1.5) yields vectors $\bar{v}$ and $\bar{a}$ as functions of time and the Eulerian coordinates, which is equivalent to setting the motion with respect to the Eulerian coordinates.

Let us now assume a motion to be set with respect to Eulerian coordinates. This means that the velocity vector $\bar{v}$ is prescribed as a function of time and spatial coordinates: $\bar{v} = \bar{v}(t, \bar{r})$. Substitution of this expression into Eq. (1.4) implies the following ordinary differential equation:

$$\frac{\partial \bar{r}}{\partial t} = \bar{v}(t, \bar{r}).$$

By solving this equation, the radius-vector $\bar{r}$ may be found as a function of time and its initial position $\bar{r}_0$. Taking components of the radius-vector $\bar{r}_0$ at the initial instant as Lagrangian coordinates ξ^i, we obtain $\bar{r}$ as a function of t and ξ, i.e. we set the motion with respect to the Lagrangian coordinates. □

1.2. Tangent Vectors

Let us consider a domain Ω with coordinates $\xi = \{\xi^i\}$. Denote by $\bar{r} = \bar{r}(\xi)$ the radius-vector of a point with coordinates ξ.

Let us fix the coordinates ξ^2 and ξ^3, and consider a line which is drawn by the radius-vector, when only the coordinate ξ^1 changes. This line is called coordinate line ξ^1. Similarly, we define coordinate lines ξ^2 and ξ^3.
Exercise 1.4. Show that the vectors

$$\bar{g}_i = \frac{\partial \bar{r}}{\partial \xi^i} \tag{1.6}$$

are *tangent* to coordinate lines ξ^i. □
Exercise 1.5. Prove that the vectors $\bar{g}_i$ are linearly independent and form a *basis*. □
Exercise 1.6. Derive the following formulas for the vectors $\bar{g}_i$ in Cartesian, cylindrical and spherical coordinates:

$$\bar{g}_1 = \bar{e}_1, \quad \bar{g}_2 = \bar{e}_2, \quad \bar{g}_3 = \bar{e}_3,$$
$$\bar{g}_1 = \bar{e}_r, \quad \bar{g}_2 = r\bar{e}_\theta, \quad \bar{g}_3 = \bar{e}_z,$$
$$\bar{g}_1 = \bar{e}_r, \quad \bar{g}_2 = r\bar{e}_\theta, \quad \bar{g}_3 = r\sin\theta\,\bar{e}_\phi. \quad □ \tag{1.7}$$

Any vector $\bar{q}$ can be expanded in tangent vectors $\bar{g}_i$

$$\bar{q} = q^i \bar{g}_i = q^1 \bar{g}_1 + q^2 \bar{g}_2 + q^3 \bar{g}_3, \tag{1.8}$$

where q^i are called *contravariant components* of vector $\bar{q}$. Here and below the summation is assumed with respect to repeating indices. The sign of the sum is omitted, but repeated indices are supposed to occupy alternately the upper and the lower positions.

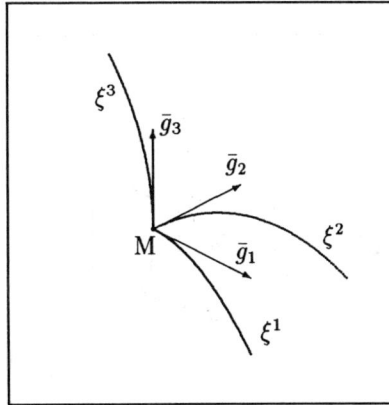

Fig. 1.2. Tangent vectors for a Lagrangian coordinate frame.

Let us calculate the differential of the radius-vector

$$d\bar{r} = \frac{\partial \bar{r}}{\partial \xi^i} d\xi^i = \bar{g}_i d\xi^i. \tag{1.9}$$

Multiplying Eq. (1.9) by itself we find the square of *the arc element*

$$ds^2 = d\bar{r} \cdot d\bar{r} = \bar{g}_i d\xi^i \cdot \bar{g}_j d\xi^j = (\bar{g}_i \cdot \bar{g}_j) d\xi^i d\xi^j = g_{ij} d\xi^i d\xi^j. \tag{1.10}$$

Here the dot stands for *the inner product* of vectors, and the quantities

$$g_{ij} = \bar{g}_i \cdot \bar{g}_j \tag{1.11}$$

are called *covariant components of the metric tensor* (a definition of tensor will be provided in Section 1.2).

Exercise 1.7. Check that the metric matrix $[g_{ij}]$ equals the unit matrix for Cartesian coordinate frames. □

Exercise 1.8. Calculate the metric matrix $[g_{ij}]$ for cylindrical and spherical coordinate frames. □

Let us construct a parallelepiped on tangent vectors $\bar{g}_i$. As is well known, its volume V is calculated as

$$V = \bar{g}_1 \cdot (\bar{g}_2 \times \bar{g}_3) = \bar{g}_2 \cdot (\bar{g}_3 \times \bar{g}_1) = \bar{g}_3 \cdot (\bar{g}_1 \times \bar{g}_2), \qquad (1.12)$$

where $\times$ stands for *the vector product*.

Exercise 1.9. Derive formulas (1.12). □

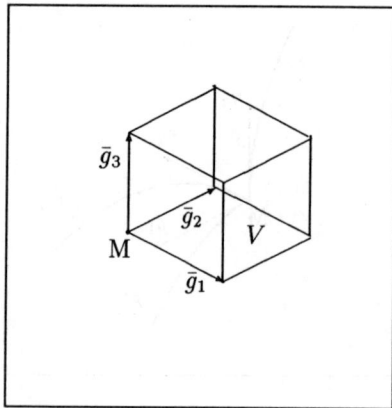

Fig. 1.3. The elementary volume.

We now introduce vectors $\bar{g}^i$ which are orthogonal to tangent vectors $\bar{g}_i$. The vectors $\bar{g}^i$ are called *dual vectors*. For any integers i and j we have

$$\bar{g}^i \cdot \bar{g}_j = \delta^i_j, \qquad (1.13)$$

where δ^i_j are *the Kronecker indices*:

$$\delta^i_j = \left\{ \begin{array}{ll} 1 & i = j, \\ 0 & i \neq j. \end{array} \right.$$

Exercise 1.10. Check that Eq. (1.13) determines vectors $\bar{g}^i$ uniquely. □

Proposition 1.2. Vectors $\bar{g}^i$ can be calculated as

$$\bar{g}^1 = \frac{\bar{g}_2 \times \bar{g}_3}{V}, \qquad \bar{g}^2 = \frac{\bar{g}_3 \times \bar{g}_1}{V}, \qquad \bar{g}^3 = \frac{\bar{g}_1 \times \bar{g}_2}{V}. \qquad (1.14)$$

6

Proof. Let us calculate the product $\bar{g}^1 \cdot \bar{g}_1$. Substitution of the first expression (1.14) into this formula implies that $\bar{g}^1 \cdot \bar{g}_1 = \bar{g}_1 \cdot (\bar{g}_2 \times \bar{g}_3)/V$. The latter expression equals 1 according to Eq. (1.12).

Let us now calculate $\bar{g}^1 \cdot \bar{g}_2$. By using a similar procedure, we find $\bar{g}^1 \cdot \bar{g}_2 = \bar{g}_2 \cdot (\bar{g}_2 \times \bar{g}_3)/V$. This expression vanishes since the vector product $\bar{g}_2 \times \bar{g}_3$ is orthogonal to $\bar{g}_2$. $\square$

Exercise 1.11. Check the other equalities (1.14). $\square$

Exercise 1.12. Derive the following formulas for dual vectors $\bar{g}^i$ for Cartesian, cylindrical and spherical coordinates:

$$\bar{g}^1 = \bar{e}_1, \qquad \bar{g}^2 = \bar{e}_2, \qquad \bar{g}^3 = \bar{e}_3,$$

$$\bar{g}^1 = \bar{e}_r, \qquad \bar{g}^2 = \frac{1}{r}\bar{e}_\theta, \qquad \bar{g}^3 = \bar{e}_z,$$

$$\bar{g}^1 = \bar{e}_r, \qquad \bar{g}^2 = \frac{1}{r}\bar{e}_\theta, \qquad \bar{g}^3 = \frac{1}{r\sin\theta}\bar{e}_\phi. \quad \square \tag{1.15}$$

Proposition 1.3. The following inverse formulas hold:

$$\bar{g}_1 = V\bar{g}^2 \times \bar{g}^3, \qquad \bar{g}_2 = V\bar{g}^3 \times \bar{g}^1, \qquad \bar{g}_3 = V\bar{g}^1 \times \bar{g}^2. \quad \square \tag{1.16}$$

Exercise 1.13. Derive Eqs. (1.16) by using the well-known formula

$$\bar{u} \times \bar{v} \times \bar{w} = \bar{v}(\bar{u} \cdot \bar{w}) - \bar{w}(\bar{u} \cdot \bar{v}). \quad \square$$

Let us now discuss another approach to constructing dual vectors. Denote by g^{ij} elements of the matrix inverse to the metric matrix $[g_{ij}]$. These elements are called contravariant components of the metric tensor. For any integers i and j

$$g^{ik}g_{kj} = \delta^i_j. \tag{1.17}$$

Exercise 1.14. Calculate contravariant components of the metric tensor for Cartesian, cylindrical, and spherical coordinates. $\square$

We set

$$\bar{g}^i = g^{ij}\bar{g}_j \tag{1.18}$$

and check that vectors (1.18) satisfy Eqs. (1.13). Indeed, $\bar{g}^i \cdot \bar{g}_j = g^{ik}\bar{g}_k \cdot \bar{g}_j$. According to Eqs. (1.11) and (1.17), this equality implies that

$$\bar{g}^i \cdot \bar{g}_j = g^{ik}g_{kj} = \delta^i_j. \quad \square$$

Proposition 1.4. The following formula is valid:

$$g^{ij} = \bar{g}^i \cdot \bar{g}^j. \tag{1.19}$$

Proof. Let us multiply vectors $\bar{g}^i$ and $\bar{g}^j$. By using Eqs. (1.11), (1.17), and (1.18), we obtain

$$\bar{g}^i \cdot \bar{g}^j = g^{ik}\bar{g}_k \cdot g^{jl}\bar{g}_l = g^{ik}g^{jl}(\bar{g}_l \cdot \bar{g}_k) = g^{ik}g^{jl}g_{lk} = g^{ik}\delta^j_k = g^{ij}. \quad \square$$

Exercise 1.15. Prove that

$$\bar{g}_i = g_{ij}\bar{g}^j. \qquad \square \tag{1.20}$$

It follows from Eqs. (1.18) and (1.20) that covariant and contravariant components of the metric tensor allow the indices of tangent vectors to be raised and lowered.

Similar to Eq. (1.8), an arbitrary vector $\bar{q}$ may be presented in the form

$$\bar{q} = q_i\bar{g}^i, \tag{1.21}$$

where q_i are *covariant components* of the vector $\bar{q}$.

Let us multiply Eq. (1.8) by $\bar{g}^i$ and use Eq. (1.13). As a result, we obtain

$$\bar{q} \cdot \bar{g}^i = q^j \bar{g}_j \cdot \bar{g}^i = q^j \delta_j^i = q^i. \tag{1.22}$$

Exercise 1.16. By using Eqs. (1.13) and (1.21), check that

$$q_i = \bar{q} \cdot \bar{g}_i. \qquad \square \tag{1.23}$$

By employing Eqs. (1.8), (1.13), and (1.21) we find that for any vectors $\bar{p}$ and $\bar{q}$

$$\bar{p} \cdot \bar{q} = p^i \bar{g}_i \cdot q_j \bar{g}^j = p^i q_j \delta_{i\cdot}^{\cdot j} = p^i q_i. \tag{1.24}$$

In particular, for $\bar{p} = \bar{q}$, Eq. (1.24) implies that

$$|\,\bar{q}\,|^2 = q^i q_i. \tag{1.25}$$

1.3. Transformation of Coordinate Frames

Let us consider two curvilinear coordinate frames: "old" $\xi = \{\xi^i\}$ and "new" $\eta = \{\eta^i\}$, which are connected by the formulas

$$\xi^i = \xi^i(\eta^1, \eta^2, \eta^3), \qquad \eta^i = \eta^i(\xi^1, \xi^2, \xi^3). \tag{1.26}$$

Differentiation of Eqs. (1.26) yields

$$d\xi^i = a_j^i d\eta^j, \qquad d\eta^i = b_j^i d\xi^j, \tag{1.27}$$

where

$$a_j^i = \frac{\partial \xi^i}{\partial \eta^j}, \qquad b_j^i = \frac{\partial \eta^i}{\partial \xi^j}. \tag{1.28}$$

Proposition 1.5. The matrices $[a_j^i]$ and $[b_j^i]$ are mutually inverse. For any integers i and j

$$a_k^i b_j^k = \delta_j^i. \tag{1.29}$$

8

Proof. Let us write $\delta_j^i = \partial \xi^i / \partial \xi^j$ and employ Eqs. (1.28). As a result, we find

$$\delta_j^i = \frac{\partial \xi^i}{\partial \xi^j} = \frac{\partial \xi^i}{\partial \eta^k} \frac{\partial \eta^k}{\partial \xi^j} = a_k^i b_j^k. \qquad \square$$

Let $\bar{g}_i$ be tangent vectors for coordinates ξ^i, and $\bar{g}_i'$ tangent vectors for coordinates η^i. By using Eq. (1.6), we write

$$\bar{g}_i' = \frac{\partial \bar{r}}{\partial \eta^i}. \qquad (1.30)$$

It follows from Eqs. (1.28) and (1.30) that

$$\bar{g}_i' = \frac{\partial \bar{r}}{\partial \eta^i} = \frac{\partial \bar{r}}{\partial \xi^j} \frac{\partial \xi^j}{\partial \eta^i} = \bar{g}_j a_i^j. \qquad (1.31)$$

Proposition 1.6. Covariant components of the metric tensor g_{ij} are transformed according to the formula

$$g_{ij}' = g_{kl} a_i^k a_j^l. \qquad \square \qquad (1.32)$$

Exercise 1.17. Derive Eq. (1.32) by employing Eqs. (1.11) and (1.31). $\square$

Proposition 1.7. Contravariant components of the metric tensor g^{ij} are transformed according to the formula

$$g^{ij\prime} = g^{kl} b_k^i b_l^j. \qquad (1.33)$$

Proof. Eq. (1.17) implies that contravariant components of the metric tensor are defined as components of a matrix which is inverse to the matrix with covariant components. Therefore, in order to derive Eq. (1.33) we should show that $g^{ik\prime} g_{kj}' = \delta_j^i$. This equality is verified with the use of Eqs. (1.17), (1.32), and (1.33) as follows:

$$g^{ik\prime} g_{kj}' = g^{mn} b_m^i b_n^k g_{pq} a_k^p a_j^q = g^{mn} g_{pq} \delta_n^p b_m^i a_j^q$$
$$= g^{mn} g_{nq} b_m^i a_j^q = \delta_q^m b_m^i a_j^q = b_m^i a_j^m = \delta_j^i. \qquad \square$$

Proposition 1.8. Dual vectors $\bar{g}^i$ are transformed according to the formula

$$\bar{g}^{i\prime} = \bar{g}^j b_j^i. \qquad (1.34)$$

Proof. It follows from Eqs. (1.18), (1.29), (1.33), and (1.34) that

$$\bar{g}^{i\prime} = g^{ik\prime} \bar{g}_k' = g^{jn} b_j^i b_n^k \bar{g}_m a_k^m = g^{jn} b_j^i \bar{g}_m \delta_n^m = g^{jm} \bar{g}_m b_j^i = \bar{g}^j b_j^i. \qquad \square$$

Definition. Quantities which are transformed by formula (1.31) are called covariant, and quantities which are transformed according to Eq. (1.34) are called contravariant.

9

Tangent vectors $\bar{g}_i$ are covariant, and dual vectors $\bar{g}^i$ are contravariant.

Definition. An object is called *invariant* if it remains unchanged at transition from one coordinate frame to another.

Example. Any scalar f is an invariant object.

Proposition 1.9. The differential of radius-vector $d\bar{r}$ is invariant.

Proof. We should show that

$$d\bar{r}' = d\bar{r}, \tag{1.35}$$

where

$$d\bar{r}' = \bar{g}'_i d\eta^i, \qquad d\bar{r} = \bar{g}_i d\xi^i. \tag{1.36}$$

Eq. (1.35) is verified by direct substitution of expressions (1.36) with the use of Eqs. (1.27), (1.29), and (1.31). $\square$

Let $\bar{q}$ be an invariant vector, $\bar{q}' = \bar{q}$. Substitution of expressions (1.8) and (1.31) into the left-hand side of this equality yields

$$\bar{q}' = q^{i'} \bar{g}'_i = q^{i'} \bar{g}_j a^j_i. \tag{1.37}$$

It follows from Eqs. (1.8) and (1.37) that $q^{i'} a^j_i = q^j$, which is equivalent to the equality

$$q^{i'} = b^i_j q^j. \tag{1.38}$$

Exercise 1.18. Prove that $q'_i = a^j_i q_j$ for any invariant vector q. $\square$

2. Tensor Algebra

In this Section we provide a brief survey of the main concepts and assertions in the tensor algebra. We discuss basic operation on tensors, define unit, inverse, orthogonal, and positive definite tensors, and derive explicit formulas for polar and triangular decomposition of a tensor.

2.1. Definition of a Tensor

In Section 1, two kinds of products for vectors were used: the inner product $\cdot$ and the vector product $\times$. The first is commutative: $\bar{q}_1 \cdot \bar{q}_2 = \bar{q}_2 \cdot \bar{q}_1$ for any $\bar{q}_1$ and $\bar{q}_2$, while the other is skew-commutative: $\bar{q}_1 \times \bar{q}_2 = -\bar{q}_2 \times \bar{q}_1$.

We now introduce the third kind of products: *tensor (diad) product*. The inner product of two vectors is a scalar, the vector product of two vectors is a vector. The tensor product of two vectors is a new object: *a diad of vectors*. The tensor product is associative, non-commutative, and distributive. For any scalars f_1 and f_2, and for any vectors $\bar{q}_1$, $\bar{q}_2$, and $\bar{q}_3$

$$(f_1 \bar{q}_1 + f_2 \bar{q}_2) \bar{q}_3 = f_1 \bar{q}_1 \bar{q}_3 + f_2 \bar{q}_2 \bar{q}_3,$$
$$\bar{q}_1 \bar{q}_2 \neq \bar{q}_2 \bar{q}_1, \qquad (\bar{q}_1 \bar{q}_2) \bar{q}_3 = \bar{q}_1 (\bar{q}_2 \bar{q}_3).$$

By calculating sums of diads of tangent vectors with scalar coefficients, we construct objects with two-component indices

$$\hat{Q} = Q^{ij}\bar{g}_i\bar{g}_j, \tag{2.1}$$

with three-component indices

$$\hat{P} = P^{ijk}\bar{g}_i\bar{g}_j\bar{g}_k, \tag{2.2}$$

etc. In Eqs. (2.1) and (2.2) both tangent vectors $\bar{g}_i$ and dual vectors $\bar{g}^i$ may be employed, e.g. we may write

$$\hat{R} = R^{ij}_{kl}\bar{g}_i\bar{g}_j\bar{g}^k\bar{g}^l. \tag{2.3}$$

Definition. *A tensor* is an invariant object with an arbitrary number of indices.
Definition. The number of indices in Eqs. (2.1) – (2.3) is called *rank* of the corresponding tensor.

A scalar f is a tensor of the zero rank, a vector $\bar{q}$ is a tensor of the first rank, an invariant object $\hat{Q} = Q^{ij}\bar{g}_i\bar{g}_j$ is a tensor of the second rank. For convenience tensors of the second and higher ranks are denoted by the superscript hat.

The main attention in the mechanics of continuum is focused on tensors of the second rank (the gradient of deformations, the deformation tensor, the strain tensor, the stress tensor), and, rarely, on tensors of the third and fourth ranks (the tensor of elastic moduli, the compliance tensor). To simplify notation we confine ourselves to tensors of the second rank, making appropriate remarks regarding tensors of higher ranks where necessary.

Let us consider in which way components of a tensor change at transition from one coordinate frame to another. For this purpose we write the condition of invariance for transformations (1.26). Using Eq. (1.31) we obtain

$$\hat{Q}' = Q^{kl\prime}\bar{g}'_k\bar{g}'_l = Q^{kl\prime}a^i_k a^j_l\bar{g}_i\bar{g}_j = Q^{ij}\bar{g}_i\bar{g}_j = \hat{Q}.$$

This equality together with Eq. (1.29) implies that

$$Q^{ij\prime} = Q^{kl}b^i_k b^j_l. \tag{2.4}$$

Exercise 2.1. Prove that for any tensor of the second rank

$$Q'_{ij} = Q_{kl}a^k_i a^l_j, \qquad Q^{i\prime}_j = Q^k_l b^i_k a^l_j. \qquad \square \tag{2.5}$$

2.2. Operations on Tensors

2.2.1. Operation on one tensor

Transpose. Tensor $\hat{Q}^T = Q_{ji}\bar{g}^i\bar{g}^j = Q_{ij}\bar{g}^j\bar{g}^i$ is called *transpose* for a tensor $\hat{Q} = Q_{ij}\bar{g}^i\bar{g}^j$.

Definition. Tensor $\hat{Q}$ is called *symmetrical* if $\hat{Q}^T = \hat{Q}$, and *skew-symmetrical* if $\hat{Q}^T = -\hat{Q}$.

Exersice 2.2. Prove that for any tensor $\hat{Q}$, tensor

$$\hat{Q}_s = \frac{1}{2}(\hat{Q} + \hat{Q}^T) \tag{2.6}$$

is symmetrical, and tensor

$$\hat{Q}_a = \frac{1}{2}(\hat{Q} - \hat{Q}^T) \tag{2.7}$$

is skew-symmetrical. $\square$

Exercise 2.3. By employing Eqs. (2.6) and (2.7), prove that any tensor $\hat{Q}$ can be presented as a sum $\hat{Q} = \hat{Q}_s + \hat{Q}_a$ of a symmetrical tensor $\hat{Q}_s$ and a skew-symmetrical tensor $\hat{Q}_a$. $\square$

2.2.2. Operations on two tensors

Multiplication by a scalar. For any tensor of the second rank $\hat{Q} = Q_{ij}\bar{g}^i\bar{g}^j$ and any scalar f their product is a tensor of the second rank

$$\widehat{fQ} = (fQ_{ij})\bar{g}^i\bar{g}^j.$$

This operation is distributive $(f_1 + f_2)\hat{Q} = f_1\hat{Q} + f_2\hat{Q}$.

Addition of tensors of the same rank. For any two tensors of the second rank $\hat{P} = P_{ij}\bar{g}^i\bar{g}^j$ and $\hat{Q} = Q_{ij}\bar{g}^i\bar{g}^j$ their sum is a tensor of the second rank

$$\widehat{P+Q} = (P_{ij} + Q_{ij})\bar{g}^i\bar{g}^j.$$

This operation has the following properties:
(i) commutativity $\hat{P} + \hat{Q} = \hat{Q} + \hat{P}$,
(ii) associativity $(\hat{P} + \hat{Q}) + \hat{R} = \hat{P} + (\hat{Q} + \hat{R})$,
(iii) distributivity $f(\hat{P} + \hat{Q}) = f\hat{P} + f\hat{Q}$.

Tensor multiplication. For any tensor $\hat{Q} = Q_{ij}\bar{g}^i\bar{g}^j$ and any vector $\bar{q} = q^k\bar{g}_k$ their tensor product is a tensor of the third rank

$$\widehat{Qq} = Q_{ij}q^k\bar{g}^i\bar{g}^j\bar{g}_k.$$

This operation is associative $(\hat{P}\hat{Q})\hat{R} = \hat{P}(\hat{Q}\hat{R})$.

Inner multiplication. For any tensor $\hat{Q} = Q_{ij}\bar{g}^i\bar{g}^j$ and vector $\bar{q} = q^k\bar{g}_k$ their inner product is a vector

$$\overline{Q \cdot q} = \hat{Q} \cdot \bar{q} = Q_{ij}\bar{g}^i\bar{g}^j \cdot q^k\bar{g}_k$$
$$= Q_{ij}q^k\bar{g}^i\bar{g}^j \cdot \bar{g}_k = Q_{ij}q^k\bar{g}^i\delta^j_{\cdot k} = Q_{ij}q^j\bar{g}^i.$$

Exercise 2.4. Show that for any tensor $\hat{Q} = Q_{ij}\bar{g}^i\bar{g}^j$ and any vector $\bar{q} = q^k\bar{g}_k$

$$\hat{Q} \cdot \bar{q} = \bar{q} \cdot \hat{Q}^T. \qquad \square \tag{2.8}$$

Exercise 2.5. Check that for any tensors $\hat{P}$ and $\hat{Q}$

$$(\hat{P} \cdot \hat{Q})^T = \hat{Q}^T \cdot \hat{P}^T. \qquad \square \tag{2.9}$$

By using the inner product we can define any positive power of a tensor $\hat{Q}$. For example, we set $\hat{Q}^2 = \hat{Q} \cdot \hat{Q}$, $\hat{Q}^3 = \hat{Q}^2 \cdot \hat{Q}$, etc. Obviously, powers of the same tensor commute with one another $\hat{Q}^m \cdot \hat{Q}^n = \hat{Q}^n \cdot \hat{Q}^m$ for any positive integers m and n and any tensor $\hat{Q}$.

Convolution or duplicate scalar product. For any two tensors $\hat{P} = P_{ij}\bar{g}^i\bar{g}^j$ and $\hat{Q} = Q^{kl}\bar{g}_k\bar{g}_l$ their *convolution* is a scalar

$$\hat{P} : \hat{Q} = P_{ij}\bar{g}^i\bar{g}^j : Q^{kl}\bar{g}_k\bar{g}_l = P_{ij}Q^{kl}\bar{g}^i\bar{g}^j : \bar{g}_k\bar{g}_l$$
$$= P_{ij}Q^{kl}\bar{g}^i \cdot \delta^j_{\cdot k}\bar{g}_l = P_{ij}Q^{kl}\delta^j_{\cdot k}\bar{g}^i \cdot \bar{g}_l = P_{ij}Q^{kl}\delta^j_{\cdot k}\delta^i_{\cdot l} = P_{ij}Q^{ji}. \tag{2.10}$$

This operation is commutative $\hat{P} : \hat{Q} = \hat{Q} : \hat{P}$.

Exercise 2.6. Show that for any tensors $\hat{P}$ and $\hat{Q}$

$$\hat{P}^T : \hat{Q}^T = \hat{P} : \hat{Q}. \qquad \square \tag{2.11}$$

Exercise 2.7. Let $\hat{P}$ be symmetrical and $\hat{Q}$ be skew-symmetrical tensor. Prove that

$$\hat{P} : \hat{Q} = 0. \qquad \square \tag{2.12}$$

The above definitions can be extended to tensors of higher ranks. The coincidence of ranks for two tensors is necessary only for their addition. Other operations can be carried out on tensors of arbitrary ranks.

2.3. The Unit Tensor

Definition. A tensor $\hat{I}$ is called *unit* if for any tensor $\hat{Q}$

$$\hat{Q} \cdot \hat{I} = \hat{I} \cdot \hat{Q} = \hat{Q}. \tag{2.13}$$

Proposition 2.1. The unit tensor is determined by the formulas

$$\hat{I} = g_{ij}\bar{g}^i\bar{g}^j = \bar{g}_i\bar{g}^i = \bar{g}^i\bar{g}_i = g^{ij}\bar{g}_i\bar{g}_j. \tag{2.14}$$

Proof. For any tensor of the second rank $\hat{Q} = Q^{ij}\bar{g}_i\bar{g}_j$ we obtain

$$\hat{Q} \cdot \hat{I} = Q^{ij}\bar{g}_i\bar{g}_j \cdot \bar{g}^k\bar{g}_k = Q^{ij}\delta^{\cdot k}_{j\cdot}\bar{g}_i\bar{g}_k = Q^{ij}\bar{g}_i\bar{g}_j = \hat{Q},$$
$$\hat{I} \cdot \hat{Q} = \bar{g}_k\bar{g}^k \cdot Q^{ij}\bar{g}_i\bar{g}_j = \delta^k_{\cdot i}Q^{ij}\bar{g}_k\bar{g}_j = Q^{ij}\bar{g}_i\bar{g}_j = \hat{Q},$$

which imply that tensor (2.14) is unit. $\square$

Exercise 2.8. Check that for any tensors of the second rank $\hat{P}$ and $\hat{Q}$

$$\hat{I} : (\hat{P} \cdot \hat{Q}) = \hat{P} : \hat{Q}. \quad \square \tag{2.15}$$

Exercise 2.9. Prove that the unit tensor is unique. $\square$

2.4. An Inverse Tensor

Definition. For any tensor of the second rank $\hat{Q}$, its *inverse* tensor $\hat{Q}^{-1}$ is defined by the equalities

$$\hat{Q} \cdot \hat{Q}^{-1} = \hat{Q}^{-1} \cdot \hat{Q} = \hat{I}.$$

Definition. A tensor $\hat{Q}$ is *non-singular* if there exists an inverse tensor $\hat{Q}^{-1}$.

Exercise 2.10. By employing appropriate assertions for matrices, show that for any two non-singular tensors $\hat{P}$ and $\hat{Q}$ the following equalities are valid:

$$(\hat{P}^T)^{-1} = (\hat{P}^{-1})^T, \quad (\hat{P}^{-1})^{-1} = \hat{P}, \quad (\hat{P} \cdot \hat{Q})^{-1} = \hat{Q}^{-1} \cdot \hat{P}^{-1}. \quad \square \tag{2.16}$$

2.5. Eigenvectors and Eigenvalues of a Tensor

Definition. A vector $\bar{e}$ is called *right eigenvector* of a tensor of the second rank $\hat{Q}$ if

$$\hat{Q} \cdot \bar{e} = \lambda \bar{e} \tag{2.17}$$

for a constant λ which is called *right eigenvalue* of $\hat{Q}$.

Definition. A vector $\bar{e}'$ is called *left eigenvector* of a tensor $\hat{Q}$ with *a left eigenvalue* λ' if

$$\bar{e}' \cdot \hat{Q} = \lambda' \bar{e}'. \tag{2.18}$$

It follows from Eqs. (2.17) and (2.18) that λ and λ' satisfy *the characteristic equation*

$$\det(\hat{Q} - \lambda \hat{I}) = 0, \tag{2.19}$$

where $\hat{I}$ is the unit tensor. Eq. (2.19) implies that λ and λ' take the same values λ_1, λ_2, and λ_3.

Denote by $\bar{e}_k$ the right eigenvector and by $\bar{e}^k$ the left eigenvector corresponding to the eigenvalue λ_k

$$\hat{Q} \cdot \bar{e}_k = \lambda_k \bar{e}_k, \qquad \bar{e}^k \cdot \hat{Q} = \lambda_k \bar{e}^k. \tag{2.20}$$

Multiplication of the first equation (2.20) by $\bar{e}^l$ yields $\bar{e}^l \cdot \hat{Q} \cdot \bar{e}_k = \lambda_k \bar{e}^l \cdot \bar{e}_k$. By employing Eq. (2.20), this equality can be written as

$$(\lambda_k - \lambda_l)\bar{e}^l \cdot \bar{e}_k = 0.$$

14

By introducing an appropriate normalization condition we obtain

$$\bar{e}^l \cdot \bar{e}_k = \delta^{l\cdot}_{\cdot k}. \tag{2.21}$$

Exercise 2.11. By using an appropriate assertion in the matrix theory, prove that any tensor $\hat{Q}$ can be presented in the form

$$\hat{Q} = \lambda_1 \bar{e}_1 \bar{e}^1 + \lambda_2 \bar{e}_2 \bar{e}^2 + \lambda_3 \bar{e}_3 \bar{e}^3. \quad \square \tag{2.22}$$

Exercise 2.12. Check that

$$\hat{Q}^T = \lambda_1 \bar{e}^1 \bar{e}_1 + \lambda_2 \bar{e}^2 \bar{e}_2 + \lambda_3 \bar{e}^3 \bar{e}_3. \quad \square \tag{2.22}$$

By using Eq. (2.22), any integer power of a tensor $\hat{Q}$ can be defined. For example, for any positive integer n we set

$$\hat{Q}^n = \lambda_1^n \bar{e}_1 \bar{e}^1 + \lambda_2^n \bar{e}_2 \bar{e}^2 + \lambda_3^n \bar{e}_3 \bar{e}^3. \tag{2.23}$$

Exercise 2.13. Show that this definition is correct. $\square$
Exercise 2.14. Prove that for any non-singular tensor $\hat{Q}$

$$\hat{Q}^{-1} = \frac{1}{\lambda_1} \bar{e}_1 \bar{e}^1 + \frac{1}{\lambda_2} \bar{e}_2 \bar{e}^2 + \frac{1}{\lambda_3} \bar{e}_3 \bar{e}^3. \quad \square \tag{2.24}$$

2.6. The Principal Invariants of a Tensor

Let us present a tensor $\hat{Q}$ in a combined basis $\hat{Q} = Q^i_j \bar{g}_i \bar{g}^j$, introduce the matrix (Q^i_j), and write the characteristic equation (2.19) in the matrix form

$$\det(Q^i_j - \lambda \delta^i_j) = 0.$$

Calculation of the determinant yields

$$\lambda^3 - I_1(\hat{Q})\lambda^2 + I_2(\hat{Q})\lambda - I_3(\hat{Q}) = 0. \tag{2.25}$$

Exercise 2.15. By direct calculations, show that

$$I_1(\hat{Q}) = \lambda_1 + \lambda_2 + \lambda_3, \quad I_2(\hat{Q}) = \lambda_1\lambda_2 + \lambda_2\lambda_3 + \lambda_3\lambda_1, \quad I_3(\hat{Q}) = \lambda_1\lambda_2\lambda_3. \quad \square \tag{2.26}$$

Definition. The coefficients $I_1(\hat{Q})$, $I_2(\hat{Q})$, $I_3(\hat{Q})$ in the characteristic equation (2.25) are called *principal invariants* of a tensor $\hat{Q}$.
Definition. Any function of the principal invariants of a tensor $\hat{Q}$ is called its *invariant*.

Exercise 2.16. Prove that for any tensor $\hat{Q}$

$$I_k(\hat{Q}) = I_k(\hat{Q}^T) \quad (k = 1, 2, 3). \quad \square \tag{2.27}$$

Exercise 2.17. Prove that for any tensor $\hat{Q}$

$$I_1(\hat{Q}) = \hat{I} : \hat{Q}, \quad I_3(\hat{Q}) = \det(Q_j^i). \quad \square \tag{2.28}$$

For any tensor $\hat{Q}$, the determinant of the matrix (Q_j^i) is called *determinant* of this tensor. Therefore, the latter equality (2.28) may be written as $I_3(\hat{Q}) = \det \hat{Q}$.

Exercise 2.18. Prove that for any two tensors $\hat{P}$ and $\hat{Q}$

$$I_1(\hat{P} \cdot \hat{Q}) = \hat{P} : \hat{Q}. \quad \square \tag{2.29}$$

Exercise 2.19. Derive the following formulas for the principal invariants of a tensor $\hat{Q}$:

$$I_2(\hat{Q}) = \frac{1}{2}[I_1^2(\hat{Q}) - I_1(\hat{Q}^2)],$$

$$I_3(\hat{Q}) = \frac{1}{3}[I_1(\hat{Q}^3) - I_1^3(\hat{Q}) + 3I_1(\hat{Q})I_2(\hat{Q})]. \quad \square \tag{2.30}$$

Exercise 2.20. Show that for any non-singular tensor $\hat{Q}$

$$I_1(\hat{Q}^{-1}) = \frac{I_2(\hat{Q})}{I_3(\hat{Q})}, \quad I_2(\hat{Q}^{-1}) = \frac{I_1(\hat{Q})}{I_3(\hat{Q})}, \quad I_3(\hat{Q}^{-1}) = \frac{1}{I_3(\hat{Q})}. \quad \square \tag{2.31}$$

Proposition 2.2 (The Caley–Hamilton theorem). Any tensor of the second rank $\hat{Q}$ satisfies its characteristic equation

$$\hat{Q}^3 - I_1(\hat{Q})\hat{Q}^2 + I_2(\hat{Q})\hat{Q} - I_3(\hat{Q})\hat{I} = 0. \quad \square \tag{2.32}$$

Exercise 2.21. Prove this theorem by employing an appropriate assertion in the matrix theory. $\square$

Eq. (2.32) allows any power (both positive and negative) of a tensor $\hat{Q}$ to be expressed in terms of $\hat{I} = \hat{Q}^0$, $\hat{Q} = \hat{Q}^1$, and $\hat{Q}^2$. For example, it follows from Eq. (2.32) that

$$\hat{Q}^{-1} = \frac{1}{I_3(\hat{Q})}[\hat{Q}^2 - I_1(\hat{Q})\hat{Q} + I_2(\hat{Q})\hat{I}]. \tag{2.33}$$

Exercise 2.22. Show that for any tensor $\hat{Q}$, tensor $\hat{Q}^T$ satisfies the same characteristic equation

$$(\hat{Q}^T)^3 - I_1(\hat{Q})(\hat{Q}^T)^2 + I_2(\hat{Q})\hat{Q}^T - I_3(\hat{Q})\hat{I} = 0. \tag{2.34}$$

16

2.7. Expansion of a Symmetrical Tensor into the Spherical and Deviatoric Parts

Definition. For any symmetrical tensor $\hat{Q}$, the product of its first principal invariant by the unit tensor is called *spherical component* of $\hat{Q}$.

Definition. For any symmetrical tensor $\hat{Q}$, the difference $\mathrm{dev}(\hat{Q}) = \hat{Q} - \frac{1}{3}I_1(\hat{Q})\hat{I}$ is called *deviatoric part* of $\hat{Q}$.

Exercise 2.23. Prove that any symmetrical tensor can be expanded into the sum of its spherical and deviatoric parts

$$\hat{Q} = \frac{1}{3}I_1(\hat{Q})\hat{I} + \mathrm{dev}(\hat{Q}). \qquad \square \qquad (2.35)$$

2.8. Positive Definite Tensors

Definition. A tensor $\hat{Q}$ is *positive definite* if for any non-zero vector $\bar{q}$

$$\bar{q} \cdot \hat{Q} \cdot \bar{q} > 0.$$

Exercise 2.24. Check that the unit tensor $\hat{I}$ is positive definite. $\square$

Proposition 2.3. For any non-singular tensor $\hat{Q}$, tensors $\hat{Q} \cdot \hat{Q}^T$ and $\hat{Q}^T \cdot \hat{Q}$ are positive definite.

Proof. It follows from Eq. (2.8) that

$$\bar{q} \cdot \hat{Q} \cdot \hat{Q}^T \cdot \bar{q} = (\bar{q} \cdot \hat{Q}) \cdot (\bar{q} \cdot \hat{Q}) = |\bar{q} \cdot \hat{Q}|^2 > 0,$$
$$\bar{q} \cdot \hat{Q}^T \cdot \hat{Q} \cdot \bar{q} = (\bar{q} \cdot \hat{Q}^T) \cdot (\bar{q} \cdot \hat{Q}^T) = |\bar{q} \cdot \hat{Q}^T|^2 > 0. \qquad \square$$

Exercise 2.25. Prove that eigenvalues λ_k of a symmetrical positive definite tensor $\hat{Q}$ are real and positive. $\square$

For a symmetrical positive definite tensor $\hat{Q}$, its fractional powers can be defined by using Eq. (2.22). In particular, the square root of a tensor $\hat{Q}$ is defined as

$$\hat{Q}^{\frac{1}{2}} = \lambda_1^{\frac{1}{2}}\bar{e}_1\bar{e}^1 + \lambda_2^{\frac{1}{2}}\bar{e}_2\bar{e}^2 + \lambda_3^{\frac{1}{2}}\bar{e}_3\bar{e}^3. \qquad (2.36)$$

Exercise 2.26. Show that $\hat{Q}^{\frac{1}{2}} \cdot \hat{Q}^{\frac{1}{2}} = \hat{Q}$. $\square$

Definition. For any tensor $\hat{Q}$, the eigenvalues $v_k(\hat{Q})$ of the tensor $(\hat{Q}^T \cdot \hat{Q})^{\frac{1}{2}}$ are called *singular values* of $\hat{Q}$.

2.9. Orthogonal Tensors

Definition. A tensor $\hat{O}$ is called *orthogonal* if

$$\hat{O}^T = \hat{O}^{-1}. \qquad (2.37)$$

Exercise 2.27. Check that for any orthogonal tensor $\hat{O}$

$$\hat{O} \cdot \hat{O}^T = \hat{O}^T \cdot \hat{O} = \hat{I}. \quad \square \tag{2.38}$$

Exercise 2.28. Show that an orthogonal transformation preserves the inner product of vectors. Prove that for any two vectors $\bar{q}_1$ and $\bar{q}_2$, and for any orthogonal tensor $\hat{O}$ we have $\bar{q}_1' \cdot \bar{q}_2' = \bar{q}_1 \cdot \bar{q}_2$, where $\bar{q}_i' = \bar{q}_i \cdot \hat{O}$. $\square$

Since $\det \hat{O}^T = \det \hat{O}$ and $\det \hat{O}^{-1} = (\det \hat{O})^{-1}$, it follows from Eq. (2.37) that $\det \hat{O} = \pm 1$ for any orthogonal tensor $\hat{O}$. We confine ourselves to orthogonal tensors with

$$\det \hat{O} = 1. \tag{2.39}$$

Exercise 2.29. By using Eqs. (2.31), (2.37), and (2.39), show that for any orthogonal tensor $\hat{O}$

$$I_1(\hat{O}) = I_2(\hat{O}), \qquad I_3(\hat{O}) = 1. \quad \square \tag{2.40}$$

2.10. Polar Decomposition

Proposition 2.4. Any non-singular tensor $\hat{Q}$ can be presented in the form

$$\hat{Q} = \hat{U}_l \cdot \hat{O}, \tag{2.41}$$

where $\hat{U}_l$ is a symmetrical positive definite tensor, and $\hat{O}$ is an orthogonal tensor. $\square$

Formula (2.41) is called the left polar decomposition of a tensor $\hat{Q}$.
Proof. We transpose Eq. (2.41) and use Eq. (2.37)

$$\hat{Q}^T = \hat{O}^T \cdot \hat{U}_l^T = \hat{O}^{-1} \cdot \hat{U}_l. \tag{2.42}$$

Multiplication of Eqs. (2.41) and (2.42) implies that

$$\hat{Q} \cdot \hat{Q}^T = \hat{U}_l \cdot \hat{O} \cdot \hat{O}^{-1} \hat{U}_l = \hat{U}_l^2.$$

It follows from this equality that

$$\hat{U}_l = (\hat{Q} \cdot \hat{Q}^T)^{\frac{1}{2}}. \tag{2.43}$$

Exercise 2.30. Show that the tensor $\hat{U}_l$ defined by Eq. (2.43) is symmetrical and positive definite. $\square$

Let us show that the tensor

$$\hat{O} = \hat{U}_l^{-1} \cdot \hat{Q} \tag{2.44}$$

18

is orthogonal. For this purpose we transpose Eq. (2.44)

$$\hat{O}^T = \hat{Q}^T \cdot (\hat{U}_l^{-1})^T = \hat{Q}^T \cdot \hat{U}_l^{-1}$$

and multiply the obtained equality by Eq. (2.44). By using Eq. (2.43), we find

$$\hat{O} \cdot \hat{O}^T = \hat{U}_l^{-1} \cdot \hat{Q} \cdot \hat{Q}^T \cdot \hat{U}_l^{-1} = \hat{U}_l^{-1} \cdot \hat{U}_l \cdot \hat{U}_l \cdot \hat{U}_l^{-1} = \hat{I}. \quad \square$$

Similar to Eq. (2.41), *the right polar decomposition* of a tensor $\hat{Q}$ can be introduced

$$\hat{Q} = \hat{O} \cdot \hat{U}_r, \tag{2.45}$$

where $\hat{U}_r$ is a symmetrical positive definite tensor, and $\hat{O}$ is an orthogonal tensor.
Exercise 2.31. Derive the following formulas for the tensors $\hat{U}_r$ and $\hat{O}$:

$$\hat{U}_r = \hat{R}^{\frac{1}{2}}, \qquad \hat{O} = \hat{Q} \cdot \hat{U}_r^{-1}, \tag{2.46}$$

where

$$\hat{R} = \hat{Q}^T \cdot \hat{Q}. \quad \square \tag{2.47}$$

Problem 2.1. Prove that the tensors $\hat{U}_l$ and $\hat{U}_r$ are unique. For a detailed discussion of this issue see Stephenson (1980). $\square$

2.11. Triangular Decomposition

In general, an explicit calculation of the matrix $\hat{U}_r$ is a complicated problem, since the principal invariants and the principal axes of $\hat{R}$ are unknown. To simplify these calculations, another (*triangular*) decomposition of $\hat{Q}$ is introduced which may be derived for any tensor $\hat{R}$ explicitly, see Souchet (1993).

Let a positive definite tensor $\hat{R} = R^{ij} \bar{g}_i \bar{g}_j$ be presented in a fixed coordinate frame in the matrix form

$$\hat{R} = \begin{bmatrix} a & \gamma & \beta \\ \gamma & b & \alpha \\ \beta & \alpha & c \end{bmatrix}. \tag{2.48}$$

Exercise 2.32. Check that for any vector $\bar{q} = q_i \bar{g}^i$

$$\bar{q} \cdot \hat{R} \cdot \bar{q} = aq_1^2 + bq_2^2 + cq_3^2 + 2\gamma q_1 q_2 + 2\beta q_1 q_3 + 2\alpha q_2 q_3$$

$$= c(q_3 + \frac{\alpha}{c}q_2 + \frac{\beta}{c}q_1)^2 + \frac{bc - \alpha^2}{c}(q_2 + \frac{\gamma c - \alpha\beta}{bc - \alpha^2}q_1)^2$$

$$+ [a - \frac{(\gamma c - \alpha\beta)^2}{c(bc - \alpha^2)} - \frac{\beta^2}{c}]q_1^2. \quad \square \tag{2.49}$$

According to Eq. (2.49), tensor $\hat{R}$ is positive definite provided

$$c > 0, \qquad \frac{bc - \alpha^2}{c} > 0, \qquad a - \frac{(\gamma c - \alpha\beta)^2}{c(bc - \alpha^2)} - \frac{\beta^2}{c} > 0. \tag{2.50}$$

These conditions are assumed to be fulfilled.

Our objective is to construct a tensor $\hat{S}$ presented by a triangular matrix

$$\hat{S} = \begin{bmatrix} s_{11} & 0 & 0 \\ s_{21} & s_{22} & 0 \\ s_{31} & s_{32} & s_{33} \end{bmatrix} \tag{2.51}$$

such that the diagonal components s_{ii} are positive and

$$\hat{S}^T \cdot \hat{S} = \hat{R}. \tag{2.52}$$

First, let us assume that such a tensor $\hat{S}$ has already been constructed. Since the diagonal elements in Eq. (2.51) are positive, the tensor $\hat{S}$ is non-singular, and an inverse tensor $\hat{S}^{-1}$ exists. Eq. (2.52) implies that $\hat{S}^T \cdot \hat{S} = \hat{Q}^T \cdot \hat{Q}$, which means that

$$(\hat{S}^T)^{-1} \cdot \hat{Q}^T \cdot \hat{Q} \cdot \hat{S}^{-1} = \hat{I}.$$

With the use of Eq. (2.9) this equality can be written as follows:

$$(\hat{Q} \cdot \hat{S}^{-1})^T \cdot (\hat{Q} \cdot \hat{S}^{-1}) = \hat{I}. \tag{2.53}$$

Comparison of Eq. (2.53) with Eq. (2.38) implies that tensor $\hat{Q} \cdot \hat{S}^{-1}$ is orthogonal: $\hat{Q} \cdot \hat{S}^{-1} = \hat{O}$. This leads to the triangular decomposition of an arbitrary non-singular tensor $\hat{Q}$ into the product of a triangular non-singular tensor and an orthogonal tensor

$$\hat{Q} = \hat{O} \cdot \hat{S}. \tag{2.54}$$

Let us now construct the tensor $\hat{S}$ explicitly. For this purpose we write Eq. (2.52) in the component form by using expressions (2.48) and (2.51). As a result, the following algebraic equations are derived:

$$s_{11}^2 + s_{22}^2 + s_{33}^2 = a, \qquad s_{22}^2 + s_{32}^2 = b, \qquad s_{33}^2 = c,$$
$$s_{32}s_{33} = \alpha, \qquad s_{31}s_{33} = \beta, \qquad s_{21}s_{22} + s_{31}s_{32} = \gamma. \tag{2.55}$$

Exercise 2.33. Check that Eqs. (2.55) have the unique solution

$$s_{11} = \sqrt{a - \frac{(\gamma c - \alpha\beta)^2}{c(bc - \alpha^2)} - \frac{\beta^2}{c}}, \qquad s_{21} = \frac{\gamma c - \alpha\beta}{\sqrt{c(bc - \alpha^2)}},$$

$$s_{22} = \sqrt{\frac{bc - \alpha^2}{c}}, \qquad s_{31} = \frac{\beta}{\sqrt{c}}, \qquad s_{32} = \frac{\alpha}{\sqrt{c}}, \qquad s_{33} = \sqrt{c}. \qquad \square \tag{2.56}$$

For a given tensor $\hat{Q}$, the tensor $\hat{R}$ may be found explicitly as a product of $\hat{Q}^T$ and $\hat{Q}$. When the components of $\hat{R}$ are calculated, Eqs. (2.56) allow the components of the triangular tensor $\hat{S}$ to be obtained explicitly. Therefore,

20

unlike the polar decomposition (2.45), the triangular decomposition (2.54) may be derived analytically.

Exercise 2.34. Derive formulas for a tensor $\hat{S}$ determining the left triangular decomposition $\hat{Q} = \hat{S} \cdot \hat{O}$ of a tensor $\hat{Q}$. $\square$

3. Tensor Analysis

This Section is concerned with basic assertions in the theory of differentiation and integration of tensor functions. We discuss properties of the Hamilton operator, derive formulas for the Christoffel symbols, introduce the covariant differentiation of tensor fields, and prove the Stokes theorem.

3.1. The Nabla-Operator

Let $\xi = \{\xi^i\}$ be *curvilinear coordinates* and $\bar{r} = \bar{r}(\xi)$ be the radius-vector of a point M with coordinates ξ. The radius-vector of a point M_1 with coordinates $\xi + d\xi = \{\xi^i + d\xi^i\}$ is denoted by $\bar{r} + d\bar{r}$. The differential $d\bar{r}$ equals

$$d\bar{r} = \frac{\partial \bar{r}}{\partial \xi^1}d\xi^1 + \frac{\partial \bar{r}}{\partial \xi^2}d\xi^2 + \frac{\partial \bar{r}}{\partial \xi^3}d\xi^3 = \frac{\partial \bar{r}}{\partial \xi^i}d\xi^i = \bar{g}_i d\xi^i, \tag{3.1}$$

where $\bar{g}_i = \partial \bar{r}/\partial \xi^i$ are tangent vectors. We multiply Eq. (3.1) by dual vectors $\bar{g}^j$ and, by using Eq. (1.9), obtain $\bar{g}^j \cdot d\bar{r} = \bar{g}^j \cdot \bar{g}_i d\xi^i = \delta_i^j d\xi^i = d\xi^j$. This formula implies that

$$d\xi^i = \bar{g}^i \cdot d\bar{r} = d\bar{r} \cdot \bar{g}^i. \tag{3.2}$$

Let $f(\xi)$ be a smooth scalar function. Differentiation of this function with the use of Eq. (3.2) yields

$$df = \frac{\partial f}{\partial \xi^i}d\xi^i = \bar{g}^i \frac{\partial f}{\partial \xi^i} \cdot d\bar{r}. \tag{3.3}$$

By introducing *the Hamilton operator* (the nabla-operator)

$$\bar{\nabla} = \bar{g}^i \frac{\partial}{\partial \xi^i}, \tag{3.4}$$

we rewrite Eq. (3.3) as follows:

$$df = \bar{\nabla}f \cdot d\bar{r}. \tag{3.5}$$

Exercise 3.1. Prove that $\bar{\nabla}$ is a vector, i.e. a tensor of the first rank. For this purpose, show that $\bar{\nabla}' = \bar{\nabla}$, where $\bar{\nabla}'$ is the nabla-operator in "new" coordinates $\eta = \{\eta^i\}$. $\square$

Let us now consider a smooth vector function $\bar{q}(\xi)$. Similar to Eq. (3.3) we obtain by using Eqs. (3.2) and (3.4)

$$d\bar{q} = \frac{\partial \bar{q}}{\partial \xi^i} d\xi^i = d\bar{r} \cdot \bar{g}^i \frac{\partial \bar{q}}{\partial \xi^i} = d\bar{r} \cdot \bar{\nabla}\bar{q}.$$

It follows from this formula and Eq. (2.8) that

$$d\bar{q} = d\bar{r} \cdot \bar{\nabla}\bar{q} = \bar{\nabla}\bar{q}^T \cdot d\bar{r}. \tag{3.6}$$

Finally, let us consider a smooth tensor function $\hat{Q}(\xi)$. By repeating a similar calculation, we arrive at the formula

$$d\hat{Q} = \frac{\partial \hat{Q}}{\partial \xi^i} d\xi^i = d\bar{r} \cdot \bar{\nabla}\hat{Q} = (\bar{\nabla}\hat{Q})^T \cdot d\bar{r}. \tag{3.7}$$

3.2. Operators Connected with the Nabla-Operator

Since $\bar{\nabla}$ is a vector, its inner products with vectors and tensors may be introduced

$$\bar{\nabla} \cdot \bar{q} = \bar{g}^i \cdot \frac{\partial \bar{q}}{\partial \xi^i}, \qquad \bar{\nabla} \cdot \hat{Q} = \bar{g}^i \cdot \frac{\partial \hat{Q}}{\partial \xi^i}. \tag{3.8}$$

Let us show that $\bar{\nabla} \cdot \bar{q}$ coincides with the standard *divergence* of a vector $\bar{q}$. For this purpose we suppose that $\xi^i = x^i$ are Cartesian coordinates with tangent vectors $\bar{e}_i$. Components of the vector $\bar{q}$ in this coordinate frame are q_{x^i}. Since the spatial derivatives of vectors $\bar{e}_i$ vanish, the first equality (3.8) implies that

$$\bar{\nabla} \cdot \bar{q} = \bar{e}_1 \cdot \frac{\partial \bar{q}}{\partial x^1} + \bar{e}_2 \cdot \frac{\partial \bar{q}}{\partial x^2} + \bar{e}_3 \cdot \frac{\partial \bar{q}}{\partial x^3} = \frac{\partial q_{x^1}}{\partial x^1} + \frac{\partial q_{x^2}}{\partial x^2} + \frac{\partial q_{x^3}}{\partial x^3} = \operatorname{div} \bar{q}. \tag{3.9}$$

Similarly, we define the divergence of a tensor $\hat{Q}$ as

$$\operatorname{div} \hat{Q} = \bar{\nabla} \cdot \hat{Q}. \tag{3.10}$$

Exercise 3.2. By using Eq. (3.8), calculate the divergence of a vector $\bar{q}$ in spherical and cylindrical coordinate frames. $\square$

The vector product of the nabla-operator with a vector function $\bar{q}$ is called its *curl or rotor*

$$\operatorname{rot} \bar{q} = \bar{\nabla} \times \bar{q} = \bar{g}^i \times \frac{\partial \bar{q}}{\partial \xi^i}. \tag{3.11}$$

Exercise 3.3. Show that in Cartesian coordinates the curl of a vector $\bar{q}$ can be presented as

$$\operatorname{rot} \bar{q} = \det \begin{bmatrix} \bar{e}_1 & \bar{e}_2 & \bar{e}_3 \\ \frac{\partial}{\partial x^1} & \frac{\partial}{\partial x^2} & \frac{\partial}{\partial x^3} \\ q_{x^1} & q_{x^2} & q_{x^3} \end{bmatrix}. \qquad \square$$

22

Exercise 3.4. Calculate the curl of a vector $\bar{q}$ in spherical and cylindrical coordinates. □

The tensor product of the nabla-operator with a vector $\bar{q}$ does not form any special object, but by using the tensor $\bar{\nabla}\bar{q}$ two important objects can be constructed: the symmetrical *strain tensor* $\hat{\epsilon}(\bar{q})$ and the skew-symmetrical *spin tensor* $\hat{\omega}(\bar{q})$

$$\hat{\epsilon}(\bar{q}) = \frac{1}{2}(\bar{\nabla}\bar{q}^T + \bar{\nabla}\bar{q}), \qquad \hat{\omega}(\bar{q}) = \frac{1}{2}(\bar{\nabla}\bar{q}^T - \bar{\nabla}\bar{q}). \tag{3.12}$$

Exercise 3.5. Check that

$$\bar{\nabla}\bar{q} = \hat{\epsilon}(\bar{q}) - \hat{\omega}(\bar{q}), \qquad \bar{\nabla}\bar{q}^T = \hat{\epsilon}(\bar{q}) + \hat{\omega}(\bar{q}). \quad □ \tag{3.13}$$

3.3. Properties of the Nabla-Operator

In this subsection we formulate several assertions regarding the characteristic features of the nabla-operator. The proof of these statments is left to the reader as an exercise.

Proposition 3.1. For any two scalar functions f_1 and f_2 and any vector function $\bar{q}$

$$\bar{\nabla}(f_1 f_2) = (\bar{\nabla}f_1)f_2 + f_1(\bar{\nabla}f_2), \qquad \bar{\nabla}(f_1\bar{q}) = (\bar{\nabla}f_1)\bar{q} + f_1(\bar{\nabla}\bar{q}). \quad □ \tag{3.14}$$

Proposition 3.2. For any two vector functions $\bar{q}_1$ and $\bar{q}_2$

$$\bar{\nabla}(\bar{q}_1 \cdot \bar{q}_2) = \bar{\nabla}\bar{q}_1 \cdot \bar{q}_2 + \bar{q}_1 \cdot (\bar{\nabla}\bar{q}_2)^T. \quad □$$

Proposition 3.3. For any vector function $\bar{q}$ and any tensor function $\hat{Q}$

$$\bar{\nabla} \cdot (\hat{Q} \cdot \bar{q}) = (\bar{\nabla} \cdot \hat{Q}) \cdot \bar{q} + \hat{Q} : (\bar{\nabla}\bar{q})^T. \quad □ \tag{3.15}$$

For a symmetrical tensor $\hat{Q}$, Eqs. (3.12) and (3.15) imply that

$$\bar{\nabla} \cdot (\hat{Q} \cdot \bar{q}) = (\bar{\nabla} \cdot \hat{Q}) \cdot \bar{q} + \hat{Q} : \hat{\epsilon}(\bar{q}). \tag{3.16}$$

Proposition 3.4. For any tensor functions $\hat{P}$ and $\hat{Q}$

$$\bar{\nabla} \cdot (\hat{P} \cdot \hat{Q}) = \hat{Q}^T \cdot (\bar{\nabla} \cdot \hat{P}) + \hat{P}^T : \bar{\nabla}\hat{Q}. \quad □ \tag{3.17}$$

3.4. Christoffel's Symbols

Let us calculate the derivatives of tangent vector $\bar{g}_i$ with respect to spatial coordinates ξ^j. Since the derivative of a vector should be a vector as well, the result may be expanded in the tangent vectors

$$\frac{\partial \bar{g}_i}{\partial \xi^j} = \Gamma_{ij}^k \bar{g}_k. \tag{3.18}$$

The quantities Γ_{ij}^k in the right-hand side of Eq. (3.18) are called *the Christoffel symbols of the second kind*. (The Christoffel symbols of the first kind will be introduced later.)

Proposition 3.5. The Christoffel symbols are symmetrical with respect to low indices

$$\Gamma_{ij}^k = \Gamma_{ji}^k. \tag{3.19}$$

Proof. Let us replace $\bar{g}_i$ in Eq. (3.18) by $\partial \bar{r}/\partial \xi^i$. As a result, we obtain

$$\frac{\partial^2 \bar{r}}{\partial \xi^j \partial \xi^i} = \Gamma_{ij}^k \bar{g}_k. \tag{3.20}$$

Similarly, we have

$$\frac{\partial^2 \bar{r}}{\partial \xi^i \partial \xi^j} = \Gamma_{ji}^k \bar{g}_k.$$

Since the left-hand sides of these relationships coincide, the right-hand sides should coincide as well, which leads us to the requested symmetry. $\square$

Let us now derive an explicit formula for the Christoffel symbols. Multiplying Eq. (3.20) by $\bar{g}_m$ and using Eq. (3.19) we find

$$\frac{\partial^2 \bar{r}}{\partial \xi^i \partial \xi^j} \cdot \bar{g}_m = \Gamma_{ij}^k g_{km}, \tag{3.21}$$

where $g_{ij} = \bar{g}_i \cdot \bar{g}_j$. Calculation of the derivative of g_{ij} with respect to ξ^k yields

$$\frac{\partial g_{ij}}{\partial \xi^k} = \frac{\partial}{\partial \xi^k}(\bar{g}_i \cdot \bar{g}_j) = \frac{\partial \bar{g}_i}{\partial \xi^k} \cdot \bar{g}_j + \bar{g}_i \cdot \frac{\partial \bar{g}_j}{\partial \xi^k} = \frac{\partial^2 \bar{r}}{\partial \xi^k \partial \xi^i} \cdot \bar{g}_j + \bar{g}_i \cdot \frac{\partial^2 \bar{r}}{\partial \xi^k \partial \xi^j}. \tag{3.22}$$

Similarly,

$$\frac{\partial g_{jk}}{\partial \xi^i} = \frac{\partial^2 \bar{r}}{\partial \xi^i \partial \xi^j} \cdot \bar{g}_k + \bar{g}_j \cdot \frac{\partial^2 \bar{r}}{\partial \xi^i \partial \xi^k}, \tag{3.23}$$

$$\frac{\partial g_{ik}}{\partial \xi^j} = \frac{\partial^2 \bar{r}}{\partial \xi^j \partial \xi^i} \cdot \bar{g}_k + \bar{g}_i \cdot \frac{\partial^2 \bar{r}}{\partial \xi^j \partial \xi^k}. \tag{3.24}$$

Summing up Eqs. (3.22) and (3.23) and subtracting Eq. (3.24) we obtain

$$\frac{\partial g_{ij}}{\partial \xi^k} + \frac{\partial g_{jk}}{\partial \xi^i} - \frac{\partial g_{ik}}{\partial \xi^j} = 2\frac{\partial^2 \bar{r}}{\partial \xi^k \partial \xi^i} \cdot \bar{g}_j.$$

This equality together with Eq. (3.21) implies that

$$\Gamma_{ik}^m g_{mj} = \Gamma_{ij,k}, \tag{3.25}$$

where

$$\Gamma_{ij,k} = \frac{1}{2}\left(\frac{\partial g_{ij}}{\partial \xi^k} + \frac{\partial g_{jk}}{\partial \xi^i} - \frac{\partial g_{ik}}{\partial \xi^j}\right) \tag{3.26}$$

are *the Christoffel symbols of the first kind*. Finally, multiplying Eq. (3.25) by g^{js} and summing up with respect to j, we find

$$\Gamma_{ik}^s = \frac{1}{2}g^{js}\left(\frac{\partial g_{ij}}{\partial \xi^k} + \frac{\partial g_{jk}}{\partial \xi^i} - \frac{\partial g_{ik}}{\partial \xi^j}\right). \tag{3.27}$$

Exercise 3.6. Calculate the Christoffel symbols for Cartesian, cylindrical, and spherical coordinate frames. □

Exercise 3.7. Show that the Christoffel symbols do not form a tensor, i.e. that they do not satisfy the formula for transformation of a tensor's components at transition from one coordinate frame to another. □

3.5. Derivatives of the Dual Vectors

Differentiation of Eq. (1.13) with respect to ξ^k yields

$$\frac{\partial \bar{g}^i}{\partial \xi^k} \cdot \bar{g}_j + \bar{g}^i \cdot \frac{\partial \bar{g}_j}{\partial \xi^k} = 0.$$

This equality together with Eq. (3.18) implies that

$$\frac{\partial \bar{g}^i}{\partial \xi^k} \cdot \bar{g}_j = -\Gamma_{jk}^m \bar{g}^i \cdot \bar{g}_m = -\Gamma_{jk}^m \delta_{\cdot m}^{i\cdot} = -\Gamma_{jk}^i. \tag{3.28}$$

It follows from Eq. (3.28) that

$$\frac{\partial \bar{g}^i}{\partial \xi^k} = -\Gamma_{jk}^i \bar{g}^j. \tag{3.29}$$

3.6. Derivative of the Elementary Volume

Denote by

$$V = \bar{g}_1 \cdot (\bar{g}_2 \times \bar{g}_3) = \bar{g}_3 \cdot (\bar{g}_1 \times \bar{g}_2) = \bar{g}_2 \cdot (\bar{g}_3 \times \bar{g}_1) \tag{3.30}$$

the volume of a parallelepiped erected on tangent vectors $\bar{g}_i$ (the elementary volume), see Eq. (1.12). By using the formula for the triple product of vectors, we write

$$V = \det \begin{bmatrix} g_{1x^1} & g_{1x^2} & g_{1x^3} \\ g_{2x^1} & g_{2x^2} & g_{2x^3} \\ g_{3x^1} & g_{3x^2} & g_{3x^3} \end{bmatrix}, \tag{3.31}$$

where g_{ix^j} are projections of vectors $\bar{g}_i$ on Cartesian axes x^j. Multiplying Eq. (3.31) by itself we find

$$V^2 = \det \begin{bmatrix} g_{1x^1} & g_{1x^2} & g_{1x^3} \\ g_{2x^1} & g_{2x^2} & g_{2x^3} \\ g_{3x^1} & g_{3x^2} & g_{3x^3} \end{bmatrix} \begin{bmatrix} g_{1x^1} & g_{2x^1} & g_{3x^1} \\ g_{1x^2} & g_{2x^2} & g_{3x^2} \\ g_{1x^3} & g_{2x^3} & g_{3x^3} \end{bmatrix}$$

$$= \det \begin{bmatrix} \bar{g}_1 \cdot \bar{g}_1 & \bar{g}_1 \cdot \bar{g}_2 & \bar{g}_1 \cdot \bar{g}_3 \\ \bar{g}_2 \cdot \bar{g}_1 & \bar{g}_2 \cdot \bar{g}_2 & \bar{g}_2 \cdot \bar{g}_3 \\ \bar{g}_3 \cdot \bar{g}_1 & \bar{g}_3 \cdot \bar{g}_2 & \bar{g}_3 \cdot \bar{g}_3 \end{bmatrix} = \det \begin{bmatrix} g_{11} & g_{12} & g_{13} \\ g_{21} & g_{22} & g_{23} \\ g_{31} & g_{32} & g_{33} \end{bmatrix} = g.$$

It follows from this equality that

$$V = \sqrt{g}. \tag{3.32}$$

Let us calculate the derivative of V with respect to ξ^k. By employing Eq. (3.30) and the cyclic rule for the triple product, we find

$$\frac{\partial \sqrt{g}}{\partial \xi^k} = \frac{\partial}{\partial \xi^k} \bar{g}_1 \cdot (\bar{g} \times \bar{g}_3)$$

$$= \frac{\partial \bar{g}_1}{\partial \xi^k} \cdot (\bar{g}_2 \times \bar{g}_3) + \bar{g}_1 \cdot \left(\frac{\partial \bar{g}_2}{\partial \xi^k} \times \bar{g}_3\right) + \bar{g}_1 \cdot \left(\bar{g}_2 \times \frac{\partial \bar{g}_3}{\partial \xi^k}\right)$$

$$= \frac{\partial \bar{g}_1}{\partial \xi^k} \cdot (\bar{g}_2 \times \bar{g}_3) + \frac{\partial \bar{g}_2}{\partial \xi^k} \cdot (\bar{g}_3 \times \bar{g}_1) + \frac{\partial \bar{g}_3}{\partial \xi^k} \cdot (\bar{g}_1 \times \bar{g}_2).$$

With the use of Eq. (3.18) this equality can be written as

$$\frac{\partial \sqrt{g}}{\partial \xi^k} = \Gamma^i_{1k} \bar{g}_i \cdot (\bar{g}_2 \times \bar{g}_3) + \Gamma^i_{2k} \bar{g}_i \cdot (\bar{g}_3 \times \bar{g}_1) + \Gamma^i_{3k} \bar{g}_i \cdot (\bar{g}_1 \times \bar{g}_2).$$

Since the triple product of tangent vectors vanishes when any two vectors coincide, and equal $\sqrt{g}$ when these vectors differ from each other, the latter equality implies that

$$\frac{\partial \sqrt{g}}{\partial \xi^k} = \sqrt{g}(\Gamma^1_{1k} + \Gamma^2_{2k} + \Gamma^3_{3k}) = \sqrt{g}\,\Gamma^i_{ik}. \tag{3.33}$$

Eq. (3.33) can be rewritten as

$$\frac{\partial \ln \sqrt{g}}{\partial \xi^k} = \Gamma^i_{ik}. \tag{3.34}$$

Exercise 3.8. By direct calculations with the use of Eq. (3.27), show that Eq. (3.34) is fulfilled in Cartesian, cylindrical, and spherical coordinates. □

3.7. Covariant Derivative of a Vector

Let $\bar{q} = q^i \bar{g}_i = q_i \bar{g}^i$ be an arbitrary smooth vector field. We calculate the derivative of $\bar{q}$ with respect to ξ^j by using Eq. (3.18) and obtain

$$\frac{\partial \bar{q}}{\partial \xi^j} = \frac{\partial q^i}{\partial \xi^j}\bar{g}_i + q^i\frac{\partial \bar{g}_i}{\partial \xi^j} = \frac{\partial q^k}{\partial \xi^j}\bar{g}_k + q^i\Gamma^k_{ij}\bar{g}_k = \nabla_j q^k \bar{g}_k, \tag{3.35}$$

where

$$\nabla_j q^k = \frac{\partial q^k}{\partial \xi^j} + q^i\Gamma^k_{ij} \tag{3.36}$$

is called *the covariant derivative* of contravariant components of $\bar{q}$.

By employing the same reasoning, and using Eq. (3.29) we find

$$\frac{\partial \bar{q}}{\partial \xi^j} = \frac{\partial q_i}{\partial \xi^j}\bar{g}^i + q_i\frac{\partial \bar{g}^i}{\partial \xi^j} = \frac{\partial q_k}{\partial \xi^j}\bar{g}^k - q_i\Gamma^i_{jk}\bar{g}^k = \nabla_j q_k \bar{g}^k, \tag{3.37}$$

where

$$\nabla_j q_k = \frac{\partial q_k}{\partial \xi^j} - q_i\Gamma^i_{jk} \tag{3.38}$$

is called the covariant derivative of covariant components of $\bar{q}$.

Multiplication of Eq. (3.35) by $\bar{g}^i$ and of Eq. (3.37) by $\bar{g}_i$ yields

$$\nabla_j q^i = \bar{g}^i \cdot \frac{\partial \bar{q}}{\partial \xi^j}, \qquad \nabla_j q_i = \bar{g}_i \cdot \frac{\partial \bar{q}}{\partial \xi^j}. \tag{3.39}$$

Tensor multiplication of Eqs. (3.35) and (3.37) by $\bar{g}^j$ implies that

$$\bar{\nabla}\bar{q} = \bar{g}^j \frac{\partial \bar{q}}{\partial \xi^j}, \tag{3.40}$$

where the tensor of the second rank

$$\bar{\nabla}\bar{q} = \nabla_j q_k \bar{g}^j \bar{g}^k = \nabla_j q^k \bar{g}^j \bar{g}_k \tag{3.41}$$

is called the covariant derivative of a vector $\bar{q}$.

3.8. Covariant Derivative of a Tensor

By employing a similar approach, we can derive expressions for the covariant derivative of a tensor

$$\hat{Q} = Q^{ij}\bar{g}_i\bar{g}_j = Q^i_{\cdot j}\bar{g}_i\bar{g}^j = Q_{ij}\bar{g}^i\bar{g}^j. \tag{3.42}$$

For example, the first expression (3.42) and Eq. (3.18) imply that

$$\frac{\partial \hat{Q}}{\partial \xi^k} = \frac{\partial}{\partial \xi^k}(Q^{ij}\bar{g}_i\bar{g}_j) = \frac{\partial Q^{ij}}{\partial \xi^k}\bar{g}_i\bar{g}_j + Q^{ij}\frac{\partial \bar{g}_i}{\partial \xi^k}\bar{g}_j + Q^{ij}\bar{g}_i\frac{\partial \bar{g}_j}{\partial \xi^k}$$

$$= (\frac{\partial Q^{ij}}{\partial \xi^k} + Q^{mj}\Gamma^i_{mk} + Q^{im}\Gamma^j_{mk})\bar{g}_i\bar{g}_j = \nabla_k Q^{ij}\bar{g}_i\bar{g}_j, \tag{3.43}$$

where

$$\nabla_k Q^{ij} = \frac{\partial Q^{ij}}{\partial \xi^k} + Q^{mj}\Gamma^i_{mk} + Q^{im}\Gamma^j_{mk} \tag{3.44}$$

is the covariant derivative of contravariant components of $\hat{Q}$.

Exercise 3.9. Derive expressions for the covariant derivatives of combined and covariant components of a tensor $\hat{Q}$. $\square$

Multiplication of Eq. (3.43) by $\bar{g}^m$ from the left and by $\bar{g}^n$ from the right yields

$$\nabla_k Q^{mn} = \bar{g}^m \cdot \frac{\partial \hat{Q}}{\partial \xi^k} \cdot \bar{g}^n. \tag{3.45}$$

Tensor multiplication of Eq. (3.43) by $\bar{g}^k$ implies that

$$\bar{\nabla}\hat{Q} = \bar{g}^k \frac{\partial \hat{Q}}{\partial \xi^k} = \nabla_k Q^{ij} \bar{g}^k \bar{g}_i \bar{g}_j. \tag{3.46}$$

Exercise 3.10. Derive formulas for the tensor $\bar{\nabla}\hat{Q}$ by using combined and covariant components of $\hat{Q}$. $\square$

Exercise 3.11. Check that for any vectors $\bar{p}$ and $\bar{q}$

$$\nabla_k(p^i q_j) = (\nabla_k p^i)q_j + p^i(\nabla_k q_j). \quad \square$$

Exercise 3.12. Derive the following formulas for the covariant derivative of a scalar function f in Cartesian, cylindrical, and spherical coordinates:

$$\bar{\nabla}f = \frac{\partial f}{\partial x^1}\bar{e}_1 + \frac{\partial f}{\partial x^2}\bar{e}_2 + \frac{\partial f}{\partial x^3}\bar{e}_3, \qquad \bar{\nabla}f = \frac{\partial f}{\partial r}\bar{e}_r + \frac{1}{r}\frac{\partial f}{\partial \theta}\bar{e}_\theta + \frac{\partial f}{\partial z}\bar{e}_z,$$

$$\bar{\nabla}f = \frac{\partial f}{\partial r}\bar{e}_r + \frac{1}{r}\frac{\partial f}{\partial \theta}\bar{e}_\theta + \frac{1}{r\sin\theta}\frac{\partial f}{\partial \phi}\bar{e}_\phi. \quad \square \tag{3.47}$$

3.9. The Ricci Theorem

Proposition 3.6 (The Ricci theorem). For any positive integers i, j and k

$$\nabla_k g_{ij} = 0. \tag{3.48}$$

Proof. In Cartesian coordinates $\{x^i\}$ with the unit vectors $\bar{e}_i$, the unit tensor $\hat{I}$ is presented as $\hat{I} = \bar{e}_1\bar{e}_1 + \bar{e}_2\bar{e}_2 + \bar{e}_3\bar{e}_3$. It follows from this expression that $\hat{I}$ is independent of spatial coordinates

$$\frac{\partial \hat{I}}{\partial \xi^k} = 0 \qquad (k = 1, 2, 3). \tag{3.49}$$

In curvilinear coordinates $\{\xi^i\}$, the unit tensor has the form $\hat{I} = g_{ij}\bar{g}^i\bar{g}^j$. Substitution of this expression into Eq. (3.49) yields

$$\nabla_k g_{ij}\bar{g}^i\bar{g}^j = 0,$$

which implies Eq. (3.48). $\square$

Exercise 3.13. Prove that for any integers i, j, and k

$$\nabla_k g^{ij} = 0. \qquad \square \tag{3.50}$$

Eqs. (3.48) and (3.50) allow us to lower and raise indices under the sign of the covariant derivative. For example, Eq. (3.50) implies that

$$\nabla_k q^i = \nabla_k(g^{ij}q_j) = \nabla_k(g^{ij})q_j + g^{ij}\nabla_k q_j = g^{ij}\nabla_k q_j.$$

3.10. The Divergence of a Vector and a Tensor

Let us calculate the divergence of a vector field $\bar{q}$. By using Eqs. (3.8), (3.9), and (3.35) we find

$$\operatorname{div} \bar{q} = \bar{\nabla} \cdot \bar{q} = \bar{g}^i \cdot \frac{\partial \bar{q}}{\partial \xi^i} = \bar{g}^i \cdot \left(\frac{\partial q^j}{\partial \xi^i} + q^k\Gamma^j_{ki}\right)\bar{g}_j$$

$$= \left(\frac{\partial q^j}{\partial \xi^i} + q^k\Gamma^j_{ki}\right)\delta^i_{\cdot j} = \frac{\partial q^i}{\partial \xi^i} + q^k\Gamma^i_{ki}.$$

By employing the symmetry of the Christoffel symbols (3.19) and formula (3.34) we obtain

$$\operatorname{div} \bar{q} = \frac{\partial q^i}{\partial \xi^i} + q^k\Gamma^i_{ik} = \frac{\partial q^k}{\partial \xi^k} + \frac{q^k}{\sqrt{g}}\frac{\partial \sqrt{g}}{\partial \xi^k} = \frac{1}{\sqrt{g}}\frac{\partial(\sqrt{g}q^k)}{\partial \xi^k}. \tag{3.51}$$

Exercise 3.14. By using formula (3.51), derive the following formulas for the divergence of a vector $\bar{q}$ in cylindrical and spherical coordinates:

$$\operatorname{div} \bar{q} = \frac{\partial q_r}{\partial r} + \frac{q_r}{r} + \frac{1}{r}\frac{\partial q_\theta}{\partial \theta} + \frac{\partial q_z}{\partial z},$$

$$\operatorname{div} \bar{q} = \frac{\partial q_r}{\partial r} + \frac{2q_r}{r} + \frac{1}{r}\left(\frac{\partial q_\theta}{\partial \theta} + \frac{\cos\theta}{\sin\theta}q_\theta + \frac{1}{\sin\theta}\frac{\partial q_\phi}{\partial \phi}\right). \qquad \square$$

By employing a similar reasoning, an explicit expression can be derived for the divergence of a tensor field of the second rank $\hat{Q} = Q^{ij}\bar{g}_i\bar{g}_j$

$$\operatorname{div} \hat{Q} = \bar{\nabla} \cdot \hat{Q} = \bar{g}^k \cdot \frac{\partial}{\partial \xi^k}(Q^{ij}\bar{g}_i\bar{g}_j)$$

$$= \bar{g}^k \cdot \left(\frac{\partial Q^{ij}}{\partial \xi^k} \bar{g}_i \bar{g}_j + Q^{ij} \Gamma^m_{ik} \bar{g}_m \bar{g}_j + Q^{ij} \bar{g}_i \frac{\partial \bar{g}_j}{\partial \xi^k} \right)$$

$$= \frac{\partial Q^{kj}}{\partial \xi^k} \bar{g}_j + Q^{ij} \Gamma^k_{ki} \bar{g}_j + Q^{kj} \frac{\partial \bar{g}_j}{\partial \xi^k}$$

$$= \frac{\partial Q^{ij}}{\partial \xi^i} \bar{g}_j + \frac{1}{\sqrt{g}} Q^{ij} \frac{\partial \sqrt{g}}{\partial \xi^i} \bar{g}_j + Q^{ij} \frac{\partial \bar{g}_j}{\partial \xi^i}$$

$$= \frac{1}{\sqrt{g}} \frac{\partial (Q^{ij} \sqrt{g} \bar{g}_j)}{\partial \xi^i} = \frac{1}{\sqrt{g}} \frac{\partial (\sqrt{g} \bar{g}^i \cdot \hat{Q})}{\partial \xi^i}. \tag{3.52}$$

Exercise 3.15. Derive expressions for the divergence of a tensor $\hat{Q}$ in cylindrical and spherical coordinates. □

3.11. The Second Covariant Derivative

In the course of calculus, the second derivative of a scalar function is defined as the derivative of the first derivative. The same approach is employed in the tensor calculus. *The second covariant derivative* of a scalar function f is defined as the covariant derivative of its first covariant derivative. By using Eq. (3.29) we obtain

$$\bar{\nabla} \bar{\nabla} f = \bar{g}^i \frac{\partial}{\partial \xi^i} \left(\bar{g}^j \frac{\partial}{\partial \xi^j} f \right) = \bar{g}^i \bar{g}^j \frac{\partial^2 f}{\partial \xi^i \partial \xi^j} + \bar{g}^i \frac{\partial \bar{g}^j}{\partial \xi^i} \frac{\partial f}{\partial \xi^j}$$

$$= \left(\frac{\partial^2 f}{\partial \xi^i \partial \xi^k} - \Gamma^j_{ik} \frac{\partial f}{\partial \xi^j} \right) \bar{g}^i \bar{g}^k. \tag{3.53}$$

Similar to Eq. (3.53), the second covariant derivative may be defined for a tensor of an arbitrary rank.

Exercise 3.16. Show that for any tensor of the second rank $\hat{Q}$ and for any vector $\bar{q}$

$$\bar{\nabla} (\bar{\nabla} \bar{q}^T) : \hat{Q} = \hat{Q}^T : \bar{\nabla} (\bar{\nabla} \bar{q}^T). \qquad \square \tag{3.54}$$

Introduce *the Laplacian* of a smooth scalar function f as

$$\Delta f = \bar{\nabla} \cdot \bar{\nabla} f. \tag{3.55}$$

Exercise 3.17. By using Eqs. (3.53) and (3.55) show that in a curvilinear coordinate frame the Laplacian of a scalar function can be presented as follows:

$$\Delta f = g^{ij} \left(\frac{\partial^2 f}{\partial \xi^i \partial \xi^j} - \Gamma^k_{ij} \frac{\partial f}{\partial \xi^k} \right). \qquad \square \tag{3.56}$$

Exercise 3.18. Check that in Cartesian coordinates Eq. (3.56) coincides with the standard formula for the Laplacian of a scalar function

$$\Delta f = \frac{\partial^2 f}{\partial (x^1)^2} + \frac{\partial^2 f}{\partial (x^2)^2} + \frac{\partial^2 f}{\partial (x^3)^2}. \qquad \square$$

Exercise 3.19. Derive the following formulas for the Laplacian of a scalar function f in the cylindrical and spherical coordinates:

$$\Delta f = \frac{\partial^2 f}{\partial r^2} + \frac{1}{r}\frac{\partial f}{\partial r} + \frac{1}{r^2}\frac{\partial^2 f}{\partial \theta^2} + \frac{\partial^2 f}{\partial z^2},$$

$$\Delta f = \frac{\partial^2 f}{\partial r^2} + \frac{2}{r}\frac{\partial f}{\partial r} + \frac{1}{r^2}\left(\frac{\partial^2 f}{\partial \theta^2} - \tan\theta\frac{\partial f}{\partial \theta} + \frac{1}{\cos^2\theta}\frac{\partial^2 f}{\partial \phi^2}\right). \quad \square$$

3.12. The Stokes Formula

Let Ω be a bounded connected domain with a smooth boundary Γ. Points of Ω refer to Cartesian coordinates $\{x^i\}$ with the unit vectors $\bar{e}_i$. Let $\bar{n} = n_{x^1}\bar{e}_1 + n_{x^2}\bar{e}_2 + n_{x^3}\bar{e}_3$ be the unit outward *normal vector* to surface Γ.

Proposition 3.7 (The Stokes theorem). For any smooth vector field $\bar{q}$

$$\int_\Omega \bar{\nabla}\cdot\bar{q}\,dV = \int_\Gamma \bar{n}\cdot\bar{q}\,dS, \tag{3.57}$$

where dV is *the volume element*, and dS is *the surface element*.

Proof. Let us expand the vector $\bar{q}$ in the unit vectors $\bar{e}_i$: $\bar{q} = q_1\bar{e}_1 + q_2\bar{e}_2 + q_3\bar{e}_3$. By using Eq. (3.9), we can write

$$\int_\Omega \bar{\nabla}\cdot\bar{q}\,dV = \int_\Omega \left(\frac{\partial q_{x^1}}{\partial x^1} + \frac{\partial q_{x^2}}{\partial x^3} + \frac{\partial q_{x^3}}{\partial x^3}\right)dV. \tag{3.58}$$

The well-known Stokes theorem states that for any sufficiently smooth scalar function f

$$\int_\Omega \frac{\partial f}{\partial x^i}dV = \int_\Gamma n_{x^i}f\,dS.$$

By applying this assertion to Eq. (3.58), we find

$$\int_\Omega \left(\frac{\partial q_{x^1}}{\partial x^1} + \frac{\partial q_{x^2}}{\partial x^3} + \frac{\partial q_{x^3}}{\partial x^3}\right)dV = \int_\Gamma (n_{x^1}q_{x^1} + n_{x^2}q_{x^2} + n_{x^3}q_{x^3})dS = \int_\Gamma \bar{n}\cdot\bar{q}\,dS.$$

This formula together with Eq. (3.58) implies Eq. (3.57). $\square$

The Stokes formula (3.57) can be extended to tensors of an arbitrary rank.

Exercise 3.20. Prove the Stokes formula for a tensor field $\hat{Q}$ of the second rank

$$\int_\Omega \bar{\nabla}\cdot\hat{Q}\,dV = \int_\Gamma \bar{n}\cdot\hat{Q}\,dS. \quad \square \tag{3.59}$$

Exercise 3.21. Show that the following equalities hold:

$$\int_\Omega \bar{\nabla}\times\bar{q}\,dV = \int_\Gamma \bar{n}\times\bar{q}\,dS, \quad \int_\Omega \bar{\nabla}\times\hat{Q}\,dV = \int_\Gamma \bar{n}\times\hat{Q}\,dS. \quad \square \tag{3.60}$$

Proposition 3.8. For a symmetrical tensor field of the second rank $\hat{Q}$ and a vector field $\bar{q}$

$$\int_\Omega (\bar{\nabla} \cdot \hat{Q}) \cdot \bar{q} dV = \int_\Gamma \bar{n} \cdot (\hat{Q} \cdot \bar{q}) dS - \int_\Omega \hat{Q} : \hat{\epsilon}(\bar{q}) dV, \qquad (3.61)$$

where tensor $\hat{\epsilon}(\bar{q})$ is determined by Eq. (3.12).
Proof. It follows from Eq. (3.16) that

$$\int_\Omega (\bar{\nabla} \cdot \hat{Q}) \cdot \bar{q} dV = \int_\Omega [\bar{\nabla} \cdot (\hat{Q} \cdot \bar{q}) - \hat{Q} : \hat{\epsilon}(\bar{q})] dV.$$

Applying the Stokes formula (3.57) to the first term in the right-hand side of this equality, we obtain Eq. (3.61). $\square$

4. Corotational Derivatives

In this Section we provide a brief survey of corotational derivatives, i.e. derivatives which transform an objective tensor (independent of superimposed rigid motions) into an objective tensor of the same range. The corotational derivatives serve as analogues of material time derivatives (which are not indifferent with respect to superimposed rigid motions) in the constitutive theory for inelastic media.

4.1. Objective Tensors

Let us consider two motions of a continuum. Points of the medium refer to Lagrangian curvilinear coordinates $\xi = \{\xi^i\}$. At current instant t, the radius-vector of a point ξ equals

$$\bar{r} = \bar{\mathcal{R}}(t, \xi) \qquad (4.1)$$

in the first motion, and equals

$$\bar{r} = \bar{\mathcal{R}}'(t, \xi) \qquad (4.2)$$

in the other one.

Fix a point $\mathcal{P}$ which is called *pole*. Let $\bar{R}_0$ and $\bar{R}_0'$ be radius-vectors of the point $\mathcal{P}$ in the first and second motions, respectively. The relative radius-vectors (with respect to the pole) are denoted as $\bar{R}$ and $\bar{R}'$

$$\bar{\mathcal{R}} = \bar{R}_0 + \bar{R}, \qquad \bar{\mathcal{R}}' = \bar{R}_0' + \bar{R}'. \qquad (4.3)$$

Definition. Two motions $\bar{\mathcal{R}}$ and $\bar{\mathcal{R}}'$ differ from each other by *a rigid motion* if $|\bar{R}'(t)| = |\bar{R}(t)|$ for any point ξ and for any instant t.

Proposition 4.1. Two motions $\bar{\mathcal{R}}$ and $\bar{\mathcal{R}}'$ differ from each other by a rigid motion if there exists an orthogonal tensor function $\hat{O} = \hat{O}(t)$ such that for any point ξ and for any instant t

$$\bar{R}' = \bar{R} \cdot \hat{O}. \qquad \square \qquad (4.4)$$

Exercise 4.1. Prove this assertion. $\square$

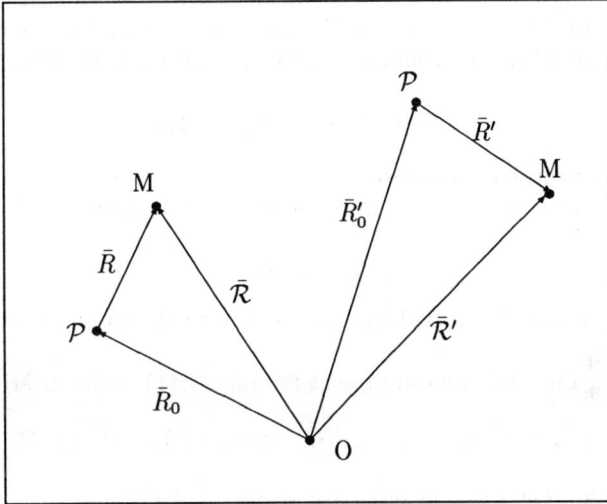

Fig. 4.1. Kinematics of two rigid motions.

Substitution of expression (4.4) into Eq. (4.3) yields

$$\bar{\mathcal{R}}'(t,\xi) = \bar{R}'_0(t) + [\bar{\mathcal{R}}(t,\xi) - \bar{R}_0(t)] \cdot \hat{O}(t). \qquad (4.5)$$

Eq. (4.5) determines a rigid motion superimposed on motion (4.1). Differentiation of Eq. (4.5) with respect to ξ^i implies the following formula for the tangent vectors:

$$\bar{g}'_i(t,\xi) = \bar{g}_i(t,\xi) \cdot \hat{O}(t) = \hat{O}^T(t) \cdot \bar{g}_i(t,\xi). \qquad (4.6)$$

Exercise 4.2. Check that dual vectors $\bar{g}^i$ and $\bar{g}^{i\prime}$ satisfy the equality

$$\bar{g}^{i\prime}(t,\xi) = \bar{g}^i(t,\xi) \cdot \hat{O}(t) = \hat{O}^T(t) \cdot \bar{g}^i(t,\xi). \qquad \square \qquad (4.7)$$

Let us consider a vector field $\bar{q}(t,\xi)$ frozen into a deformable medium. Denote by q^i and $q^{i\prime}$ contravariant components of vector $\bar{q}$

$$\bar{q} = q^i \bar{g}_i, \qquad \bar{q}' = q^{i\prime} \bar{g}'_i. \qquad (4.8)$$

33

Definition. A vector field $\bar{q}$ is called *objective* or *indifferent with respect to rigid motions* if

$$q^{i\prime} = q^i.$$

It follows from this equality together with Eq. (4.6) that

$$\bar{q}' = q^{i\prime}(\bar{g}_i \cdot \hat{O}) = q^i(\hat{O}^T \cdot \bar{g}_i) = \hat{O}^T \cdot \bar{q}. \tag{4.9}$$

Eq. (4.9) provides a coordinate-independent definition of an objective vector.

Using a similar approach, an objective tensor of an arbitrary rank may be defined. For simplicity we confine ourselves to a tensor field of the second rank

$$\hat{Q} = Q^{ij}\bar{g}_i\bar{g}_j, \qquad \hat{Q} = Q^{ij\prime}\bar{g}_i'\bar{g}_j', \tag{4.10}$$

frozen into a deformable medium.

Definition. A tensor field $\hat{Q}$ is called objective or indifferent with respect to a rigid motion if

$$Q^{ij\prime} = Q^{ij}, \tag{4.11}$$

i.e. if its components are independent of any rigid motion superimposed on motion (4.1).

By using Eq. (4.6) we can present formula (4.11) in the invariant form

$$\hat{Q}' = Q^{ij\prime}(\bar{g}_i \cdot \hat{O})(\bar{g}_j \cdot \hat{O}) = Q^{ij}(\hat{O}^T \cdot \bar{g}_i)(\bar{g}_j \cdot \hat{O}) = \hat{O}^T \cdot \hat{Q} \cdot \hat{O}. \tag{4.12}$$

Exercise 4.3. Let $\bar{q}$ be an objective vector, and $\hat{Q}$ be an objective tensor. Prove that $\bar{q} \cdot \hat{Q}$ is an objective vector. $\square$

Exercise 4.4. Let $\hat{P}$ and $\hat{Q}$ be objective tensors. Show that $\hat{P} \cdot \hat{Q}$ is an objective tensor. $\square$

Let us now introduce invariant and isotropic tensors.

Definition. A tensor $\hat{Q}$ is called *invariant with respect to rigid motion* if

$$\hat{Q}' = \hat{Q}, \tag{4.13}$$

i.e. if tensor $\hat{Q}$ does not respond to superimposed rigid motions.

Definition. A tensor $\hat{Q}$ is called *isotropic with respect to rigid motions* if

$$\hat{Q} = Q^{ij}\bar{g}_i\bar{g}_j = Q^{ij}\bar{g}_i'\bar{g}_j', \tag{4.14}$$

i.e. if its components in nonperturbed and perturbed bases coincide.

Exercise 4.5. Prove the following assertions:

(i) any scalar is an isotropic tensor of zero rank;

(ii) there are no isotropic vectors;

(iii) the only isotropic tensor of the second rank is the unit tensor $\hat{I}$;

34

(iv) the only isotropic tensor of the third rank is *the Levi–Civita tensor*

$$\hat{\mathfrak{z}} = -\hat{I} \times \hat{I}. \quad \square$$

4.2. Velocity and Its Gradient

Denote by $\bar{v}$ and $\bar{v}'$ the velocity vectors corresponding to motions (4.1) and (4.2)

$$\bar{v} = \frac{\partial \bar{R}}{\partial t}, \qquad \bar{v}' = \frac{\partial \bar{R}'}{\partial t}. \tag{4.15}$$

Transform the latter formula (4.15) with the use of Eq. (4.5) as follows:

$$\bar{v}' = \frac{d\bar{R}'_0}{dt} + (\bar{v} - \frac{d\bar{R}_0}{dt}) \cdot \hat{O} + (\bar{R} - \bar{R}_0) \cdot \frac{d\hat{O}}{dt}. \tag{4.16}$$

Differentiation of Eq. (2.38) with respect to time yields

$$\frac{d\hat{O}}{dt} \cdot \hat{O}^T + \hat{O} \cdot \frac{d\hat{O}^T}{dt} = 0,$$

which implies that

$$\frac{d\hat{O}}{dt} = -\hat{O} \cdot \frac{d\hat{O}^T}{dt} \cdot \hat{O}. \tag{4.17}$$

It follows from Eqs. (4.5), (4.16), and (4.17) that

$$\begin{aligned}
\bar{v}' &= \frac{d\bar{R}'_0}{dt} + (\bar{v} - \frac{d\bar{R}_0}{dt}) \cdot \hat{O} - (\bar{R} - \bar{R}_0) \cdot \hat{O} \cdot \frac{d\hat{O}^T}{dt} \cdot \hat{O} \\
&= \frac{d\bar{R}'_0}{dt} + (\bar{v} - \frac{d\bar{R}_0}{dt}) \cdot \hat{O} - (\bar{R}' - \bar{R}'_0) \cdot \frac{d\hat{O}^T}{dt} \cdot \hat{O}.
\end{aligned} \tag{4.18}$$

Definition. The tensor

$$\hat{\Omega} = \frac{d\hat{O}^T}{dt} \cdot \hat{O} \tag{4.19}$$

is called *spin tensor*.

Exercise 4.6. By using Eq. (4.17), check that the spin tensor $\hat{\Omega}$ is skew-symmetrical

$$\hat{\Omega}^T = -\hat{\Omega}. \quad \square \tag{4.20}$$

Exercise 4.7. Show that tensors $\hat{O}$ and $\hat{O}^T$ satisfy the differential equations

$$\frac{d\hat{O}}{dt} = -\hat{O} \cdot \hat{\Omega}, \qquad \frac{d\hat{O}^T}{dt} = \hat{\Omega} \cdot \hat{O}^T. \quad \square \tag{4.21}$$

It follows from Eqs. (4.18) – (4.20) that

$$\bar{v}' = \frac{d\bar{R}'_0}{dt} + (\bar{v} - \frac{d\bar{R}_0}{dt}) \cdot \hat{O} - (\bar{R}' - \bar{R}'_0) \cdot \hat{\Omega}$$

$$= \frac{d\bar{R}'_0}{dt} + (\bar{v} - \frac{d\bar{R}_0}{dt}) \cdot \hat{O} + \hat{\Omega} \cdot (\bar{R}' - \bar{R}'_0). \qquad (4.22)$$

Let us calculate the covariant derivative of vector $\bar{v}'$. Eqs. (3.40), (4.7), and (4.22) imply that

$$\bar{\nabla}' \bar{v}' = \bar{g}^{i\prime} \frac{\partial \bar{v}'}{\partial \xi^i} = \hat{O}^T \cdot \bar{g}^i \frac{\partial \bar{v}'}{\partial \xi^i} = \hat{O}^T \cdot \bar{g}^i (\frac{\partial \bar{v}}{\partial \xi^i} \cdot \hat{O} - \frac{\partial \bar{R}'}{\partial \xi^i} \cdot \hat{\Omega})$$

$$= \hat{O}^T \cdot (\bar{g}^i \frac{\partial \bar{v}}{\partial \xi^i} \cdot \hat{O} - \bar{g}^i \bar{g}'_i \cdot \hat{\Omega}) = \hat{O}^T \cdot (\bar{\nabla} \bar{v} \cdot \hat{O} - \bar{g}^i \bar{g}_i \cdot \hat{O} \cdot \hat{\Omega})$$

$$= \hat{O}^T \cdot \bar{\nabla} \bar{v} \cdot \hat{O} - \hat{O}^T \cdot \hat{I} \cdot \hat{O} \cdot \hat{\Omega} = \hat{O}^T \cdot \bar{\nabla} \bar{v} \cdot \hat{O} - \hat{\Omega}. \qquad (4.23)$$

We transpose Eq. (4.23) and use Eq. (4.20). As a result, we find

$$(\bar{\nabla}' \bar{v}')^T = \hat{O}^T \cdot \bar{\nabla} \bar{v}^T \cdot \hat{O} + \hat{\Omega}. \qquad (4.24)$$

According to Eqs. (4.16) and (4.23), both the velocity vector and its gradient are not indifferent with respect to superposition of rigid motions.

4.3. The Zorawski Formula

Introduce *the rate-of-strain tensor* $\hat{D}$ and *the vorticity tensor* $\hat{Y}$ as follows:

$$\hat{D} = \frac{1}{2}(\bar{\nabla} \bar{v}^T + \bar{\nabla} \bar{v}), \quad \hat{Y} = \frac{1}{2}(\bar{\nabla} \bar{v}^T - \bar{\nabla} \bar{v}). \qquad (4.25)$$

The tensor $\hat{D}$ is also called the rate-of-deformation tensor, the tensor of the strain rate, and the velocity gradient.

Let us calculate tensors $\hat{D}$ and $\hat{Y}$ for the perturbed motion (4.2). Substitution of expressions (4.23) and (4.24) into Eq. (4.25) yields

$$\hat{D}' = \hat{O}^T \cdot \hat{D} \cdot \hat{O}, \qquad \hat{Y}' = \hat{O}^T \cdot \hat{Y} \cdot \hat{O} + \hat{\Omega}. \qquad (4.26)$$

The latter equation (4.26) can be presented in the form

$$\hat{\Omega} = \hat{Y}' - \hat{O}^T \cdot \hat{Y} \cdot \hat{O}. \qquad (4.27)$$

Eqs. (4.26) are called *Zorawski's formulas*. According to them, the rate-of-strain tensor is objective, whereas the vorticity tensor is not indifferent with respect to superposition of rigid motions. This assertion is used for constructing *corotational derivatives*, i.e. derivatives of objective tensors which are indifferent with respect to superposition of rigid motions.

4.4. The Jaumann Derivative

Let $\bar{q}$ be an objective vector field

$$\bar{q}' = \bar{q} \cdot \hat{O}. \tag{4.28}$$

We calculate the derivative of $\bar{q}$ with respect to time for fixed Lagrangian coordinates. This derivative is called *material derivative* of $\bar{q}$. Differentiation of Eq. (4.28) with the use of Eq. (4.21) yields

$$\frac{\partial \bar{q}'}{\partial t} = \frac{\partial \bar{q}}{\partial t} \cdot \hat{O} + \bar{q} \cdot \frac{d\hat{O}}{dt} = \frac{\partial \bar{q}}{\partial t} \cdot \hat{O} - \bar{q} \cdot \hat{O} \cdot \hat{\Omega}.$$

Substitution of expression (4.27) into this equality implies that

$$\frac{\partial \bar{q}'}{\partial t} = \frac{\partial \bar{q}}{\partial t} \cdot \hat{O} - \bar{q} \cdot \hat{O} \cdot (\hat{Y}' - \hat{O}^T \cdot \hat{Y} \cdot \hat{O})$$

$$= \frac{\partial \bar{q}}{\partial t} \cdot \hat{O} - \bar{q} \cdot \hat{O} \cdot \hat{Y}' + \bar{q} \cdot \hat{Y} \cdot \hat{O}. \tag{4.29}$$

It follows from Eqs. (4.28) and (4.29) that

$$\frac{\partial \bar{q}'}{\partial t} + \bar{q}' \cdot \hat{Y}' = (\frac{\partial \bar{q}}{\partial t} + \bar{q}' \cdot \hat{Y}) \cdot \hat{O}. \tag{4.30}$$

According to Eq. (4.30), the vector

$$\bar{q}^{\circ} = \frac{\partial \bar{q}}{\partial t} + \bar{q} \cdot \hat{Y} \tag{4.31}$$

is indifferent with respect to superposed rigid motions. It is called *the Jaumann derivative* of an objective vector $\bar{q}$.

By employing a similar reasoning, the Jaumann derivative may be defined for an objective tensor of the second rank $\hat{Q}$. By using Eq. (4.12) and repeating the above calculations, we find

$$\frac{\partial \hat{Q}'}{\partial t} = \frac{d\hat{O}^T}{dt} \cdot \hat{Q} \cdot \hat{O} + \hat{O}^T \cdot \frac{\partial \hat{Q}}{\partial t} \cdot \hat{O} + \hat{O}^T \cdot \hat{Q} \cdot \frac{d\hat{O}}{dt}$$

$$= \hat{\Omega} \cdot \hat{O}^T \cdot \hat{Q} \cdot \hat{O} + \hat{O}^T \cdot \frac{\partial \hat{Q}}{\partial t} \cdot \hat{O} - \hat{O}^T \cdot \hat{Q} \cdot \hat{O} \cdot \hat{\Omega}$$

$$= (\hat{Y}' - \hat{O}^T \cdot \hat{Y} \cdot \hat{O}) \cdot \hat{O}^T \cdot \hat{Q} \cdot \hat{O} + \hat{O}^T \cdot \frac{\partial \hat{Q}}{\partial t} \cdot \hat{O}$$

$$\qquad - \hat{O}^T \cdot \hat{Q} \cdot \hat{O} \cdot (\hat{Y}' - \hat{O}^T \cdot \hat{Y} \cdot \hat{O})$$

$$= \hat{Y}' \cdot \hat{Q}' - \hat{O}^T \cdot \hat{Y} \cdot \hat{Q} \cdot \hat{O} + \hat{O}^T \cdot \frac{\partial \hat{Q}}{\partial t} \cdot \hat{O}$$

$$\qquad - \hat{Q}' \cdot \hat{Y}' + \hat{O}^T \cdot \hat{Q} \cdot \hat{Y} \cdot \hat{O}.$$

Therefore,

$$\frac{\partial \hat{Q}'}{\partial t} - \hat{Y}' \cdot \hat{Q}' + \hat{Q}' \cdot \hat{Y}' = \hat{O}^T \cdot \left(\frac{\partial \hat{Q}}{\partial t} - \hat{Y} \cdot \hat{Q} + \hat{Q} \cdot \hat{Y}\right) \cdot \hat{O}.$$

This equality means that the Jaumann derivative of an objective tensor $\hat{Q}$

$$\hat{Q}^\circ = \frac{\partial \hat{Q}}{\partial t} + \hat{Q} \cdot \hat{Y} - \hat{Y} \cdot \hat{Q} \tag{4.32}$$

is indifferent with respect to superposed rigid motions.

By using *the Lie brackets* $\{\hat{Q}, \hat{Y}\} = \hat{Q} \cdot \hat{Y} - \hat{Y} \cdot \hat{Q}$, Eq. (4.32) can be presented as follows:

$$\hat{Q}^\circ = \frac{\partial \hat{Q}}{\partial t} + \{\hat{Q}, \hat{Y}\}. \tag{4.33}$$

Proposition 4.2. The Jaumann derivative of the inner product of two objective tensors $\hat{P}$ and $\hat{Q}$ obeys *the Leibnitz formula*

$$(\hat{P} \cdot \hat{Q})^\circ = \hat{P}^\circ \cdot \hat{Q} + \hat{P} \cdot \hat{Q}^\circ. \qquad \square \tag{4.34}$$

Exercise 4.8. By direct calculations, derive Eq. (4.34) from Eq. (4.32). $\square$
Proposition 4.3. Let $\hat{Q}$ be an objective tensor with

$$\hat{Q}^\circ = 0. \tag{4.35}$$

Then all the principal invariants $I_k(\hat{Q})$ are independent of time.
Proof. It follows from Eqs. (4.32) and (4.35) that

$$\frac{\partial \hat{Q}}{\partial t} = \hat{Y} \cdot \hat{Q} - \hat{Q} \cdot \hat{Y}. \tag{4.36}$$

Let us calculate the derivative with respect to time of the first principal invariant $I_1(\hat{Q}) = \hat{I} : \hat{Q}$. By using Eq. (4.36), we find

$$\frac{\partial I_1(\hat{Q})}{\partial t} = \frac{\partial}{\partial t}(\hat{I} : \hat{Q}) = \hat{I} : \frac{\partial \hat{Q}}{\partial t} = \hat{I} : (\hat{Y} \cdot \hat{Q} - \hat{Q} \cdot \hat{Y}) = \hat{Y} : \hat{Q} - \hat{Q} : \hat{Y} = 0. \tag{4.37}$$

Eqs. (4.34) and (4.35) imply that $(\hat{Q}^2)^\circ = 0$ and $(\hat{Q}^3)^\circ = 0$. Carrying out calculations similar to Eq. (4.37), we obtain

$$\frac{\partial I_1(\hat{Q}^2)}{\partial t} = 0, \qquad \frac{\partial I_1(\hat{Q}^3)}{\partial t} = 0. \tag{4.38}$$

The required assertion follows from these equalities together with Eqs. (2.30) and (4.37). □

4.5. The Oldroyd Derivatives

It follows from Eqs. (4.25) that

$$\hat{Y} = \hat{D} - \bar{\nabla}\bar{v}, \qquad \hat{Y} = -\hat{D} + \bar{\nabla}\bar{v}^T. \tag{4.39}$$

Let $\bar{q}$ be an objective vector field with the Jaumann derivative (4.31). Substitution of the first expression (4.38) into (4.31) yields

$$\bar{q}^{\circ} = \frac{\partial \bar{q}}{\partial t} + \bar{q} \cdot \hat{D} - \bar{q} \cdot \bar{\nabla}\bar{v} = \bar{q}^{\triangle} + \bar{q} \cdot \hat{D}, \tag{4.40}$$

where

$$\bar{q}^{\triangle} = \frac{\partial \bar{q}}{\partial t} - \bar{q} \cdot \bar{\nabla}\bar{v}. \tag{4.41}$$

Since $\bar{q}$, $\hat{D}$, and $\bar{q}^{\circ}$ are objective vectors and tensors, it follows from Eq. (4.39) that $\bar{q}^{\triangle}$ is objective as well. This vector is called *contravariant (upper) convected derivative* of a vector $\bar{q}$, see Oldroyd (1950).

Let us consider now an objective tensor field $\hat{Q}$ with the Jaumann derivative (4.32). Substitution of the first expression (4.38) into Eq. (4.32) yields

$$\hat{Q}^{\circ} = \frac{\partial \hat{Q}}{\partial t} + \hat{Q} \cdot \hat{D} - \hat{D} \cdot \hat{Q} - \hat{Q} \cdot \bar{\nabla}\bar{v} + \bar{\nabla}\bar{v} \cdot \hat{Q}. \tag{4.42}$$

It follows from Eq. (4.25) that $\bar{\nabla}\bar{v} = 2\hat{D} - \bar{\nabla}\bar{v}^T$. Substituting this expression into Eq. (4.41) we obtain

$$\hat{Q}^{\circ} = \frac{\partial \hat{Q}}{\partial t} + \hat{Q} \cdot \hat{D} - \hat{Q} \cdot \bar{\nabla}\bar{v} + \hat{D} \cdot \hat{Q} - \bar{\nabla}\bar{v}^T \cdot \hat{Q} = \hat{Q}^{\triangle} + \hat{Q} \cdot \hat{D} + \hat{D} \cdot \hat{Q}. \tag{4.43}$$

Here

$$\hat{Q}^{\triangle} = \frac{\partial \hat{Q}}{\partial t} - \hat{Q} \cdot \bar{\nabla}\bar{v} - \bar{\nabla}\bar{v}^T \cdot \hat{Q} \tag{4.44}$$

is the contravariant (upper) convected derivative of an objective tensor $\hat{Q}$. According to Eq. (4.42), tensor $\hat{Q}^{\triangle}$ is indifferent with respect to superposed rigid motions.

By employing the latter expression (4.38) and repeating similar calculations, we may introduce *covariant (lower) convected derivatives* of an objective vector $\bar{q}$ and an objective tensor $\hat{Q}$, see Oldroyd (1950),

$$\bar{q}^{\triangledown} = \frac{\partial \bar{q}}{\partial t} + \bar{q} \cdot \bar{\nabla}\bar{v}^T, \qquad \hat{Q}^{\triangledown} = \frac{\partial \hat{Q}}{\partial t} + \hat{Q} \cdot \bar{\nabla}\bar{v}^T + \bar{\nabla}\bar{v} \cdot \hat{Q}. \tag{4.45}$$

Exercise 4.9. Check that $\bar{q}^{\nabla}$ is an objective vector and $\hat{Q}^{\nabla}$ is an objective tensor. □

Exercise 4.10. Find a connection between the Jaumann and the Oldroyd covariant derivatives. □

Exercise 4.11. Show that the Leibnitz formula (4.34) is not valid for the Oldroyd covariant and contravariant derivatives. □

Exercise 4.12 Find such an objective tensor $\hat{Q}$ that the Oldroyd contravariant derivative of it equals zero, while its principal invariants do not vanish. □

Exercise 4.13. Find a similar example of an objective tensor $\hat{Q}$ for the Oldroyd covariant derivative. □

Spriggs et al. (1966) introduced *the general corotational derivative*

$$\hat{Q}^{\square} = \hat{\sigma}^{\circ} + a_1(\hat{Q} \cdot \hat{D} + \hat{D} \cdot \hat{Q}) + a_2(\hat{Q} : \hat{D})\hat{I} + a_3 I_1(\hat{Q})\hat{D}, \qquad (4.46)$$

where a_k $(k = 1, 2, 3)$ are arbitrary constants.

Exercise 4.14. Check that the general corotational derivative of an objective tensor is indifferent with respect to superposed rigid motions. □

Exercise 4.15. Show that for any tensor $\hat{Q}$, the Jaumann and the Oldroyd corotational derivatives are particular cases of the general corotational derivative. □

4.6. The Rivlin–Ericksen Tensors

Eq. (4.25) determines the rate-of-strain tensor $\hat{D}$ which is an objective tensor. By using formulas for the Oldroyd covariant derivatives, we can construct the second, the third, etc., time derivatives of the rate-of-strain tensor which are objective tensors as well. These tensors introduced by Oldroyd (1950) and employed by Rivlin & Ericksen (1955) are known as *the Rivlin–Ericksen tensors*

$$\hat{A}_0 = \hat{I}, \qquad \hat{A}_1 = 2\hat{D} = \bar{\nabla}\bar{v} + \bar{\nabla}\bar{v}^T,$$

$$\hat{A}_{n+1} = \hat{A}_n^{\nabla} = \frac{\partial \hat{A}_n}{\partial t} + \hat{A}_n \cdot \bar{\nabla}\bar{v}^T + \bar{\nabla}\bar{v} \cdot \hat{A}_n \qquad (n = 1, 2, \ldots). \qquad (4.47)$$

Instead of the Oldroyd covariant derivatives, we may employ other corotational derivatives. For example, by using the Oldroyd contravariant derivatives, we obtain *the White–Metzner tensors*, see White & Metzner (1963),

$$\hat{B}_0 = -\hat{I}, \qquad \hat{B}_1 = 2\hat{D} = \bar{\nabla}\bar{v} + \bar{\nabla}\bar{v}^T,$$

$$\hat{B}_{n+1} = \hat{B}_n^{\triangle} = \frac{\partial \hat{B}_n}{\partial t} - \hat{B}_n \cdot \bar{\nabla}\bar{v}^T - \bar{\nabla}\bar{v} \cdot \hat{B}_n \qquad (n = 1, 2, \ldots). \qquad (4.48)$$

The Rivlin-Ericksen tensors (4.46) were used by Coleman et al. (1966), while the White-Metzner tensors (4.47) were employed by Astarita & Marrucci (1974) and Huilgol (1979) to develop constitutive equations in nonlinear mechanics of continua.

5. Tensor Functions

In this Section we discuss some features of objective and isotropic tensor functions of tensor variables, prove the Rivlin-Ericksen theorem, introduce derivatives of scalar functions with respect to tensor arguments, derive the Finger formula, and establish some elementary properties of convex and polyconvex functions.

5.1. The Rivlin–Ericksen Theorem

Let $\xi = \{\xi^i\}$ be Lagrangian coordinates with tangent vectors $\bar{g}_i$ and $\hat{Q}(\xi)$ be an arbitrary tensor field. For definiteness, we confine ourselves to tensors of the second rank, while most of the results derived are also true for tensors of an arbitrary rank.

We consider two motions of a medium which differ from each other by a superimposed rigid motion characterized by an orthogonal tensor $\hat{O} = \hat{O}(t)$. In Section 4 we introduced objective, isotropic, and invariant tensors. With reference to their features, we provide the following

Definition. A tensor function $\hat{F}(\hat{Q})$ is called *invariant* if its values coincide for two motions which differ from each other by a rigid motion.

Definition. A tensor function $\hat{F}(\hat{Q})$ is called *objective* if its values are transformed according to the formula

$$\hat{F}(\hat{Q} \cdot \hat{O}) = \hat{O}^T \cdot \hat{F}(\hat{Q}) \cdot \hat{O}. \tag{5.1}$$

Definition. A tensor function $\hat{F}(\hat{Q})$ is called *isotropic* if its values satisfy the equality

$$\hat{F}(\hat{O} \cdot \hat{Q}) = \hat{F}(\hat{Q}). \tag{5.2}$$

Our objective now is to establish a representation of tensor functions of tensor variables, see Rivlin & Ericksen (1955).

Proposition 5.1 (The Rivlin–Ericksen theorem). Let $\hat{F}(\hat{Q})$ be a smooth, objective, and isotropic tensor function. Then
(i) there is a smooth function $\hat{G}$ of a tensor variable such that for any tensor $\hat{Q}$

$$\hat{F}(\hat{Q}) = \hat{G}(\hat{Q}^T \cdot \hat{Q}); \tag{5.3}$$

(ii) there are three scalar functions ψ_k of the principal invariants $I_k(\hat{P})$ ($k = 1, 2, 3$) of a tensor $\hat{P}$ such that

$$\hat{G}(\hat{P}) = \psi_0 \hat{I} + \psi_1 \hat{P} + \psi_2 \hat{P}^2, \tag{5.4}$$

41

where $\hat{I}$ is the unit tensor.

Proof. We divide the proof into four steps.

Step 1. Let us show that equality (5.3) is fulfilled. According to the polar decomposition formula (2.45), any tensor $\hat{Q}$ can be presented in the form $\hat{Q} = \hat{O}_* \cdot \hat{U}_r$, where $\hat{O}_*$ is an orthogonal tensor, and $\hat{U}_r$ is a symmetrical positive definite tensor. It follows from the isotropicity of function $\hat{F}(\hat{Q})$ that for any orthogonal tensor $\hat{O}$

$$\hat{F}(\hat{Q}) = \hat{F}(\hat{O} \cdot \hat{Q}) = \hat{F}(\hat{O} \cdot \hat{O}_* \cdot \hat{U}_r).$$

Taking $\hat{O} = \hat{O}_*^T$ and using the equality $\hat{O}_*^T = \hat{O}_*^{-1}$, we obtain

$$\hat{F}(\hat{Q}) = \hat{F}(\hat{U}_r). \tag{5.5}$$

Exercise 5.1. Show that for any smooth function $\hat{F}(\hat{U})$ of a symmetrical positive definite tensor $\hat{U}$ there is a smooth function $\hat{G}$ such that

$$\hat{F}(\hat{U}) = \hat{G}(\hat{U}^2). \quad \square \tag{5.6}$$

By using Eq. (5.6) we can rewrite Eq. (5.5) as $\hat{F}(\hat{Q}) = \hat{G}(\hat{U}_r^2)$. This equality together with Eqs. (2.46) and (2.47) implies the required formula (5.3).

Step 2. Let us show that for any orthogonal tensor $\hat{O}$, the function $\hat{G}$ satisfies the equation

$$\hat{G}(\hat{O}^T \cdot \hat{P} \cdot \hat{O}) = \hat{O}^T \cdot \hat{G}(\hat{P}) \cdot \hat{O}. \tag{5.7}$$

Any symmetrical positive definite tensor $\hat{P}$ can be presented in the form $\hat{P} = \hat{P}^{\frac{1}{2}} \cdot \hat{P}^{\frac{1}{2}} = (\hat{P}^{\frac{1}{2}})^T \cdot \hat{P}^{\frac{1}{2}}$. By employing this equality, we write the left-hand side of Eq. (5.7) as

$$\hat{G}(\hat{O}^T \cdot \hat{P} \cdot \hat{O}) = \hat{G}(\hat{O}^T \cdot (\hat{P}^{\frac{1}{2}})^T \cdot \hat{P}^{\frac{1}{2}} \cdot \hat{O}) = \hat{G}((\hat{P}^{\frac{1}{2}} \cdot \hat{O})^T \cdot (\hat{P}^{\frac{1}{2}} \cdot \hat{O})).$$

This expression together with Eq. (5.3) implies that

$$\hat{G}(\hat{O}^T \cdot \hat{P} \cdot \hat{O}) = \hat{F}(\hat{P}^{\frac{1}{2}} \cdot \hat{O}). \tag{5.8}$$

By applying Eq. (5.1) to function $\hat{F}(\hat{P}^{\frac{1}{2}})$, we find

$$\hat{F}(\hat{P}^{\frac{1}{2}} \cdot \hat{O}) = \hat{O}^T \cdot \hat{F}(\hat{P}^{\frac{1}{2}}) \cdot \hat{O}.$$

We replace $\hat{F}(\hat{P}^{\frac{1}{2}})$ by $\hat{G}(\hat{P})$ and obtain

$$\hat{F}(\hat{P}^{\frac{1}{2}} \cdot \hat{O}) = \hat{O}^T \cdot \hat{G}(\hat{P}) \cdot \hat{O}. \tag{5.9}$$

Eq. (5.7) follows from Eqs. (5.8) and (5.9).

Step 3. Any tensor function of a tensor variable can be treated as a mapping which transforms one tensor into another. Our objective now is to describe tensor functions with the use of scalar functions of real variables. We provide a definition which allows functions mapping one real number into another to be extended to functions which map one tensor into another. We restrict ourselves to functions of symmetrical positive definite tensors $\hat{P}$.

According to Eq. (2.22), any tensor $\hat{P}$ can be presented in the form

$$\hat{P} = \lambda_1^{(P)}\bar{e}_1\bar{e}^1 + \lambda_2^{(P)}\bar{e}_2\bar{e}^2 + \lambda_3^{(P)}\bar{e}_3\bar{e}^3, \tag{5.10}$$

where $\lambda_k^{(P)}$ are eigenvalues, $\bar{e}_k$ and $\bar{e}^k$ are right and left eigenvectors of tensor $\hat{P}$. Referring to the connections between tensors and matrices and to the well-known properties of symmetrical matrices, we can state that for any symmetrical tensor $\hat{P}$ and for any tangent vectors $\bar{g}_i$ there is an orthogonal tensor $\hat{O}$ such that the tensor $\hat{O}^T \cdot \hat{P} \cdot \hat{O}$ can be written as

$$\hat{O}^T \cdot \hat{P} \cdot \hat{O} = \lambda_1^{(P)}\bar{g}_1\bar{g}^1 + \lambda_2^{(P)}\bar{g}_2\bar{g}^2 + \lambda_3^{(P)}\bar{g}_3\bar{g}^3. \tag{5.11}$$

Exercise 5.2. Check that

$$\bar{g}_i = \hat{O}^T \cdot \bar{e}_i, \qquad \bar{g}^i = \bar{e}^i \cdot \hat{O}. \qquad \square \tag{5.12}$$

Let $G(\lambda)$ be a smooth scalar function of a real variable λ. A tensor function $\hat{G}(\hat{P})$ is defined in the eigenbasis $\bar{e}_k$ of a tensor $\hat{P}$ as a mapping of the matrix $diag\ (\lambda_1^{(P)}, \lambda_2^{(P)}, \lambda_3^{(P)})$ into the matrix $diag\ (G(\lambda_1^{(P)}), G(\lambda_2^{(P)}), G(\lambda_3^{(P)}))$. To calculate the function $\hat{G}(\hat{P})$ we should transform tensor $\hat{P}$ into its diagonal form, find the values $G(\lambda_k^{(P)})$, and set

$$\hat{G}(\hat{P}) = G(\lambda_1^{(P)})\bar{e}_1\bar{e}^1 + G(\lambda_2^{(P)})\bar{e}_2\bar{e}^2 + G(\lambda_3^{(P)})\bar{e}_3\bar{e}^3. \tag{5.13}$$

Exercise 5.3. Check that for the power functions $G(\lambda) = \lambda^n$ this definition is consistent with Eq. (2.23). $\square$

Substituting expressions (5.12) into Eq. (5.13) we obtain the function $\hat{G}(\hat{P})$ for an arbitrary basis $\bar{g}_i$

$$\hat{G}(\hat{P}) = \hat{O} \cdot [G(\lambda_1^{(P)})\bar{g}_1\bar{g}^1 + G(\lambda_2^{(P)})\bar{g}_2\bar{g}^2 + G(\lambda_3^{(P)})\bar{g}_3\bar{g}^3] \cdot \hat{O}^T. \tag{5.14}$$

It follows from Eqs. (5.11) and Eq. (5.13) that

$$\hat{G}(\hat{O}^T \cdot \hat{P} \cdot \hat{O}) = G(\lambda_1^{(P)})\bar{g}_1\bar{g}^1 + G(\lambda_2^{(P)})\bar{g}_2\bar{g}^2 + G(\lambda_3^{(P)})\bar{g}_3\bar{g}^3. \tag{5.15}$$

Exercise 5.4. By using formulas (5.14) and (5.15), check that Eq. (5.7) is fulfilled. $\square$

43

Any smooth function $G(\lambda)$ can be expanded in *the Taylor series*

$$G(\lambda) = \sum_{n=0}^{\infty} a_n \lambda^n, \tag{5.16}$$

where

$$a_n = \frac{1}{n!} \frac{d^n G}{d\lambda^n}(0).$$

For simplicity, series (5.16) is assumed to converge for any real λ.

By using Eqs. (2.23), (5.13), and (5.16), we find

$$\hat{G}(\hat{P}) = \sum_{n=0}^{\infty} a_n [(\lambda_1^{(P)})^n \bar{e}_1 \bar{e}^1 + (\lambda_2^{(P)})^n \bar{e}_2 \bar{e}^2 + (\lambda_3^{(P)})^n \bar{e}_3 \bar{e}^3] = \sum_{n=0}^{\infty} a_n \hat{P}^n. \tag{5.17}$$

Exercise 5.5. Show that any function (5.17) satisfies Eq. (5.7). □

Step 4. In order to derive Eq. (5.4) we apply the Caley–Hamilton theorem to formula (5.17). According to Eq. (2.32), the third power of any tensor $\hat{P}$ can be expressed in terms of tensors $\hat{I}$, $\hat{P}$, and $\hat{P}^2$

$$\hat{P}^3 = I_1(\hat{P}) \hat{P}^2 - I_2(\hat{P}) \hat{P} + I_3(\hat{P}) \hat{I}.$$

Multiplying this equality by $\hat{P}$, we can express tensor $\hat{P}^4$ in terms of the tensors $\hat{I}$, $\hat{P}$, and $\hat{P}^2$, etc. Finally, we obtain that any power $\hat{P}^n$ ($n = 3, 4, \ldots$) of $\hat{P}$ can be expressed in terms of $\hat{I}$, $\hat{P}$, and $\hat{P}^2$

$$\hat{P}^n = \mu_{2,n} \hat{P}^2 + \mu_{1,n} \hat{P} + \mu_{0,n} \hat{I}, \tag{5.18}$$

where coefficients $\mu_{k,n}$ depend on the principal invariants of $\hat{P}$. Substitution of expression (5.18) into Eq. (5.17) implies Eq. (5.4). □

Our proof is based on an assumption regarding the smoothness of a function $G(\lambda)$ which allows this function to be expanded into series (5.17). The Rivlin-Ericksen theorem is valid for continuous functions as well. An appropriate proof is left to the reader with reference to Rivlin & Ericksen (1955) and Ciarlet (1988).

By using a similar reasoning, we can derive the following

Proposition 5.2. Let $F(\hat{Q})$ be a scalar, smooth, invariant, and isotropic function of a tensor $\hat{Q}$. Then

(i) there is a smooth function G of a tensor variable such that for any tensor $\hat{Q}$

$$F(\hat{Q}) = G(\hat{Q}^T \cdot \hat{Q}); \tag{5.19}$$

(ii) there is a smooth function Ψ of the principal invariants $I_k(\hat{P})$ ($k = 1, 2, 3$) of a tensor $\hat{P}$ such that

$$G(\hat{P}) = \Psi(I_1(\hat{P}), I_2(\hat{P}), I_3(\hat{P})). \tag{5.20}$$

44

Proof. We provide a sketch of the proof leaving details to the reader as an exercise. Eq. (5.19) is a consequence of the isotropicity of function F. To derive formula (5.20) we show that function G satisfies the condition

$$G(\hat{P}) = G(\hat{O}^T \cdot \hat{P} \cdot \hat{O}) \tag{5.21}$$

for any symmetrical positive definite tensor $\hat{P}$ and for any orthogonal tensor $\hat{O}$. Choosing tensor $\hat{O}$ in an appropriate way, we find that in the current basis $\bar{g}_i$, tensor $\hat{O}^T \cdot \hat{P} \cdot \hat{O}$ has the diagonal form (5.11). Any scalar function of this tensor is determined by its eigenvalues $\lambda_k^{(P)}$ only. Eq. (5.21) implies that for an arbitrary basis, function G is determined by these eigenvalues, or, which is equivalent, by the principal invariants $I_k(\hat{P})$ expressed uniquely in terms of the eigenvalues $\lambda_k^{(P)}$. $\square$

5.2. Derivatives of a Scalar Function with Respect to a Tensor Variable

In this subsection we introduce *the derivative* of smooth functions *with respect to tensor variables*. For simplicity we confine ourselves to scalar functions. The general case may be studied by using a similar reasoning.

Let $F(\hat{Q})$ be a smooth scalar function of a tensor $\hat{Q}$. In a fixed basis $\bar{g}_i$ we can write

$$F(\hat{Q}) = F(Q^{11}, Q^{12}, \ldots, Q^{33}), \tag{5.22}$$

where $\hat{Q} = Q^{ij}\bar{g}_i\bar{g}_j$.

Definition. A tensor $F_{\hat{Q}}$ is called the derivative of a scalar function F with respect to a tensor $\hat{Q}$ if for any $\delta\hat{Q}$ we have

$$F(\hat{Q} + \delta\hat{Q}) = F(\hat{Q}) + F_{\hat{Q}} : \delta\hat{Q}^T + o(\delta\hat{Q}). \tag{5.23}$$

Here

$$\lim_{\|\delta\hat{Q}\|\to 0} \frac{\|o(\delta\hat{Q})\|}{\|\delta\hat{Q}\|} = 0,$$

where $\|\cdot\|$ means *the Euclidean norm* of a tensor.

Proposition 5.3. For fixed curvilinear coordinates

$$F_{\hat{Q}} = \frac{\partial F}{\partial Q^{ij}}\bar{g}^i\bar{g}^j. \tag{5.24}$$

Proof. In a basis $\bar{g}_i$, Eq. (5.22) allows us to treat function F as a function of several variables $\bar{Q}^{ij}$. It follows from the formula for the differential of a scalar function that

$$F(\hat{Q} + \delta\hat{Q}) = F(\hat{Q}) + \frac{\partial F}{\partial Q^{ij}}\delta Q^{ij} + o(\delta\hat{Q}).$$

45

Formula (5.24) follows from this expression and Eq. (5.23). □

Exercise 5.6. By using Eq. (5.24) prove that for any symmetrical tensor $\hat{Q}$

$$F_{\hat{Q}^T} = F_{\hat{Q}}. \qquad \square \tag{5.25}$$

Exercise 5.7. Prove that any sufficiently smooth functions F_1 and F_2 obey the Leibnitz formula

$$(F_1 F_2)_{\hat{Q}} = (F_1)_{\hat{Q}} F_2 + F_1 (F_2)_{\hat{Q}}. \qquad \square \tag{5.26}$$

Exercise 5.8. Let $F(\hat{Q})$ be a smooth function and $\hat{Q}(\xi)$ be a smooth tensor field. Prove that

$$\bar{\nabla} F(\hat{Q}) = \bar{\nabla} \hat{Q}^T : F_{\hat{Q}}. \qquad \square \tag{5.27}$$

5.2.1. Derivatives of the principal invariants

Let us calculate the derivatives of the principal invariants of a tensor $\hat{Q}$. We begin with the first invariant $I_1(\hat{Q}) = \hat{I} : \hat{Q}$. It follows from Eqs. (2.27) and (2.28) that

$$I_1(\hat{Q} + \delta\hat{Q}) - I_1(\hat{Q}) = \hat{I} : (\hat{Q} + \delta\hat{Q}) - \hat{I} : \hat{Q} = \hat{I} : \delta\hat{Q} = \hat{I} : \delta\hat{Q}^T.$$

This equality together with Eq. (5.23) implies that

$$[I_1(\hat{Q})]_{\hat{Q}} = \hat{I}. \tag{5.28}$$

By using Eqs. (2.28) and (2.29) we calculate the derivative of the first principal invariant of $\hat{Q}^2$ as follows:

$$I_1((\hat{Q} + \delta\hat{Q})^2) - I_1(\hat{Q}^2) = \hat{I} : (\hat{Q}^2 + \hat{Q} \cdot \delta\hat{Q} + \delta\hat{Q} \cdot \hat{Q} + \delta\hat{Q}^2) - \hat{I} : \hat{Q}^2$$
$$= \hat{I} : (\hat{Q} \cdot \delta\hat{Q}) + \hat{I} : (\delta\hat{Q} \cdot \hat{Q}) + o(\delta\hat{Q}) = \hat{Q} : \delta\hat{Q} + \delta\hat{Q} : \hat{Q} + o(\delta\hat{Q})$$
$$= 2\hat{Q} : \delta\hat{Q} + o(\delta\hat{Q}) = 2\hat{Q}^T : \delta\hat{Q}^T + o(\delta\hat{Q}).$$

It follows from this formula and Eq. (5.23) that

$$[I_1(\hat{Q}^2)]_{\hat{Q}} = 2\hat{Q}^T. \tag{5.29}$$

Exercise 5.9. Check that

$$[I_1(\hat{Q}^3)]_{\hat{Q}} = 3(\hat{Q}^T)^2. \qquad \square \tag{5.30}$$

Exercise 5.10. Show that for any positive integer n

$$[I_1(\hat{Q}^n)]_{\hat{Q}} = n(\hat{Q}^T)^{n-1}. \qquad \square$$

In order to calculate the derivatives of the second principal invariant of tensor $\hat{Q}$ we employ formulas (2.30), (5.28), and (5.29). As a result, we obtain

$$[I_2(\hat{Q})]_{\hat{Q}} = \frac{1}{2}\{2I_1(\hat{Q})[I_1(\hat{Q})]_{\hat{Q}} - [I_1(\hat{Q}^2)]_{\hat{Q}}\}$$

$$= \frac{1}{2}[2I_1(\hat{Q})\hat{I} - 2\hat{Q}^T] = I_1(\hat{Q})\hat{I} - \hat{Q}^T. \qquad (5.31)$$

By using Eqs. (2.30), (5.28), (5.30), and (5.31), we find

$$[I_3(\hat{Q})]_{\hat{Q}} = \frac{1}{3}\{3(\hat{Q}^T)^2 - 3I_1^2(\hat{Q})\hat{I} + 3I_2(\hat{Q})\hat{I} + 3I_1(\hat{Q})[I_1(\hat{Q})\hat{I} - \hat{Q}^T]\}$$

$$= I_2(\hat{Q})\hat{I} - I_1(\hat{Q})\hat{Q}^T + (\hat{Q}^T)^2. \quad (5.32)$$

To derive another presentation for the derivative $I_3(\hat{Q})$ we multiply the right-hand side of Eq. (5.32) by $\hat{Q}^T$ and find

$$[I_3(\hat{Q})]_{\hat{Q}} = [I_2(\hat{Q})\hat{Q}^T - I_1(\hat{Q})(\hat{Q}^T)^2 + (\hat{Q}^T)^3] \cdot (\hat{Q}^T)^{-1}.$$

By employing Eq. (2.34), we arrive at the formula

$$[I_3(\hat{Q})]_{\hat{Q}} = I_3(\hat{Q})(\hat{Q}^T)^{-1}. \qquad (5.33)$$

5.2.2. Finger's formula

Let us derive an expression for the derivative with respect to a tensor $\hat{Q}$ of a scalar function $F(I_1, I_2, I_3)$ of the principal invariants $I_k(\hat{Q})$. By using the formula for the derivative of a composite function, we find

$$[F(I_1(\hat{Q}), I_2(\hat{Q}), I_3(\hat{Q}))]_{\hat{Q}} = \frac{\partial F}{\partial I_1}[I_1(\hat{Q})]_{\hat{Q}} + \frac{\partial F}{\partial I_2}[I_2(\hat{Q})]_{\hat{Q}} + \frac{\partial F}{\partial I_3}[I_3(\hat{Q})]_{\hat{Q}}. \quad (5.34)$$

Substitution of expressions (5.28), (5.31), and (5.32) into Eq. (5.34) yields

$$[F(I_1(\hat{Q}), I_2(\hat{Q}), I_3(\hat{Q}))]_{\hat{Q}} = [\frac{\partial F}{\partial I_1} + I_1(\hat{Q})\frac{\partial F}{\partial I_2} + I_2(\hat{Q})\frac{\partial F}{\partial I_3}]\hat{I}$$

$$-[\frac{\partial F}{\partial I_2} + I_1(\hat{Q})\frac{\partial F}{\partial I_3}]\hat{Q}^T + \frac{\partial F}{\partial I_3}(\hat{Q}^T)^2. \qquad (5.35)$$

Eq. (5.35) is called *the Finger formula* for the derivative of a scalar function with respect to a tensor argument. Another form of the Finger formula can be obtained if we substitute expression (5.33), instead of (5.32), into Eq. (5.34).

Exercise 5.11. Derive the following formula

$$[F(I_1(\hat{Q}), I_2(\hat{Q}), I_3(\hat{Q}))]_{\hat{Q}} = [\frac{\partial F}{\partial I_1} + I_1(\hat{Q})\frac{\partial F}{\partial I_2}]\hat{I}$$
$$-\frac{\partial F}{\partial I_2}\hat{Q}^T + I_3(\hat{Q})\frac{\partial F}{\partial I_3}(\hat{Q}^T)^{-1}. \qquad \square \quad (5.36)$$

5.2.3. Derivatives with respect to the inverse tensor

Let us express the derivative of a scalar function F with respect to the inverse tensor $\hat{Q}^{-1}$ in terms of the derivative of F with respect to a tensor $\hat{Q}$. For this purpose, we calculate the differential of the equality $\hat{Q} \cdot \hat{Q}^{-1} = \hat{I}$ and find

$$\delta\hat{Q} \cdot \hat{Q}^{-1} + \hat{Q} \cdot \delta\hat{Q}^{-1} = 0.$$

It follows from this formula that

$$\delta\hat{Q}^{-1} = -\hat{Q}^{-1} \cdot \delta\hat{Q} \cdot \hat{Q}^{-1}. \qquad (5.37)$$

According to the definition of the derivative of a function F with respect to a tensor $\hat{Q}^{-1}$ we have $\delta F = F_{\hat{Q}^{-1}} : (\delta\hat{Q}^{-1})^T + o(\delta\hat{Q}^{-1})$. Substitution of expression (5.37) into this formula yields

$$\begin{aligned} \delta F &= -F_{\hat{Q}^{-1}} : (\hat{Q}^{-1} \cdot \delta\hat{Q} \cdot \hat{Q}^{-1})^T + o(\delta\hat{Q}^{-1}) \\ &= -F_{\hat{Q}^{-1}} : (\hat{Q}^{-1})^T \cdot \delta\hat{Q}^T \cdot (\hat{Q}^{-1})^T + o(\delta\hat{Q}^{-1}) \\ &= -(\hat{Q}^{-1})^T \cdot F_{\hat{Q}^{-1}} \cdot (\hat{Q}^{-1})^T : \delta\hat{Q}^T + o(\delta\hat{Q}^{-1}). \end{aligned} \qquad (5.38)$$

Exercise 5.12. Check that $o(\delta\hat{Q}^{-1}) = o(\delta\hat{Q})$ for any non-singular tensor $\hat{Q}$. $\square$
Comparison of expressions (5.23) and (5.38) implies that

$$F_{\hat{Q}} = -(\hat{Q}^T)^{-1} \cdot F_{\hat{Q}^{-1}} \cdot (\hat{Q}^T)^{-1},$$

which leads to the desired formula

$$F_{\hat{Q}^{-1}} = -\hat{Q}^T \cdot F_{\hat{Q}} \cdot \hat{Q}^T. \qquad (5.39)$$

Exercise 5.13. By employing a similar reasoning, show that

$$(I_1(\hat{Q}))_{\hat{Q}^2} = \frac{1}{2}(\hat{Q}^T)^{-1}, \qquad (I_2(\hat{Q}))_{\hat{Q}^2} = \frac{1}{2}[I_1(\hat{Q})(\hat{Q}^T)^{-1} - \hat{I}],$$
$$(I_3(\hat{Q}))_{\hat{Q}^2} = \frac{1}{2}I_3(\hat{Q})(\hat{Q}^T)^{-2}. \qquad \square$$

5.2.4. The second derivatives of a scalar function with respect to a tensor variable

By using Eq. (5.23) we can define *the second derivative* of a scalar function *with respect to a tensor variable* as the derivative of the first derivative. We employ the coordinate presentation (5.24) and find that for a smooth function $F(\hat{Q})$, its second derivative with respect to a tensor $\hat{Q} = Q^i_j \bar{g}_i \bar{g}^j$ is equal to the following tensor of the fourth rank:

$$F_{\hat{Q}\hat{Q}}(\hat{Q}) = \bar{g}^j \bar{g}_l \frac{\partial^2 F}{\partial Q^j_l \partial Q^i_k} \bar{g}^i \bar{g}_k. \tag{5.40}$$

Exercise 5.14. Show that for any two tensors $\hat{Q}$ and $\delta\hat{Q}$

$$F(\hat{Q} + \delta\hat{Q}) = F(\hat{Q}) + F_{\hat{Q}}(\hat{Q}) : \delta\hat{Q}^T$$

$$+ \frac{1}{2}\delta\hat{Q}^T : F_{\hat{Q}\hat{Q}}(\hat{Q}) : \delta\hat{Q}^T + o(\|\delta\hat{Q}\|^2). \qquad \Box \tag{5.41}$$

One of the most important inequalities in the calculus of variations is *the Legendre–Hadamard condition* (ellipticity condition) which is formulated in terms of the second derivative of a scalar function.

Definition. A scalar function $F(\hat{Q})$ is *elliptic* (satisfies the Legendre–Hadamard condition) in a $\hat{Q}$ if for any vectors $\bar{a}$ and $\bar{b}$ we have

$$\bar{b}\bar{a} : F_{\hat{Q}\hat{Q}} : \bar{b}\bar{a} \geq 0. \tag{5.42}$$

Definition. A scalar function $F(\hat{Q})$ is *strongly elliptic* in a $\hat{Q}$ if for any nonzero vectors $\bar{a}$ and $\bar{b}$ we have

$$\bar{b}\bar{a} : F_{\hat{Q}\hat{Q}} : \bar{b}\bar{a} > 0. \tag{5.43}$$

Exercise 5.15. Write inequality (5.42) in the component form

$$\frac{\partial^2 F}{\partial Q^i_k \partial Q^j_l} a^i a^j b_k b_l \geq 0. \qquad \Box$$

5.3. Convex Functions

In this subsection we introduce convex subsets of the set of tensors of the second rank and convex scalar functions of tensor variables. The concepts provided in this subsection will be used in Chapter 3 in order to formulate the characteristic features of strain energy densities for hyperelastic materials.

Let $\mathcal{V}$ be *a vector space*, and $V \subset \mathcal{V}$ be a subset of $\mathcal{V}$.

Definition. A subset V is called *convex* if for any elements $\hat{P}, \hat{Q} \subset V$, it contains also the entire interval $s\hat{P} + (1-s)\hat{Q}$, where $s \in [0,1]$.

Definition. A *convex hull* Co V of a subset V is an intersection of all the convex subsets of $\mathcal{V}$ which contain subset V.

Evidently, Co $V = V$ for any convex subset V. For a non-convex subset V, its convex hull can be defined as a set Co $V = \{\sum_{n=1}^{N} \nu_n \hat{Q}_n\}$, where N is a positive integer, $\hat{Q}_n \in V$, and scalars $\nu_n \geq 0$ satisfy the equality $\sum_{n=1}^{N} \nu_n = 1$.

Let $\mathcal{M}$ be a set of real tensors of the second rank (3×3-matrices), and $M \subset \mathcal{M}$ be a subset of $\mathcal{M}$ consisting of tensors with positive determinants.

For any non-singular tensor $\hat{Q}$ we denote by adj $\hat{Q} = \hat{Q}^{-1} \det \hat{Q}$ its *adjugate* tensor. Let $\tilde{M}$ be the set of triples $\mathcal{Q} = (\hat{Q}, \operatorname{adj} \hat{Q}, \det \hat{Q})$ with $\hat{Q} \in M$.

Proposition 5.4. The following equalities are valid:

$$\text{Co } M = \mathcal{M}, \qquad \text{Co } \tilde{M} = \mathcal{M} \times \mathcal{M} \times (0, \infty). \qquad \square \qquad (5.44)$$

We leave the proof of the above assertion to the reader as an exercise, referring to Ciarlet (1988).

Corollary 5.1. Set M is not convex.

Definition. A function $F(\hat{Q})$ defined on a convex subset V of a vector space $\mathcal{V}$ is called *convex* if for any $\hat{P}, \hat{Q} \subset V$ and for any $s \in [0,1]$ we have

$$F(s\hat{P} + (1-s)\hat{Q}) \leq sF(\hat{P}) + (1-s)F(\hat{Q}). \qquad (5.45)$$

Definition. A function $F(\hat{Q})$ defined on a convex subset V of a vector space $\mathcal{V}$ is called *strictly convex* if for any $\hat{P}, \hat{Q} \subset V$, $\hat{P} \neq \hat{Q}$, and for any $s \in [0,1]$ we have

$$F(s\hat{P} + (1-s)\hat{Q}) < sF(\hat{P}) + (1-s)F(\hat{Q}). \qquad (5.46)$$

Exercise 5.16. Check that Eq. (5.46) can be written as

$$F(\hat{Q} + s(\hat{P} - \hat{Q})) \leq sF(\hat{P}) + (1-s)F(\hat{Q}).$$

By introducing the notation $\hat{R} = \hat{P} - \hat{Q}$, rewrite this inequality as

$$F(\hat{Q} + s\hat{R}) \leq sF(\hat{R} + \hat{Q}) + (1-s)F(\hat{Q}). \qquad \square \qquad (5.47)$$

Definition. A function $F(\hat{Q})$ is called *rank-one-convex* at $\hat{Q} \subset \mathcal{M}$ if for any vectors $\bar{a}$ and $\bar{b}$ and for any $s \in [0,1]$ we have

$$F(\hat{Q} + s\bar{a}\bar{b}) \leq sF(\hat{Q} + \bar{a}\bar{b}) + (1-s)F(\hat{Q}). \qquad (5.48)$$

Definition. A function $F(\hat{Q})$ is called *strictly rank-one-convex* at $\hat{Q}$ if for any non-zero vectors $\bar{a}$ and $\bar{b}$, and for any $s \in (0,1)$ we have

$$F(\hat{Q} + s\bar{a}\bar{b}) < sF(\hat{Q} + \bar{a}\bar{b}) + (1-s)F(\hat{Q}). \qquad (5.49)$$

It is worth noting an important difference between the concepts of convexity and rank-one-convexity: a function may be convex on a set, while it may be rank-one-convex at a point. On the other hand, these concepts have several similar features. For example, the convexity of a smooth function is very close to the positivity of its second derivative, while the concept of rank-one-convexity is extremely close to the concept of ellipticity. A precise assertion is provided in the following propositions.

Proposition 5.5. Let a function $F(\hat{Q})$ be twice continuously differentiable on a convex subset V of a vector space $\mathcal{V}$.

(i) If for any $\hat{P}, \hat{Q} \subset V$, $\hat{P} \neq \hat{Q}$

$$(\hat{P} - \hat{Q}) : F_{\hat{Q}\hat{Q}}(\hat{Q}) : (\hat{P} - \hat{Q}) > 0,$$

then the function $F(\hat{Q})$ is strictly convex on V.

(ii) Function $F(\hat{Q})$ is convex on V if and only if

$$(\hat{P} - \hat{Q}) : F_{\hat{Q}\hat{Q}}(\hat{Q}) : (\hat{P} - \hat{Q}) \geq 0$$

for any $\hat{P}, \hat{Q} \subset V$. $\square$

Proposition 5.6. Let a function $F(\hat{Q})$ be twice continuously differentiable in the vicinity of $\hat{Q} \subset \mathcal{M}$.

(i) If function F is strongly elliptic, then it is strictly rank-one-convex;

(ii) Function F is rank-one convex if and only if the Legendre–Hadamard condition (5.42) is fulfilled. $\square$

Problem 5.1. Prove these theorems. For a detailed proof, see Ciarlet (1988).

Exercise 5.17. By using Proposition 5.5, check that the function $F(\hat{Q}) = \|\hat{Q}\|^2$ is strictly convex on $\mathcal{M}$. $\square$

According to Corollary 5.1, several sets of interest for applications are not convex. In order to extend the concept of convexity to functions defined on *non-convex* sets, we introduce the following

Definition. Let V be a non-convex subset of a vector space $\mathcal{V}$, and Co V be its convex hull. A function $F(\hat{Q})$ defined on V is called convex if there is a convex function $F^*(\hat{Q})$ defined on Co V such that $F(\hat{Q}) = F^*(\hat{Q})$ for any $\hat{Q} \in V$.

A criterion of convexity for functions defined on arbitrary subsets is provided by Busemann et al. (1963).

Proposition 5.7. Let V be an arbitrary subset of a finite-dimensional vector space $\mathcal{V}$. A function $F(\hat{Q})$ is convex on V if and only if

(i) there is a linear function $G(\hat{Q})$ such that $F(\hat{Q}) \geq G(\hat{Q})$ for any $\hat{Q} \in V$;

(ii) for any positive integer N, any $\hat{Q}_n \in V$, and any non-negative coefficients ν_n $(n = 1, \ldots, N)$ such that $\sum_{n=1}^{N} \nu_n = 1$ we have

$$F(\sum_{n=1}^{N} \nu_n \hat{Q}_n) \leq \sum_{n=1}^{N} \nu_n F(\hat{Q}_n). \quad \square \tag{5.50}$$

Problem 5.2. Prove this assertion. □

The following examples formulated as exercises show that we cannot confine ourselves to convex functions, since some functions of special interest for applications are not convex. For a detailed analysis of these examples, see Ciarlet (1988).

Exercise 5.18. Show that the function $F(\hat{Q}) = \|\text{adj}\hat{Q}\|^2$ is not convex on set M. □

Exercise 5.19. Show that the function $F(\hat{Q}) = \det \hat{Q}$ is not convex on set M. □

A generalization of the concept of convex functions, which allows the above functions to be included into consideration, was suggested by Ball (1977). It is based on the idea of polyconvexity.

Definition. A function $F(\hat{Q})$ defined on the set M is called *polyconvex* if there is a convex function $F^*(Q)$ defined on Co $\tilde{M}$ such that for any $\hat{Q} \in M$

$$F(\hat{Q}) = F^*(\hat{Q}, \text{adj } \hat{Q}, \det \hat{Q}). \tag{5.51}$$

Exercise 5.20. Check that functions $F(\hat{Q}) = \|\text{adj}\hat{Q}\|^2$ and $F(\hat{Q}) = \det \hat{Q}$ are polyconvex. □

Exercise 5.21. Let $\psi(s)$ be a convex function defined on $(0, \infty)$, a and b be positive constants. Show that the function $F(\hat{Q}) = a\|\hat{Q}\|^2 + b\|\text{adj } \hat{Q}\|^2 + \psi(\det \hat{Q})$ is polyconvex. □

To provide more sophisticated examples of polyconvex functions we should refer to the following

Proposition 5.8. Let $\hat{Q}$ be an arbitrary element of M and v_k be the singular values of tensor $\hat{Q}$. Suppose that a function $\Phi(x_1, x_2, x_3, y_1, y_2, y_3, z)$ is convex on the set of non-negative x_k, y_l, and z and satisfies the following conditions:
(i) for any vectors $\bar{x} = \{x_k\}$, $\bar{y} = \{y_l\}$ with non-negative components, for any non-negative z, and for any *permutations* $\mathcal{P}_x$ and $\mathcal{P}_y$ of components of vectors $\bar{x}$ and $\bar{y}$ we have $\Phi(\mathcal{P}_x\bar{x}, \mathcal{P}_y\bar{y}, z) = \Phi(\bar{x}, \bar{y}, z)$;
(ii) function $\Phi(x_1, x_2, x_3, y_1, y_2, y_3, z)$ is non-decreasing in each x_k and y_l.
Then the function

$$F(\hat{Q}) = \Phi(v_1, v_2, v_3, v_2 v_3, v_3 v_1, v_1 v_2, v_1 v_2 v_3) \tag{5.52}$$

is polyconvex. □

The proof of this theorem is left to the reader as an exercise with reference to Ball (1977).

Exercise 5.22. Let N_1 and N_2 be positive integers, $a_i > 0$, $b_j > 0$, $\alpha_i \geq 1$, and $\beta_j \geq 1$ positive constants, and $\psi(s)$ a convex function. By using Proposition 5.8, show that the function

$$F(\hat{Q}) = \sum_{i=1}^{N_1} a_i(v_1^{\alpha_i} + v_2^{\alpha_i} + v_3^{\alpha_i}) + \sum_{j=1}^{N_2} b_j[(v_2 v_3)^{\beta_j} + (v_3 v_1)^{\beta_j} + (v_1 v_2)^{\beta_j}]$$

$$+\psi(v_1 v_2 v_3) \quad (5.53)$$

is polyconvex. □

Functions (5.53) are called *Ogden's functions*. These functions (and some others similar to them) demonstrate fair fitting of experimental data for a number of hyperelastic media, see Ogden (1972).

Chapter 2

Mechanics of continua

This Chapter is concerned with the kinematics and dynamics of continua with finite strains. In Section 1, the main kinematic concepts are introduced: the deformation gradients, the deformation tensors, and the strain tensors. Section 2 provides a classification of forces. We formulate the main principles in the mechanics of continua: the mass conservation law, the principle of linear momentum and the principle of angular momentum, introduce the Cauchy and Piola stress tensors, and derive the motion equations and the boundary conditions in stresses. Finally, in Section 3 basic axioms of the constitutive theory are formulated and discussed.

1. Kinematics of Continua

In this Section we introduce the main kinematics concepts in the nonlinear mechanics of continua.

1.1. The Deformation Gradient

Let us consider a medium that occupies a domain Ω_0 in the initial configuration, and a domain $\Omega(t)$ in the actual configuration at instant t. Denote by $\xi = \{\xi^i\}$ Lagrangian curvilinear coordinates. Fix a point M with coordinates ξ and denote by $\bar{r}_0(\xi)$ and by $\bar{r}(t, \xi)$ its radius-vectors in the initial and actual configurations. Introduce *the displacement vector* $\bar{u}(t, \xi)$ according to the formula

$$\bar{r}(t, \xi) = \bar{r}_0(\xi) + \bar{u}(t, \xi). \tag{1.1}$$

For the description of motion some generally accepted requirements are imposed on admissible displacement fields:

- the map $\bar{r} = \bar{r}(t, \bar{r}_0)$ is twice continuously differentiable;

54

- the map $\bar{r} = \bar{r}(t, \bar{r}_0)$ is globally one-to-one and it preserves the orientation.

The first restriction is introduced mainly for convenience and simplicity of exposition. It can and should be violated for the analysis of crack propagation and shock waves in deformable media.

The first part of the other restriction means that two distinct material points cannot occupy the same position simultaneously. Therefore, the map $\bar{r}(t, \bar{r}_0)$ is globally invertible. This assertion excludes such phenomena as e.g. the collapse of a cavity or the attachment of strips. The other part of this restriction means that orientation of any three non-coplanar vectors does not change during the motion.

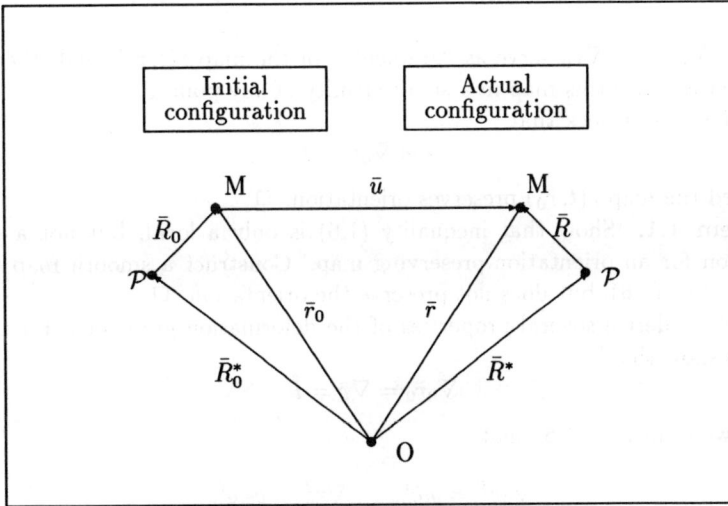

Fig. 1.1. The radius-vectors and the displacement vector.

Denote by $\bar{R}_0(\xi)$ and $\bar{R}(t, \xi)$ radius-vectors of a point M with respect to a pole $\mathcal{P}$, and by $\bar{R}_0^*$ and $\bar{R}^*(t)$ radius-vectors of the point $\mathcal{P}$ with respect to some Euler coordinate frame in the initial and actual configurations, see Fig. 1.1,

$$\bar{r}_0(\xi) = \bar{R}_0^* + \bar{R}_0(\xi), \qquad \bar{r}(t, \xi) = \bar{R}^*(t) + \bar{R}(t, \xi). \tag{1.2}$$

We differentiate Eq. (1.2) with respect to ξ^i and use Eq. (1.1). As a result, we obtain

$$\bar{g}_i = \frac{\partial \bar{R}}{\partial \xi^i} = \frac{\partial \bar{r}}{\partial \xi^i} = \frac{\partial \bar{r}_0}{\partial \xi^i} + \frac{\partial \bar{u}}{\partial \xi^i} = \frac{\partial \bar{R}_0}{\partial \xi^i} + \frac{\partial \bar{u}}{\partial \xi^i} = \bar{g}_{0i} + \frac{\partial \bar{u}}{\partial \xi^i}. \tag{1.3}$$

Since

$$\frac{\partial \bar{u}}{\partial \xi^i} = \bar{g}_{0i} \cdot \bar{\nabla}_0 \bar{u} = \bar{g}_i \cdot \bar{\nabla} \bar{u},$$

it follows from Eq. (1.3) that

$$\bar{g}_i = \bar{g}_{0i} + \bar{g}_{0i} \cdot \bar{\nabla}_0 \bar{u} = \bar{g}_{0i} \cdot (\hat{I} + \bar{\nabla}_0 \bar{u}) = (\hat{I} + \bar{\nabla}_0 \bar{u}^T) \cdot \bar{g}_{0i},$$
$$\bar{g}_{0i} = \bar{g}_i - \bar{g}_i \cdot \bar{\nabla} \bar{u} = \bar{g}_i \cdot (\hat{I} - \bar{\nabla} \bar{u}) = (\hat{I} - \bar{\nabla} \bar{u}^T) \cdot \bar{g}_i. \tag{1.4}$$

Let $\bar{g}_0^i(\xi)$ and $\bar{g}^i(t, \xi)$ be dual vectors in the initial and actual configurations. We introduce two *deformation gradients* $\bar{\nabla}_0 \bar{r}$ and $\bar{\nabla} \bar{r}_0$

$$\bar{\nabla}_0 \bar{r} = \bar{g}_0^i \frac{\partial \bar{r}}{\partial \xi^i} = \bar{g}_0^i \bar{g}_i, \qquad \bar{\nabla} \bar{r}_0 = \bar{g}^i \frac{\partial \bar{r}_0}{\partial \xi^i} = \bar{g}^i \bar{g}_{0i}. \tag{1.5}$$

Tensors $\bar{\nabla}_0 \bar{r}$ and $\bar{\nabla} \bar{r}_0$ serve as "gradients" of the map $\bar{r}(t, \bar{r}_0)$, and, therefore, they characterize this map in a small vicinity of any point ξ.

Exercise 1.1. Check that

$$\det \bar{\nabla}_0 \bar{r} > 0 \tag{1.6}$$

provided the map $\bar{r}(t, \bar{r}_0)$ preserves orientation. □

Problem 1.1. Show that inequality (1.6) is only a local, but not a global condition for an orientation-preserving map. Construct a smooth map which satisfies Eq. (1.6), but does not preserve the orientation. □

Let us derive several properties of the deformation gradients. First, it is easy to show that

$$\bar{\nabla}_0 \bar{r}_0 = \bar{\nabla} \bar{r} = \hat{I}.$$

It follows from Eq. (1.5) that

$$\bar{\nabla}_0 \bar{r}^T = \bar{g}_i \bar{g}_0^i, \qquad \bar{\nabla} \bar{r}_0^T = \bar{g}_{0i} \bar{g}^i. \tag{1.7}$$

Exercise 1.2. Derive from Eqs. (1.4) and (1.5) the following equalities:

$$\bar{\nabla}_0 \bar{r} = \hat{I} + \bar{\nabla}_0 \bar{u}, \qquad \bar{\nabla} \bar{r} = \hat{I} - \bar{\nabla} \bar{u}. \quad \square \tag{1.8}$$

Multiplying equations (1.5) we obtain

$$\bar{\nabla}_0 \bar{r} \cdot \bar{\nabla} \bar{r}_0 = \hat{I}.$$

This equality implies that

$$\bar{\nabla}_0 \bar{r} = \bar{\nabla} \bar{r}_0^{-1}. \tag{1.9}$$

Multiplying the first equality (1.5) by $\bar{g}_{0i}$ and the other equality (1.5) by $\bar{g}_0^i$ we find

$$\bar{g}_{0i} \cdot \bar{\nabla}_0 \bar{r} = \bar{g}_{0i} \cdot \bar{g}_0^j \bar{g}_j = \bar{g}_i, \qquad \bar{\nabla} \bar{r}_0 \cdot \bar{g}_0^i = \bar{g}^j \bar{g}_{0j} \cdot \bar{g}_0^i = \bar{g}^i. \tag{1.10}$$

It follows from these formulas that

$$\bar{\nabla} = \bar{g}^i \frac{\partial}{\partial \xi^i} = \bar{\nabla}\bar{r}_0 \cdot \bar{g}_0^i \frac{\partial}{\partial \xi^i} = \bar{\nabla}\bar{r}_0 \cdot \bar{\nabla}_0. \tag{1.11}$$

Eqs. (1.9) and (1.11) imply that

$$\bar{\nabla}_0 = \bar{\nabla}\bar{r}_0^{-1} \cdot \bar{\nabla} = \bar{\nabla}_0\bar{r} \cdot \bar{\nabla}. \tag{1.12}$$

Let us consider a vector $d\bar{r}_0 = \bar{g}_{0i} d\xi^i$ in the initial configuration and its image $d\bar{r} = \bar{g}_i d\xi^i$ in the actual configuration. By using Eq. (1.10) we find

$$d\bar{r} = d\bar{r}_0 \cdot \bar{\nabla}_0\bar{r} = \bar{\nabla}_0\bar{r}^T \cdot d\bar{r}_0. \tag{1.13}$$

Eq. (1.13) together with (1.9) implies that

$$d\bar{r}_0 = d\bar{r} \cdot \bar{\nabla}\bar{r}_0 = \bar{\nabla}\bar{r}_0^T \cdot d\bar{r}. \tag{1.14}$$

1.2. Deformation Tensors and Strain Tensors

In this subsection we introduce the main deformation tensors employed in the mechanics of continua. The geometrical meaning of these tensors will be discussed in subsection 1.8.

Denote by ds_0 and ds the arc elements in the initial and actual configurations

$$ds_0^2 = d\bar{r}_0 \cdot d\bar{r}_0, \quad ds^2 = d\bar{r} \cdot d\bar{r}. \tag{1.15}$$

Substitution of expressions (1.13) into the second formula (1.15) yields

$$ds^2 = d\bar{r} \cdot d\bar{r} = d\bar{r}_0 \cdot \bar{\nabla}_0\bar{r} \cdot \bar{\nabla}_0\bar{r}^T \cdot d\bar{r}_0 = d\bar{r}_0 \cdot \hat{g} \cdot d\bar{r}_0, \tag{1.16}$$

where

$$\hat{g} = \bar{\nabla}_0\bar{r} \cdot \bar{\nabla}_0\bar{r}^T \tag{1.17}$$

is called *the Cauchy deformation tensor*. It follows from Eqs. (1.8) and (1.17) that

$$\hat{g} = \hat{I} + 2\hat{\epsilon}_0(\bar{u}) + \bar{\nabla}_0\bar{u} \cdot \bar{\nabla}_0\bar{u}^T, \tag{1.18}$$

where

$$\hat{\epsilon}_0(\bar{u}) = \frac{1}{2}(\bar{\nabla}_0\bar{u} + \bar{\nabla}_0\bar{u}^T) \tag{1.19}$$

is *the first (Cauchy) infinitesimal strain tensor*.

A reciprocal deformation measure may be derived if we substitute expression (1.14) into the first formula (1.15)

$$ds_0^2 = d\bar{r}_0 \cdot d\bar{r}_0 = d\bar{r} \cdot \bar{\nabla}\bar{r}_0 \cdot \bar{\nabla}\bar{r}_0^T \cdot d\bar{r} = d\bar{r} \cdot \hat{g}_0 \cdot d\bar{r}, \tag{1.20}$$

57

where

$$\hat{g}_0 = \bar{\nabla}\bar{r}_0 \cdot \bar{\nabla}\bar{r}_0^T \qquad (1.21)$$

is called *the Almansi deformation tensor*. It follows from Eqs. (1.8) and (1.21) that

$$\hat{g}_0 = \hat{I} - 2\hat{\epsilon}(\bar{u}) + \bar{\nabla}\bar{u} \cdot \bar{\nabla}\bar{u}^T, \qquad (1.22)$$

where

$$\hat{\epsilon}(\bar{u}) = \frac{1}{2}(\bar{\nabla}\bar{u} + \bar{\nabla}\bar{u}^T) \qquad (1.23)$$

is *the second (Swainger) infinitesimal strain tensor*.

Substitution of expressions (1.5) into Eqs. (1.17) and (1.21) yields

$$\hat{g} = g_{ij}\bar{g}_0^i\bar{g}_0^j, \qquad \hat{g}_0 = g_{0ij}\bar{g}^i\bar{g}^j. \qquad (1.24)$$

According to Eqs. (1.16) and (1.20), the Cauchy and Almansi deformation tensors indicate changes in the arc element for the transition from the initial to the actual configuration.

By multiplying the deformation gradients $\bar{\nabla}_0\bar{r}$ and $\bar{\nabla}\bar{r}_0$ we can construct four symmetrical tensors. In addition to the Cauchy and Almansi tensors, we introduce *the Finger deformation tensor*

$$\hat{F} = \bar{\nabla}_0\bar{r}^T \cdot \bar{\nabla}_0\bar{r} \qquad (1.25)$$

and *the Piola deformation tensor*

$$\hat{F}_0 = \bar{\nabla}\bar{r}_0^T \cdot \bar{\nabla}\bar{r}_0. \qquad (1.26)$$

Exercise 1.3. Check that

$$\hat{F} = \hat{g}_0^{-1}, \qquad \hat{F}_0 = \hat{g}^{-1}. \qquad \Box \qquad (1.27)$$

Exercise 1.4. By using Eqs. (1.5), (1.25), and (1.26), show that

$$\hat{F} = g_0^{ij}\bar{g}_i\bar{g}_j, \qquad \hat{F}_0 = g^{ij}\bar{g}_{0i}\bar{g}_{0j}. \qquad \Box$$

Exercise 1.5. Prove that

$$I_k(\hat{F}) = I_k(\hat{g}), \qquad I_k(\hat{F}_0) = I_k(\hat{g}_0), \qquad (1.28)$$

where I_k ($k = 1, 2, 3$) stands for the principal invariants of tensors. $\Box$

Exercise 1.6. By using Eqs. (1.8), (1.25), and (1.26) check that

$$\hat{F} = \hat{I} + 2\hat{\epsilon}_0(\bar{u}) + \bar{\nabla}_0\bar{u}^T \cdot \bar{\nabla}_0\bar{u},$$
$$\hat{F}_0 = \hat{I} - 2\hat{\epsilon}(\bar{u}) + \bar{\nabla}\bar{u}^T \cdot \bar{\nabla}\bar{u}. \qquad \Box \qquad (1.29)$$

By using the above deformation tensors, other deformation tensors may be introduced as functions of the Cauchy or Finger measures. For example, *the Hencky deformation tensor* is defined as, see e.g. Fitzgerald (1980),

$$\hat{H} = \frac{1}{2} \ln \hat{F}. \tag{1.30}$$

In general, in order to construct the Hencky tensor $\hat{H}$ the eigenvalues and eigenvectors of the Finger deformation tensor $\hat{F}$ are necessary. Their explicit determination is not a simple problem, and the transition from the Finger deformation tensor to the Hencky tensor leads to additional cumbersome calculations.

It follows from Eqs. (1.15) and (1.16) that

$$ds^2 - ds_0^2 = d\bar{r}_0 \cdot \hat{g} \cdot d\bar{r}_0 - d\bar{r}_0 \cdot \hat{I} \cdot d\bar{r}_0 = 2d\bar{r}_0 \cdot \hat{C} \cdot d\bar{r}_0, \tag{1.31}$$

where

$$\hat{C} = \frac{1}{2}(\hat{g} - \hat{I}) = \hat{\epsilon}_0(\bar{u}) + \frac{1}{2}\bar{\nabla}_0\bar{u} \cdot \bar{\nabla}_0\bar{u}^T \tag{1.32}$$

is called *the Cauchy strain tensor*.

Carrying out similar calculations, we obtain from Eqs. (1.15) and (1.20) that

$$ds^2 - ds_0^2 = d\bar{r} \cdot \hat{I} \cdot d\bar{r} - d\bar{r} \cdot \hat{g}_0 \cdot d\bar{r} = 2d\bar{r} \cdot \hat{A} \cdot d\bar{r}, \tag{1.33}$$

where

$$\hat{A} = \frac{1}{2}(\hat{I} - \hat{g}_0) = \hat{\epsilon}(\bar{u}) - \frac{1}{2}\bar{\nabla}\bar{u} \cdot \bar{\nabla}\bar{u}^T \tag{1.34}$$

is called *the Almansi strain tensor*.

Substitution of expressions (1.24) into Eqs. (1.32) and (1.34) implies that

$$\hat{C} = \frac{1}{2}(g_{ij} - g_{0ij})\bar{g}_0^i\bar{g}_0^j, \qquad \hat{A} = \frac{1}{2}(g_{ij} - g_{0ij})\bar{g}^i\bar{g}^j. \tag{1.35}$$

It follows from Eqs. (1.35) that the Cauchy and Almansi strain tensors have the same covariant components, but in different bases. Obviously, their contravariant and mixed components can differ from each other.

Similar to Eqs. (1.32) and (1.34), we can introduce *the Finger strain tensor*

$$\hat{E}_F = \frac{1}{2}(I_3(\hat{g}_0)\hat{F} - \hat{I}) \tag{1.36}$$

and *the Piola strain tensor*

$$\hat{E}_{F_0} = \frac{1}{2}(\hat{I} - I_3(\hat{g})\hat{F}_0). \tag{1.37}$$

The terminology used in the nonlinear mechanics has not yet been fixed. The deformation gradient is also called *the distortion tensor*. The Cauchy strain tensor is also called the Cauchy-Green strain tensor and the Green strain tensor. The Cauchy deformation tensor is also called the left Cauchy tensor, while the Almansi deformation tensor is called the Green tensor, the right Cauchy tensor, and the Euler strain tensor.

1.3. Stretch Tensors

By using the polar decomposition formula (1.2.41), we can present the deformation gradient $\bar{\nabla}_0 \bar{r}$ in the form

$$\bar{\nabla}_0 \bar{r} = \hat{U}_l \cdot \hat{O}, \tag{1.38}$$

where $\hat{U}_l$ is a symmetrical positive definite tensor, and $\hat{O}$ is an orthogonal tensor.
Definition. The tensor $\hat{U}_l$ is called *the left stretch tensor*, while the tensor $\hat{O}$ is called *the rotation tensor*.

Substitution of expression (1.38) into Eq. (1.17) with the use of Eq. (1.2.37) implies that

$$\hat{g} = \hat{U}_l \cdot \hat{O} \cdot \hat{O}^{-1} \cdot \hat{U}_l = \hat{U}_l^2.$$

It follows from this equality that

$$\hat{U}_l = \hat{g}^{1/2}. \tag{1.39}$$

Another important relation may be derived from the right polar decomposition of the deformation gradient

$$\bar{\nabla}_0 \bar{r} = \hat{O} \cdot \hat{U}_r, \tag{1.40}$$

where $\hat{U}_r$ is a symmetrical positive definite tensor, and $\hat{O}$ is an orthogonal tensor.
Definition. The tensor $\hat{U}_r$ is called *the right stretch tensor*, while the tensor $\hat{O}$ is called *the rotation tensor*.
Exercise 1.7. Show that

$$\hat{U}_r = \hat{F}^{1/2}. \quad \square \tag{1.41}$$

Exercise 1.8. Check that

$$\hat{H} = \ln \hat{U}_r, \tag{1.42}$$

where $\hat{H}$ is the Hencky deformation tensor (1.30). $\square$
Exercise 1.9. Prove that the eigenvalues of tensors $\hat{U}_l$ and $\hat{U}_r$ coincide. $\square$
Definition. Eigenvalues v_1, v_2, v_3 of the stretch tensors are called *the principal stretches*.
Exercise 1.10. Derive the following equalities:

$$I_1(\hat{F}) = v_1^2 + v_2^2 + v_3^3, \qquad I_2(\hat{F}) = v_1^2 v_2^2 + v_1^2 v_3^2 + v_2^2 v_3^2,$$
$$I_3(\hat{F}) = v_1^2 v_2^2 v_3^2. \quad \square \tag{1.43}$$

1.4. Relative Deformation Tensors

The deformation tensors describe transformations from the initial (at $t = 0$) to the actual (at the current instant) configurations. For the analysis of inelastic media, it is important to characterize the entire history of deformations on the interval $[0, t]$. For this purpose, the so-called *relative deformation tensors* are introduced.

Let us fix two arbitrary instants $0 \leq s \leq t$ and consider transition from the actual configuration at moment s to the actual configuration at moment t. The corresponding deformation gradients $\bar{\nabla}_s \bar{r}(t)$ and $\bar{\nabla}_t \bar{r}(s)$ are called *relative deformation gradients*

$$\bar{\nabla}_s \bar{r}(t) = \bar{\nabla}_s \bar{r}_0 \cdot \bar{\nabla}_0 \bar{r}(t) = \bar{g}^i(s)\bar{g}_i(t),$$
$$\bar{\nabla}_t \bar{r}(s) = \bar{\nabla}_t \bar{r}_0 \cdot \bar{\nabla}_0 \bar{r}(s) = \bar{g}^i(t)\bar{g}_i(s). \tag{1.44}$$

Exercise 1.11. Check that for any $0 \leq s \leq t < \infty$

$$\bar{\nabla}_s \bar{r}(t) = (\bar{\nabla}_0 \bar{r}(s))^{-1} \cdot \bar{\nabla}_0 \bar{r}(t). \qquad \square \tag{1.45}$$

By using Eqs. (1.17), (1.25), and (1.44), we can introduce *the relative Cauchy and Finger deformation tensors* as follows:

$$\hat{g}_*(t,s) = \bar{\nabla}_s \bar{r}(t) \cdot \bar{\nabla}_s \bar{r}(t)^T = \bar{\nabla}_s \bar{r}_0 \cdot \hat{g}(t) \cdot \bar{\nabla}_s \bar{r}_0^T$$
$$= g_{ij}(t)\bar{g}^i(s)\bar{g}^j(s),$$
$$\hat{F}_*(t,s) = \bar{\nabla}_s \bar{r}^T(t) \cdot \bar{\nabla}_s \bar{r}(t) = \bar{\nabla}_0 \bar{r}^T(t) \cdot \hat{g}^{-1}(s) \cdot \bar{\nabla}_0 \bar{r}(t)$$
$$= g^{ij}(s)\bar{g}_i(t)\bar{g}_j(t). \tag{1.46}$$

Exercise 1.12. Check that for any $t \geq 0$

$$\hat{g}_*(t,t) = \hat{I}, \qquad \hat{F}_*(t,t) = \hat{I}. \qquad \square \tag{1.47}$$

Exercise 1.13. Derive the following formula for the relative Cauchy deformation tensor:

$$\hat{g}_*(t,s) = (\bar{\nabla}_0 \bar{r}(s))^{-1} \cdot \bar{\nabla}_0 \bar{r}(t) \cdot \bar{\nabla}_0 \bar{r}^T(t) \cdot (\bar{\nabla}_0 \bar{r}^T(s))^{-1}. \qquad \square \tag{1.48}$$

Formulas (1.27) show that only two deformation tensors (e.g. the Cauchy and Finger tensors) are independent, the other two tensors are inverse to these tensors. The following exercise demonstrates that only one relative deformation tensor is independent, while all the others are expressed in terms of it.

61

Exercise 1.14. Show that for any $0 \leq s \leq t < \infty$

$$\hat{F}_*(t, s) = \hat{g}_*^{-1}(s, t). \quad \square \tag{1.49}$$

Our objective now is to establish a connection between the time derivative of the relative Cauchy deformation tensor and the rate-of-strain tensor (1.4.25).

Proposition 1.1. For any instant $t \geq 0$

$$\hat{D}(t) = \frac{1}{2} \frac{\partial}{\partial t} F_*(t, s)|_{s=t}. \tag{1.50}$$

Proof. Let us calculate the time derivative of the deformation gradient. Using Eqs. (1.1.4) and (1.5) we obtain

$$\frac{\partial}{\partial t} \bar{\nabla}_0 \bar{r}(t) = \frac{\partial}{\partial t} \bar{g}_0^i \frac{\partial \bar{r}(t)}{\partial \xi^i} = \bar{g}_0^i \frac{\partial}{\partial \xi^i} \frac{\partial \bar{r}(t)}{\partial t} = \bar{g}_0^i \frac{\partial \bar{v}(t)}{\partial \xi^i} = \bar{\nabla}_0 \bar{v}(t), \tag{1.51}$$

where $\bar{v}(t)$ is the velocity vector. Eq. (1.51) together with Eq. (1.12) yields

$$\frac{\partial}{\partial t} \bar{\nabla}_0 \bar{r}(t) = \bar{\nabla}_0 \bar{r}(t) \cdot \bar{\nabla}_t \bar{v}(t). \tag{1.52}$$

Differentiation of the identity

$$(\bar{\nabla}_0 \bar{r}(t))^{-1} \cdot \bar{\nabla}_0 \bar{r}(t) = \hat{I},$$

implies that

$$\frac{\partial}{\partial t} (\bar{\nabla}_0 \bar{r}(t))^{-1} = -(\bar{\nabla}_0 \bar{r}(t))^{-1} \cdot \frac{\partial}{\partial t} \bar{\nabla}_0 \bar{r}(t) \cdot (\bar{\nabla}_0 \bar{r}(t))^{-1}. \tag{1.53}$$

Substituting expression (1.52) into Eq. (1.53) we obtain

$$\frac{\partial}{\partial t} (\bar{\nabla}_0 \bar{r}(t))^{-1} = -\bar{\nabla}_t \bar{v}(t) \cdot (\bar{\nabla}_0 \bar{r}(t))^{-1}. \tag{1.54}$$

Exercise 1.15. Check that

$$\frac{\partial}{\partial t} (\bar{\nabla}_0 \bar{r}^T(t))^{-1} = -(\bar{\nabla}_0 \bar{r}^T(t))^{-1} \cdot \bar{\nabla}_t \bar{v}^T(t). \quad \square \tag{1.55}$$

We differentiate expression (1.48) with respect to s and employ formulas (1.54) and (1.55). As a result, we find

$$\frac{\partial}{\partial s} \hat{g}_*(t, s) = -[\frac{\partial}{\partial s} (\bar{\nabla}_0 \bar{r}(s))^{-1}] \cdot \bar{\nabla}_0 \bar{r}(t) \cdot \bar{\nabla}_0 \bar{r}^T(t) \cdot (\bar{\nabla}_0 \bar{r}^T(s))^{-1}$$

$$- (\bar{\nabla}_0 \bar{r}(s))^{-1} \cdot \bar{\nabla}_0 \bar{r}(t) \cdot \bar{\nabla}_0 \bar{r}^T(t) \cdot [\frac{\partial}{\partial s} (\bar{\nabla}_0 \bar{r}^T(s))^{-1}]$$

$$= -\bar{\nabla}_s \bar{v}(s) \cdot \hat{g}_*(t, s) - \hat{g}_*(t, s) \cdot \bar{\nabla}_s \bar{v}^T(s). \tag{1.56}$$

Setting $t = s$ in Eq. (1.56) and using Eq. (1.47) we obtain

$$\frac{\partial}{\partial s}\hat{g}_*(t,s)|_{s=t} = -(\bar{\nabla}_t\bar{v}(t) + \bar{\nabla}_t\bar{v}^T(t)) = -2\hat{D}(t), \qquad (1.57)$$

where $\hat{D}(t)$ is the rate-of-strain tensor.

Exercise 1.16. By using Eq. (1.49), show that

$$\frac{\partial}{\partial t}\hat{F}_*(t,s) = -\hat{F}_*(t,s) \cdot [\frac{\partial}{\partial t}\hat{g}_*(s,t)] \cdot \hat{F}_*(t,s). \qquad \Box \qquad (1.58)$$

Eqs. (1.47), (1.57), and (1.58) imply the desired relationship (1.50). $\Box$

By employing formulas (1.32), (1.34), (1.36), and (1.37), we can define the relative strain tensors $\hat{C}_*(t,s)$, $\hat{A}_*(t,s)$, etc.

For the analysis of viscoelastic media, the so-called *difference histories of strains* are employed together with the relative strain tensors, see e.g. Coleman & Noll (1961). The difference history of the Cauchy strain $\hat{C}_d(t,s)$ is defined as

$$\hat{C}_d(t,s) = \hat{C}(t) - \hat{C}(t-s), \qquad (1.59)$$

where $\hat{C}(t)$ is the Cauchy strain tensor for the transition from the initial to the actual configuration.

Exercise 1.17. By using Eqs. (1.16), (1.31), and (1.59), show that the difference history of strains determines changes in the arc element for transition from the actual configuration at instant $t - s$ to the actual configuration at instant t

$$ds^2(t) - ds^2(t-s) = 2d\bar{r}_0 \cdot \hat{C}_d(t,s) \cdot d\bar{r}_0. \qquad \Box$$

Let us now calculate corotational derivatives of the Finger deformation tensor $\hat{F}(t)$.

Proposition 1.2. For any instant $t \geq 0$

$$\hat{F}^\triangle(t) = 0, \qquad (1.60)$$

where $\hat{F}^\triangle(t)$ is the contravariant Oldroyd derivative, see Eq. (1.4.43).

Proof. Differentiation of Eq. (1.25) with the use of Eq. (1.52) yields

$$\frac{\partial}{\partial t}\hat{F}(t) = (\frac{\partial}{\partial t}\bar{\nabla}_0\bar{r}(t))^T \cdot \bar{\nabla}_0\bar{r}(t) + \bar{\nabla}_0\bar{r}^T(t) \cdot \frac{\partial}{\partial t}\bar{\nabla}_0\bar{r}(t)$$
$$= \bar{\nabla}_t\bar{v}^T(t) \cdot \bar{\nabla}_0\bar{r}(t) \cdot \bar{\nabla}_0\bar{r}(t) + \bar{\nabla}_0\bar{r}^T(t) \cdot \bar{\nabla}_0\bar{r}(t) \cdot \bar{\nabla}_t\bar{v}(t)$$
$$= \bar{\nabla}_t\bar{v}^T(t) \cdot \hat{F}(t) + \hat{F}(t) \cdot \bar{\nabla}_t\bar{v}(t). \qquad (1.61)$$

This formula together with Eq. (1.4.43) implies the required equality (1.60). $\Box$

It follows from Eqs. (1.4.25) that

$$\bar{\nabla}_t\bar{v}(t) = \hat{D}(t) - \hat{Y}(t), \qquad \bar{\nabla}_t\bar{v}^T(t) = \hat{D}(t) + \hat{Y}(t),$$

where $\hat{Y}(t)$ is the vorticity tensor. Substituting these expressions into Eq. (1.61) we find

$$\hat{F}^{\circ}(t) = \hat{D}(t) \cdot \hat{F}(t) + \hat{F}(t) \cdot \hat{D}(t), \qquad (1.62)$$

where $\hat{F}^{\circ}(t)$ is the Jaumann derivative of the Finger deformation tensor, see Eq. (1.4.32).

Exercise 1.18. By using Eqs. (1.4.32) and (1.62), check that

$$(\hat{F}^{-1}(t))^{\circ} + \hat{F}^{-1}(t) \cdot \hat{D}(t) + \hat{D}(t) \cdot \hat{F}^{-1}(t) = 0. \quad \square \qquad (1.63)$$

Eq. (1.62) may be treated as an algebraic equation for determining the rate-of-strain tensor $\hat{D}$ for a given deformation tensor $\hat{F}$.

Problem 1.2. Derive *Zubov's formula*

$$\hat{D} = \frac{1}{2}(I_1 I_2 - I_3)^{-1}[(I_1^2 + I_2)\hat{F}^{\circ} + I_1 I_3 \hat{F}^{-1} \cdot \hat{F}^{\circ} \hat{F}^{-1} + \hat{F} \cdot \hat{F}^{\circ} \hat{F}$$
$$-I_3(\hat{F}^{-1} \cdot \hat{F}^{\circ} + \hat{F}^{\circ} \cdot \hat{F}^{-1}) - I_1(\hat{F} \cdot \hat{F}^{\circ} + \hat{F}^{\circ} \cdot \hat{F})], \qquad (1.64)$$

where I_k ($k = 1, 2, 3$) are principal invariants of the Finger deformation tensor. For a detailed discussion of Eq. (1.64), see Leonov (1976). $\square$

Problem 1.3. The Finger deformation tensor is connected with the Hencky deformation tensor by Eq. (1.30). Replace tensor $\hat{F}$ by the Hencky tensor $\hat{H}$ in Zubov's formula (1.64). The required formula was established in another way by Reinhardt & Dubey (1995). $\square$

Problem 1.4. Calculate the Jaumann derivatives of the Hencky deformation tensor. For a detailed analysis of the corotational derivatives of the Hencky tensor, see Gurtin & Spear (1983). $\square$

1.5. Generalized Strain Tensors

Four strain tensors $\hat{A}$, $\hat{C}$, $\hat{E}_F$, and $\hat{E}_{F_0}$, are expressed in terms of the deformation tensors $\hat{g}$, $\hat{g}_0$, $\hat{F}$, and $\hat{F}_0$. It is natural to generalize this approach and to treat any "admissible" tensor function $\hat{L}$ of the deformation tensors as a new, generalized measure of strains. Several *generalized strain tensors* were introduced by Seth (1964). Conditions of "admissibility" for generalized strain tensors were formulated by Hill (1968):

- the generalized strain tensor should coincide for two motions which differ from each other by a rigid motion;

- the generalized strain tensor must vanish for the identical transformation from the initial to the actual configuration, $\hat{L}(0) = 0$;

- $\hat{L}$ must be a tensor of the second rank;

- it should be an isotropic invertible tensor function of either the Cauchy deformation tensor $\hat{g}$ or the Finger deformation tensor $\hat{F}$. For example, by assuming that $\hat{L}$ is a function of the Cauchy strain tensor $\hat{C}$, we obtain

$$\hat{L} = l_0\hat{I} + l_1\hat{C} + l_2\hat{C}^2, \qquad \hat{C} = c_0\hat{I} + c_1\hat{L} + c_2\hat{L}^2, \qquad (1.65)$$

where c_k and l_k $(k = 0, 1, 2)$ are functions of the principal invariants of tensors $\hat{L}$ and $\hat{C}$, respectively;

- the derivative of the generalized strain tensor with respect to the strain tensor should vanish in the initial configuration, $\hat{L}_{\hat{C}}|_{\hat{C}=0} = 0$;

- for infinitesimal strains, the generalized strain tensor should coincide with the first and second infinitesimal strain tensors;

- the eigenvalues of the generalized strain tensor should be positive provided the corresponding principal stretches are greater than unity.

Two generalized strain tensors are widely used in applications:
– the *Eulerian m-tensor of strains*

$$\hat{\mathcal{E}}_E^{(m)} = \begin{cases} \frac{1}{m}(\hat{g}^{\frac{m}{2}} - \hat{I}) & m \neq 0 \\ \frac{1}{2}\ln\hat{g} & m = 0 \end{cases}, \qquad (1.66)$$

– the *Lagrangian m-tensor of strains*

$$\hat{\mathcal{E}}_L^{(m)} = \begin{cases} \frac{1}{m}(\hat{I} - \hat{F}^{-\frac{m}{2}}) & m \neq 0 \\ \frac{1}{2}\ln\hat{F} & m = 0 \end{cases}. \qquad (1.67)$$

Problem 1.5. Derive the following formulas for the fractional powers of the Cauchy and Finger tensors:

$$\hat{g}^{\frac{m}{2}} = \sum_{k=1}^{3} v_k^m \frac{\hat{g}^2 - (I_1 - v_k^2)\hat{g} + I_3 v_k^{-2}\hat{I}}{2v_k^4 - I_1 v_k^2 + I_3 v_k^{-2}},$$

$$\hat{F}^{\frac{m}{2}} = \sum_{k=1}^{3} v_k^m \frac{\hat{F}^2 - (I_1 - v_k^2)\hat{F} + I_3 v_k^{-2}\hat{I}}{2v_k^4 - I_1 v_k^2 + I_3 v_k^{-2}},$$

where v_k are the principal stretches, and I_k are the principal invariants of the Cauchy and Finger deformation tensors, see Morman (1986). $\square$

1.6. Rigid Motions

Let us assume that transition from the initial to the actual configuration coincides with a rigid motion

$$\bar{R}(t, \xi) = \bar{R}_0(\xi) \cdot \hat{O}(t), \qquad (1.68)$$

where $\hat{O}(t)$ is an orthogonal tensor function. It follows from Eqs. (1.2) and (1.68) that

$$\bar{r}(t, \xi) = \bar{R}_0^*(t) + [\bar{r}_0(\xi) - \bar{R}_0^*] \cdot \hat{O}(t). \qquad (1.69)$$

Differentiation of (1.69) with respect to ξ^i yields $\bar{g}_i = \bar{g}_{0i} \cdot \hat{O}$. This equality together with Eqs. (1.5) and (1.8) implies that

$$\bar{\nabla}_0 \bar{r} = \hat{O}, \quad \bar{\nabla} \bar{r}_0 = \hat{O}^T, \quad \bar{\nabla}_0 \bar{u} = \hat{O} - \hat{I}, \quad \bar{\nabla} \bar{u} = \hat{I} - \hat{O}^T.$$

Substitution of these expressions into Eqs. (1.17), (1.19), (1.21), (1.23), (1.25), (1.26), (1.32), (1.34), (1.36), and (1.37) yields

$$\hat{g} = \hat{g}_0 = \hat{F} = \hat{F}_0 = \hat{I}, \quad \hat{A} = \hat{C} = \hat{E}_F = \hat{E}_{F_0} = 0,$$

$$\hat{\epsilon}_0 = \frac{1}{2}(\hat{O} + \hat{O}^T) - \hat{I} \neq 0, \quad \hat{\epsilon} = \hat{I} - \frac{1}{2}(\hat{O} + \hat{O}^T) \neq 0. \qquad (1.70)$$

Exercise 1.19. Derive formulas (1.70). □

According to (1.70), tensors $\hat{A}$, $\hat{C}$, $\hat{E}_F$, and $\hat{E}_{F_0}$ determine "pure" deformation of a medium: they vanish for any rigid motion. The infinitesimal strain tensors $\hat{\epsilon}_0$ and $\hat{\epsilon}$ do not possess this property: they do not vanish for rigid motions with rotations. This means that only strain tensors $\hat{A}$, $\hat{C}$, $\hat{E}_F$, and $\hat{E}_{F_0}$ (and their functions) describe correctly the deformation of a continuum, but not tensors $\hat{\epsilon}_0$ and $\hat{\epsilon}$.

1.7. Volume Deformation

Let us write Eq. (1.24) as $\hat{g} = g_{ij}\bar{g}_0^i\bar{g}_0^j = g_{ik}g_0^{kj}\bar{g}_0^i\bar{g}_{0\,j}$. It follows from this equality that

$$I_3(\hat{g}) = \det \hat{g} = \det[g_{ik}g_0^{kj}] = \det[g_{ik}]\det[g_{0kj}^{-1}] = \frac{\det[g_{ik}]}{\det[g_{0kj}]} = \frac{g}{g_0}. \qquad (1.71)$$

On the other hand, Eq. (1.17) yields

$$\det \hat{g} = \det \bar{\nabla}_0 \bar{r} \, \det \bar{\nabla}_0 \bar{r}^T = (\det \bar{\nabla}_0 \bar{r})^2.$$

Comparison of these equalities implies that

$$\det \bar{\nabla}_0 \bar{r} = \sqrt{\frac{g}{g_0}}. \qquad (1.72)$$

Exercise 1.20. Show that

$$\det \bar{\nabla} \bar{r}_0 = \sqrt{\frac{g_0}{g}}. \quad □$$

Denote by dV_0 and dV the volume elements (volumes of the elementary parallelepipeds erected on tangent vectors $\bar{g}_{0\,i}$ and $\bar{g}_i$ with sides $d\xi^i$) in the initial and actual configurations. It follows from Eq. (1.3.32) that

$$dV_0 = \sqrt{g_0}\,d\xi^1 d\xi^2 d\xi^3, \qquad dV = \sqrt{g}\,d\xi^1 d\xi^2 d\xi^3. \tag{1.73}$$

Substitution of expression (1.71) into Eqs. (1.73) with the use of Eq. (1.28) implies that

$$\frac{dV}{dV_0} = \sqrt{\frac{g}{g_0}} = \sqrt{I_3(\hat{g})} = \sqrt{I_3(\hat{F})}. \tag{1.74}$$

The volume deformation is defined as

$$\varphi = \frac{dV - dV_0}{dV_0}.$$

Substitution of expression (1.74) into this formula yields

$$\varphi = \sqrt{\frac{g}{g_0}} - 1 = \sqrt{I_3(\hat{g})} - 1 = \sqrt{I_3(\hat{F})} - 1. \tag{1.75}$$

1.8. Deformation of the Surface Element

Let us consider a rectangular surface element in the initial configuration erected on vectors $d\bar{r}_0'$ and $d\bar{r}_0''$. Denote by $\bar{n}_0$ the unit normal vector to the surface element and by dS_0 its area

$$\bar{n}_0 dS_0 = d\bar{r}_0' \times d\bar{r}_0''. \tag{1.76}$$

In the actual configuration, vectors $d\bar{r}_0'$, $d\bar{r}_0''$, and $\bar{n}_0$ are transformed into $d\bar{r}'$, $d\bar{r}''$, and $\bar{n}$ and formula (1.76) is written as

$$\bar{n}dS = d\bar{r}' \times d\bar{r}''. \tag{1.77}$$

Our objective is to derive an explicit expression for the ratio dS/dS_0. For this purpose we apply formula (1.13) to $d\bar{r}_0'$ and $d\bar{r}_0''$ and find

$$d\bar{r}' = d\bar{r}_0' \cdot \bar{\nabla}_0 \bar{r}, \qquad d\bar{r}'' = \bar{\nabla}_0 \bar{r}^T \cdot d\bar{r}_0''. \tag{1.78}$$

Eqs. (1.77) and (1.78) imply that

$$\bar{n}dS = d\bar{r}_0' \cdot \bar{\nabla}_0 \bar{r} \times \bar{\nabla}_0 \bar{r}^T \cdot d\bar{r}_0''. \tag{1.79}$$

We now transform the expression $\bar{\nabla}_0 \bar{r} \times \bar{\nabla}_0 \bar{r}^T$ by employing Eqs. (1.5) and (1.7)

$$\bar{\nabla}_0 \bar{r} \times \bar{\nabla}_0 \bar{r}^T = \bar{g}_0^i \bar{g}_i \times \bar{g}_j \bar{g}_0^j. \tag{1.80}$$

It follows from Eq. (1.1.14) that

$$\bar{g}_i \times \bar{g}_j = V\bar{g}^k, \tag{1.81}$$

where indices i, j and k create an even permutation of 1, 2 and 3. Substitution of expression (1.81) into Eq. (1.80) with the use of Eq. (1.10) yields

$$\bar{\nabla}_0\bar{r} \times \bar{\nabla}_0\bar{r}^T = V\bar{g}_0^i\bar{g}^k\bar{g}_0^j = V\bar{g}_0^i\bar{\nabla}\bar{r}_0 \cdot \bar{g}_0^k\bar{g}_0^j.$$

Let us return to the indices i and j by applying formula (1.81) to the vector $\bar{g}_0^k$.

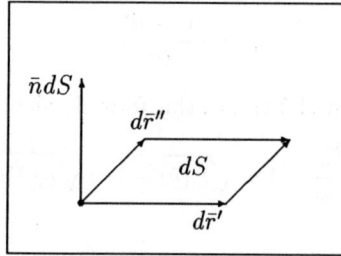

Fig. 1.2. The surface element.

By using Eq. (1.74), we find

$$\bar{\nabla}_0\bar{r} \times \bar{\nabla}_0\bar{r}^T = \frac{V}{V_0}\bar{g}_0^i\bar{\nabla}\bar{r}_0 \cdot (\bar{g}_{0i} \times \bar{g}_{0j})\bar{g}_0^j = \sqrt{\frac{g}{g_0}}\bar{g}_0^i\bar{\nabla}\bar{r}_0 \cdot (\bar{g}_{0i} \times \bar{g}_{0j})\bar{g}_0^j. \tag{1.82}$$

Eqs. (1.79) and (1.82) imply that

$$\bar{n}dS = \sqrt{\frac{g}{g_0}}d\bar{r}_0' \cdot \bar{g}_0^i\bar{\nabla}\bar{r}_0 \cdot (\bar{g}_{0i} \times \bar{g}_{0j})\bar{g}_0^j \cdot d\bar{r}_0''$$

$$= \sqrt{\frac{g}{g_0}}(d\bar{r}_0')^i\bar{\nabla}\bar{r}_0 \cdot (\bar{g}_{0i} \times \bar{g}_{0j})(d\bar{r}_0'')^j = \sqrt{\frac{g}{g_0}}\bar{\nabla}\bar{r}_0 \cdot (d\bar{r}_0' \times d\bar{r}_0'').$$

It follows from this equality and Eqs. (1.76) and (1.77) that

$$\bar{n}dS = \sqrt{\frac{g}{g_0}}\bar{\nabla}\bar{r}_0 \cdot \bar{n}_0 dS_0 = \sqrt{\frac{g}{g_0}}\bar{n}_0 \cdot \bar{\nabla}\bar{r}_0^T dS_0. \tag{1.83}$$

We multiply Eq. (1.83) by itself, use the condition $\bar{n} \cdot \bar{n} = 1$, and obtain

$$dS^2 = \frac{g}{g_0}\bar{n}_0 \cdot \bar{\nabla}\bar{r}_0^T \cdot \bar{\nabla}\bar{r}_0 \cdot \bar{n}_0 dS_0^2.$$

68

This equality together with Eqs. (1.9) and (1.17) implies that

$$\frac{dS}{dS_0} = \sqrt{\frac{g}{g_0}} (\bar{n}_0 \cdot \hat{g}^{-1} \cdot \bar{n}_0)^{1/2}. \tag{1.84}$$

By using Eq. (1.27), we obtain

$$\frac{dS}{dS_0} = \sqrt{\frac{g}{g_0}} (\bar{n}_0 \cdot \hat{F}_0 \cdot \bar{n}_0)^{1/2}. \tag{1.85}$$

Another expression for this ratio can be derived if we present Eq. (1.83) as follows:

$$\bar{n}_0 dS_0 = \sqrt{\frac{g_0}{g}} \bar{\nabla}_0 \bar{r} \cdot \bar{n} dS = \sqrt{\frac{g_0}{g}} \bar{n} \cdot \bar{\nabla}_0 \bar{r}^T dS. \tag{1.86}$$

We multiply Eq. (1.86) by itself, use Eq. (1.25) and the condition $\bar{n}_0 \cdot \bar{n}_0 = 1$, and find

$$\frac{dS_0}{dS} = \sqrt{\frac{g_0}{g}} (\bar{n} \cdot \hat{F} \cdot \bar{n})^{1/2}. \tag{1.87}$$

Eqs. (1.16), (1.20), (1.84), and (1.87) demonstrate the geometrical meaning of various deformation tensors. The Cauchy tensor $\hat{g}$ and the Almansi tensor $\hat{g}_0$ determine changes in the arc element, while the Finger tensor $\hat{F}$ and the Piola tensor $\hat{F}_0$ characterize changes of the area element for transition from the initial to the actual configuration.

2. Dynamics of Continua

In this Section we discuss some features of external and internal forces, formulate the mass conservation law and the principles of linear and angular momenta, and derive motion equations and boundary conditions in stresses.

2.1. Forces

Let us consider a medium which occupies a domain Ω_0 with a smooth boundary Γ_0 in the initial configuration. After application of loads, the medium transforms into the actual configuration where it occupies a domain Ω with a boundary Γ.

The main concepts in the dynamics of continua are *mass* and *forces*. We cannot provide precise definitions for them. Mass m is treated as a measure of material which the body contains. A force is considered as a measure of interactions between the body and its environment.

The dimension of mass is kg, the dimension of force is N, $1\text{N}=1\text{ kg} \cdot \text{m/s}^2$.

69

A mass distribution is characterized by *mass density* ρ, which is defined as mass per unit volume

$$\rho = \frac{dm}{dV},$$

where dV is an elementary volume and dm is the mass of this volume. Function $\rho = \rho(t, \xi)$, where t stands for time, and $\xi = \{\xi^i\}$ denotes Lagrangian coordinates, is assumed to be sufficiently smooth. The dimension of density is kg/m^3.

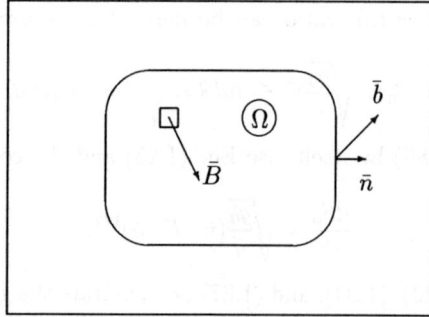

Fig. 2.1. Body and surface forces.

For any part $\tilde{V}$ of a medium, its mass m is defined as

$$m = \int_{\tilde{V}} \rho dV. \tag{2.1}$$

Mass m and density ρ are scalar quantities. Forces are vector quantities, which are characterised by their values, directions and points of application.

All the forces are divided into two classes:

- *concentrated* forces applied at an isolated point;

- *distributed* forces applied along a line, a surface, or a volume.

To avoid additional complications we confine ourselves to distributed forces only.

Forces applied to a subdomain $\tilde{V}$ with a boundary $\partial\tilde{V}$ are divided into:

- *body* load $\bar{B}$;

- *surface* load $\bar{b}$.

70

A body load is a force applied to a unit mass (or to a unit volume) of $\tilde{V}$. The force acting on a volume element dV equals $\rho\bar{B}dV$. The dimension of body load is N/kg $=$ m/s^2. Classical examples of body loads are the gravity force and electro-magnetic forces.

A surface traction is a force applied to a unit area of the boundary $\partial\tilde{V}$. The force acting on a surface element dS with the unit normal $\bar{n}$ equals $\bar{b}_{\bar{n}}dS$. The dimension of surface traction is N/m^2. Pressure is a typical example of surface traction.

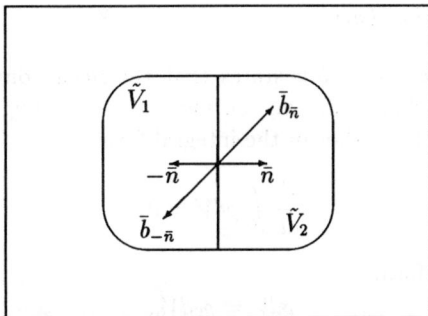

Fig. 2.2. Surface traction along an interface.

Let us consider two volumes $\tilde{V}_1$ and $\tilde{V}_2$ with a common boundary S. Denote by $\bar{n}_1 = \bar{n}$ the unit normal to S outward with respect to $\tilde{V}_1$. The unit outward normal to $\tilde{V}_2$ is $\bar{n}_2 = -\bar{n}$. The surface traction applied to volume $\tilde{V}_1$ from volume $\tilde{V}_2$ is $\bar{b}_{\bar{n}_1} = \bar{b}_{\bar{n}}$. On the other hand, the surface traction applied to volume $\tilde{V}_2$ from $\tilde{V}_1$ is $\bar{b}_{\bar{n}_2} = \bar{b}_{-\bar{n}}$. These vectors are connected by the relationship

$$\bar{b}_{-\bar{n}} = -\bar{b}_{\bar{n}}, \tag{2.2}$$

which means that two bodies act on one another with forces which are equal in their values and turned in opposite directions.

The resultant force acting on a domain $\tilde{V}$ is

$$\bar{B} = \int_{\tilde{V}} \rho\bar{B}dV + \int_{\partial\tilde{V}} \bar{b}dS. \tag{2.3}$$

For simplicity, index $\bar{n}$ is omitted.

Let $\bar{r}(t,\xi)$ be radius-vector, and $\bar{v}(t,\xi)$ velocity of a point ξ in the actual configuration. *The resultant moment* acting on a domain $\tilde{V}$ is defined as

$$\bar{M} = \int_{\tilde{V}} \rho\bar{r} \times \bar{B}dV + \int_{\partial\tilde{V}} \bar{r} \times \bar{b}dS. \tag{2.4}$$

71

Definition. *The total linear momentum* of a domain $\tilde{V}$ is

$$\int_{\tilde{V}} \rho \bar{v} dV, \tag{2.5}$$

and *the total angular momentum of $\tilde{V}$* is

$$\int_{\tilde{V}} \rho \bar{r} \times \bar{v} dV. \tag{2.6}$$

2.2. Mass Conservation Law

The mass conservation law states that the mass contained in a moving volume $\tilde{V}$ does not change during the motion. According to Eq. (2.1), this assertion can be written either in the integral form

$$\frac{d}{dt} \int_{\tilde{V}} \rho dV = 0, \tag{2.7}$$

or in the differential form

$$\rho dV = \rho_0 dV_0, \tag{2.8}$$

where ρ_0 is mass density in the initial configuration.

By using Eq. (1.74) we write the mass conservation law (2.8) as follows:

$$\rho = \rho_0 \sqrt{\frac{g_0}{g}}. \tag{2.9}$$

Eq. (2.9) can be presented in the form

$$\rho = \frac{\rho_0}{\sqrt{I_3(\hat{g})}} = \frac{\rho_0}{\sqrt{I_3(\hat{F})}}. \tag{2.10}$$

2.3. Principle of Linear Momentum

The principle of linear momentum states that for an arbitrary subdomain $\tilde{V}$, the rate of change of the total linear momentum equals the resultant force applied to this domain. This assertion with the use of Eqs. (2.3) and (2.5) implies that

$$\frac{d}{dt} \int_{\tilde{V}} \rho \bar{v} dV = \int_{\tilde{V}} \rho \bar{B} dV + \int_{\partial \tilde{V}} \bar{b} dS. \tag{2.11}$$

Eq. (2.11) is also called the Cauchy law of motion for domain $\tilde{V}$.

Let us transform Eq. (2.11) by using Eq. (2.8). For this purpose we rewrite the expression in the left-hand side of Eq. (2.11) as follows:

$$\frac{d}{dt}\int_{\tilde{V}} \rho \bar{v} dV = \frac{d}{dt}\int_{\tilde{V}_0} \rho_0 \bar{v} dV_0 = \int_{\tilde{V}_0} \rho_0 \frac{\partial \bar{v}}{\partial t} dV_0 = \int_{\tilde{V}_0} \rho_0 \bar{a} dV_0 = \int_{\tilde{V}} \rho \bar{a} dV,$$

where $\bar{a}$ is acceleration, and $\tilde{V}_0$ is the subdomain under consideration in the initial configuration. Substitution of this expression into Eq. (2.11) yields

$$\int_{\tilde{V}} \rho \bar{a} dV = \int_{\tilde{V}} \rho \bar{B} dV + \int_{\partial \tilde{V}} \bar{b} dS. \tag{2.12}$$

2.4. The Cauchy Stress Tensor

In this subsection we introduce *the Cauchy stress tensor* $\hat{\sigma}$. For this purpose, we employ some artificial construction choosing a domain $\tilde{V}$ of a special form. Namely, we fix a point M and consider a tetrahedron $\tilde{V}$ whose lateral surfaces coincide with coordinate planes of Cartesian coordinates $\{x^i\}$ with the origin at point M and basic vectors $\bar{e}_i$.

Denote by

$$\bar{n} = n_i \bar{e}_i$$

the unit outward normal to triangle ABC, see Fig. 2.3. Areas of triangle ABC and appropriate lateral triangles are denoted by $\tilde{S}$ and $\tilde{S}_i$, respectively.

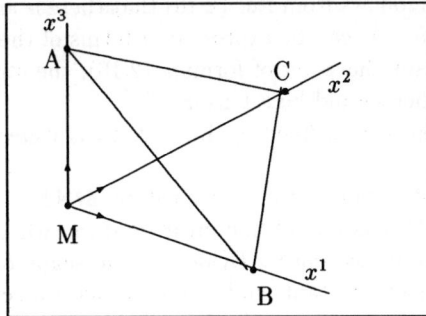

Fig. 2.3. A tetrahedron under the action of body and surface loads.

Exercise 2.1. Check that

$$\tilde{S}_i = n_i \tilde{S}. \quad \square \tag{2.13}$$

The volume of domain $\tilde{V}$ equals

$$| \tilde{V} | = \frac{1}{3} \tilde{S} h, \qquad (2.14)$$

where h is the length of the perpendicular to triangle ABC lowered from the point M.

By applying formula (2.12) to tetrahedron $\tilde{V}$, we obtain up to higher order terms compared with the volume $|\tilde{V}|$

$$\rho \bar{a} |\tilde{V}| = \rho \bar{B} |\tilde{V}| + \bar{b}_{\bar{n}} \tilde{S} + \bar{b}_{-\bar{e}_1} \tilde{S}_1 + \bar{b}_{-\bar{e}_2} \tilde{S}_2 + \bar{b}_{-\bar{e}_3} \tilde{S}_3.$$

Substitution of expressions (2.13) and (2.14) into this equality with the use of Eq. (2.2) yields

$$\frac{1}{3} \rho \bar{a} \tilde{S} h = \frac{1}{3} \rho \bar{B} \tilde{S} h + \bar{b}_{\bar{n}} \tilde{S} - \bar{b}_{\bar{e}_1} \tilde{S} n_1 - \bar{b}_{\bar{e}_2} \tilde{S} n_2 - \bar{b}_{\bar{e}_3} \tilde{S} n_3.$$

Approaching $h \to 0$ and assuming the boundedness for forces and acceleration we find

$$\bar{b}_{\bar{n}} = \bar{b}_{\bar{e}_1} n_1 + \bar{b}_{\bar{e}_2} n_2 + \bar{b}_{\bar{e}_3} n_3 = \bar{b}^i n_i, \qquad (2.15)$$

where $\bar{b}^i = \bar{b}_{\bar{e}_i}$. Vectors $\bar{b}^i$ can be expanded as follows: $\bar{b}^i = \sigma^{ij} \bar{e}_j$. Substitution of this expression into Eq. (2.15) implies that

$$\bar{b}_{\bar{n}} = n_i \sigma^{ij} \bar{e}_j = \bar{n} \cdot \hat{\sigma}, \qquad (2.16)$$

where $\hat{\sigma} = \sigma^{ij} \bar{e}_i \bar{e}_j$. It follows from Eq. (2.16) that there is an object $\hat{\sigma}$ such that the surface traction $\bar{b} = \bar{b}_{\bar{n}}$ can be expressed in terms of the normal unit vector $\bar{n}$. Owing to the tensor character of formula (2.16), the object $\hat{\sigma}$ transforming one vector into another should be a tensor.

Definition. The tensor $\hat{\sigma}$ defined by Eq. (2.16) is called the Cauchy stress tensor.

The above construction with a tetrahedron $MABC$ can be found practically in all the textbooks on continuum mechanics, where it is employed to introduce the Cauchy stress tensor. Evidently, the shape of the domain under consideration is not so important, and a more general assertion should be the reason for the linear relationship (2.16). We formulate a mathematical assertion which implies the existence of the Cauchy tensor leaving the proof to the reader as an exercise (with reference to Gurtin & Martins (1976) where an appropriate proof can be found). For an extension of this statement to an n-dimensional case, see Valent (1988).

Proposition 2.1. Let $\tilde{V}$ be an arbitrary polyhedron with a boundary $\tilde{\Gamma}$, and $\bar{n}$ be the unit outward normal to $\tilde{\Gamma}$. Suppose that $\bar{b} = \bar{b}(\xi, \bar{e})$ is a vector function of

Lagrangian coordinates ξ and unit vectors $\bar{e}$ such that for any $\bar{e}$ it is continuous in ξ. Additionally, we assume that for any polyhedron $\tilde{V}$

$$| \int_{\tilde{\Gamma}} \bar{b}(\xi, \bar{n}(\xi))dS| \leq C|\tilde{V}|,$$

where $|\tilde{V}|$ is the volume of the domain $\tilde{V}$, and the constant $C > 0$ is independent of $\bar{b}$. Then there is a tensor function $\hat{\sigma}(\xi)$ such that for any ξ and for any $\bar{e}$

$$\bar{b}(\xi, \bar{e}) = \hat{\sigma}(\xi) \cdot \bar{e}. \quad \Box$$

2.5. Equation of Motion

Let us substitute expression (2.16) into Eq. (2.12)

$$\int_{\tilde{V}} \rho \bar{a} dV = \int_{\tilde{V}} \rho \bar{B} dV + \int_{\partial \tilde{V}} \bar{n} \cdot \hat{\sigma} dS. \tag{2.17}$$

The second term in the right-hand side of Eq. (2.17) can be transformed by using the Stokes formula (1.3.59)

$$\int_{\partial \tilde{V}} \bar{n} \cdot \hat{\sigma} dS = \int_{\tilde{V}} \bar{\nabla} \cdot \hat{\sigma} dV.$$

Substituting this expression into Eq. (2.17) we obtain

$$\int_{\tilde{V}} (\rho \bar{a} - \rho \bar{B} - \bar{\nabla} \cdot \hat{\sigma})dV = 0. \tag{2.18}$$

Since Eq. (2.18) is valid for an arbitrary elementary volume $\tilde{V}$, it implies that

$$\rho \bar{a} = \rho \bar{B} + \bar{\nabla} \cdot \hat{\sigma}. \tag{2.19}$$

Eq. (2.19) is called *the motion equation*.

To write a boundary condition for Eq. (2.19) we use Eq. (2.16), where surface traction $\bar{b}$ is treated as a given vector

$$\bar{n} \cdot \hat{\sigma} |_{\Gamma} = \bar{b}. \tag{2.20}$$

A particular case of the motion equation is *the equilibrium equation* which follows from Eq. (2.19) provided the inertia term in the left-hand side vanishes

$$\bar{\nabla} \cdot \hat{\sigma} + \rho \bar{B} = 0. \tag{2.21}$$

The part of the mechanics of continua which deals with motion equation (2.19) is called *dynamics*, whereas the part of the mechanics which is concerned with the study of equilibrium equation (2.21) is called *statics*.

For fixed curvilinear coordinates $\{\xi^i\}$, Eq. (2.19) can be presented in the form

$$\rho a^i = \nabla_j \sigma^{ji} + \rho B^i.$$

Exercise 2.2. By using Eq. (1.3.52), write Eq. (2.21) in the form

$$\rho a^i \bar{g}_i = \frac{1}{\sqrt{g}} \frac{\partial(\sqrt{g} \sigma^{ij} \bar{g}_j)}{\partial \xi^i} + \rho B^i \bar{g}_i. \quad \square \tag{2.22}$$

2.6. Principle of Angular Momentum

The objective of this subsection is to prove that the Cauchy stress tensor is symmetrical

$$\hat{\sigma} = \hat{\sigma}^T. \tag{2.23}$$

For this purpose we return to a Cartesian coordinate system $\{x^i\}$ and employ *the principle of angular momentum* for a domain $\tilde{V}$

$$\frac{d}{dt} \int_{\tilde{V}} \rho \bar{r} \times \bar{v} dV = \int_{\tilde{V}} \rho \bar{r} \times \bar{B} dV + \int_{\partial \tilde{V}} \bar{r} \times \bar{b} dS. \tag{2.24}$$

We transform the term in the left-hand side of Eq. (2.24) with the use of Eq. (2.10)

$$\frac{d}{dt} \int_{\tilde{V}} \rho \bar{r} \times \bar{v} dV = \frac{d}{dt} \int_{\tilde{V}_0} \rho_0 \bar{r} \times \bar{v} dV_0$$

$$= \int_{\tilde{V}_0} \rho_0 \frac{\partial}{\partial t} (\bar{r} \times \bar{v}) dV_0 = \int_{\tilde{V}_0} \rho_0 (\frac{\partial \bar{r}}{\partial t} \times \bar{v} + \bar{r} \times \frac{\partial \bar{v}}{\partial t}) dV_0$$

$$= \int_{\tilde{V}_0} \rho_0 (\bar{v} \times \bar{v} + \bar{r} \times \bar{a}) dV_0 = \int_{\tilde{V}_0} \rho_0 \bar{r} \times \bar{a} dV_0 = \int_{\tilde{V}} \rho \bar{r} \times \bar{a} dV. \tag{2.25}$$

Here we employ the formula $\bar{v} \times \bar{v} = 0$.

We now transform the second term in the right-hand side of Eq. (2.24) by employing expansion (2.15) and the Stokes formula

$$\int_{\partial \tilde{V}} \bar{r} \times \bar{b} dS = \int_{\partial \tilde{V}} n_i \bar{r} \times \bar{b}^i dS = \int_{\tilde{V}} \frac{\partial}{\partial x^i} (\bar{r} \times \bar{b}^i) dV$$

$$= \int_{\tilde{V}} (\frac{\partial \bar{r}}{\partial x^i} \times \bar{b}^i + \bar{r} \times \frac{\partial \bar{b}^i}{\partial x^i}) dV = \int_{\tilde{V}} (\bar{e}_i \times \sigma^{ij} \bar{e}_j + \bar{r} \times \frac{\partial \sigma^{ij}}{\partial x^i} \bar{e}_j) dV$$

$$= \int_{\tilde{V}} [\sigma^{ij} \bar{e}_i \times \bar{e}_j + \bar{r} \times (\bar{\nabla} \cdot \hat{\sigma})] dV. \tag{2.26}$$

Substitution of expressions (2.25) and (2.26) into Eq. (2.24) yields

$$\int_{\tilde{V}} \sigma^{ij} \bar{e}_i \times \bar{e}_j dV = \int_{\tilde{V}} \bar{r} \times (\rho \bar{a} - \rho \bar{B} - \bar{\nabla} \cdot \hat{\sigma}) dV.$$

It follows from this equality and the motion equation (2.19) that

$$\int_{\tilde{V}} \sigma^{ij} \bar{e}_i \times \bar{e}_j dV = 0.$$

Since this equality is valid for an arbitrary domain $\tilde{V}$, we find

$$\sigma^{ij} \bar{e}_i \times \bar{e}_j = 0, \qquad (2.27)$$

which implies the symmetry condition (2.23).

Exercise 2.3. Derive Eq. (2.23) from formula (2.27). □

We now employ Eq. (2.23) in order to obtain the equilibrium equation in the coordinate form. Let $\{x^i\}$ be Cartesian coordinates in the actual configuration with unit vectors $\bar{e}_i$. Expand tensor $\hat{\sigma}$ and vector $\bar{B}$ in unit vectors $\bar{e}_i$

$$\hat{\sigma} = \sigma_{x^1 x^1} \bar{e}_1 \bar{e}_1 + \sigma_{x^2 x^2} \bar{e}_2 \bar{e}_2 + \sigma_{x^3 x^3} \bar{e}_3 \bar{e}_3$$
$$+ \sigma_{x^1 x^2}(\bar{e}_1 \bar{e}_2 + \bar{e}_2 \bar{e}_1) + \sigma_{x^1 x^3}(\bar{e}_1 \bar{e}_3 + \bar{e}_3 \bar{e}_1) + \sigma_{x^2 x^3}(\bar{e}_2 \bar{e}_3 + \bar{e}_3 \bar{e}_2),$$
$$\bar{B} = B_{x^1} \bar{e}_1 + B_{x^2} \bar{e}_2 + B_{x^3} \bar{e}_3.$$

Since

$$\bar{\nabla} = \bar{e}_1 \frac{\partial}{\partial x^1} + \bar{e}_2 \frac{\partial}{\partial x^2} + \bar{e}_3 \frac{\partial}{\partial x^3},$$

we find

$$\bar{\nabla} \cdot \hat{\sigma} = \bar{e}_1 \frac{\partial \sigma_{x^1 x^1}}{\partial x^1} + \bar{e}_2 \frac{\partial \sigma_{x^1 x^2}}{\partial x^1} + \bar{e}_3 \frac{\partial \sigma_{x^1 x^3}}{\partial x^1}$$
$$+ \bar{e}_1 \frac{\partial \sigma_{x^1 x^2}}{\partial x^2} + \bar{e}_2 \frac{\partial \sigma_{x^2 x^2}}{\partial x^2} + \bar{e}_3 \frac{\partial \sigma_{x^2 x^3}}{\partial x^2}$$
$$+ \bar{e}_1 \frac{\partial \sigma_{x^1 x^3}}{\partial x^3} + \bar{e}_2 \frac{\partial \sigma_{x^2 x^3}}{\partial x^3} + \bar{e}_3 \frac{\partial \sigma_{x^3 x^3}}{\partial x^3}.$$

Substitution of these expressions into Eq. (2.21) implies the following equilibrium equations:

$$\frac{\partial \sigma_{x^1 x^1}}{\partial x^1} + \frac{\partial \sigma_{x^1 x^2}}{\partial x^2} + \frac{\partial \sigma_{x^1 x^3}}{\partial x^3} + \rho B_{x^1} = 0,$$
$$\frac{\partial \sigma_{x^1 x^2}}{\partial x^1} + \frac{\partial \sigma_{x^2 x^2}}{\partial x^2} + \frac{\partial \sigma_{x^2 x^3}}{\partial x^3} + \rho B_{x^2} = 0,$$
$$\frac{\partial \sigma_{x^1 x^3}}{\partial x^1} + \frac{\partial \sigma_{x^2 x^3}}{\partial x^2} + \frac{\partial \sigma_{x^3 x^3}}{\partial x^3} + \rho B_{x^3} = 0. \qquad (2.28)$$

By employing a similar algorithm and using Eqs. (1.1.1) and (1.1.2) we can derive the equilibrium equations in cylindrical and spherical coordinates.

Exercise 2.4. Let $\{r, \theta, z\}$ be cylindrical coordinates in the actual configuration with unit vectors $\bar{e}_r$, $\bar{e}_\theta$, $\bar{e}_z$, and

$$\hat{\sigma} = \sigma_{rr}\bar{e}_r\bar{e}_r + \sigma_{\theta\theta}\bar{e}_\theta\bar{e}_\theta + \sigma_{zz}\bar{e}_z\bar{e}_z$$
$$+\sigma_{r\theta}(\bar{e}_r\bar{e}_\theta + \bar{e}_\theta\bar{e}_r) + \sigma_{rz}(\bar{e}_r\bar{e}_z + \bar{e}_z\bar{e}_r) + \sigma_{\theta z}(\bar{e}_\theta\bar{e}_z + \bar{e}_z\bar{e}_\theta),$$
$$\bar{B} = B_r\bar{e}_r + B_\theta\bar{e}_\theta + B_z\bar{e}_z.$$

Then the equilibrium equations are

$$\frac{\partial \sigma_{rr}}{\partial r} + \frac{1}{r}\frac{\partial \sigma_{r\theta}}{\partial \theta} + \frac{\partial \sigma_{rz}}{\partial z} + \frac{\sigma_{rr} - \sigma_{\theta\theta}}{r} + \rho B_r = 0,$$

$$\frac{\partial \sigma_{r\theta}}{\partial r} + \frac{1}{r}\frac{\partial \sigma_{\theta\theta}}{\partial \theta} + \frac{\partial \sigma_{\theta z}}{\partial z} + \frac{2\sigma_{r\theta}}{r} + \rho B_\theta = 0,$$

$$\frac{\partial \sigma_{rz}}{\partial r} + \frac{1}{r}\frac{\partial \sigma_{\theta z}}{\partial \theta} + \frac{\partial \sigma_{zz}}{\partial z} + \frac{\sigma_{rz}}{r} + \rho B_z = 0. \qquad \square \qquad (2.29)$$

Exercise 2.5. Let $\{r, \theta, \phi\}$ be spherical coordinates in the actual configuration with unit vectors $\bar{e}_r$, $\bar{e}_\theta$, $\bar{e}_\phi$, and

$$\hat{\sigma} = \sigma_{rr}\bar{e}_r\bar{e}_r + \sigma_{\theta\theta}\bar{e}_\theta\bar{e}_\theta + \sigma_{\phi\phi}\bar{e}_\phi\bar{e}_\phi$$
$$+\sigma_{r\theta}(\bar{e}_r\bar{e}_\theta + \bar{e}_\theta\bar{e}_r) + \sigma_{r\phi}(\bar{e}_r\bar{e}_\phi + \bar{e}_\phi\bar{e}_r) + \sigma_{\theta\phi}(\bar{e}_\theta\bar{e}_\phi + \bar{e}_\phi\bar{e}_\theta),$$
$$\bar{B} = B_r\bar{e}_r + B_\theta\bar{e}_\theta + B_\phi\bar{e}_\phi.$$

Then the equilibrium equations are

$$\frac{\partial \sigma_{rr}}{\partial r} + \frac{1}{r}\frac{\partial \sigma_{r\theta}}{\partial \theta} + \frac{1}{r\sin\theta}\frac{\partial \sigma_{r\phi}}{\partial \phi} + \frac{1}{r}\left(2\sigma_{rr} - \sigma_{\theta\theta} - \sigma_{\phi\phi} + \frac{\cos\theta}{\sin\theta}\sigma_{r\theta}\right)$$
$$+\rho B_r = 0,$$

$$\frac{\partial \sigma_{r\theta}}{\partial r} + \frac{1}{r}\frac{\partial \sigma_{\theta\theta}}{\partial \theta} + \frac{1}{r\sin\theta}\frac{\partial \sigma_{\theta\phi}}{\partial \phi} + \frac{1}{r}\left[3\sigma_{r\theta} + \frac{\cos\theta}{\sin\theta}(\sigma_{\theta\theta} - \sigma_{\phi\phi})\right]$$
$$+\rho B_\theta = 0,$$

$$\frac{\partial \sigma_{r\phi}}{\partial r} + \frac{1}{r}\frac{\partial \sigma_{\theta\phi}}{\partial \theta} + \frac{1}{r\sin\theta}\frac{\partial \sigma_{\phi\phi}}{\partial \phi} + \frac{1}{r}\left(3\sigma_{r\phi} + 2\frac{\cos\theta}{\sin\theta}\sigma_{\theta\phi}\right)$$
$$+\rho B_\phi = 0. \qquad \square \qquad (2.30)$$

2.7. Energy Balance Equation

In this subsection we develop *the energy balance equation* from motion equation (2.19). For this purpose we multiply Eq. (2.19) by $\bar{v}$ and integrate it over a domain V occupied by a body in the actual configuration. As a result, we obtain

$$\int_V \rho\bar{v} \cdot \bar{a}dV = \int_V \bar{v} \cdot (\bar{\nabla} \cdot \hat{\sigma})dV + \int_V \rho\bar{v} \cdot \bar{B}dV. \qquad (2.31)$$

78

Transform the term in the left-hand side of Eq. (2.31) as follows:

$$\int_V \rho \bar{v} \cdot \bar{a} dV = \int_{V_0} \rho_0 \bar{v} \cdot \frac{\partial \bar{v}}{\partial t} dV_0 = \frac{1}{2} \int_{V_0} \rho_0 \frac{\partial v^2}{\partial t} dV_0$$

$$= \frac{d}{dt} \int_{V_0} \frac{\rho_0 v^2}{2} dV_0 = \frac{d}{dt} \int_V \frac{\rho v^2}{2} dV. \qquad (2.32)$$

In order to transform the first term in the right-hand side of Eq. (2.31) we use Eqs. (1.3.61), (1.4.25), and (2.16)

$$\int_V (\bar{\nabla} \cdot \hat{\sigma}) \cdot \bar{v} dV = \int_\Gamma \bar{n} \cdot \hat{\sigma} \cdot \bar{v} dS - \int_V \hat{\sigma} : \hat{D} dV = \int_\Gamma \bar{b} \cdot \bar{v} dS - \int_V \hat{\sigma} : \hat{D} dV. \quad (2.33)$$

Substitution of expressions (2.32) and (2.33) into Eq. (2.31) yields the energy balance equation

$$\frac{d}{dt} \int_V \frac{\rho v^2}{2} dV + \int_V \hat{\sigma} : \hat{D} dV = \int_V \rho \bar{v} \cdot \bar{B} dV + \int_\Gamma \bar{b} \cdot \bar{v} dS. \qquad (2.34)$$

The first term in the left-hand side of Eq. (2.34) is called *kinetic energy*, the second term is called *stress power*, while the sum in the right-hand side of this equality is called *rate of working*.

2.8. The Piola Stress Tensor

It follows from Eq. (2.16) that the surface load applied to the surface element in the actual configuration equals

$$d\bar{b} = \bar{n} \cdot \hat{\sigma} dS. \qquad (2.35)$$

Let us replace the surface element in the actual configuration dS by the surface element in the initial configuration dS_0 by using Eq. (1.83)

$$d\bar{b} = \bar{n}_0 \cdot \sqrt{\frac{g}{g_0}} \bar{\nabla} \bar{r}_0^T \cdot \hat{\sigma} dS_0 = \bar{n}_0 \cdot \hat{P} dS_0, \qquad (2.36)$$

where tensor

$$\hat{P} = \sqrt{\frac{g}{g_0}} \bar{\nabla} \bar{r}_0^T \cdot \hat{\sigma} \qquad (2.37)$$

is called *the Piola stress tensor*. As common practice in the nonlinear mechanics of continua, different authors provide different names to the same object. The Piola tensor is also called the first Piola–Kirchhoff tensor and the first Kirchhoff tensor.

Eq. (2.37) together with Eq. (1.9) implies that

$$\hat{\sigma} = \sqrt{\frac{g_0}{g}} \bar{\nabla}_0 \bar{r}^T \cdot \hat{P}. \tag{2.38}$$

Exercise 2.6. Check that the Piola stress tensor is not symmetrical: in general, $\hat{P}^T \neq \hat{P}$. □

In some situations, instead of the non-symmetrical Piola stress tensor, *the Kirchhoff stress tensor* $\hat{\Sigma}$ is employed

$$\hat{\Sigma} = \sqrt{\frac{g}{g_0}} \bar{\nabla} \bar{r}_0^T \cdot \hat{\sigma} \cdot \bar{\nabla} \bar{r}_0. \tag{2.39}$$

Tensor $\hat{\Sigma}$ is also called the second Piola-Kirchhoff tensor.

Exercise 2.7. Check that the Kirchhoff stress tensor is symmetrical. □

Exercise 2.8. Establish a connection between the Piola and Kirchhoff stress tensors. □

Let us now derive the motion equation in terms of the Piola tensor $\hat{P}$. For this purpose, we rewrite Eq. (2.17) by using (2.10) and (2.36)

$$\int_{\tilde{V}_0} \rho_0 \bar{a} \, dV_0 = \int_{\tilde{V}_0} \rho_0 \bar{B} \, dV_0 + \int_{\partial \tilde{V}_0} \bar{n}_0 \cdot \hat{P} \, dS_0.$$

Applying the Stokes formula (1.3.59) to this equality we obtain

$$\int_{\tilde{V}_0} (\rho_0 \bar{a} - \rho_0 \bar{B} - \bar{\nabla}_0 \cdot \hat{P}) \, dV_0 = 0.$$

Since this equality is valid for an arbitrary volume $\tilde{V}_0$, it implies the motion equation in the basis of the initial configuration

$$\rho_0 \bar{a} = \bar{\nabla}_0 \cdot \hat{P} + \rho_0 \bar{B}. \tag{2.40}$$

The corresponding equilibrium equation is

$$\bar{\nabla}_0 \cdot \hat{P} + \rho_0 \bar{B} = 0. \tag{2.41}$$

According to (2.35), a boundary condition for Eqs. (2.40) and (2.41) is written as

$$\bar{n}_0 \cdot \hat{P} \mid_{\Gamma_0} = \bar{b} \frac{dS}{dS_0}. \tag{2.42}$$

As an example, let us present Eq. (2.41) in the component form for Cartesian coordinates $\{x^i\}$ in the initial configuration with unit vectors $\bar{e}_i$. We have

$$\bar{\nabla}_0 = \bar{e}_1 \frac{\partial}{\partial x^1} + \bar{e}_2 \frac{\partial}{\partial x^2} + \bar{e}_3 \frac{\partial}{\partial x^3}, \quad \bar{B} = B_{x^1} \bar{e}_1 + B_{x^2} \bar{e}_2 + B_{x^3} \bar{e}_3,$$

$$\hat{P} = P_{x^1 x^1} \bar{e}_1 \bar{e}_1 + P_{x^2 x^2} \bar{e}_2 \bar{e}_2 + P_{x^3 x^3} \bar{e}_3 \bar{e}_3 + P_{x^1 x^2} \bar{e}_1 \bar{e}_2 + P_{x^2 x^1} \bar{e}_2 \bar{e}_1$$
$$+ P_{x^1 x^3} \bar{e}_1 \bar{e}_3 + P_{x^3 x^1} \bar{e}_3 \bar{e}_1 + P_{x^2 x^3} \bar{e}_2 \bar{e}_3 + P_{x^3 x^2} \bar{e}_3 \bar{e}_2. \tag{2.43}$$

It follows from Eq. (2.43) that

$$
\begin{aligned}
\bar{\nabla}_0 \cdot \hat{P} = \bar{e}_1 \frac{\partial P_{x^1 x^1}}{\partial x^1} + \bar{e}_2 \frac{\partial P_{x^1 x^2}}{\partial x^1} + \bar{e}_3 \frac{\partial P_{x^1 x^3}}{\partial x^1} \\
+ \bar{e}_1 \frac{\partial P_{x^2 x^1}}{\partial x^2} + \bar{e}_2 \frac{\partial P_{x^2 x^2}}{\partial x^2} + \bar{e}_3 \frac{\partial P_{x^2 x^3}}{\partial x^2} \\
+ \bar{e}_1 \frac{\partial P_{x^3 x^1}}{\partial x^3} + \bar{e}_2 \frac{\partial P_{x^3 x^2}}{\partial x^3} + \bar{e}_3 \frac{\partial P_{x^3 x^3}}{\partial x^3}.
\end{aligned}
\tag{2.44}
$$

Substitution of expressions (2.43) and (2.44) into Eq. (2.41) yields

$$
\begin{aligned}
\frac{\partial P_{x^1 x^1}}{\partial x^1} + \frac{\partial P_{x^2 x^1}}{\partial x^2} + \frac{\partial P_{x^3 x^1}}{\partial x^3} + \rho_0 B_{x^1} = 0, \\
\frac{\partial P_{x^1 x^2}}{\partial x^1} + \frac{\partial P_{x^2 x^2}}{\partial x^2} + \frac{\partial P_{x^3 x^2}}{\partial x^3} + \rho_0 B_{x^2} = 0, \\
\frac{\partial P_{x^1 x^3}}{\partial x^1} + \frac{\partial P_{x^2 x^3}}{\partial x^2} + \frac{\partial P_{x^3 x^3}}{\partial x^3} + \rho_0 B_{x^3} = 0.
\end{aligned}
\tag{2.45}
$$

The equilibrium equations (2.45) for the Piola stress tensor in Cartesian coordinates coincide with the equilibrium equations for linear media with infinitesimal strains.

Exercise 2.9. Derive the equilibrium equations for the Piola stress tensor in cylindrical and spherical coordinates. □

3. Constitutive Equations

The motion equation (2.19) together with the mass conservation law (2.10), and expressions (1.1.5) for the acceleration vector $\bar{a}$ and (2.1.25) for the Finger deformation tensor $\hat{F}$ are fulfilled for any medium, both solid and liquid. This system of differential equations is not closed, i.e. it contains more unknown functions than the number of equations. In order to close it, additional relations should be introduced, which connect the stress tensor $\hat{\sigma}$ with some deformation tensor, e.g. with the Finger tensor. These relations are called *the constitutive equations* (constitutive laws), since they constitute the physical state of the medium under consideration.

3.1. Axioms of the Constitutive Theory

In this subsection we formulate basic axioms of the constitutive theory for *isothermal processes*, when the temperature is assumed to be constant. The constitutive theory for non-isothermal processes will be the subject of the second volume of this textbook.

Causality. The displacement field $\bar{u}$ determines entirely stresses in a medium:

$$
\hat{\sigma}(t, \xi) = \hat{\mathcal{F}}(\bar{u}(\tilde{t}, \tilde{\xi})),
\tag{3.1}
$$

where t and $\tilde{t}$ are moments of time, ξ and $\tilde{\xi}$ are spatial coordinates and $\hat{\mathcal{F}}$ is *a tensor valued functional.*

Determinism. Stresses at instant t depend on the displacement fields at previous moments of time. This means that $\tilde{t}$ should be less than or equal to t in Eq. (3.1).

Axiom of neighborhood. Displacements $\bar{u}$ at points $\tilde{\xi}$ far from a point ξ do not affect stress $\hat{\sigma}$ at the point ξ.

To provide the correct mathematical meaning to the latter statement, we fix a vector $\Delta\bar{r}$ from the point ξ to a point $\tilde{\xi}$ and expand function $\bar{u}$ into *the Taylor series* with respect to $\Delta\bar{r}$

$$\bar{u}(\tilde{t},\tilde{\xi}) = \bar{u}(\tilde{t},\xi) + \Delta\bar{r}\cdot\bar{\nabla}_0\bar{u}(\tilde{t},\xi) + \frac{1}{2}\Delta\bar{r}\cdot\bar{\nabla}_0\bar{\nabla}_0\bar{u}(\tilde{t},\xi)\cdot\Delta\bar{r} + \dots. \qquad (3.2)$$

Substitution of expression (3.2) into Eq. (3.1) implies that

$$\hat{\mathcal{F}}(\bar{u}(\tilde{t},\tilde{\xi})) = \hat{\mathcal{F}}(\bar{u}(\tilde{t},\xi),\bar{\nabla}_0\bar{u}(\tilde{t},\xi),\bar{\nabla}_0\bar{\nabla}_0\bar{u}(\tilde{t},\xi),\dots). \qquad (3.3)$$

Materials are distinguished depending on the number of derivatives which are taken into account in Eq. (3.3).

Definition. A material is called *simple* if the right-hand side of Eq. (3.3) contains the displacement field $\bar{u}$ and its gradient $\bar{\nabla}_0\bar{u}$ only.

Definition. A material is called *non-simple of gradient type* if the functional $\hat{\mathcal{F}}$ depends on higher derivatives. The maximal order of these derivatives determines the order of a non-simple material.

We confine ourselves to simple materials, when Eq. (3.3) may be written as

$$\hat{\sigma}(t,\xi) = \hat{\mathcal{F}}(\bar{u}(\tilde{t},\tilde{\xi})),\bar{\nabla}_0\bar{u}(\tilde{t},\xi)). \qquad (3.4)$$

Since $\bar{u}(t,\xi) = \bar{r}(t,\xi) - \bar{r}_0(\xi)$ and $\bar{\nabla}_0\bar{u}(t,\xi) = \bar{\nabla}_0\bar{r}(t,\xi) - \hat{I}$, Eq. (3.4) implies that

$$\hat{\sigma}(t,\xi) = \hat{\mathcal{F}}(\xi,\bar{r}(\tilde{t},\xi),\bar{\nabla}_0\bar{r}(\tilde{t},\xi)). \qquad (3.5)$$

Tensor-valued functionals in the right-hand side of Eqs. (3.4) and (3.5) differ from each other, but the same notation is preserved for simplicity.

Definition. A material is called *homogeneous* if the tensor-valued functional $\hat{\mathcal{F}}$ does not depend explicitly on Lagrangian coordinates ξ.

For homogeneous materials Eq. (3.4) implies that

$$\hat{\sigma}(t,\xi) = \hat{\mathcal{F}}(\bar{r}(\tilde{t},\xi),\bar{\nabla}_0\bar{r}(\tilde{t},\xi)). \qquad (3.6)$$

Axiom of objectivity (Principle of material frame indifference). The stress tensor $\hat{\sigma}$ is indifferent with respect to superposition of rigid motions:

$$\hat{\sigma}'(t,\xi) = \hat{O}^T(t)\cdot\hat{\sigma}(t,\xi)\cdot\hat{O}(t) \qquad (3.7)$$

for any motion $\bar{\mathcal{R}}'$ which differs from motion $\bar{\mathcal{R}}$ by a rigid motion, see Eq. (1.4.5).

Substitution of expressions (1.4.6) and (1.4.7) into Eq. (1.5) implies that

$$\bar{\nabla}_0 \bar{r}'(t,\xi) = \bar{\nabla}_0 \bar{r}(t,\xi) \cdot \hat{O}(t). \tag{3.8}$$

It follows from Eqs. (3.6) and (3.8) that

$$\hat{\sigma}'(t,\xi) = \hat{\mathcal{F}}(\bar{R}'_0(\tilde{t}) + (\bar{r}(\tilde{t},\xi) - \bar{R}_0(\tilde{t})) \cdot \hat{O}(\tilde{t}), \bar{\nabla}_0 \bar{r}(\tilde{t},\xi) \cdot \hat{O}(\tilde{t})).$$

Substitution of this expression and Eq. (3.6) into Eq. (3.7) yields

$$\hat{\mathcal{F}}(\bar{R}'_0(\tilde{t}) + (\bar{r}(\tilde{t},\xi) - \bar{R}_0(\tilde{t})) \cdot \hat{O}(\tilde{t}), \bar{\nabla}_0 \bar{r}(\tilde{t},\xi) \cdot \hat{O}(\tilde{t}))$$
$$= \hat{O}^T(t) \cdot \hat{\mathcal{F}}(\bar{r}(\tilde{t},\xi), \bar{\nabla}_0 \bar{r}(\tilde{t},\xi)) \cdot \hat{O}(t).$$

This equality is valid for arbitrary functions $\bar{R}_0(t)$ and $\bar{R}'_0(t)$ provided the tensor-valued functional $\hat{\mathcal{F}}$ is independent of $\bar{r}$. Therefore, we find

$$\hat{\sigma}(t,\xi) = \hat{\mathcal{F}}(\bar{\nabla}_0 \bar{r}(\tilde{t},\xi)), \qquad \hat{\sigma}'(t,\xi) = \hat{\mathcal{F}}(\bar{\nabla}_0 \bar{r}(\tilde{t},\xi) \cdot \hat{O}(\tilde{t})). \tag{3.9}$$

It follows from Eqs. (3.7) and (3.9) that

$$\hat{\mathcal{F}}(\bar{\nabla}_0 \bar{r}(\tilde{t},\xi) \cdot \hat{O}(\tilde{t})) = \hat{O}^T(t) \cdot \hat{\sigma}(t,\xi) \cdot \hat{O}(t). \tag{3.10}$$

By using the polar decomposition formula (1.38) in the form $\bar{\nabla}_0 \bar{r} = \hat{U}_l \cdot \hat{O}_*$, we find

$$\hat{\sigma}(t,\xi) = \hat{O}(t) \cdot \hat{\mathcal{F}}(\hat{U}_l(\tilde{t},\xi) \cdot \hat{O}_*(\tilde{t}) \cdot \hat{O}(\tilde{t})) \cdot \hat{O}^T(t).$$

This equality is fulfilled for any orthogonal tensor $\hat{O}$, in particular, for $\hat{O} = \hat{O}_*^T$. Substituting $\hat{O}_*^T$ instead of $\hat{O}$, we obtain

$$\hat{\sigma}(t,\xi) = \hat{O}_*^T(t) \cdot \hat{\mathcal{F}}(\hat{U}_l(\tilde{t},\xi)) \cdot \hat{O}_*(t). \tag{3.11}$$

Eq. (3.11) is called *the Truesdell formula* for the Cauchy stress tensor.

Since $\hat{U}_l = \hat{g}^{1/2}$, Eq. (3.11) can be written as

$$\hat{\sigma}(t,\xi) = \hat{O}_*^T(t) \cdot \hat{\mathcal{F}}(\hat{g}(\tilde{t},\xi)) \cdot \hat{O}_*(t). \tag{3.12}$$

For simplicity, the same notation is employed for the tensor-valued functionals in the right-hand sides of Eqs. (3.11) and (3.12).

Another formula for the stress tensor can be derived if the orthogonal tensor $\hat{O}_*$ is replaced by $\hat{U}_l^{-1} \cdot \bar{\nabla}_0 \bar{r}$. This implies that

$$\hat{\sigma}(t,\xi) = \bar{\nabla}_0 \bar{r}^T(t,\xi) \cdot \hat{\mathcal{G}}(\hat{U}_l(\tilde{t},\xi)) \cdot \bar{\nabla}_0 \bar{r}(t,\xi), \tag{3.13}$$

where $\hat{\mathcal{G}}(\hat{U}_l(\tilde{t}, \xi)) = \hat{U}_l^{-1}(t, \xi) \cdot \hat{\mathcal{F}}(\hat{U}_l(\tilde{t}, \xi)) \cdot \hat{U}_l^{-1}(t, \xi)$. Replacing $\hat{U}_l$ by $\hat{g}^{1/2}$, we obtain from Eq. (3.13) that

$$\hat{\sigma}(t, \xi) = \bar{\nabla}_0 \bar{r}^T(t, \xi) \cdot \hat{\mathcal{G}}(\hat{g}(\tilde{t}, \xi)) \cdot \bar{\nabla}_0 \bar{r}(t, \xi). \tag{3.14}$$

For simplicity, the same notation is preserved in the right-hand sides of Eqs. (3.13) and (3.14).

Formulas (3.12) and (3.14) serve as basic tools for the analysis of the constitutive equations with finite strains. It is convenient to distinguish explicitly the current instant t and previous moments of time $\tilde{t} \in [0, t)$. This leads us to the relationships

$$\hat{\sigma}(t, \xi) = \hat{O}_*^T(t) \cdot \hat{\mathcal{F}}(\hat{g}(t, \xi), \hat{g}(\tilde{t}, \xi)) \cdot \hat{O}_*(t),$$
$$\hat{\sigma}(t, \xi) = \bar{\nabla}_0 \bar{r}^T(t, \xi) \cdot \hat{\mathcal{G}}(\hat{g}(t, \xi), \hat{g}(\tilde{t}, \xi)) \cdot \bar{\nabla}_0 \bar{r}(t, \xi).$$

By using the new variable $s = t - \tilde{t} \in [0, t)$, these equalities can be presented as

$$\hat{\sigma}(t, \xi) = \hat{O}_*^T(t) \cdot \hat{\mathcal{F}}(\hat{g}(t, \xi), \hat{g}(t - s, \xi)) \cdot \hat{O}_*(t),$$
$$\hat{\sigma}(t, \xi) = \bar{\nabla}_0 \bar{r}^T(t, \xi) \cdot \hat{\mathcal{G}}(\hat{g}(t, \xi), \hat{g}(t - s, \xi)) \cdot \bar{\nabla}_0 \bar{r}(t, \xi). \tag{3.15}$$

Finally, replacing the Cauchy deformation tensor $\hat{g}$ by the Cauchy strain tensor $\hat{C}$ with the use of Eq. (1.32) we find

$$\hat{\sigma}(t, \xi) = \hat{O}_*^T(t) \cdot \hat{\mathcal{F}}(\hat{C}(t, \xi), \hat{C}(t - s, \xi)) \cdot \hat{O}_*(t),$$
$$\hat{\sigma}(t, \xi) = \bar{\nabla}_0 \bar{r}^T(t, \xi) \cdot \hat{\mathcal{G}}(\hat{C}(t, \xi), \hat{C}(t - s, \xi)) \cdot \bar{\nabla}_0 \bar{r}(t, \xi). \tag{3.16}$$

The second argument of the functionals $\hat{\mathcal{F}}$ and $\hat{\mathcal{G}}$ can be replaced by the difference history of the Cauchy strain $\hat{C}_d(t, s, \xi)$, see Eq. (1.59). As a result, we obtain (for simplicity, we preserve the same notation for the functionals)

$$\hat{\sigma}(t, \xi) = \hat{O}_*^T(t) \cdot \hat{\mathcal{F}}(\hat{C}(t, \xi), \hat{C}_d(t, s, \xi)) \cdot \hat{O}_*(t),$$
$$\hat{\sigma}(t, \xi) = \bar{\nabla}_0 \bar{r}^T(t, \xi) \cdot \hat{\mathcal{G}}(\hat{C}(t, \xi), \hat{C}_d(t, s, \xi)) \cdot \bar{\nabla}_0 \bar{r}(t, \xi). \tag{3.17}$$

The axiom of objectivity imposes limitations on the constitutive equations which follow from the replacement of one actual configuration by another. Let us now consider a transition from one initial configuration to another. We denote by $\bar{g}_0^{(1)i}$ and $\bar{g}_0^{(2)i}$ the corresponding vectors of dual bases, and assume that these vectors are connected by the formula

$$\bar{g}_0^{(2)i} = \hat{O} \cdot \bar{g}_0^{(1)i}, \tag{3.18}$$

where $\hat{O}$ is an orthogonal tensor of rotation corresponding to a "rigid motion" in the initial configuration. Eqs. (1.5) and (3.18) imply that

$$\bar{\nabla}_0^{(2)} \bar{r} = \hat{O} \cdot \bar{\nabla}_0^{(1)} \bar{r}. \tag{3.19}$$

Definition. A material is called *isotropic* if its constitutive equations are invariant with respect to orthogonal transformations of the initial configuration

$$\hat{\mathcal{F}}(\bar{\nabla}_0^{(2)}\bar{r}) = \hat{\mathcal{F}}(\bar{\nabla}_0^{(1)}\bar{r}). \tag{3.20}$$

We confine ourselves to isotropic media only. Substitution of expression (3.19) into Eq. (3.20) implies that for any orthogonal tensor $\hat{O}$

$$\hat{\mathcal{F}}(\bar{\nabla}_0\bar{r}) = \hat{\mathcal{F}}(\hat{O} \cdot \bar{\nabla}_0\bar{r}). \tag{3.21}$$

By using the polar decomposition formula (1.40), we write $\bar{\nabla}_0\bar{r} = \hat{O}_* \cdot \hat{U}_r$, where $\hat{U}_r$ is the right stretch tensor, and present Eq. (3.21) as follows:

$$\hat{\mathcal{F}}(\bar{\nabla}_0\bar{r}) = \hat{\mathcal{F}}(\hat{O} \cdot \hat{O}_* \cdot \hat{U}_r).$$

This relation is fulfilled for any orthogonal $\hat{O}$, in particular, for $\hat{O} = \hat{O}_*^T$. In the latter case, we obtain

$$\hat{\sigma}(t,\xi) = \hat{\mathcal{F}}(\hat{U}_r(\tilde{t},\xi)). \tag{3.22}$$

According to Eq. (3.22), the Cauchy stress tensor is a tensor-valued functional on the history of stretch tensors.

Axiom of memory. The deformation gradient $\bar{\nabla}_0\bar{r}(\tilde{t},\xi)$ at distant past from the current instant t does not affect appreciably the stress tensor $\hat{\sigma}(t,\xi)$.

To present this axiom in the mathematically correct manner, we should introduce a precise measure for evaluating the effect of the deformation history on the material response at the current instant. This will be accomplished below, while we now discuss the difference between elastic and inelastic solids.

Definition. A material is called *elastic* if the Cauchy stress tensor at the current instant t depends only on the deformation gradient at this moment of time.

For elastic media, the constitutive equation (3.14) is written as

$$\hat{\sigma}(t,\xi) = \bar{\nabla}_0\bar{r}^T(t,\xi) \cdot \hat{\mathcal{G}}(\hat{g}(t,\xi)) \cdot \bar{\nabla}_0\bar{r}(t,\xi), \tag{3.23}$$

and the constitutive equation (3.22) implies that

$$\hat{\sigma}(t,\xi) = \hat{\mathcal{F}}(\hat{U}_r(t,\xi)). \tag{3.24}$$

According to Eq. (3.24), the Cauchy stress tensor is a function of the right stretch tensor. By using Eq. (1.41), we can present Eq. (3.24) in the form

$$\hat{\sigma}(t,\xi) = \hat{\mathcal{F}}(\hat{F}(t,\xi)), \tag{3.25}$$

where the same notation for the function in the right-hand side is preserved for simplicity.

It follows from Eqs. (3.10) and (3.21) that function $\hat{\mathcal{F}}(\bar{\nabla}_0\bar{r})$ is objective and isotropic. By applying the Rivlin–Ericksen theorem, see Section 1.5, we obtain the following constitutive equation:

$$\hat{\sigma}(t,\xi) = \phi_0\hat{I} + \phi_1\hat{F}(t,\xi) + \phi_2\hat{F}^2(t,\xi), \qquad (3.26)$$

where ϕ_k are functions of the principal invariants of the Finger tensor $\hat{F}(t,\xi)$. Eq. (3.26) means that the mechanical behavior of an isotropic elastic material is determined by three material functions ϕ_k of three principal invariants I_k. An approach based on the constitutive equation (3.26) is called *the Cauchy elasticity*.

Definition. A material is called *viscoelastic* if the Cauchy stress tensor at the current instant t depends on the entire history of the deformation gradient on the interval $[0, t]$, where $t = 0$ is the instant when external loads are applied.

3.2. Principles of Fading Memory

A decay in the effect of the deformation history on stresses at the current instant is described by a special theory of fading memory, see e.g. Coleman (1964) and Coleman & Mizel (1968). We do not intend to provide a detailed exposition of this theory, referring to the original works, and confine ourselves to several basic definitions.

Definition. The set $\mathcal{C}(t) = \{\hat{C}(s)\ (0 \le s \le t)\}$ is called *the strain history* up to a moment t, and the set $\mathcal{C}_*(t) = \{\hat{C}(s)\ (0 \le s < t)\}$ is called *the past strain history* up to a moment t.

Definition. A function $\beta(t)$ is called *influence function* of an order $\kappa > 1$ if the following conditions are fulfilled:

1. $\beta(t) > 0$ for any $t \ge 0$;

2. $\beta(0) = 1$;

3. $\int_0^\infty \beta(t)dt < \infty$;

4. there is a $\kappa > 1$ such that $\lim_{t\to\infty} t^\kappa \beta(t) = 0$.

Definition. For a fixed instant $t > 0$ and a fixed influence function $\beta(t)$, a set $\mathcal{W}_t$ is called *fading memory space* up to the moment t if it contains all the pairs $(\hat{C}(t), \mathcal{C}(t))$ with the standard operations of addition and multiplication by a real number and with the inner product

$$\langle(\hat{C}_1(t), \mathcal{C}_1(t)), (\hat{C}_2(t), \mathcal{C}_2(t))\rangle = \hat{C}_1(t) : \hat{C}_2(t) + \int_0^\infty \beta(s)\hat{C}_1(t-s) : \hat{C}_2(t-s)ds.$$

Exercise 3.1. Show that for any $t > 0$, the fading memory space $\mathcal{W}_t$ is a *Hilbert space* with the norm

$$\|(\hat{C}(t), \mathcal{C}(t))\|^2_{\mathcal{W}_t} = \hat{C}(t) : \hat{C}(t) + \int_0^\infty \beta(s)\hat{C}(t-s) : \hat{C}(t-s)ds.$$

The zero element of this space corresponds to rigid motions with $\hat{C}(t) = 0$. $\square$

Let us consider a tensor valued functional $\hat{\mathcal{H}}$ mapping $\mathcal{W}_t$ into the set of symmetrical tensors of the second rank.

Definition. A functional $\hat{\mathcal{H}}$ is called *continuous* at a point $(\hat{C}(t), \mathcal{C}(t))$ provided

$$\lim \|\hat{\mathcal{H}}(\hat{C}(t) + \hat{C}_1(t), \mathcal{C}(t) + \mathcal{C}_1(t)) - \hat{\mathcal{H}}(\hat{C}(t), \mathcal{C}_1(t))\| = 0$$

as $\|(\hat{C}_1(t), \mathcal{C}_1(t))\|^2_{\mathcal{W}_t} \to 0$, where $\| \cdot \|$ is the Euclidean norm of a tensor.

Definition. A functional $\hat{\mathcal{H}}$ is called continuous on a subset of $\mathcal{W}_t$ if it is continuous at any point of this subset.

Two *principles of fading memory* should be distinguished: weak and strong.

- **Weak principle of fading memory.** There is an influence function $\beta(t)$ of an order $\kappa > 1$ such that for any $t \geq 0$ the functionals $\hat{\mathcal{F}}$ and $\hat{\mathcal{G}}$ in Eqs. (3.16) are continuous in a neighborhood of the zero element of $\mathcal{W}_t$.

This assertion means that two histories of deformation which differ from each other by a distant past, but are close in the recent past lead to similar stresses at the current instant.

The weak principle of fading memory ensures only the existence of a function $\beta(t)$, but does not guarantee its uniqueness. Moreover, this assertion does not allow concrete expressions for functionals $\hat{\mathcal{F}}$ and $\hat{\mathcal{G}}$ to be developed.

- **Strong principle of fading memory.** There is an influence function $\beta(t)$ of an order $\kappa > n + 1$ such that for any $t \geq 0$ the functionals $\hat{\mathcal{F}}$ and $\hat{\mathcal{G}}$ in Eqs. (3.16) are n times differentiable in a neighborhood of the zero element of $\mathcal{W}_t$.

The strong principle of fading memory allows the constitutive functionals to be expanded in series in multiple integrals of the deformation history similar to the Taylor series for a function of several variables. This approach to constructing constitutive equations goes back to Volterra for nonlinear viscoelastic media with infinitesimal strains, and to Green & Rivlin (1957) and Coleman & Noll (1960) for nonlinear viscoelastic media with finite strains.

Chapter 3

Constitutive equations in finite elasticity

This Chapter is concerned with constitutive models in the elasticity theory with finite strains. In Section 1, we discuss basic features of elastic media and provide some experimental data. Section 2 is concerned with constitutive equations in finite elasticity. We consider both the Cauchy and Green theories of elastic media, but concentrate on hyperelastic materials. Finally, in Section 3 we formulate boundary-value problems in finite elasticity and discuss the existence and uniqueness of their solutions.

1. Elastic Behavior of Materials

This Section is concerned with a discussion of some characteristic features of elastic materials with finite strains.

Elastic media are defined in Chapter 2 as materials where stresses depend on the current strains only, and where the history of deformation can be neglected in the constitutive relations. Despite the (mathematically) precise character of this definition, questions arise as to (i) whether such materials really exist, (ii) if they do exist, whether they demonstrate large deformations, (iii) if they can bear finite deformations, which range of strains is typical of them. An additional question of interest concerns the accuracy of the above constitutive assumption regarding independence of the past: what is the order of error (compared with "purely" elastic deformations) when the dependence on the deformation history (*hysteresis phenomena*) is neglected.

Apparently, any conscious answer to these questions may be obtained from experimental data only. Thus, we begin our analysis of elastic solids with a discussion of some experimental results.

Let us consider a specimen in the form of a rectilinear rod which is in its

88

natural (stress-free) configuration. At instant $t = 0$, tensile forces P are applied to the ends of the rod, see Fig. 1.1. Under their action the rod deforms. The forces P are assumed to be sufficiently smooth functions of time, which change so slowly that any accidental vibrations decay. This kind of loading is called *quasi-static*. An alternative kind of loading is called *dynamic*.

To characterize the rod's deformation we introduce the so-called *proof strain*

$$\epsilon = \frac{l - l_0}{l_0}, \tag{1.1}$$

where l_0 and l are initial and current lengths of the rod.

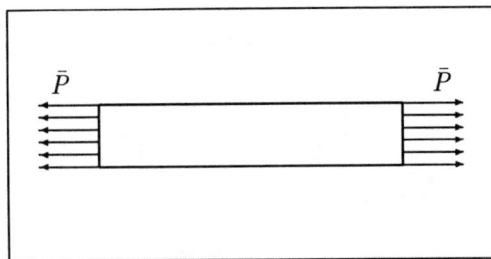

Fig. 1.1. A specimen under uniaxial load.

The strain ϵ is not the only measure employed in experiments. For example, instead of it the following characteristic may be used:

$$\tilde{\epsilon} = \frac{l - l_0}{l}.$$

Other measures of stretches are *the extension ratio*

$$\tilde{\lambda} = \frac{l}{l_0}, \tag{1.2}$$

and *the logarithmic (natural) strain*

$$e = \ln \frac{l}{l_0}. \tag{1.3}$$

Exercise 1.1. Show that the proof and logarithmic strains are connected by the formula

$$e = \ln(1 + \epsilon). \quad \square$$

Exercise 1.2. Prove that up to the second order terms compared with the strain ϵ, the following equalities are true:

$$\tilde{\epsilon} = \epsilon, \qquad \tilde{\lambda} = 1 + \epsilon, \qquad e = \epsilon. \qquad \square$$

As common practice, the proof strain ϵ is used when the corresponding elongations are rather small (e.g. less than 0.01 to 0.1). For deformations which exceed 0.1, either the extension ratio λ or the logarithmic strain is employed.

An important merit of the proof strains is an opportunity to add them in order to calculate the total strain. Let us assume that a rod is stretched, first, from its initial length l_0 to a length l_1, and, afterward, it is stretched from the length l_1 to a length l_2. Then the total strain can be found as

$$\epsilon = \epsilon_1 + \epsilon_2, \tag{1.4}$$

where

$$\epsilon_1 = \frac{l_1 - l_0}{l_0}, \qquad \epsilon_2 = \frac{l_2 - l_1}{l_1},$$

provided $|l_1 - l_0| \ll l_0$ and $|l_2 - l_0| \ll l_0$.

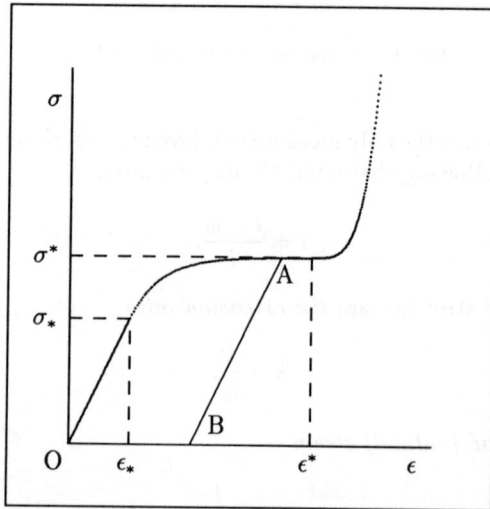

Fig. 1.2. Schematic plot of the stress-strain dependence for an elastic solid.

Exercise 1.3. Show that for a small uniaxial deformation, the proof strain coincides with the only non-zero component of the Cauchy infinitesimal strain tensor, while the logarithmic strain coincides with the only non-zero component of the Hencky strain tensor. □

As a measure of stresses, we choose *the engineering stress* $p = P/S_0$, where S_0 is the cross-section area in the initial configuration. An alternative measure of stresses is the so-called *true stress* $\sigma = P/S$, where S is the cross-section area in the actual configuration.

Exercise 1.4. Check that the engineering stress corresponds to the Piola stress tensor, while the true stress corresponds to the Cauchy stress tensor. □

A typical stress-strain dependence is plotted in Fig. 1.2 (curve OA). In the interval $[0, \epsilon_*]$ the material behaves linearly, whereas for $\epsilon > \epsilon_*$ its response becomes nonlinear. *The ultimate strain* ϵ_* (or the corresponding stress σ_*) characterizes a boundary between regions of linear and nonlinear elasticity.

For a number of materials, ϵ_* value is rather low, of the order of from 0.001 to 0.01. This is the standard ultimate strain for metals at room temperature, concrete, stones, semicrystalline plastics, etc. On the other hand, some media demonstrate a linear stress-strain dependence for deformations up to 0.05 – 0.1, e.g. several types of rubber and biological tissues.

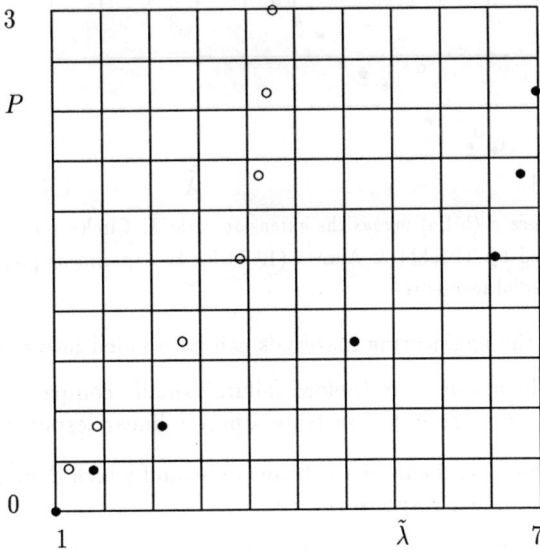

Fig. 1.3. Tensile force P (kg) *versus* the extension ratio $\tilde{\lambda}$. Circles correspond to experimental data presented by Mooney (1940) for tread stock (unfilled circles) and gum stock (filled circles).

Two basic sources of nonlinearity may be distinguished in the mechanics of continua:

- geometrical nonlinearity due to large deformations, when the linear rule of summation (1.4) is not valid;

- physical nonlinearity due to the nonlinear response to applied loads.

According to this classification, the geometrically nonlinear and physically nonlinear problems should be distinguished.

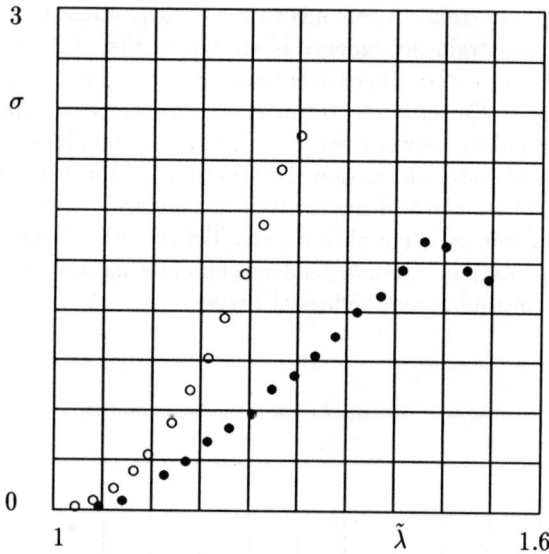

Fig. 1.4. Tensile stress σ (MPa) *versus* the extension ratio $\tilde{\lambda}$. Circles correspond to experimental data obtained by Tissakht & Ahmed (1995) for two specimens prepared from the distal layer of the medial meniscus.

Formally, all the engineering materials can be divided into four groups:

1. materials which can bear (before failure) small (compared with unity) strains only, and which demonstrate a purely linear response;

2. materials which can bear small strains only, and which demonstrate both linear and nonlinear responses;

3. materials which allow both small and large (of the order of unity) deformations, and which show a linear stress-strain dependence up to their failure;

4. materials which allow both small and large deformations, and which show a linear stress-strain dependence for small strains and a nonlinear dependence for finite deformations.

The first group of materials is the subject of the linear theory of elasticity. The second group is considered in the geometrically linear and physically nonlinear theory of elasticity. These groups are rather far from the main subject of our study.

To the best of our knowledge, materials of the third group do not exist at all.

Finally, materials of the fourth group which demonstrate both physical and geometrical nonlinearity are in the focus of our attention.

Nonlinearities in the material response are demonstrated in Figs. 1.3 and 1.4, where the stress-strain dependencies are plotted for vulcanized rubber and for the knee meniscus.

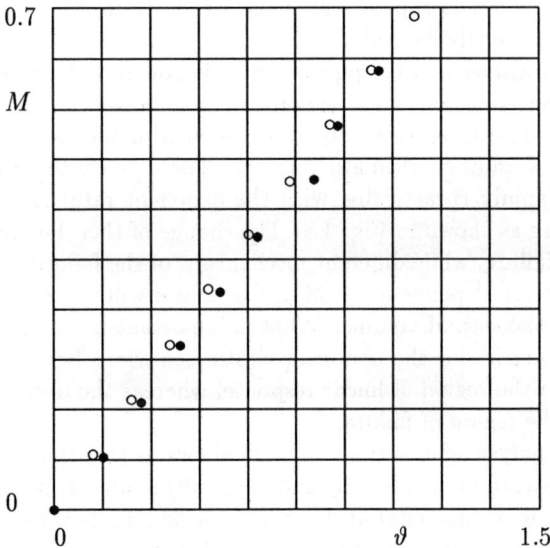

Fig. 1.5. Torque M (N·m) *versus* twist angle ϑ (rad). Circles correspond to experimental data obtained by Rivlin & Saunders (1951) for vulcanized rubber. Unfilled circles correspond to loading, while filled circles correspond to unloading.

The curve presented in Fig. 1.2 may belong to a nonlinear elastic material as well as to an elastoplastic material. The difference between a nonlinear elastic medium and an elastoplastic medium can be revealed at the unloading process only, when the applied load P decreases and tends to zero.

For a nonlinear elastic material, any reduction of stresses leads to a decrease in strains, whereas stresses and strains obey the same stress-strain relation both for loading and unloading, see curve OA in Fig. 1.2.

93

On the contrary, for an elastoplastic medium, a reduction of stresses leads to a decrease in strains, but the stress-strain diagrams for loading and unloading differ from each other. As common practice, the loading curve OA is essentially nonlinear, whereas the unloading curve AB is linear. The line AB is parallel to the initial (linear) portion of curve OA.

The theory of elastoplasticity will be considered in detail in the second volume of this textbook. In sequel, we will analyze the nonlinear elastic behavior only. To convince the reader that some materials do demonstrate the elastic behavior at finite strains, we present experimental data for vulcanized rubber in torsion, see Fig. 1.5. For our purposes, it is important to stress that in torsional tests the torque M is a measure of applied forces, while the twist angle ϑ plays the role of a measure of deformation. Experimental data plotted in Fig. 1.5 show that for large twists (up to 1 rad) the stress-strain diagrams for loading and unloading practically coincide.

Let us now consider another phenomenon demonstrated in Fig. 1.2. When the strain ϵ exceeds the ultimate strain ϵ_*, the stress-strain curves increase monotonically and sufficiently slowly (the region of nonlinear elasticity) until the strain reaches another ultimate value ϵ^*. For $\epsilon > \epsilon^*$, these curves begin either to grow rapidly (practically, with the exponent rate) as shown in Fig. 1.3, or to decrease as shown in Fig. 1.4. The change of their behavior is caused by the material failure, while different mechanisms of the failure imply different types of stress-strain dependencies. Material failure will be the subject of our consideration in the second volume. What is important now is the presence of geometrical boundaries for the nonlinear elastic behavior. The lower boundary ϵ_* corresponds to the region of linear response, whereas the upper boundary ϵ^* corresponds to the region of failure.

The above analysis of experimental data allows us to formulate the subject of finite elasticity as the geometrically and physically nonlinear behavior of solids (without hysteresis phenomena) at deformations ranging between the ultimate strains for the linear response and the strains of failure.

2. Constitutive Equations in Finite Elasticity

This Section is concerned with constitutive equations for elastic media. We suggest several constitutive equations for elastic materials with finite strains which are reduced to the constitutive equation of a linear elastic solid for infinitesimal strains. The main attention is focused on hyperelastic media, i.e. elastic materials which possess strain energy densities. We suggest some expressions for strain energy densities both for compressible and incompressible materials and discuss basic restrictions imposed on these functions by the constitutive theory.

2.1. Elasticity and Hyperelasticity

According to the constitutive theory, see Section 2.3, in order to formulate constitutive equations in finite elasticity we should introduce three functions ϕ_k ($k = 0, 1, 2$) of three principal invariants $I_k(\hat{F})$ of the Finger deformation tensor, see Eq. (2.3.26). Although there is no rational way to obtain these functions, some attempts are undertaken to postulate constitutive equations for isotropic elastic materials with finite strains. The theory developed (which is called *the Cauchy elasticity*) is based on the assumption that constitutive relations with large deformations may be derived as generalizations of the linear constitutive equation with infinitesimal strains

$$\hat{\sigma} = \lambda I_1(\hat{\epsilon}_0)\hat{I} + 2\mu\hat{\epsilon}_0. \tag{2.1}$$

Here $\hat{\sigma}$ is the stress tensor, $\hat{\epsilon}_0$ is the Cauchy infinitesimal strain tensor, $\hat{I}$ is the unit tensor, and λ and μ are constants (*the Lame parameters*), and I_k stands for the kth principal invariant.

For infinitesimal strains, all the strain tensors coincide with each other. The same is true for the stress tensors as well. For finite strains, we should determine which stress and strain tensors are employed in Eq. (2.1). Depending on the choice, we derive different constitutive models which may be called "*linear*" constitutive models in finite elasticity. One of the first linear models was suggested by Seth (1935). According to *the Seth model*, tensor $\hat{\sigma}$ coincides with the Cauchy stress tensor, while tensor $\hat{\epsilon}_0$ is be replaced by the Almansi strain tensor $\hat{A}$. As a result, we obtain the constitutive equation

$$\hat{\sigma} = \lambda I_1(\hat{A})\hat{I} + 2\mu\hat{A}. \tag{2.2}$$

The Saint-Venant–Kirchhoff model, see Ciarlet (1988), is another example of linear elastic media with finite strains. According to this model, tensor $\hat{\sigma}$ in Eq. (2.1) is replaced by the Kirchhoff stress tensor $\hat{\Sigma}$, while tensor $\hat{\epsilon}_0$ is replaced by the Cauchy strain tensor $\hat{C}$. The constitutive equation of the Saint-Venant–Kirchhoff model is written as

$$\hat{\Sigma} = \lambda I_1(\hat{C})\hat{I} + 2\mu\hat{C}. \tag{2.3}$$

The third example is provided by *the "linear" Signorini model*, see Signorini (1943). According to this model, tensor $\hat{\sigma}$ coincides with the Cauchy stress tensor, tensor $\hat{\epsilon}_0$ is replaced by the Almansi strain tensor $\hat{A}$, but, unlike the Seth model, coefficients λ and μ are assumed to be functions of the principal invariants of tensor $\hat{A}$

$$\hat{\sigma} = [\lambda I_1(\hat{A}) + \frac{1}{2}(\lambda + \mu)I_1^2(\hat{A})]\hat{I} + 2[\mu - (\lambda + \mu)I_1(\hat{A})]\hat{A}. \tag{2.4}$$

95

In order to derive nonlinear constitutive equations in finite elasticity, the second power of an appropriate strain tensor is included into the constitutive equation, the coefficients are expanded into the Taylor series, and only a few terms in the expansions are taken into account. An example of the nonlinear constitutive equation is provided by *the Mernon model*, see Ciarlet (1988),

$$\hat{\Sigma} = \lambda I_1(\hat{C})\hat{I} + 2\mu\hat{C} + \nu_1 I_1(\hat{C}^2)\hat{I} + \nu_2 I_1^2(\hat{C})\hat{I} + \nu_3 I_1(\hat{C})\hat{C} + \nu_4\hat{C}^2, \qquad (2.5)$$

where λ, μ, ν_k ($k = 1, 2, 3, 4$) are material parameters. By neglecting in Eq. (2.5) terms of the second order compared with the Cauchy strain tensor, we return to the Saint-Venant–Kirchhoff model (2.3).

Apparently, there is no conscious procedure to find the number of material constants which ensures fair prediction of experimental data, as well as to determine which nonlinear terms are important for constructing constitutive equations. Therefore, it is quite natural to reduce the number of unknown functions which should be found by fitting experimental data to the minimum. Evidently, the minimal number of material functions is unity.

Definition. The constitutive theory which describes the mechanical behavior of elastic solids with the use of one material function is called hyperelasticity (*the Green elasticity*).

To be accurate, we should say that *the hyperelasticity theory* implies the existence of a smooth function $W(\bar{\nabla}_0\bar{r})$, which characterizes *the mechanical energy* of material. The function W is called *strain energy density* (per unit volume in the initial configuration). For isothermal processes, when the temperature is fixed, the function W coincides (up to a constant) with the specific *free energy* of material. For *adiabatic processes*, when energy remains unchanged, this function coincides (up to a constant) with the specific *internal energy* of material.

2.2. The First Law of Thermodynamics

In this subsection we formulate *the first law of thermodynamics* and derive (by using this assertion) constitutive equations of a hyperelastic medium. Let a hyperelastic medium occupy a domain $\tilde{V}_0$ with a smooth boundary $\tilde{\Gamma}_0$ in the initial configuration and a domain $\tilde{\Omega}$ with a boundary $\tilde{\Gamma}$ in the actual configuration under the action of a body force $\bar{B}$ and a surface traction $\bar{b}$.

Definition. *The total strain energy* of a medium is determined as

$$W^\dagger = \int_{\tilde{V}_0} W(\bar{\nabla}_0\bar{r})dV_0, \qquad (2.6)$$

where dV_0 is the volume element.

Let us consider two displacement fields for transition from the initial to the actual configuration: $\bar{u}_1 = \bar{u}$ and $\bar{u}_2 = \bar{u} + \delta\bar{u}$, where $\delta\bar{u}$ is a perturbation of the displacement field $\bar{u}$.

Definition. *The elementary work of a body force* $\bar{B}$ *and a surface traction* $\bar{b}$ equals

$$\delta A = \int_{\tilde{V}} \rho \bar{B} \cdot \delta \bar{u} dV + \int_{\tilde{\Gamma}} \bar{b} \cdot \delta \bar{u} dS, \qquad (2.7)$$

where dV and dS are the volume and surface elements, respectively, and ρ is mass density in the actual configuration.

To transform Eq. (2.7) we replace the integrals over domains in the actual configuration by the integrals over appropriate domains in the initial configuration. By using mass conservation law (2.2.8) and boundary condition (2.2.42), we obtain

$$\delta A = \int_{\tilde{V}_0} \rho_0 \bar{B} \cdot \delta \bar{u} dV_0 + \int_{\tilde{\Gamma}_0} \bar{n}_0 \cdot \hat{P} \cdot \delta \bar{u} dS_0,$$

where $\hat{P}$ is the Piola stress tensor, ρ_0 is mass density, and dS_0 is the surface element in the initial configuration. By applying the Stokes formula (1.3.59) and Eq. (1.3.15), we find

$$\delta A = \int_{\tilde{V}_0} [\rho_0 \bar{B} \cdot \delta \bar{u} + (\bar{\nabla}_0 \cdot \hat{P}) \cdot \delta \bar{u} + \hat{P} : \bar{\nabla}_0 \delta \bar{u}^T] dV_0. \qquad (2.8)$$

We confine ourselves to static deformations and assume that the medium is in equilibrium in the actual configuration. Eq. (2.8) together with the equilibrium equation (2.2.41) implies that

$$\delta A = \int_{\tilde{V}_0} \hat{P} : \bar{\nabla}_0 \delta \bar{u}^T dV_0.$$

Finally, by using Eq. (2.1.1) we replace perturbation $\delta \bar{u}$ of the displacement vector by perturbation $\delta \bar{r}$ of the radius-vector, and obtain

$$\delta A = \int_{\tilde{V}_0} \hat{P} : \bar{\nabla}_0 \delta \bar{r}^T dV_0. \qquad (2.9)$$

The first law of thermodynamics. For isothermal and adiabatic processes, variation of the total strain energy $W^\dagger$ equals the elementary work of external loads.

This principle together with Eqs. (2.6) and (2.9) yields

$$\int_{\tilde{V}_0} \delta W dV_0 = \int_{\tilde{V}_0} \hat{P} : \bar{\nabla}_0 \delta \bar{r}^T dV_0.$$

Since the equality is valid for an arbitrary volume $\tilde{V}_0$, it implies that

$$\delta W = \hat{P} : \bar{\nabla}_0 \delta \bar{r}^T = \hat{P} : \delta \bar{\nabla}_0 \bar{r}^T.$$

Bearing in mind definition (1.5.23) of the derivative of a scalar function with respect to a tensor variable, this equality is written as

$$(W_{\bar{\nabla}_0 \bar{r}} - \hat{P}) : \delta \bar{\nabla}_0 \bar{r}^T = 0. \qquad (2.10)$$

97

Since tensor $\delta\bar{\nabla}_0\bar{r}^T$ is arbitrary, Eq. (2.10) implies the following constitutive equation of a hyperelastic material:

$$\hat{P} = W_{\bar{\nabla}_0\bar{r}}. \tag{2.11}$$

According to the principle of material frame indifference, function W should be objective. We confine ourselves to isotropic media and assume additionally that strain energy density W is isotropic. It follows from Proposition 1.5.2 that function W has the form

$$W(\bar{\nabla}_0\bar{r}) = W(I_1(\hat{F}), I_2(\hat{F}), I_3(\hat{F})), \tag{2.12}$$

where $I_k(\hat{F})$ are the principal invariants of the Finger deformation tensor $\hat{F}$. For simplicity, we will write below I_k instead of $I_k(\hat{F})$.

2.3. Finger's Formula

Let us transform the constitutive equation (2.11) by replacing the differentiation with respect to the deformation gradient $\bar{\nabla}_0\bar{r}$ by the differentiation with respect to the Finger deformation tensor $\hat{F}$. For this purpose, we differentiate Eq. (2.1.25) and find

$$\delta\hat{F} = \delta\bar{\nabla}_0\bar{r}^T \cdot \bar{\nabla}_0\bar{r} + \bar{\nabla}_0\bar{r}^T \cdot \delta\bar{\nabla}_0\bar{r}. \tag{2.13}$$

The derivative of function W with respect to the symmetrical tensor $\hat{F}$ is defined as $\delta W = W_{\hat{F}} : \delta\hat{F}^T = W_{\hat{F}} : \delta\hat{F}$. Substitution of expression (2.13) into this formula yields

$$\delta W = W_{\hat{F}} : (\delta\bar{\nabla}_0\bar{r}^T \cdot \bar{\nabla}_0\bar{r} + \bar{\nabla}_0\bar{r}^T \cdot \delta\bar{\nabla}_0\bar{r}) = 2\bar{\nabla}_0\bar{r} \cdot W_{\hat{F}} : \delta\bar{\nabla}_0\bar{r}^T.$$

On the other hand, we may write $\delta W = W_{\bar{\nabla}_0\bar{r}} : \delta\bar{\nabla}_0\bar{r}^T$. Comparison of these equalities implies that

$$W_{\bar{\nabla}_0\bar{r}} = 2\bar{\nabla}_0\bar{r} \cdot W_{\hat{F}}. \tag{2.14}$$

It follows from Eqs. (2.11) and (2.14) that $\hat{P} = 2\bar{\nabla}_0\bar{r} \cdot W_{\hat{F}}$. Substitution of this expression into Eq. (2.2.38) yields the constitutive equation of a hyperelastic medium

$$\hat{\sigma} = \frac{2}{\sqrt{I_3(\hat{F})}} \hat{F} \cdot W_{\hat{F}}. \tag{2.15}$$

Exercise 2.1. By using similar transformations, show that

$$\hat{\sigma} = \frac{2}{\sqrt{I_3(\hat{g})}} \bar{\nabla}_0\bar{r}^T \cdot W_{\hat{g}} \cdot \bar{\nabla}_0\bar{r}. \quad \Box \tag{2.16}$$

By applying Finger's formula (1.5.36) to Eq. (2.12), we find

$$W_{\hat{F}} = [\frac{\partial W}{\partial I_1} + I_1 \frac{\partial W}{\partial I_2}]\hat{I} - \frac{\partial W}{\partial I_2}\hat{F} + I_3 \frac{\partial W}{\partial I_3}\hat{F}^{-1}. \tag{2.17}$$

Substitution of expression (2.17) into Eq. (2.15) implies *the Finger formula*

$$\hat{\sigma} = \frac{2}{\sqrt{I_3}}(\psi_0 \hat{I} + \psi_1 \hat{F} + \psi_2 \hat{F}^2), \tag{2.18}$$

where

$$\psi_0 = I_3(\hat{F})\frac{\partial W}{\partial I_3}, \qquad \psi_1 = \frac{\partial W}{\partial I_1} + I_1(\hat{F})\frac{\partial W}{\partial I_2}, \qquad \psi_2 = -\frac{\partial W}{\partial I_2}. \tag{2.19}$$

It follows from Eq. (1.2.33) that $\hat{F}^2 = I_1\hat{F} - I_2\hat{I} + I_3\hat{F}^{-1}$. Substitution of this expression into Eq. (2.18) yields the other version of the Finger formula

$$\hat{\sigma} = \frac{2}{\sqrt{I_3}}[(I_2\frac{\partial W}{\partial I_2} + I_3\frac{\partial W}{\partial I_3})\hat{I} + \frac{\partial W}{\partial I_1}\hat{F} - I_3\frac{\partial W}{\partial I_2}\hat{F}^{-1}]. \tag{2.20}$$

2.4. Constitutive Restrictions

Let us assume that the initial configuration is stress-free, i.e. that the Cauchy stress tensor $\hat{\sigma}$ vanishes in the initial configuration. Our objective is to formulate restrictions imposed on strain energy density $W = W(I_1, I_2, I_3)$:
1. *Strain energy density W vanishes in the initial configuration.* According to Eqs. (2.18) and (2.19), when we add an arbitrary constant to strain energy density, the stress tensor remains unchanged. Therefore, this restriction may be fulfilled for any strain energy density by adding to function W an appropriate constant. Since in the initial configuration

$$\hat{F} = \hat{F}^2 = \hat{I}, \qquad I_1(\hat{F}) = 3, \qquad I_2(\hat{F}) = 3, \qquad I_3(\hat{F}) = 1, \tag{2.21}$$

this condition implies

$$W\mid_{I_1=3,\ I_2=3,\ I_3=1} = 0. \tag{2.22}$$

2. *The Cauchy stress tensor $\hat{\sigma}$ vanishes in the initial configuration.*
Exercise 2.2. By using Eqs. (2.18), (2.19), and (2.21), show that the Cauchy stress tensor in the initial configuration equals

$$\sigma = 2(\frac{\partial W}{\partial I_1} + 2\frac{\partial W}{\partial I_2} + \frac{\partial W}{\partial I_3})\mid_{I_1=3,\ I_2=3,\ I_3=1} \hat{I}. \qquad \square$$

It follows from this formula that in the stress-free initial configuration

$$(\frac{\partial}{\partial I_1} + 2\frac{\partial}{\partial I_2} + \frac{\partial}{\partial I_3})W \mid_{I_1=3, \ I_2=3, \ I_3=1} = 0. \tag{2.23}$$

3. *For infinitesimal strains, constitutive equation (2.18) turns into the constitutive relation of an isotropic linear elastic material (2.1).* To derive the corresponding condition in terms of strain energy density W, we, first, substitute expression (2.1.29) into Eq. (2.18). Neglecting terms of the second order compared with $|\bar{\nabla}_0\tilde{u}|$, we obtain

$$\hat{\sigma} = \frac{2}{\sqrt{I_3(\hat{F})}}[(\psi_0 + \psi_1 + \psi_2)\hat{I} + 2(\psi_1 + 2\psi_2)\hat{\epsilon}_0]. \tag{2.24}$$

Exercise 2.3. By using formulas (1.2.30) and (2.1.29), show that up to the second order terms

$$I_1(\hat{F}) = 3 + 2I_1(\hat{\epsilon}_0), \quad I_2(\hat{F}) = 3 + 4I_1(\hat{\epsilon}_0), \quad I_3(\hat{F}) = 1 + 2I_1(\hat{\epsilon}_0). \quad \square \tag{2.25}$$

Substituting expressions (2.25) into Eq. (2.20) and using condition (2.23), we obtain with the desired level of accuracy

$$\psi_1 + 2\psi_2 = (\psi_1 + 2\psi_2)|_{I_1=3, \ I_2=3, \ I_3=1} = (\frac{\partial W}{\partial I_1} + \frac{\partial W}{\partial I_2})|_{I_1=3, \ I_2=3, \ I_3=1},$$

$$\psi_0 + \psi_1 + \psi_2 = [(I_1 - 3)\frac{\partial W}{\partial I_2} + (I_3 - 1)\frac{\partial W}{\partial I_3}] + (\frac{\partial W}{\partial I_1} + 2\frac{\partial W}{\partial I_2} + \frac{\partial W}{\partial I_3})$$

$$= 2(\frac{\partial}{\partial I_2} + \frac{\partial}{\partial I_3})W|_{I_1=3, \ I_2=3, \ I_3=1}I_1(\hat{\epsilon}_0)$$

$$+ (\frac{\partial}{\partial I_1} + 2\frac{\partial}{\partial I_2} + \frac{\partial}{\partial I_3})W|_{I_1=3, \ I_2=3, \ I_3=1}$$

$$+ 2(\frac{\partial}{\partial I_1} + 2\frac{\partial}{\partial I_2} + \frac{\partial}{\partial I_3})^2W|_{I_1=3, \ I_2=3, \ I_3=1}I_1(\hat{\epsilon}_0)$$

$$= 2[(\frac{\partial}{\partial I_2} + \frac{\partial}{\partial I_3})W + (\frac{\partial}{\partial I_1} + 2\frac{\partial}{\partial I_2} + \frac{\partial}{\partial I_3})^2W]|_{I_1=3, \ I_2=3, \ I_3=1}I_1(\hat{\epsilon}_0). \tag{2.26}$$

Substitution of expressions (2.25) and (2.26) into Eq. (2.24) implies that up to the second order terms

$$\hat{\sigma} = 4[(\frac{\partial}{\partial I_2} + \frac{\partial}{\partial I_3})W + (\frac{\partial}{\partial I_1} + 2\frac{\partial}{\partial I_2} + \frac{\partial}{\partial I_3})^2W]|_{I_1=3, \ I_2=3, \ I_3=1}I_1(\hat{\epsilon}_0)\hat{I}$$

$$+ 4(\frac{\partial}{\partial I_1} + \frac{\partial}{\partial I_2})W|_{I_1=3, \ I_2=3, \ I_3=1}\hat{\epsilon}_0. \tag{2.27}$$

Comparison of Eqs. (2.1) and (2.27) yields

$$4[(\frac{\partial}{\partial I_2} + \frac{\partial}{\partial I_3})W + (\frac{\partial}{\partial I_1} + 2\frac{\partial}{\partial I_2} + \frac{\partial}{\partial I_3})^2 W]|_{I_1=3,\ I_2=3,\ I_3=1} = \lambda,$$

$$2(\frac{\partial}{\partial I_1} + \frac{\partial}{\partial I_2})W|_{I_1=3,\ I_2=3,\ I_3=1} = \mu. \qquad (2.28)$$

It follows from Eq. (2.23) and the latter equality (2.28) that

$$(\frac{\partial}{\partial I_1} + \frac{\partial}{\partial I_2})W|_{I_1=3,\ I_2=3,\ I_3=1} = -(\frac{\partial}{\partial I_2} + \frac{\partial}{\partial I_3})W|_{I_1=3,\ I_2=3,\ I_3=1} = \frac{\mu}{2}. \qquad (2.29)$$

This expression together with the first equality (2.28) implies that

$$4(\frac{\partial}{\partial I_1} + 2\frac{\partial}{\partial I_2} + \frac{\partial}{\partial I_3})^2 W|_{I_1=3,\ I_2=3,\ I_3=1} = \lambda + 2\mu. \qquad (2.30)$$

4. *Infinitely large strains correspond to infinitely large values of strain energy density.* To provide an accurate formulation of this restriction it is convenient to present strain energy density W as a function of principal stretches v_1, v_2, and v_3, see Antman (1983). For this purpose we substitute expressions (2.1.43) into function $W(I_1, I_2, I_3)$ and arrive at the new strain energy density

$$\overset{\smile}{W}(v_1, v_2, v_3) = W(v_1^2 + v_2^2 + v_3^2,\ v_1^2 v_2^2 + v_1^2 v_3^2 + v_2^2 v_3^2,\ v_1^2 v_2^2 v_3^2).$$

We assume that for any integers $l \neq m \neq n$, and for arbitrary v_m, $v_n \in (0, \infty)$

$$\overset{\smile}{W} \to \infty \qquad \text{if either } v_l \to 0 \quad \text{or} \quad v_l \to \infty. \qquad (2.31)$$

Eq. (2.31) means that an infinite amount of energy is needed to stretch any fiber to infinite length or to compress it to zero length.

Let us transform Eq. (2.31) as follows. Since $\det \hat{F} = v_1^2 v_2^2 v_3^2$, condition $v_l \to 0$ implies that $\det \hat{F} \to 0$. Thus, the first condition (2.31) is a consequence of the following equality:

$$\lim_{\det \hat{F} \to 0} W = \infty. \qquad (2.32)$$

For any matrix $\hat{A} = [a_{ij}]$, *the Hilbert–Schmidt norm* of $\hat{A}$ is defined by the formula $\|\hat{A}\|_*^2 = \sum_{i,j} a_{ij}^2$.

Exercise 2.4. Prove that the Euclidean norm of any matrix is less than or equal to its Hilbert–Schmidt norm. ☐

Exercise 2.5. Let tensors $\hat{F}$ and $\mathrm{adj}\hat{F} = \hat{F}^{-1} \det \hat{F}$ be presented in their eigenbases. Check that their Hilbert–Schmidt norms equal $\|\hat{F}\|_*^2 = v_1^2 + v_2^2 + v_3^2$ and $\|\mathrm{adj}\hat{F}\|_*^2 = v_1^2 v_2^2 + v_2^2 v_3^2 + v_3^2 v_1^2$. ☐

It follows from Exercises 2.4 and 2.5 that the latter condition (2.31) follows from the equality

$$\lim_{\|\hat{F}\|+\|\text{adj }\hat{F}\|+\det \hat{F}\to\infty} W = \infty. \tag{2.33}$$

Eqs. (2.32) and (2.33) impose limitations on strain energy density W for very large deformations. Bearing in mind that such deformations do not occur in reality (the maximal extension ratio observed in experiments is about 10, which means that the maximal principal stretches are less than 4), it is convenient to replace Eqs. (2.32) and (2.33) by *the coercivity condition*, see Ciarlet (1988), which states that there are positive constants a, b, α, β, and γ such that for any Finger's tensor $\hat{F}$

$$W \geq a[\|\hat{F}\|^\alpha + \|\text{adj}\hat{F}\|^\beta + (\det \hat{F})^\gamma] + b. \tag{2.34}$$

2.5. Non-convexity and Polyconvexity of Strain Energy Densities

The above restrictions on strain energy densities are rather weak. However, they allow several functions W to be excluded from our consideration.

For example, the Saint-Venant–Kirchhoff medium (2.3) is a hyperelastic material with the strain energy density

$$W = \frac{3\lambda + 2\mu}{2}(I_1 - 3) + \frac{\lambda + \mu}{4}(I_1 - 3)^2 - \frac{\mu}{2}(I_2 - 3). \tag{2.35}$$

Exercise 2.6. By using Eqs. (2.18), (2.19), and (2.2.39), check that expression (2.35) implies the constitutive relation (2.3). □

It follows from Eq. (2.35) that the Saint-Venant–Kirchhoff model does not satisfy restriction (2.32): function (2.35) does not tend to infinity when $\det \hat{F} \to 0$, since it is independent of the third principal invariant of the Finger deformation tensor.

The following assertion characterizes an important feature of strain energy densities.

Proposition 2.1. A strain energy density which satisfies equality (2.32) is non-convex.

Proof. Let $\mathcal{M}$ be the set of tensors of the second rank, and $M \subset \mathcal{M}$ be the subset of tensors with positive determinants. According to Corollary 1.5.1, the subset M is non-convex. This means that there are tensors $\hat{Q}_0$ and $\hat{Q}_1$ and a parameter $s \in (0, 1)$ such that

$$\hat{Q}_0 \in M, \quad \hat{Q}_1 \in M, \quad s\hat{Q}_0 + (1 - s)\hat{Q}_1 \notin M. \tag{2.36}$$

Let us assume that a strain energy density $W = W(\hat{F})$ is convex on M. This means that there is a convex function $W^*(\hat{F})$ defined on $\text{Co }M = \mathcal{M}$ such that $W(\hat{F}) = W^*(\hat{F})$ for any $\hat{F} \in M$. It follows from the convexity of W^* that for any $\tau \in [0, 1]$

$$w(\tau) = W^*(\tau \hat{Q}_0 + (1 - \tau)\hat{Q}_1) \leq \tau W^*(\hat{Q}_0) + (1 - \tau)W^*(\hat{Q}_1)$$
$$= \tau W(\hat{Q}_0) + (1 - \tau)W(\hat{Q}_1) \leq W(\hat{Q}_0) + W(\hat{Q}_1). \quad (2.37)$$

On the other hand, condition (2.36) implies that there is a $T \in (0, s)$ such that for any $\tau \in [0, T)$ we have $\det[\tau \hat{Q}_0 + (1-\tau)\hat{Q}_1] > 0$ and $\det[T\hat{Q}_0 + (1-T)\hat{Q}_1] = 0$. It follows from these conditions and Eq. (2.32) that

$$\lim_{\tau \to T} w(\tau) = \infty. \quad (2.38)$$

The contradiction between Eqs. (2.37) and (2.38) shows that our assumption regarding the convexity of strain energy density is false. $\square$

Convexity of strain energy densities is extremely convenient to study the existence and uniqueness of solutions in nonlinear problems of elastostatics, since it allows powerful methods of convex analysis to be employed, see Section 3 for details. According to Proposition 2.1, function W is not convex, but a weaker assumption may be imposed regarding its polyconvexity, see Section 1.5. We postulate polyconvexity as an additional restriction imposed on strain energy densities in hyperelasticity.

2.6. Examples of Strain Energy Densities

In this subsection we present several strain energy densities employed in finite elasticity. Not all of these functions satisfy the above restrictions. However, they are widely used in applied problems owing to (i) their simplicity for engineering calculations, (ii) historical traditions, and (iii) their connections with physical models of materials.

A formal, but general way to prescribe a strain energy density consists in expanding function $W(I_1, I_2, I_3)$ in the Taylor series

$$W(I_1, I_2, I_3) = \sum_{k,l,m=0}^{\infty} c_{klm}(I_1 - 3)^k (I_2 - 3)^l (I_3 - 1)^m \quad (2.39)$$

and neglecting terms of higher orders compared with the differences $I_1 - 3$, $I_2 - 3$, and $I_3 - 1$.

Exercise 2.7. Show that condition (2.22) implies that $c_{000} = 0$. $\square$

2.6.1. Strain energy densities in the form of truncated Taylor series

We begin with the "linear" approximation of a strain energy density

$$W = c_{100}(I_1 - 3) + c_{010}(I_2 - 3) + c_{001}(I_3 - 1). \quad (2.40)$$

103

Exercise 2.8. By using condition (2.23), show that

$$c_{100} + 2c_{010} + c_{001} = 0. \qquad \square \qquad (2.41)$$

Exercise 2.9. By using Eqs. (2.28), check that

$$\lambda = 4(c_{010} + c_{001}), \qquad \mu = 2(c_{100} + c_{010}). \qquad \square \qquad (2.42)$$

Eqs. (2.41) and (2.42) imply that a "linear" strain energy density cannot describe adequately the material behavior even for relatively small deformations, since expression (2.40) leads to the non-realistic condition $\lambda + 2\mu = 0$. It is quite natural to "improve" this expression by introducing nonlinearities in Eq. (2.40). The simplest way is to assume that function W depends on two principal invariants linearly, while the nonlinearity is concentrated in a function of one variable only. Thus, we obtain the following three models:
 – *Haughton's solid*, see Haughton (1987),

$$W = W_1(I_1) + c_{010}(I_2 - 3) + c_{001}(I_3 - 1), \qquad (2.43)$$

 – *Carroll's solid*, see Carroll (1988),

$$W = c_{100}(I_1 - 3) + W_2(I_2) + c_{011}(I_3 - 1), \qquad (2.44)$$

 – *the Mooney–Rivlin compressible solids*, see Varga (1966) and Ciarlet & Geymonat (1982),

$$W = c_{100}(I_1 - 3) + c_{010}(I_2 - 3) + W_3(I_3). \qquad (2.45)$$

In Eqs. (2.43) – (2.45), W_k are sufficiently smooth functions satisfying the condition $W_k(3) = 0$ $(k = 1, 2, 3)$.
Exercise 2.10. By using Eqs. (1.2.31), check that $I_2(\hat{F}) = I_1(\mathrm{adj}\hat{F})$ and rewrite Eq. (2.45) as follows:

$$W = c_{100}[I_1(\hat{F}) - 3] + c_{010}[I_1(\mathrm{adj}\hat{F}) - 3] + W_3(I_3). \qquad \square \qquad (2.46)$$

A particular case of the Mooney–Rivlin medium is *neo-Hookean compressible material*, see Blatz (1971), with the strain energy density

$$W = c_{100}(I_1 - 3) + W_3(I_3). \qquad (2.47)$$

Bearing in mind Eqs. (2.42), we rewrite Eq. (2.47) as

$$W = \frac{\mu}{2}(I_1 - 3) + W_3(I_3). \qquad (2.48)$$

A particular case of the neo-Hookean medium is *an elastic liquid* with

$$W = W_3(I_3). \qquad (2.49)$$

Let ρ and ρ_0 be mass densities in the actual and initial configurations, respectively. Substitution of expression (2.2.10) into Eq. (2.49) implies that the strain energy density of an elastic liquid depends on mass density ρ only

$$W = W_*(\rho). \qquad (2.50)$$

Exercise 2.11. By using Eqs. (2.18), (2.19), and (2.50), show that in an elastic liquid the Cauchy stress tensor $\hat{\sigma}$ is spherical

$$\hat{\sigma} = -p(\rho)\hat{I}, \qquad p(\rho) = \frac{\rho^2}{\rho_0}\frac{\partial W_*}{\partial \rho}(\rho). \qquad \Box \qquad (2.51)$$

To simplify Eq. (2.48), a particular form of the nonlinear function $W_3(I_3)$ should be suggested. For example, by choosing the power-law function

$$W_3(I_3) = \frac{\mu}{2k}(I_3^{-k} - 1), \qquad (2.52)$$

we arrive at *the Agarwal solid*, see Agarwal (1979), Burgess & Levinson (1972), and Simpson & Spector (1984),

$$W = \frac{\mu}{2}[(I_1 - 3) + \frac{1}{k}(I_3^{-k} - 1)]. \qquad (2.53)$$

By assuming function $W_3(I_3)$ to be a sum of power-law functions depending on two material parameters k and m, we obtain *the Atkin–Fox solid*, see Atkin & Fox (1980),

$$W = \frac{\mu}{2}(I_1 - 3I_3^{1/3}) + \frac{k}{m}(I_3^{1/2} + \frac{I_3^{(1-m)/2}}{m-1} - \frac{m}{m-1}). \qquad (2.54)$$

A generalization of the Mooney–Rivlin compressible material is *the Knowles solid*, see Knowles (1977), which is characterized by three smooth functions of one variable

$$W = \frac{\mu}{2}[(I_1 - 3)W_1(I_3) + (I_2 - 3)W_2(I_3) + W_3(I_3)]. \qquad (2.55)$$

Exercise 2.12. By using Eqs. (2.22), (2.23), and (2.28), derive restrictions on the Knowles functions $W_k(I_3)$ ($k = 1, 2, 3$). $\Box$

As was noted, there is no rational procedure to determine which nonlinear terms in the Taylor series (2.39) should be included, and which terms may be neglected. We do not intend to concentrate on this approach and refer to two models only. The first is *the Murnaghan solid*, which is well known owing to its applications in elastodynamics of metals, see Lurie (1990) for experimental data. The Murnaghan model is based on the following

Proposition 2.2. Let W be a smooth, objective, and isotropic function of the deformation gradient. Then the following expansion is valid:

$$W = \frac{\lambda}{2}I_1^2(\hat{C}) + \mu I_1(\hat{C}^2) + \frac{1}{6}[\nu_1 I_1^3(\hat{C}) + 6\nu_2 I_1(\hat{C})I_1(\hat{C}^2) + 8\nu_3 I_1(\hat{C}^3)] + o(\|\hat{C}\|^3),$$

where λ, μ, ν_k ($k = 1, 2, 3$) are constants, and

$$\lim_{\|\hat{C}\| \to 0} \frac{o(\|\hat{C}\|)}{\|\hat{C}\|} = 0. \quad \square$$

Problem 2.1. Prove this assertion. For details, see Murnaghan (1951) and Novozhilov (1953). $\square$

Neglecting terms of the fourth order compared with the norm of the Cauchy strain tensor $\hat{C}$, we obtain the Murnaghan strain energy density

$$W = \frac{\lambda}{2}I_1^2(\hat{C}) + \mu I_1(\hat{C}^2) + \frac{1}{6}[\nu_1 I_1^3(\hat{C}) + 6\nu_2 I_1(\hat{C})I_1(\hat{C}^2) + 8\nu_3 I_1(\hat{C}^3)]. \quad (2.56)$$

Exercise 2.13. Show that function (2.56) can be presented in terms of the principal invariants $I_k = I_k(\hat{F})$ as follows:

$$W = \frac{1}{4}[(-3\lambda - 2\mu + \frac{9l}{2} + \frac{n}{2})I_1 + \frac{1}{2}(\lambda + 2\mu - 3l - 3m)I_1^2$$
$$+ (-2\mu + 3m - \frac{n}{2})I_2 - mI_1I_2 + \frac{1}{6}(l + 2m)I_1^3 + \frac{n}{2}(I_3 - 1)]. \quad (2.57)$$

Here l, m, and n are *the Murnaghan parameters*: $l = \frac{1}{2}\nu_1 + \nu_2$, $m = \nu_2 + 2\nu_3$, $n = 4\nu_3$. $\square$

Another class of nonlinear constitutive models is based on *the separability assumption*, see Zdunek (1992),

$$W(I_1, I_2, I_3) = \Phi(I_1, I_2) + \Psi(I_3), \quad (2.58)$$

where Φ and Ψ are smooth functions. For weakly compressible materials, we may set $\Psi(I_3) = K(I_3 - 1)^2$, where K is the bulk modulus in the linear elasticity. Expanding function Φ into the Taylor series in $I_1 - 3$ and $I_2 - 3$ and neglecting terms of the third order compared with the norm of the Cauchy strain tensor $\hat{C}$, we obtain

$$\Phi(I_1, I_2) = c_{10}(I_1 - 3) + c_{01}(I_2 - 1) + c_{20}(I_1 - 3)^2.$$

Substitution of these expressions into Eq. (2.58) implies *the Zdunek model*

$$W(I_1, I_2, I_3) = c_{10}(I_1 - 3) + c_{01}(I_2 - 1) + c_{20}(I_1 - 3)^2 + K(I_3 - 1)^2 \quad (2.59)$$

with four material parameters c_{10}, c_{20}, c_{01}, and K.

We now return to "linear" strain energy densities and consider another class of constitutive models which may be developed if we (i) assume that function W depends on the principal invariants of the Almansi deformation measure $\hat{g}_0 = \hat{F}^{-1}$, (ii) expand the obtained function into the Taylor series (2.39) with respect to the differences $I_1(\hat{g}_0) - 3$, $I_2(\hat{g}_0) - 3$, $I_3(\hat{g}_0) - 1$, and (iii) neglect the nonlinear terms. As a result, we obtain

$$W = \tilde{c}_{100}[I_1(\hat{g}_0) - 3] + \tilde{c}_{010}[I_2(\hat{g}_0) - 3] + \tilde{c}_{011}[I_3(\hat{g}_0) - 1]. \quad (2.60)$$

Exercise 2.14. By using Eqs. (1.2.31), show that Eq. (2.60) may be written as

$$W = \tilde{c}_{100}(\frac{I_2}{I_3} - 3) + \tilde{c}_{010}(\frac{I_1}{I_3} - 3) + \tilde{c}_{011}(\frac{1}{I_3} - 1). \quad \square \quad (2.61)$$

Exercise 2.15. Derive conditions on coefficients $\tilde{c}_{100}$, $\tilde{c}_{010}$, and $\tilde{c}_{001}$ which are imposed by the constitutive restrictions (2.23) and (2.28). $\square$

Similar to Eqs. (2.43) – (2.45), a nonlinearity may be introduced into the constitutive model (2.60) by replacing a linear term in the right-hand side of Eq. (2.60) by a nonlinear function of one variable. As a result, we arrive at the models

$$W = W_1(I_1(\hat{g}_0)) + \tilde{c}_{010}[I_2(\hat{g}_0) - 3] + \tilde{c}_{001}[I_3(\hat{g}_0) - 1], \quad (2.62)$$
$$W = \tilde{c}_{100}[I_1(\hat{g}_0) - 3] + W_2(I_2(\hat{g}_0)) + \tilde{c}_{011}[I_3(\hat{g}_0) - 1], \quad (2.63)$$
$$W = \tilde{c}_{100}[I_1(\hat{g}_0) - 3] + \tilde{c}_{010}[I_2(\hat{g}_0) - 3] + W_3(I_3(\hat{g}_0)), \quad (2.64)$$

where W_k $(k = 1, 2, 3)$ are sufficiently smooth functions.

A particular case of Eq. (2.64) is obtained if we set $\tilde{c}_{100} = \frac{\mu}{2}$, $c_{010} = 0$, and use nonlinear function (2.52), cf. Eq. (2.53),

$$W = \frac{\mu}{2}\{[I_1(\hat{g}_0) - 3] + \frac{1}{k}[I_3^{-k}(\hat{g}_0) - 1]\} = \frac{\mu}{2}[(\frac{I_2}{I_3} - 3) + \frac{1}{k}(I_3^k - 1)], \quad (2.65)$$

where $I_k = I_k(\hat{F})$. Model (2.65) corresponding to $k = \frac{1}{2}$ is called *the Blatz–Ko solid*, see Blatz & Ko (1962).

Exercise 2.16. Show that the Blatz–Ko strain energy density has the form

$$W = \frac{\mu}{2}(\frac{I_2}{I_3} + 2\sqrt{I_3} - 5). \quad \square \quad (2.66)$$

In order to construct new nonlinear strain energy densities, expressions (2.43) – (2.45) and (2.62) – (2.64) are combined. For example, taking a linear combination of strain energy densities for the Agarwal and for the Blatz–Ko models, we obtain the "generalized" Blatz–Ko medium

$$W = \frac{\mu}{2}\{a[(I_1 - 3) + \frac{1}{k}(I_3^{-k} - 1)] + (1 - a)[(\frac{I_2}{I_3} - 3) + \frac{1}{k}(I_3^k - 1)]\}, \quad (2.67)$$

where a, μ, and k are material constants.

A similar approach is suggested by Levinson & Burgess (1971). Strain energy density W is treated as a function of $I_1(\hat{F})$, $I_1(\hat{g}_0)$, and $I_3(\hat{F})$. By expanding this function into the Taylor series with respect to $I_1(\hat{F}) - 3$ and $I_1(\hat{g}_0) - 3$, and neglecting nonlinear terms, we find

$$W = c'_{10}[I_1(\hat{F}) - 3] + c'_{01}(I_1(\hat{g}_0) - 3) + W_3(I_3(\hat{F})), \quad (2.68)$$

where c'_{10} and c'_{01} are material parameters. Setting $c'_{10} = \frac{\mu}{2}a$, $c'_{01} = \frac{\mu}{2}(1 - a)$, we obtain the model

$$W = \frac{\mu}{2}[a(I_1 - 3) + (1 - a)(\frac{I_2}{I_3} - 3)] + W_3(I_3). \quad (2.69)$$

Model (2.69) is of particular interest for weakly compressible elastic media. By expanding function $W_3(I_3)$ into the Taylor series with respect to $\sqrt{I_3} - 1$ and neglecting terms of the third order, we obtain *the polynomial model*

$$W = \frac{\mu}{2}[a(I_1-3)+(1-a)(\frac{I_2}{I_3}-3)+2(1-2a)(\sqrt{I_3}-1)+(2a+b)(\sqrt{I_3}-1)^2], \quad (2.70)$$

where a, b, and μ are material parameters. The quantity $\sqrt{I_3}$ is chosen, since it determines the ratio of the volume element in the actual configuration to that in the initial configuration.

In particular, for $a = 1$ and $b = (-1 + 4\nu)(1 - 2\nu)^{-1}$, Eq. (2.70) implies that

$$W = \frac{\mu}{2}[(I_1 - 3) - 2(\sqrt{I_3} - 1) + \frac{1}{1 - 2\nu}(\sqrt{I_3} - 1)^2]. \quad (2.71)$$

Eq. (2.71) was employed by Vorp et al. (1995) to describe the elastic response of blood vessel segments.

2.6.2. Strain energy densities in the form of truncated series in powers of the principal invariants

The above models are based on the concept of linearity, when all the nonlinear terms (or some part of them) in the Taylor expansion of a strain energy density are neglected. Another procedure for constructing strain energy densities is based of the idea of replacing linear terms by fractional powers. For

108

example, taking the Mooney–Rivlin model (2.46) and replacing tensors $\hat{F}$ and $\text{adj}\hat{F}$ by their powers, we obtain *the Ogden solid*, see Ogden (1972),

$$W = c_{100}[I_1(\hat{F}^{\frac{\alpha}{2}}) - 3] + c_{010}[I_1(\text{adj}\hat{F}^{\frac{\beta}{2}}) - 3] + W_3(I_3), \qquad (2.72)$$

where c_{100}, c_{010}, α, and β are material parameters, and $W_3(I_3)$ is a material function. *The generalized Ogden model* may be derived by summing up expressions (2.72) with different powers

$$W = \sum_{i=1}^{N_1} a_i[I_1(\hat{F}^{\frac{\alpha_i}{2}}) - 3] + \sum_{j=1}^{N_2} b_j[I_1(\text{adj}\hat{F}^{\frac{\beta_j}{2}}) - 3] + W_3(I_3), \qquad (2.73)$$

where N_1, N_2 are positive integers, a_i, b_j, α_i, β_j are material parameters, and $W_3(I_3)$ is a material function.

Exercise 2.17. Check that expression (2.73) coincides with the Ogden function (1.5.53) provided $W_3 = \psi$. By using Exercise 1.5.22, show that function (2.73) is polyconvex. $\square$

A particular case of Ogden's model is *the John solid*, see John (1960). Instead of introducing this constitutive model formally, we expose some reasons which lead to the John model. A linear isotropic elastic material (2.1) is characterized by the strain energy density

$$W = \frac{\lambda}{2}I_1^2(\hat{\epsilon}_0) + \mu I_1(\hat{\epsilon}_0^2). \qquad (2.74)$$

The symmetrical tensor $\hat{\epsilon}_0$ has an eigenbasis where it may be presented by a diagonal matrix with quantities ϵ_{kk} on the diagonal. Therefore, Eq. (2.74) may be written as

$$W = \frac{\lambda}{2}(\epsilon_{11} + \epsilon_{22} + \epsilon_{33})^2 + \mu(\epsilon_{11}^2 + \epsilon_{22}^2 + \epsilon_{33}^2),$$

where ϵ_{kk} are proof strains along the eigenvectors of tensor $\hat{\epsilon}_0$. By using Eqs. (1.1) and (1.2), we present this equality as follows:

$$W = \frac{\lambda}{2}[(\tilde{\lambda}_1 - 1) + (\tilde{\lambda}_2 - 1) + (\tilde{\lambda}_3 - 1)]^2 + \mu[(\tilde{\lambda}_1 - 1)^2 + (\tilde{\lambda}_2 - 1)^2 + (\tilde{\lambda}_3 - 1)^2],$$

where $\tilde{\lambda}_k$ are the extension ratios along the eigenvectors of $\hat{\epsilon}_0$. Since $\tilde{\lambda}_k$ coincide with the principal stretches v_k, we obtain the following expression for the strain energy density of the John solid:

$$W = \frac{\lambda}{2}[(v_1-1)+(v_2-1)+(v_3-1)]^2+\mu[(v_1-1)^2+(v_2-1)^2+(v_3-1)^2]$$

$$= \frac{\lambda}{2}I_1^2(\hat{F} - \hat{I}) + \mu I_1((\hat{F} - \hat{I})^2). \qquad (2.75)$$

109

Exercise 2.18. Prove that function (2.75) belongs to the class of Ogden functions (1.5.53). For this purpose, show that strain energy density (2.75) can be presented as

$$W = \frac{\lambda + 2\mu}{2}(v_1^2 + v_2^2 + v_3^2 - 3)$$
$$+\lambda(v_1 v_2 + v_1 v_3 + v_2 v_3 - 3) - (3\lambda + 2\mu)(v_1 + v_2 + v_3 - 3). \quad \square \quad (2.76)$$

Exercise 2.19. Check that function (2.76) does not satisfy constitutive restriction (2.32). $\square$

A special case of the constitutive model (2.68) with $W_3(I_3) = -c \ln I_3$ was proposed by Blatz (1960)

$$W = c_1(I_1 - 3) + c_2(\frac{I_2}{I_3} - 3) - c \ln I_3.$$

Replacing I_3 in the second term by its power $I_3^{1-\alpha}$, we arrive at *the Flory–Tatara solid*, see Flory & Tatara (1975),

$$W = c_1(I_1 - 3) + c_2(I_2 I_3^{\alpha-1} - 3) - c \ln I_3, \qquad (2.77)$$

where c, c_1, c_2, and α are material parameters.

Another example of "power" strain energy functions can be derived if we use the Agarwal solid (2.53) with the strain energy density

$$W = \frac{\mu}{2}[2I_1(\hat{C}) + \frac{1}{k}(I_3^{-k} - 1)] = \frac{\mu}{2}[I_1(\hat{C}) + I_1(\hat{C}) + \frac{1}{k}(I_3^{-k} - 1)].$$

Replacing one of the first invariants the Cauchy strain tensor $\hat{C}$ by the first invariant of a power of tensor $\hat{C}$, we arrive at the formula

$$W = \frac{\mu}{2}[I_1(\hat{C}) + I_1(\hat{C}^\beta) + \frac{1}{k}(I_3^{-k} - 1)].$$

Finally, replacing the first invariants by their powers, we obtain *the Antman solid*, see Antman (1979),

$$W = \frac{\mu}{2}[I_1^\alpha(\hat{C}) + I_1^\alpha(\hat{C}^\beta) + \frac{1}{k}(I_3^{-k} - 1)]. \qquad (2.78)$$

2.6.3. Strain energy densities as symmetrical functions

Together with the concepts of linearization and fractional powers, it is worth noting the concept of symmetry. According to this concept, a nonlinear function of three variables W is replaced by a combination of functions of one variable

110

only, see Stafford (1969). As common practice, additive and multiplicative combinations are employed. We confine ourselves to additive combinations and mention two models: *the Rivlin–Saunders solid*, and *the Valanis–Landel solid*.

According to the Rivlin–Saunders model, see Rivlin & Saunders (1951), strain energy density W may be presented in the form

$$W = W_1(I_1) + W_2(I_2) + W_3(I_3), \qquad (2.79)$$

where $I_k = I_k(\hat{F})$ and functions W_k $(k = 1, 2, 3)$ are found from experimental data. For comparison of theoretical predictions with experimental data, see Section 4.1.

According to the Valanis–Landel model, see Valanis & Landel (1967), strain energy density W is presented in the form

$$W = w(v_1) + w(v_2) + w(v_3), \qquad (2.80)$$

where v_k are principal stretches, and function w is found by fitting experimental data. As a rule, the power function $w(v) = av^n$ is employed. Strain energy density (2.80) provides fair prediction of experimental data for rubbers, see e.g. Glucklich & Landel (1977), but fails to predict correctly the mechanical response in polymeric melts, see Feigl et al. (1993).

2.7. Constitutive Equations of Particular Materials

In order to derive the constitutive relation (stress-strain dependence) of a hyperelastic material, we should substitute strain energy density W into Eqs. (2.18) and (2.19). In this subsection we provide two examples of the stress-strain relations in finite elasticity. We begin with the Blatz–Ko material (2.66). Calculating the derivatives of strain energy density, we obtain

$$\frac{\partial W}{\partial I_1} = 0, \qquad \frac{\partial W}{\partial I_2} = \frac{\mu}{2I_3}, \qquad \frac{\partial W}{\partial I_3} = \frac{\mu}{2I_3^{1/2}}\left(1 - \frac{I_2}{I_3^{3/2}}\right).$$

Substitution of these expressions into Eqs. (2.18) and (2.19) yields

$$\hat{\sigma} = \frac{\mu}{I_3^{3/2}}[(I_3^{3/2} - I_2)\hat{I} + I_1\hat{F} - \hat{F}^2]. \qquad (2.81)$$

Exercise 2.20. By using similar calculations, derive the constitutive equation for the Agarwal solid (2.53). □

Exercise 2.21. Calculate the Cauchy stress tensor for the Levinson–Burgess solid (2.71). □

As another example, we derive a stress-strain relation when strain energy density W is a function of the principal invariants $J_k = I_k(\hat{g}_0)$ of the Almansi deformation tensor $\hat{g}_0 = \hat{F}^{-1}$. According to Eq. (1.5.39), $W_{\hat{F}} = -\hat{g}_0 \cdot W_{\hat{g}_0} \cdot \hat{g}_0$. This equality and Eq. (2.15) imply that

$$\hat{\sigma} = -\frac{2}{\sqrt{I_3(\hat{F})}} W_{\hat{g}_0} \cdot \hat{g}_0 = -2\sqrt{J_3} W_{\hat{g}_0} \cdot \hat{g}_0. \tag{2.82}$$

It follows from the Finger formula (1.5.36) that

$$W_{\hat{g}_0} = [\frac{\partial W}{\partial J_1} + J_1 \frac{\partial W}{\partial J_2}]\hat{I} - \frac{\partial W}{\partial J_2}\hat{g}_0 + J_3 \frac{\partial W}{\partial J_3}\hat{g}_0^{-1}.$$

This expression together with Eqs. (2.82) and (2.1.71) yields

$$\hat{\sigma} = -2\sqrt{J_3}[J_3 \frac{\partial W}{\partial J_3}\hat{I} + (\frac{\partial W}{\partial J_1} + J_1 \frac{\partial W}{\partial J_2})\hat{g}_0 - \frac{\partial W}{\partial J_2}\hat{g}_0^2]. \tag{2.83}$$

As an example, we consider the Signorini material, see Signorini (1943), with the strain energy density

$$W = \frac{1}{2\sqrt{J_3}}[\mu(J_1 - 3) + \frac{\lambda + \mu}{4}(J_1 - 3)^2] + \mu(\frac{1}{\sqrt{J_3}} - 1). \tag{2.84}$$

Differentiation of Eq. (2.84) implies that

$$\frac{\partial W}{\partial J_1} = \frac{1}{4\sqrt{J_3}}[(\lambda + \mu)(J_1 - 3) + 2\mu], \qquad \frac{\partial W}{\partial J_2} = 0,$$

$$\frac{\partial W}{\partial J_3} = -\frac{1}{4\sqrt{J_3^3}}[2\mu + \mu(J_1 - 3) + \frac{\lambda + \mu}{4}(J_1 - 3)^2].$$

We substitute these expressions into Eq. (2.83) and use Eq. (2.1.34). After simple algebra we derive formula (2.4). Eq. (2.4) is a linear (with respect to the Almansi strain tensor $\hat{A}$) constitutive equation of a hyperelastic medium with coefficients depending on the Almansi strain tensor.

Problem 2.2. Prove that there are no linear constitutive equations with constant coefficients either for the Cauchy or for the Piola stress tensor. The impossibility of a linear elasticity theory with finite strains follows from the axiom of material frame indifference, see Fosdick & Serrin (1979). This assertion is based on the assumption that the initial configuration is stress-free. For a detailed discussion of linear constitutive models in hyperelasticity, see Podio-Guidugli (1987). □

2.8. Materials with Kinematic Constraints

Definition. *A kinematic constraint is a restriction on the displacement field in a medium.*

112

We should mention two example of constrains:

(i) an incompressible solid where the volume element does not change during the deformation

$$dV = dV_0. \tag{2.85}$$

It follows from Eqs. (2.85) and (2.1.74) that

$$I_3(\hat{F}) = I_3(\hat{g}) = 1. \tag{2.86}$$

Eq. (2.86) and mass conservation law (2.2.8) imply that

$$\rho = \rho_0. \tag{2.87}$$

The incompressibility condition means that mass density remains unchanged;

(ii) an elastic sheet with rigid fibers. The fibers are assumed to be located regularly, in parallel to each other, in an $\bar{e}$ direction. They do not resist to bending or to shear of the sheet, and do not allow the sheet elongation in the $\bar{e}$ direction. The kinematic constraint can be written as

$$ds = ds_0, \tag{2.88}$$

where ds and ds_0 are the arc elements in the fiber direction.

Any constraint is described by the equation

$$\alpha(\bar{\nabla}_0 \bar{r}) = 0, \tag{2.89}$$

where α is a sufficiently smooth function. The constrain equation should be objective. Confining ourselves to isotropic constrains, we apply Proposition 1.5.2 and find

$$\alpha(\bar{\nabla}_0 \bar{r}) = \alpha(I_1(\hat{F}), I_2(\hat{F}), I_3(\hat{F})). \tag{2.90}$$

By differentiating Eq. (2.89) and using Eq. (2.14), we find

$$0 = \delta\alpha(\bar{\nabla}_0 \bar{r}) = \alpha_{\bar{\nabla}_0 \bar{r}} : \delta\bar{\nabla}_0 \bar{r}^T = 2\bar{\nabla}_0 \bar{r} \cdot \alpha_{\hat{F}} : \delta\bar{\nabla}_0 \bar{r}^T. \tag{2.91}$$

For an arbitrary variation of the deformation gradient $\delta\bar{\nabla}_0\bar{r}$, Eq. (2.10) implies constitutive equation (2.11). For materials with constraints this is not true, since variations of the deformation gradient are not arbitrary, but obey Eq. (2.91). We use the Lagrange method and introduce an additional variable (the Lagrange multiplier) λ, multiply Eq. (2.91) by λ, and add to Eq. (2.10). By using Eq. (2.14), we obtain

$$[2\bar{\nabla}_0\bar{r} \cdot (W_{\hat{F}} + \lambda\alpha_{\hat{F}}) - \hat{P}] : \delta\bar{\nabla}_0\bar{r}^T = 0,$$

where $\delta\bar{\nabla}_0\bar{r}$ may be treated as an arbitrary variation. This leads to the constitutive equation

$$\hat{P} = 2\bar{\nabla}_0\bar{r} \cdot (W_{\hat{F}} + \lambda\alpha_{\hat{F}}). \tag{2.92}$$

Substitution of this expression into formula (2.2.38) yields

$$\hat{\sigma} = \hat{\sigma}_1 + \hat{\sigma}_2, \tag{2.93}$$

where

$$\hat{\sigma}_1 = \frac{2}{\sqrt{I_3(\hat{F})}}\hat{F} \cdot W_{\hat{F}}, \qquad \hat{\sigma}_2 = \frac{2\lambda}{\sqrt{I_3(\hat{F})}}\hat{F} \cdot \alpha_{\hat{F}}. \tag{2.94}$$

Exercise 2.22. Derive formulas similar to Eq. (2.94) when m kinematic constrains $\alpha_k(\bar{\nabla}_0\bar{r}) = 0$ $(k = 1, \ldots, m)$ are imposed. □

For *an incompressible material*, $\alpha = I_3(\hat{F}) - 1$. We differentiate this expression with the use of Eq. (1.5.33) and find $\alpha_{\hat{F}} = I_3(\hat{F})\hat{F}^{-1}$. By employing Eq. (2.86), we obtain $\alpha_{\hat{F}} = \hat{F}^{-1}$. Substitution of this expression into the latter equality (2.94) yields $\hat{\sigma}_2 = 2\lambda\hat{I}$. Finally, introducing the notation $p = -2\lambda$, where the scalar p is called hydrostatic pressure, we present this expression as

$$\hat{\sigma}_2 = -p\hat{I}. \tag{2.95}$$

Exercise 2.23. By substituting expressions (2.94) and (2.95) into Eq. (2.93) and using condition (2.86), show that the constitutive equation of an incompressible hyperelastic medium has the form

$$\hat{\sigma} = -p\hat{I} + 2\hat{F} \cdot W_{\hat{F}}. \qquad \Box \tag{2.96}$$

It follows from Eq. (2.86) that strain energy density W of an incompressible hyperelastic medium depends on $I_1(\hat{F})$ and $I_2(\hat{F})$ only

$$W = W(I_1(\hat{F}), I_2(\hat{F})). \tag{2.97}$$

Substituting expression (2.97) into Eq. (2.19) we find

$$\psi_0 = 0, \qquad \psi_1 = \frac{\partial W}{\partial I_1} + I_1\frac{\partial W}{\partial I_2}, \qquad \psi_2 = -\frac{\partial W}{\partial I_2}. \tag{2.98}$$

Eqs. (2.18) and (2.98) imply that

$$\hat{\sigma}_1 = 2(\psi_1\hat{F} + \psi_2\hat{F}^2).$$

Substitution of this expression and Eq. (2.95) into Eq. (2.93) yields the Finger formula for the Cauchy stress tensor in an incompressible hyperelastic medium

$$\hat{\sigma} = -p\hat{I} + 2(\psi_1\hat{F} + \psi_2\hat{F}^2). \tag{2.99}$$

Exercise 2.24. Check that for an incompressible hyperelastic medium, Eq. (2.20) implies that

$$\hat{\sigma}_1 = 2(I_2\frac{\partial W}{\partial I_2}\hat{I} + \frac{\partial W}{\partial I_1}\hat{F} - \frac{\partial W}{\partial I_2}\hat{F}^{-1}).$$

114

By substituting this expression and (2.95) into Eq. (2.93), derive the following constitutive equation:

$$\hat{\sigma} = -p\hat{I} + 2(\frac{\partial W}{\partial I_1}\hat{F} - \frac{\partial W}{\partial I_2}\hat{F}^{-1}). \quad \Box \qquad (2.100)$$

Exercise 2.25. By substituting Eq. (2.99) into Eq. (2.2.37), develop the following expression for the Piola stress tensor $\hat{P}$:

$$\hat{P} = -p(\bar{\nabla}_0\bar{r}^T)^{-1} + 2(\psi_1\hat{I} + \psi_2\hat{g}) \cdot \bar{\nabla}_0\bar{r}. \quad \Box \qquad (2.101)$$

2.9. Examples of Strain Energy Densities for Incompressible Materials

In this subsection we provide several expressions for strain energy densities of incompressible hyperelastic materials which are widely used in applications.

In the general case, function (2.97) may be presented as the Taylor series in $I_1 - 3$ and $I_2 - 3$

$$W = \sum_{m,n=0}^{\infty} c_{mn}(I_1 - 3)^m(I_2 - 3)^n \qquad (2.102)$$

with $c_{00} = 0$.

A material with the only non-zero coefficient c_{10} is called *neo-Hookean*. Treloar (1958) suggested the strain energy density

$$W = c_{10}(I_1 - 3) = \frac{\mu}{2}(I_1 - 3) \qquad (2.103)$$

for the description of the mechanical response in rubbers based on the concept of polymeric chains.

Exercise 2.26. Derive the following constitutive equation of the neo-Hookean material:

$$\hat{\sigma} = -p\hat{I} + \mu\hat{F}. \quad \Box$$

The model with only $c_{10} \neq 0$ and $c_{01} \neq 0$ was suggested by Mooney, see Mooney (1940). It is also called *the Mooney–Rivlin material*

$$W = c_{10}(I_1 - 3) + c_{01}(I_2 - 3). \qquad (2.104)$$

Although model (2.104) predicts adequately experimental data for some rubbers for uniaxial extension (with the extension ratio less than 3) and simple shear, it demonstrates poor correlation with observations for biaxial extension and pure shear, see Rivlin & Saunders (1951).

Exercise 2.27. Derive the following constitutive equation for the Mooney–Rivlin material:

$$\hat{\sigma} = -p\hat{I} + 2(c_{10} + c_{01}I_1)\hat{F} - 2c_{01}\hat{F}^2. \quad \square$$

Taking into account several nonlinear terms in the Taylor expansion (2.102), we may derive nonlinear constitutive models. For example, by assuming the linear dependence of W on $I_2 - 3$, and restricting ourselves to a truncated Taylor series in $I_1 - 3$, we obtain *the Isihara–Hashitsume–Tatibana model*, see Isihara et al. (1951),

$$W = c_{10}(I_1 - 3) + c_{01}(I_2 - 3) + c_{20}(I_1 - 3)^2 \tag{2.105}$$

and *the Biderman model*

$$W = c_{10}(I_1 - 3) + c_{01}(I_2 - 3) + c_{20}(I_1 - 3)^2 + c_{30}(I_1 - 3)^3, \tag{2.106}$$

cf. Eq. (2.59). Model (2.105) provides fair approximation of experimental data for equi-biaxial experiments, however, its prediction of data in uniaxial tests is rather poor. Model (2.106) demonstrates excellent agreement with experimental data for sulfur rubber for uniaxial deformations and pure shear, but indicates poor correlation with observations in equi-biaxial tests, see Alexander (1968).

Based on a series of experiments for vulcanized rubbers, Rinvin & Saunders (1951) suggested strain energy density W in the form

$$W(I_1, I_2) = c_{10}(I_1 - 3) + W_2(I_2), \tag{2.107}$$

where $W_2(I_2)$ is a function to be found by fitting experimental data. Apparently, there is no rational procedure to construct an explicit expression for the function W_2. As common practice, it is presented as a linear combination of polynomials, exponents, and logarithms of the second principal invariant of the Finger tensor. We confine ourselves to several formulas:
– Gent & Thomas (1958) assumed that

$$\frac{dW_2}{dI_2} = \frac{C}{I_2}, \tag{2.108}$$

where c is a material parameter.

Exercise 2.28. By integrating Eq. (2.108) and using condition (2.22), derive the following formula for *the Gent–Thomas strain energy density*

$$W = c_{10}(I_1 - 3) + C \ln \frac{I_2}{3}. \quad \square \tag{2.109}$$

Eq. (2.109) provides fair prediction of experimental data for vulcanized rubber.

– Hart-Smith (1966) refined the Gent–Thomas model by replacing the first term in the right-hand side of Eq. (2.107) by a nonlinear function $W_1(I_1)$ and supposing the following approximation for its derivative

$$\frac{dW_1}{dI_1} = c_1 \exp[k(I_1 - 3)^2],$$

where c_1 and k are material parameters.

Exercise 2.29. Derive the following formula for *the Hart-Smith strain energy density*:

$$W = c_1 \int_3^{I_1} \exp[k(I - 3)^2]dI + C \ln \frac{I_2}{3}. \tag{2.110}$$

The Hart-Smith model (2.110) provides excellent agreement with experimental data for sulphur rubber under uniaxial and equi-biaxial tension with the extension ratio less than 3.

– Alexander (1968) refined the Gent–Thomas model by assuming a more sophisticated expression for the derivative of function W_2. To fit experimental data for neoprene film, Alexander supposed that

$$\frac{dW_2}{dI_2} = c_{01} + \frac{C}{(I_2 - 3) + c}, \tag{2.111}$$

where c_{01}, C and c are material parameters.

Exercise 2.30. By integrating Eq. (2.111) and using condition (2.22), derive the following formula for *the Alexander strain energy density*

$$W = c_{10}(I_1 - 3) + c_{01}(I_2 - 3) + C \ln \frac{I_2 - 3 + c}{c}. \quad \square \tag{2.112}$$

– Hutchinson et al. (1965) proposed to approximate the derivative of function W_2 by a truncated series in the exponents (*the Prony series*)

$$\frac{dW_2}{dI_2} = \frac{C_1}{k_1} \exp[k_1(I_2 - 3)] + \frac{C_2}{k_2} \exp[k_2(I_2 - 3)],$$

where C_1, C_2 and k_1, k_2 are material parameters. We integrate this equality and substitute the obtained expression into Eq. (2.105) replacing the second term in the right-hand side. As a result, we arrive at the strain energy density for *the Hutchinson–Becker–Landel medium*:

$$W = c_{10}(I_1 - 3) + c_{20}(I_1 - 3)^2 + C_1[1 - \exp(k_1(I_2 - 3))]$$
$$+ C_2[1 - \exp(k_2(I_2 - 3))]. \tag{2.113}$$

Knowles & Sternberg (1981) introduced incompressible hyperelastic media with

$$W = W(I_1). \tag{2.114}$$

117

A particular case of Eq. (2.114) for *the Knowles medium* is the power-law material

$$W = \frac{\mu}{2b}\{[1 + \frac{b}{n}(I_1 - 3)]^n - 1\}, \tag{2.115}$$

where μ, b and n are positive parameters. For $n = 1$ and an arbitrary $b \neq 0$, Eq. (2.115) is reduced to the constitutive equation of the neo-Hookean material (2.103).

Exercise 2.31. By using Eqs. (2.98) and (2.101), show that the Piola stress tensor in the Knowles medium (2.114) equals

$$\hat{P} = -p(\bar{\nabla}_0\bar{r}^T)^{-1} + 2\frac{dW}{dI_1}\bar{\nabla}_0\bar{r}. \quad \square \tag{2.116}$$

By setting $I_3 = 1$ in Eq. (2.72), we obtain *the Ogden incompressible solid*. The simplest version of the Ogden model may be derived when we set $c_{100} = 2\mu m^{-2}$, $c_{010} = 0$, $\alpha = m$, and $W_3(I_3) = 0$

$$W = \frac{2\mu}{m^2}[I_1(\hat{F}^{\frac{m}{2}}) - 3]. \tag{2.117}$$

Exercise 2.32. By using Eq. (2.96) and Exercise 1.5.10, derive the constitutive equation for the Ogden solid (2.117)

$$\hat{\sigma} = -p\hat{I} + \frac{2\mu}{m}\hat{F}^{\frac{m}{2}}. \quad \square \tag{2.118}$$

Bloch et al. (1978), Chang et al. (1976), and Morman (1988) suggested an approach to constructing constitutive equations based on replacing the principal invariants of the Finger strain tensor by the principal invariants of the generalized strain tensors. By replacing the Cauchy strain tensor $\hat{C}$ in formula (2.103) for the strain energy density of the neo-Hookean medium $W = \frac{\mu}{2}I_1(\hat{C})$ by the Eulerian m-tensor of strains (2.1.66), we obtain *the Morman solid*

$$W = \frac{\mu}{2}I_1(\hat{\mathcal{E}}_E^{(m)}) = \frac{\mu}{2m}[I_1(\hat{g}^{\frac{m}{2}}) - 3]. \tag{2.119}$$

Exercise 2.33. Derive the constitutive equation for the Morman material (2.119). $\square$

3. Boundary Value Problems in Finite Elasticity

In this Section we formulate boundary value problems in finite elastostatics and elastodynamics and discuss the existence and uniqueness of their solutions.

3.1. Problems in the Nonlinear Theory of Elasticity

The basic problem in the elasticity theory with finite strains may be formulated as follows. A medium is in its natural state and occupies a domain

Ω_0 with a smooth boundary Γ_0. Surface Γ_0 is divided into two parts. On the first part, $\Gamma_0^{(u)}$, the displacement vector is given as a function $\bar{w}$ of time t and Lagrangian coordinates $\xi = \{\xi^i\}$ (as a particular case, this part of the boundary may be clamped). On the other part, $\Gamma_0^{(\sigma)}$, surface traction $\bar{b}$ is prescribed.

At instant $t = 0$, surface forces $\bar{b}$ and body forces $\bar{B}$ (dependent on time t and coordinates ξ) are applied. Our objective is to find the displacement vector $\bar{u}(t, \xi)$ and the Cauchy stress tensor $\hat{\sigma}(t, \xi)$ at any instant t and at any point ξ of the medium.

For this purpose, we write out the governing equations which consist of the motion equation (2.2.19)

$$\rho \bar{a} = \bar{\nabla} \cdot \hat{\sigma} + \rho \bar{B}, \qquad \bar{a} = \frac{\partial^2 \bar{u}}{\partial t^2}, \tag{3.1}$$

the constitutive equation (2.18)

$$\hat{\sigma} = \frac{2}{\sqrt{I_3(\hat{F})}} (\psi_0 \hat{I} + \psi_1 \hat{F} + \psi_2 \hat{F}^2) \tag{3.2}$$

and formula (2.1.25) for the Finger deformation tensor

$$\hat{F} = \bar{\nabla}_0 \bar{r}^T \cdot \bar{\nabla}_0 \bar{r}, \qquad \bar{\nabla}_0 \bar{r} = \hat{I} + \bar{\nabla}_0 \bar{u}. \tag{3.3}$$

Here ρ is mass density, $\hat{\sigma}$ is the Cauchy stress tensor, $\hat{I}$ is the unit tensor.

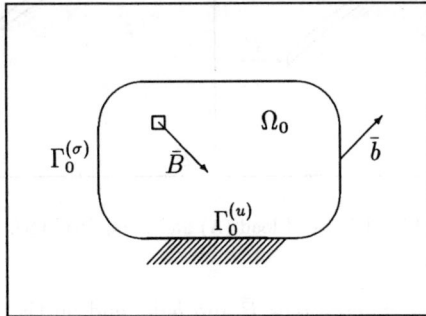

Fig. 3.1. An elastic body under external loads.

To close this system, we should add the boundary conditions

$$\bar{n} \cdot \hat{\sigma} \big|_{\xi \in \Gamma_0^{(\sigma)}} = \bar{b}, \qquad \bar{u} \big|_{\xi \in \Gamma_0^{(u)}} = \bar{w}, \tag{3.4}$$

and the initial data

$$\bar{u}|_{t=0} = \bar{u}_0, \qquad \frac{\partial \bar{u}}{\partial t}|_{t=0} = \bar{v}_0, \qquad (3.5)$$

where $\bar{n}$ is the unit outward normal to boundary Γ in the actual configuration, $\bar{u}_0$ and $\bar{v}_0$ are initial displacements and velocities.

To be precise, we should determine arguments of "external" functions $\bar{B}$, $\bar{b}$, $\bar{w}$, $\bar{u}_0$, and $\bar{v}_0$. Vectors $\bar{u}_0$ and $\bar{v}_0$ are given functions of Lagrangian coordinates ξ. Vector $\bar{w}$ is a given function of time t and coordinates ξ. In linear elasticity, vectors $\bar{b}$ and $\bar{B}$ are assumed to be given functions of time t and coordinates ξ. In finite elasticity, two opportunity arise: either to treat vectors $\bar{b}$ and $\bar{B}$ as functions of t and ξ, or to consider them as functions of the displacement vector $\bar{u}$ and its spatial derivatives. The former case corresponds to the so-called *dead forces*.

Definition. Body forces $\bar{B}$ and surface loads $\bar{b}$ are called dead if there are functions $\bar{B}_0(t,\xi)$ and $\bar{b}_0(t,\xi)$ such that for any instant t

$$\rho \bar{B} dV = \rho_0 \bar{B}_0(t,\xi) dV_0, \qquad \bar{b} dS = \bar{b}_0(t,\xi) dS_0, \qquad (3.6)$$

where ρ and ρ_0 are mass densities, dV and dV_0 are volume elements, and dS and dS_0 are surface elements in the actual and initial configurations, respectively.

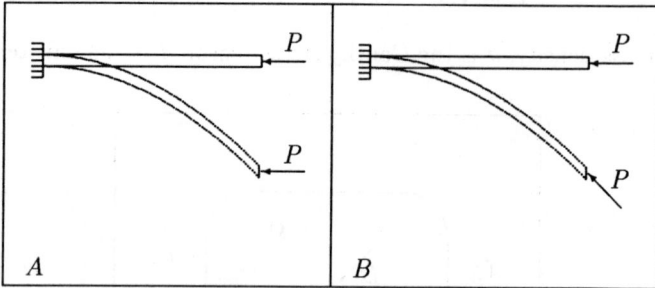

Fig. 3.2. Dead load (A) and living load (B).

In the latter case, when forces $\bar{B}$ and $\bar{b}$ depend on the displacement vector and its derivatives, these loads are called living. We do not intend to discuss *living forces* here, taking into account that their classification is far from being complete, and refer to original works by Sewell (1967) and Batra (1972). It is worth noting an important particular case of living forces: *a follower traction*, which preserves the angle between its direction and the normal $\bar{n}$ to boundary Γ, see Fig. 3.2. Formally, living forces can depend on the displacement field $\bar{u}$ non-locally, however, we confine ourselves to "simple" loads (similar to simple media

120

in the constitutive theory, see Section 2.3) which depend on the displacement vector and on the deformation gradient only.

3.2. Existence of Solutions in Finite Elasticity

Up to date there is no general theory concerning the existence and uniqueness of solutions to the initial-boundary problem (3.1) – (3.5) in finite elasticity. The existence and uniqueness theorems for linear elastic media with infinitesimal strains are derived by Duvaut & Lions (1976) and Fichera (1972). For an elastic medium with finite strains, the existence problem is essentially more complicated. In Section 4.5 we will demonstrate that even for particular types of motion and for extremely smooth initial conditions, dynamic problem (3.1) – (3.5) has a classical solution only on a finite interval of time. For a nonlinear medium, with the growth of time, any classical solution is transformed into a discontinuous solution corresponding to a shock wave. This example shows that we cannot expect the existence of classical solutions "in large", for an arbitrary instant t.

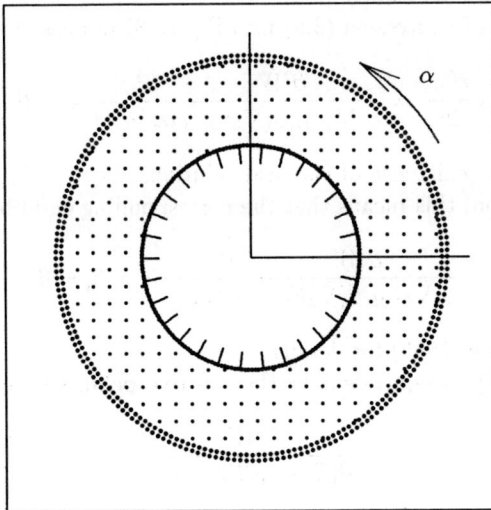

Fig. 3.3. An elastic medium between two rigid cylinders.

In order to consider the existence problem for small intervals of time, it is convenient to write the governing equations in terms of the Piola tensor $\hat{P}$. Substituting formula (2.11) for the Piola tensor into the motion equation

121

(2.2.40), we obtain

$$\rho_0 \frac{\partial^2 \bar{\varphi}}{\partial t^2} = \bar{\nabla}_0 \cdot W_{\bar{\nabla}_0 \bar{r}} + \rho_0 \bar{B}, \tag{3.7}$$

where $\bar{r} = \bar{\varphi}(t, \xi)$ is a law of motion to be found. Although Eq. (3.7) has a tensor structure, we confine ourselves to the component presentation of this equality. By choosing Cartesian coordinates $\{x^i\}$ with unit vectors $\bar{e}_i$, we write Eq. (3.7) as follows:

$$\rho_0 \frac{\partial^2 \varphi_i}{\partial t^2} = \frac{\partial}{\partial x^k} \left(\frac{\partial W}{\partial (\bar{\nabla}_0 \bar{r})_{ki}} \right) + \rho_0 B_i. \tag{3.8}$$

Transforming the first term in the right-hand side of Eq. (3.8) we obtain

$$\frac{\partial}{\partial x^k} \left(\frac{\partial W}{\partial (\bar{\nabla}_0 \bar{r})_{ki}} \right) = \frac{\partial^2 W}{\partial (\bar{\nabla}_0 \bar{r})_{ki} \partial (\bar{\nabla}_0 \bar{r})_{mn}} \frac{\partial (\bar{\nabla}_0 \bar{r})_{mn}}{\partial x^k}$$

$$= \frac{\partial^2 W}{\partial (\bar{\nabla}_0 \bar{r})_{ki} \partial (\bar{\nabla}_0 \bar{r})_{mn}} \frac{\partial^2 \varphi_n}{\partial x^k \partial x^m}. \tag{3.9}$$

Exercise 3.1. Derive equality (3.9). $\square$

Substitution of expression (3.9) into Eq. (3.8) implies that

$$\rho_0 \frac{\partial^2 \varphi_i}{\partial t^2} = \frac{\partial^2 W}{\partial (\bar{\nabla}_0 \bar{r})_{ki} \partial (\bar{\nabla}_0 \bar{r})_{mn}} \frac{\partial^2 \varphi_n}{\partial x^k \partial x^m} + \rho_0 B_i. \tag{3.10}$$

To ensure the existense of classical solutions, Eq. (3.10) should be *hyperbolic*. By definition, this means that the corresponding equilibrium equation

$$\frac{\partial^2 W}{\partial (\bar{\nabla}_0 \bar{r})_{ki} \partial (\bar{\nabla}_0 \bar{r})_{mn}} \frac{\partial^2 \varphi_n}{\partial x^k \partial x^m} + \rho_0 B_i = 0 \tag{3.11}$$

is *elliptic*, see Lurie (1990) for details.

The ellipticity is equivalent to the positive definiteness of the matrix of coefficients

$$\frac{\partial^2 W}{\partial (\bar{\nabla}_0 \bar{r})_{ki} \partial (\bar{\nabla}_0 \bar{r})_{mn}},$$

which, in turn, is equivalent to the strong ellipticity of function $W(\bar{\nabla}_0 \bar{r})$, see Eq. (1.5.43).

Exercise 3.2. Check that the ellipticity of Eq. (3.11) is equivalent to inequality (1.5.43). $\square$

Problem 3.1. Prove that the linearized dynamic problem has a unique classical solution provided strain energy density W is strongly elliptic at $\bar{\nabla}_0 \bar{r} = \hat{I}$. For a detailed proof, see Duvaut & Lions (1976). $\square$

For the essentially nonlinear problems of elastodynamics, this assertion is not true, and a criterion of the existence of classical solutions is absent.

To simplify the problem of existence, we confine ourselves to static problems, when the inertia term in the left-hand side of Eq. (3.1) as well as initial conditions (3.5) can be neglected.

Two basic approaches to the existence problem may be distinguished in finite elastostatics. According to the first approach, a nonlinear problem is treated as a perturbation of a linear one with the use of *the implicit function theorem*. The results derived here impose feeble limitations on the strain energy density W (mainly connected with its smoothness) and rigid restrictions on the amplitude of body forces and surface loads. Results typical of this method are formulated as follows: if intensities of external loads are sufficiently small, then there is a smooth displacement field satisfying equations (3.1) – (3.4), see e.g. Valent (1988).

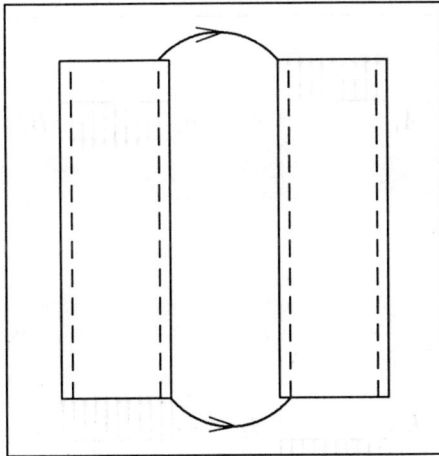

Fig. 3.4. Turning a tube inside out.

An alternative approach is based on replacing Eqs. (3.1) – (3.4) by a problem of minimization for an energy functional which equals a sum of integrals depending on strain-energy density and external loads. We will derive such a problem in Section 5.1 concerned with variational principles in finite elastostatics.

As common practice, variational problems are based on the assumption regarding the convexity of integrands. Regretfully, for hyperelastic media such a condition cannot be imposed, see Proposition 2.1. Therefore, a weaker restriction should be introduced. The first attempt in this direction was made by Ball

(1977), who applied the polyconvexity condition to the existence problems in finite elasticity. On one hand, this condition does not contradict basic hypotheses in continuum mechanics. On the other hand, it is sufficient to establish the existence theorem for appropriate energy functionals.

We do not intend to discuss the existence problem referring to Ciarlet (1988), where a detailed exposition of this question can be found.

3.3. Non-uniqueness of Solutions in Finite Elasticity

Any "classical" mathematical theory implies the uniqueness of solutions to governing equations. However, several applied problems show that we cannot expect the uniqueness of solutions in finite elasticity. Our objective now is to present three examples of static problems in the absence of body forces, where solutions are non-unique.

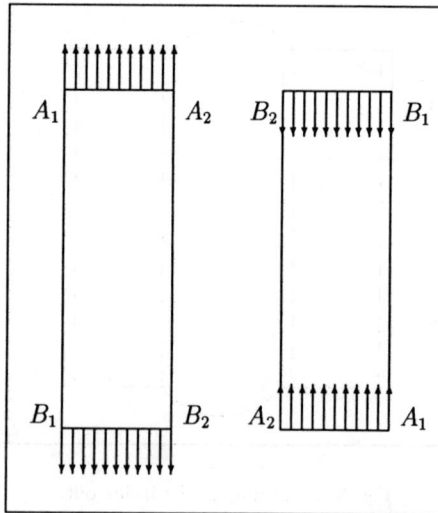

Fig. 3.5. Turning a rod upside down.

1. Following John (1964), let us consider an elastic cylinder located between two co-axial, rigid cylinders, see Fig. 3.3. The inner cylinder is fixed, the outer one rotates around their common axis with some angle α. Evidently, any solution to this problem is not unique, since the boundary conditions are the same for the angle of rotation α as well as for the angle of rotation $\alpha + 2\pi k$ for any integer k.

In the above example, the set of solutions is countable (solutions are numerated by the index k). By replacing two cylinders by two spheres, we obtain a non-countable set of solutions (any solution is characterized by an axis of rotation and a number k).

2. Following Truesdell (1978), let us consider a cylindrical tube made of an elastic material. Evidently, the initial configuration is a solution to the equilibrium equations in the absence of external forces. If we turn the tube inside out and replace the load, we obtain another position where the tube is in equilibrium, see Fig. 3.4. A similar non-uniqueness can also occur due to inversion of a spherical shell, see Ericksen (1955).

3. Following Ericksen (1955), let us consider an elastic rod under tensile dead forces $\bar{P}$ applied to its ends. Evidently, uni-axial elongation of the rod is a solution to problem (3.1) – (3.4). Let us imagine now that we turn the rod upside down so that its ends exchange their positions. Since dead forces preserve their directions in space, after turning the rod over we obtain the same rod, but under the action of compressive loads, see Fig. 3.5. The corresponding solution is uni-axial compression of the rod. Therefore, the Ericksen problem with dead loads allows both compression and tension of a rod to exist.

Chapter 4

Boundary problems in finite elasticity

This Chapter is concerned with solutions to several boundary-value problems in the elasticity theory with finite strains. Our objective is to demonstrate techniques employed in solving boundary-value problems in finite elasticity, and to show some physical effects which are typical of large deformations. Section 1 deals with the so-called universal solutions to problems in finite elastostatics. The main attention is focused on the uniaxial and biaxial tension of an incompressible elastic medium, as well as on the correspondence between experimental data and their theoretical prediction. Section 2 is concerned with the simple shear of an elastic medium. It is shown that for large deformation, the simple shear should be accompanied by normal stresses (the Weissenberg effect), and the first difference of the normal stresses is proportional to the square of the shear stress (the Lodge–Meissner formula). In Section 3, we discuss torsion of a circular cylinder made of an incompressible elastic material. We compare the Rivlin solution to this problem with experimental data and demonstrate the Poynting effect: elongation of an elastic bar under the action of torques applied to its ends. Torsion of a circular bar is one of the most interesting (from the applied point of view) problems, and we will return to it in Chapter 7 to demonstrate the separability principle in finite viscoelasticity. Section 4 deals with the discussion of the Saint-Venant principle in finite elasticity. This principle states that the effect of surface load has a local character: far away from the region where surface traction is applied, stresses are determined by the resultant force and the resultant moment, and are practically independent of the load distribution. We provide a detailed proof of the Saint-Venant principle (i) to define correctly the locality property for surface traction, and (ii) to demonstrate the tools employed in the mathematical theory of elasticity. Section 5 is concerned with one-dimensional motions of an elastic medium: elongation and shear. We demonstrate a significant difference between these motions: any classical solution to the dynamic problem for elongation motion exists on a finite interval of time only, and it is transformed into a shock wave independently of the smoothness of initial conditions. This behavior is typical of a number of conservative dynamic systems. Shear motion is an exception from this rule of thumb: its classical solutions may exist on an arbitrary interval of time provided the initial conditions are sufficiently smooth. Finally, Section 6 deals with nonlinear transversal oscillations of a rod. We show that vibrations of an elastic rod are described by the modified Korteweg–de Vries equation, derive the Hirota method for the construction of particular solutions in the form of solitary waves, and discuss some physical properties of n-soliton solutions.

1. Universal Solutions

In this Section we consider basic static problems for a hyperelastic medium in the absence of body forces. Our objective is to derive explicit solutions corresponding to particular *homogeneous deformations* which are of special interest in applications.

1.1. The Ericksen Theorem

Definition. A solution $\bar{r} = \bar{r}(\xi)$ of the governing equations is called *universal* if the equilibrium equation in the absence of body forces

$$\bar{\nabla} \cdot \hat{\sigma} = 0 \qquad (1.1)$$

is fulfilled for any strain energy density W. Here $\hat{\sigma}$ is the Cauchy stress tensor, $\bar{\nabla}$ is the gradient operator in the actual configuration.

An *affine deformation* from the initial to the actual configuration is determined by the formula

$$\bar{r} = \bar{r}_0 \cdot \hat{\Lambda}, \qquad (1.2)$$

where $\det \hat{\Lambda} \neq 0$ is a constant tensor.

Proposition 1.1 (The Ericksen theorem). For a compressible hyperelastic material, any affine deformation provides a universal solution, and there are no other universal solutions. $\square$

Proof. This statement was formulated and proved by Ericksen (1955). We do not intend to prove the entire assertion, and confine ourselves to the first part of it. Namely, we demonstrate that any affine deformation (1.2) determines a universal solution.

Differentiation of (1.2) implies that $\bar{\nabla}_0 \bar{r} = \hat{\Lambda}$. Substitution of this expression into Eq. (2.1.25) yields $\hat{F} = \hat{\Lambda}^T \cdot \hat{\Lambda}$. It follows from this formula together with the constitutive equation (3.2.18) that

$$\hat{\sigma} = \frac{2}{\det \hat{\Lambda}} [\psi_0 \hat{I} + \psi_1 \hat{\Lambda}^T \cdot \hat{\Lambda} + \psi_2 (\hat{\Lambda}^T \cdot \hat{\Lambda})^2], \qquad (1.3)$$

where $\hat{I}$ is the unit tensor, and functions ψ_i are determined by Eqs. (3.2.19).

It follows from Eq. (1.3) that $\hat{\sigma}$ is constant. Substitution of expression (1.3) into Eq. (1.1) implies that the equilibrium equation is fulfilled for any function W. $\square$

Our objective now is to describe particular affine transformations (1.2).

1.2. Homogeneous Deformation

Let $\{X^i\}$ and $\{x^i\}$ be Cartesian coordinates in the initial and actual configurations, respectively. Radius-vectors of an arbitrary point are calculated as

$$\bar{r}_0 = X^1 \bar{e}_1 + X^2 \bar{e}_2 + X^3 \bar{e}_3, \qquad \bar{r} = x^1 \bar{e}_1 + x^2 \bar{e}_2 + x^3 \bar{e}_3, \qquad (1.4)$$

where $\bar{e}_1$, $\bar{e}_2$ and $\bar{e}_3$ are unit vectors of the Cartesian coordinate frame in the initial configuration.

A homogeneous deformation is determined by the formulas

$$x^1 = \lambda_1 X^1, \qquad x^2 = \lambda_2 X^2, \qquad x^3 = \lambda_1 X^3, \qquad (1.5)$$

where λ_i are unknown coefficients (the extension ratios).

Eqs. (1.4) and (1.5) together with Eq. (2.1.3) imply that

$$\bar{g}_{10} = \bar{e}_1, \quad \bar{g}_{20} = \bar{e}_2, \quad \bar{g}_{30} = \bar{e}_3, \quad \bar{g}_1 = \lambda_1 \bar{e}_1, \quad \bar{g}_2 = \lambda_2 \bar{e}_2, \quad \bar{g}_3 = \lambda_3 \bar{e}_3. \qquad (1.6)$$

Substituting expressions (1.6) into Eqs. (2.1.5) and (2.1.25), we obtain

$$\bar{\nabla}_0 \bar{r} = \lambda_1 \bar{e}_1 \bar{e}_1 + \lambda_2 \bar{e}_2 \bar{e}_2 + \lambda_3 \bar{e}_3 \bar{e}_3, \qquad \hat{F} = \lambda_1^2 \bar{e}_1 \bar{e}_1 + \lambda_2^2 \bar{e}_2 \bar{e}_2 + \lambda_3^2 \bar{e}_3 \bar{e}_3. \qquad (1.7)$$

Exercise 1.1. By using Eqs. (1.7), show that

$$I_1(\hat{F}) = \lambda_1^2 + \lambda_2^2 + \lambda_3^2, \quad I_2(\hat{F}) = \lambda_1^2 \lambda_2^2 + \lambda_1^2 \lambda_3^2 + \lambda_2^2 \lambda_3^2, \quad I_3(\hat{F}) = \lambda_1^2 \lambda_2^2 \lambda_3^2. \quad \square \ (1.8)$$

Substitution of Eqs. (1.7) and (1.8) into Eq. (1.3) implies that

$$\hat{\sigma} = \sigma_1 \bar{e}_1 \bar{e}_1 + \sigma_2 \bar{e}_2 \bar{e}_2 + \sigma_3 \bar{e}_3 \bar{e}_3, \qquad (1.9)$$

where

$$\sigma_j = \frac{2}{\lambda_1 \lambda_2 \lambda_3} (\psi_0 + \lambda_j^2 \psi_1 + \lambda_j^4 \psi_2). \qquad (1.10)$$

1.3. Dilatation

Dilatation is a particular case of homogeneous deformations with

$$\lambda_1 = \lambda_2 = \lambda_3 = \lambda. \tag{1.11}$$

It follows from Eqs. (1.9) – (1.11) that for dilatation

$$\hat{\sigma} = -p(\lambda)\hat{I}, \tag{1.12}$$

where

$$p(\lambda) = -2\lambda^{-3}(\psi_0 + \lambda^2 \psi_1 + \lambda^4 \psi_2).$$

Exercise 1.2. By using Eqs. (3.2.19), show that

$$p(\lambda) = -\frac{2}{\lambda}\left(\frac{\partial W}{\partial I_1} + 2\lambda^2 \frac{\partial W}{\partial I_2} + \lambda^4 \frac{\partial W}{\partial I_3}\right), \tag{1.13}$$

where $W(I_1, I_2, I_3)$ is a strain energy density. $\square$

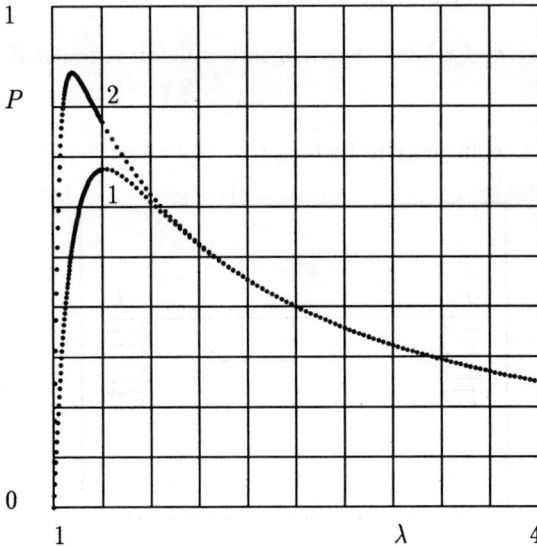

Fig. 1.1. The dimensionless pressure $P = -p/\mu$ *versus* the extension ratio λ for the Agarwal material. Curve 1 corresponds to $k = 1$, curve 2 corresponds to $k = 5$.

Eq. (1.12) and boundary condition (2.2.20) imply that

$$\bar{b} = -p\bar{n}, \tag{1.14}$$

129

where $\bar{b}$ is a surface traction, and $\bar{n}$ is the unit outward normal to the boundary in the actual configuration. Eq. (1.14) means that dilatation of an elastic medium occurs under normal surface traction only.

As an example, let us consider dilatation of the Agarwal material (3.2.53) with parameters k and μ.

Exercise 1.3. Show that for dilatation of the Agarwal solid

$$p(\lambda) = -\frac{\mu}{\lambda}(1 - \frac{1}{\lambda^{6k+2}}). \quad \square$$

The dimensionless pressure $P = -p/\mu$ *versus* the extension ratio λ is plotted in Fig 1.1. The obtained results demonstrate a typical danger in the nonlinear mechanics of solids: the Agarwal model with appropriate parameters satisfies all the restrictions imposed on strain energy densities, but the sphere of its applicability is extremely narrow. Even for small k values, e.g. for $k = 1$, this model leads to unrealistic results (decrease of stress P with the growth of the extension ratio λ) when $\lambda > \lambda_* = 1.3$. The ultimate extension ratio λ_* decreases in k, and becomes close to 1.1 for $k = 5$.

1.4. Uniaxial Tension of a Compressible Bar

Uniaxial tension of a bar is a homogeneous deformation with

$$\lambda_1 = \lambda, \qquad \lambda_2 = \lambda_3 = \alpha\lambda, \qquad (1.15)$$

where α and λ are coefficients to be found.

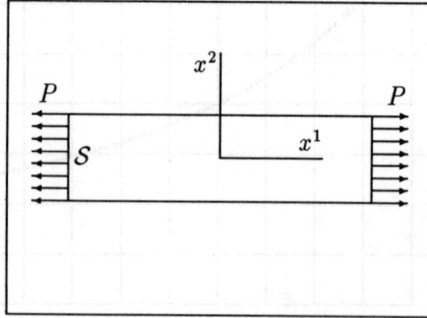

Fig. 1.2. Uniaxial tension of an elastic bar.

Exercise 1.4. By using Eqs. (1.7) and (1.8), check that for uniaxial tension

$$\hat{F} = \lambda^2[\bar{e}_1\bar{e}_1 + \alpha^2(\bar{e}_2\bar{e}_2 + \bar{e}_3\bar{e}_3)], \qquad (1.16)$$

$$I_1(\hat{F}) = (1 + 2\alpha^2)\lambda^2, \quad I_2(\hat{F}) = (2 + \alpha^2)\alpha^2\lambda^4, \quad I_3(\hat{F}) = \alpha^4\lambda^6. \quad \square \qquad (1.17)$$

For an arbitrary strain energy density W, explicit formulas cannot be derived for parameters α and λ. Thus, we confine ourselves to a particular case of the Blatz–Ko material (3.2.66).

Exercise 1.5. By substitution of expressions (1.15), (1.17), and (3.2.66) into Eqs. (1.10), show that

$$\sigma_1 = \frac{\mu(\alpha^2\lambda^5 - 1)}{\alpha^2\lambda^5}, \qquad \sigma_2 = \sigma_3 = \frac{\mu(\alpha^4\lambda^5 - 1)}{\alpha^4\lambda^5}. \qquad \square \qquad (1.18)$$

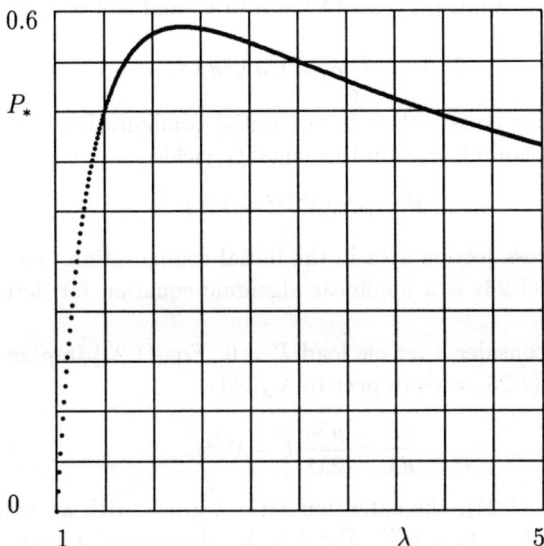

Fig. 1.3. The dimensionless tensile load $P_* = P/(\mu S_0)$ *versus* the extension ratio λ for the Blatz–Ko material.

Since surface traction on the lateral surface vanishes, we may write the boundary condition

$$\sigma_2 = \sigma_3 = 0. \qquad (1.19)$$

Stresses σ_2 and σ_3 are independent of spatial coordinates. Therefore, Eq. (1.19) should be fulfilled at any point of the bar. Eqs. (1.18) and (1.19) imply that

$$\alpha = \lambda^{-5/4}. \qquad (1.20)$$

Substitution of expression (1.20) into Eqs. (1.18) yields

$$\sigma_1 = \mu(1 - \lambda^{-5/2}). \qquad (1.21)$$

131

According to Eqs. (1.9) and (1.19), σ_1 is the only non-zero component of the Cauchy tensor. To find it, we write the boundary condition at the ends in the integral form

$$\bar{P} = \int_S \bar{n} \cdot \hat{\sigma} dx^2 dx^3. \qquad (1.22)$$

Here $\bar{P}$ is a tensile load, and S is the bar cross-section in the actual configuration. Since $\bar{P} = P\bar{e}_1$, $\bar{n} = \bar{e}_1$, $\hat{\sigma} = \sigma_1 \bar{e}_1 \bar{e}_1$, Eq. (1.22) implies that

$$P = \int_S \sigma_1 dx^2 dx^3.$$

Taking into account that $dx^2 = \lambda_2 dX^2 = \alpha \lambda dX^2$ and $dx^3 = \lambda_3 dX^3 = \alpha \lambda dX^3$, we obtain

$$P = \int_{S_0} \sigma_1 \alpha^2 \lambda^2 dX^2 dX^3,$$

where S_0 is the bar cross-section in the initial configuration. Substitution of expressions (1.20) and (1.21) into this equality yields

$$P = \mu S_0 (\lambda^{-1/2} - \lambda^{-3}), \qquad (1.23)$$

where S_0 is the cross-section area in the initial configuration. For a given tensile load P, Eq. (1.23) is a nonlinear algebraic equation for determining the extension ratio λ.

First, let us consider a tensile load $P > 0$. Eq. (1.23) implies that $\lambda > 1$. Differentiation of (1.23) with respect to λ yields

$$\frac{dP}{d\lambda} = \frac{\mu S_0}{2\lambda^4}(6 - \lambda^{5/2}). \qquad (1.24)$$

According to Eq. (1.24), the extension ratio λ grows with an increase in the tensile force until $\lambda = \lambda_* = 6^{2/5}$. For $\lambda > \lambda_*$, the extension ratio λ decreases with the growth of P, see Fig. 1.3. Thus, for any tensile load the extension ratio cannot exceed λ_*. Since this result contradicts experimental data, the Blatz–Ko model may be employed only for relatively small elongations, when $\lambda < \lambda_*$.

Let us now consider a compressive load $P < 0$. Setting $P = -\tilde{P}$, $\lambda = 1/\tilde{\lambda}$, we rewrite Eq. (1.23) as

$$\tilde{P} = \mu S_0 (\tilde{\lambda}^3 - \tilde{\lambda}^{1/2}). \qquad (1.25)$$

It follows from Eq. (1.25) that $\tilde{\lambda} > 1$ for $\tilde{P} > 0$. Differentiation of (1.25) yields

$$\frac{d\tilde{P}}{d\tilde{\lambda}} = \frac{\mu S_0}{2\tilde{\lambda}^{1/2}}(6\tilde{\lambda}^{5/2} - 1).$$

This inequality together with the estimate $\tilde{\lambda} > 1$ implies that

$$\frac{d\tilde{P}}{d\tilde{\lambda}} \geq \frac{5\mu S_0}{2\tilde{\lambda}^{1/2}} > 0.$$

This means that $\tilde{\lambda}$ increases monotonically with the growth of compressive load $\tilde{P}$.

Exercise 1.6. Analyze the dependence of the extension ratio λ on tensile force P for the Levinson–Burgess solid (3.2.71) and for the Antman solid (3.2.78). □

1.5. Uniaxial Tension of an Incompressible Bar

Explicit solutions to the problem of uniaxial tension are obtained only for a narrow class of compressible materials. In this subsection, by assuming the material incompressibility, an explicit solution is derived for an arbitrary strain energy density W.

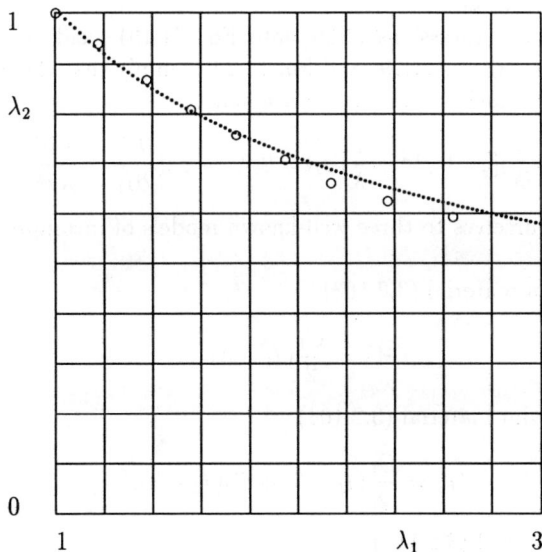

Fig. 1.4. The extension ratio λ_2 *versus* the extension ratio λ_1 for uniaxial tension of styrene butadiene rubber. Large circles correspond to experimental data obtained by Glucklich & Landel (1977). Small circles correspond to Eq. (1.27).

Substituting expression (1.17) into the incompressibility condition $I_3(\hat{F}) = 1$ we obtain

$$\alpha = \lambda^{-\frac{3}{2}}. \tag{1.26}$$

Expression (1.26) together with Eq. (1.15) implies that for uniaxial tension of an incompressible bar

$$\lambda_1 = \lambda, \qquad \lambda_2 = \lambda_3 = \lambda^{-\frac{1}{2}}. \tag{1.27}$$

It is of special interest to verify these relationships. Comparison of experimental data with their theoretical prediction shows that the incompressibility assumption is valid for a wide range of axial elongations, see Fig. 1.4.

Exercise 1.7. By using Eqs. (1.16), (1.17), and (1.26) show that

$$\hat{F} = \lambda^2 \bar{e}_1 \bar{e}_1 + \lambda^{-1}(\bar{e}_2 \bar{e}_2 + \bar{e}_3 \bar{e}_3),$$

$$I_1(\hat{F}) = \lambda^2 + 2\lambda^{-1}, \qquad I_2(\hat{F}) = 2\lambda + \lambda^{-2}. \quad \square \tag{1.28}$$

Exercise 1.8. By substituting expressions (1.28) into Eq. (3.2.99), derive formula (1.9) with

$$\sigma_1 = -p + 2\lambda^2(\psi_1 + \lambda^2 \psi_2), \qquad \sigma_2 = \sigma_3 = -p + 2\lambda^{-1}(\psi_1 + \lambda^{-1}\psi_2), \tag{1.29}$$

where p is a pressure. $\square$

Substitution of expressions (1.29) into Eq. (1.19) yields $p = 2\lambda^{-1}(\psi_1 + \lambda^{-1}\psi_2)$. This equality, Eq. (1.29) and Eq. (3.2.98) imply that the only non-zero component of the Cauchy stress tensor $\hat{\sigma}$ equals

$$\sigma_1 = 2(\lambda^2 - \frac{1}{\lambda})[\psi_1 + (\lambda^2 + \frac{1}{\lambda})\psi_2] = 2(\lambda^2 - \frac{1}{\lambda})(\frac{\partial W}{\partial I_1} + \frac{1}{\lambda}\frac{\partial W}{\partial I_2}). \tag{1.30}$$

We confine ourselves to three well-known models of incompressible hyperelastic media:
– the neo-Hookean material (3.2.103)

$$W_1 = \frac{C_1}{2}(I_1 - 3), \tag{1.31}$$

– the Mooney–Rivlin material (3.2.104)

$$W_1 = \frac{C_1}{2}(I_1 - 3) + C_2(I_2 - 3), \tag{1.32}$$

– the Knowles material (3.2.115)

$$W_1 = \frac{C_1}{2b}\{[1 + \frac{b}{m}(I_1 - 3)]^m - 1\}. \tag{1.33}$$

Parameters C_1, C_2, b and m in expressions (1.31) – (1.33) are found by fitting experimental data by the least square method.

Exercise 1.9. Check that for the neo-Hookean material

$$\sigma_1 = C_1(\lambda^2 - \frac{1}{\lambda}), \tag{1.34}$$

for the Mooney–Rivlin material

$$\sigma_1 = (C_1 + \frac{2C_2}{\lambda})(\lambda^2 - \frac{1}{\lambda}), \tag{1.35}$$

134

and for the Knowles material

$$\sigma_1 = C_1[1 + \frac{b}{m}(\lambda^2 + \frac{2}{\lambda} - 3)]^{m-1}(\lambda^2 - \frac{1}{\lambda}). \qquad \square \qquad (1.36)$$

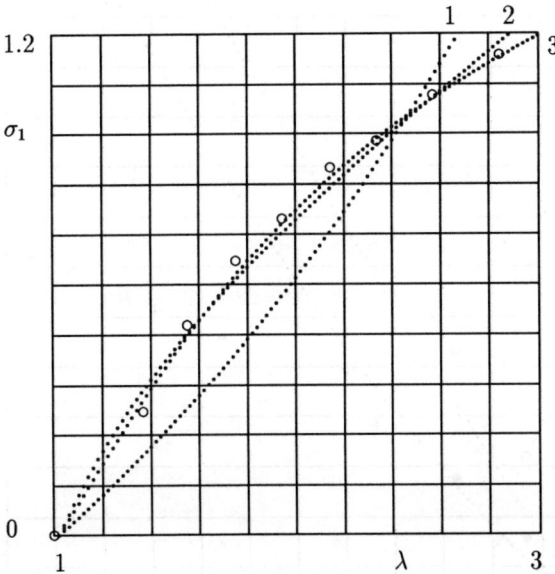

Fig. 1.5. Stress σ_1 (MPa) *versus* the extension ratio λ for styrene butadiene rubber. Large circles correspond to experimental data obtained by Glucklich & Landel (1977). Small circles correspond to the approximations of experimental data. Curve 1: the neo-Hookean model (1.31) with $C_1 = 0.180$ (MPa); curve 2: the Mooney–Rivlin model (1.32) with $C_1 = 0$ (MPa) and $C_2 = 0.219$ (MPa); curve 3: the Knowles model (1.33) with $C_1 = 0.322$ (MPa), $b = 0.06$, and $m = 0.21$.

Experimental data obtained by Glucklich & Landel (1977) for styrene butadiene rubber after 10 (min) of loading together with their theoretical predictions by using formulas (1.34) – (1.36) are plotted in Fig. 1.5. These results lead us to the following conclusions:

- the neo-Hookean model does not describe adequately the experimental data. For example, experimental points demonstrate a concave dependence of the longitudinal stress on the extension ratio, while the neo-Hookean model implies a convex dependence;

- both the Mooney–Rivlin and the Knowles models demonstrate excellent agreement between observations and their predictions. Despite the difference between these models (the former model depends on the second

135

principal invariant I_2, while the latter depends on the first principal invariant I_1), the material responses predicted by these models are extremely close to each other.

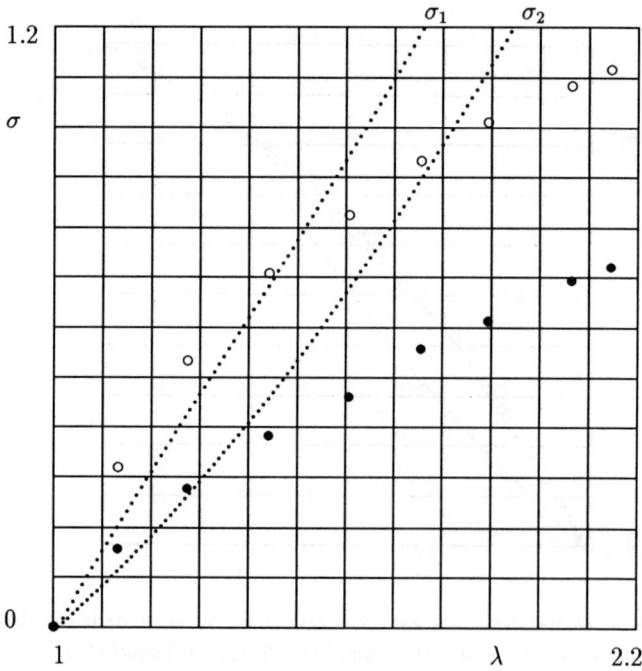

Fig. 1.6. Stresses σ_1 (unfilled circles) and σ_2 (filles circles) (MPa) *versus* the extension ratio λ_1 for styrene butadiene rubber. Large circles correspond to experimental data for $\lambda_2 = 1$ obtained by Glucklich & Landel (1977). Small circles correspond to the approximation of experimental data by the Mooney–Rivlin model (1.32).

The latter means that experimental data for uniaxial tension are not sufficient to determine an adequate model, and additional experiments are necessary. We choose biaxial tension of an elastic sheet to verify the constitutive models.

1.6. Biaxial tension of an incompressible elastic sheet

Let us consider *biaxial tension* of an elastic plate with length l_1, width l_2, and thickness h. The plate is in its natural state and occupies the domain

$$\{0 \leq X_1 \leq l_1, \quad 0 \leq X_2 \leq l_2, \quad -\frac{h}{2} \leq X_3 \leq \frac{h}{2}\}.$$

Distributed loads are applied to edges of the plate and cause its purely homogeneous deformation (1.5).

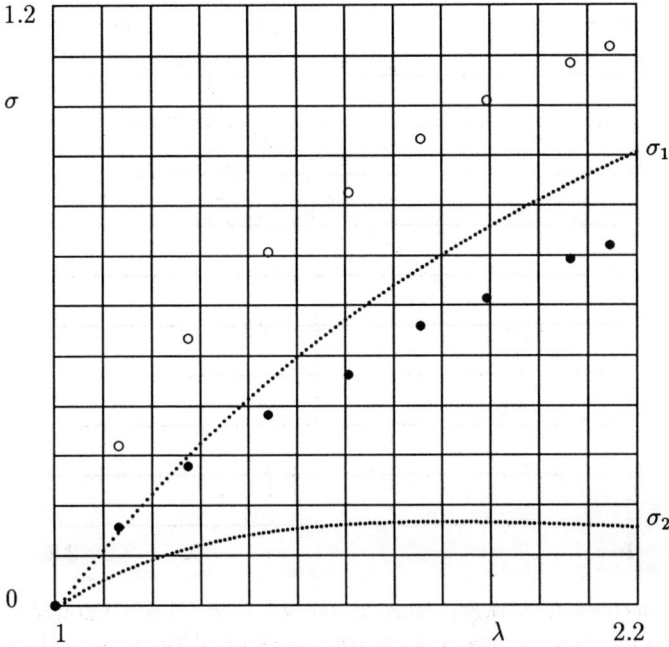

Fig. 1.7. Stresses σ_1 (unfilled circles) and σ_2 (filled circles) (MPa) *versus* the extension ratio λ_1 for styrene butadiene rubber. Large circles correspond to experimental data for $\lambda_2 = 1$ obtained by Glucklich & Landel (1977). Small circles correspond to the approximation of experimental data by the Knowles model (1.33).

It follows from Eq. (1.8) and the incompressibility condition $I_3(\hat{F}) = 1$ that

$$\lambda_3 = \frac{1}{\lambda_1 \lambda_2}. \tag{1.37}$$

Substitution of expression (1.37) into Eqs. (1.7) and (1.8) implies that

$$\hat{F} = \lambda_1^2 \bar{e}_1 \bar{e}_1 + \lambda_2^2 \bar{e}_2 \bar{e}_2 + \frac{1}{\lambda_1^2 \lambda_2^2} \bar{e}_3 \bar{e}_3,$$

$$I_1(\hat{F}) = \lambda_1^2 + \lambda_2^2 + \lambda_1^{-2} \lambda_2^{-2}, \qquad I_2(\hat{F}) = \lambda_1^{-2} + \lambda_2^{-2} + \lambda_1^2 \lambda_2^2. \tag{1.38}$$

Exercise 1.10. By substituting expressions (1.38) into the constitutive equation (3.2.99), derive Eq. (1.9) with

$$\sigma_1 = -p + 2\lambda_1^2 (\psi_1 + \lambda_1^2 \psi_2), \qquad \sigma_2 = -p + 2\lambda_2^2 (\psi_1 + \lambda_2^2 \psi_2),$$

137

$$\sigma_3 = -p + \frac{2}{\lambda_1^2\lambda_2^2}(\psi_1 + \frac{1}{\lambda_1^2\lambda_2^2}\psi_2). \qquad \Box \qquad (1.39)$$

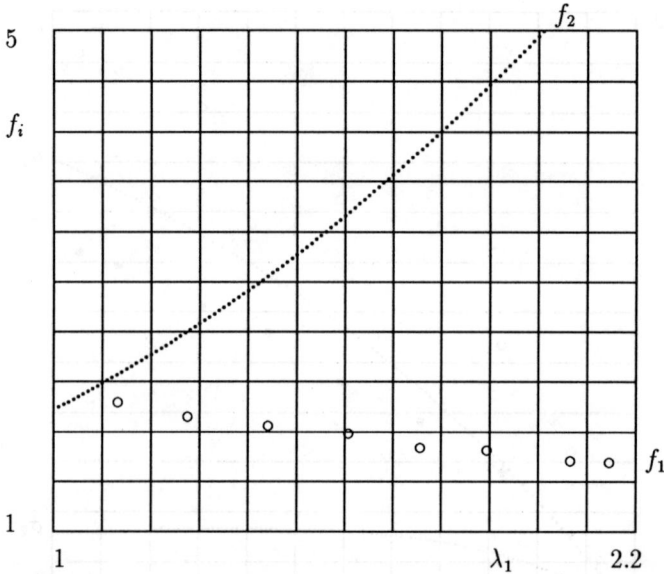

Fig. 1.8. Functions $f_1 = \sigma_1/\sigma_2$ (large circles) and $f_2 = (\lambda_1^4\lambda_2 - 1)/(\lambda_1^2\lambda_2^4 - 1)$ (small circles) *versus* the extension ratio λ_1 for styrene butadiene rubber. Large circles correspond to experimental data for $\lambda_2 = 1$ obtained by Glucklich & Landel (1977).

It follows from the boundary conditions at surfaces $X_3 = \pm\frac{h}{2}$ that $\sigma_3 = 0$. This equality and Eqs. (1.39) imply that

$$p = \frac{2}{\lambda_1^2\lambda_2^2}(\psi_1 + \frac{1}{\lambda_1^2\lambda_2^2}\psi_2).$$

We substitute this expression into Eqs. (1.39) and use Eqs. (3.2.98) and (1.38). As a result, we obtain

$$\sigma_1 = 2(\frac{\partial W}{\partial I_1} + \lambda_2^2\frac{\partial W}{\partial I_2})(\lambda_1^2 - \frac{1}{\lambda_1^2\lambda_2^2}), \quad \sigma_2 = 2(\frac{\partial W}{\partial I_1} + \lambda_1^2\frac{\partial W}{\partial I_2})(\lambda_2^2 - \frac{1}{\lambda_1^2\lambda_2^2}). \quad (1.40)$$

For given extension ratios λ_1 and λ_2, Eqs. (1.40) determine the non-zero components of the Cauchy stress tensor $\hat{\sigma}$.

In order to fit experimental data for biaxial tension of the plate, it is quite natural to begin with strain energy densities (1.31) – (1.33) with the material parameters found for uniaxial tension. The corresponding results are

138

plotted in Figs. 1.6 and 1.7 for the Mooney–Rivlin and Knowles materials. The figures demonstrate significant discrepancies between experimental data and their theoretical predictions. The same result is also true for the neo-Hookean model (1.31), while an appropriate figure is omitted.

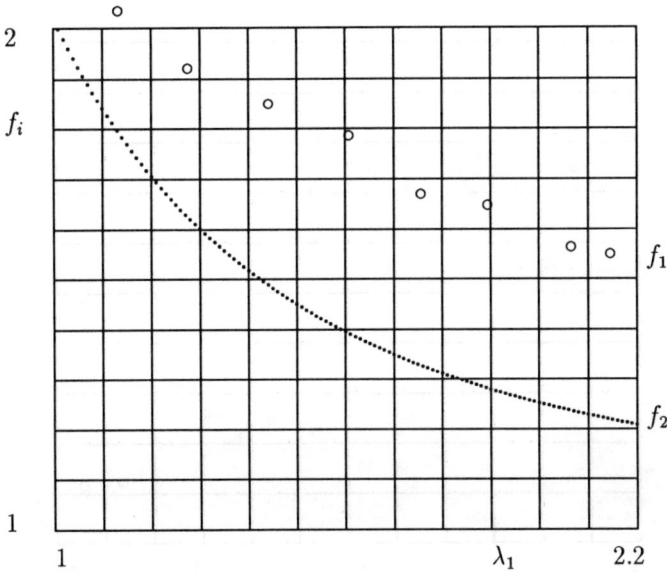

Fig. 1.9. Functions $f_1 = \sigma_1/\sigma_2$ (large circles) and $f_2 = \lambda_2^2(\lambda_1^4\lambda_2 - 1)/[\lambda_1^2(\lambda_1^2\lambda_2^4 - 1)]$ (small circles) *versus* the extension ratio λ_1 for styrene butadiene rubber. Large circles correspond to experimental data for $\lambda_2 = 1$ obtained by Glucklich & Landel (1977).

Figs. 1.6 and 1.7 show that:
(i) simple, uniaxial experiments are not sufficient to validate constitutive models with finite strains;
(ii) the standard constitutive equations do not describe adequately both uniaxial and biaxial tension, and more sophisticated models should be developed.

We do not intend to suggest a new explicit expression for the strain energy density $W(I_1, I_2)$. Following Rivlin (1956), we assume that this function of two variables may be reduced to several functions of one variable, which are found by fitting experimental data. Four different versions of such a reduction are proposed:
(i) W depends on the first principal invariant I_1 only;
(ii) W depends on the second principal invariant I_2 only;
(iii) W depends on the ratio I_1/I_2^ν only, where ν is a material constant (*the self-similarity condition*);

139

(iv) W is presented as a sum

$$W(I_1, I_2) = W_1(I_1) + W_2(I_2),\qquad (1.41)$$

where W_k depends on the kth principal invariant I_k only.

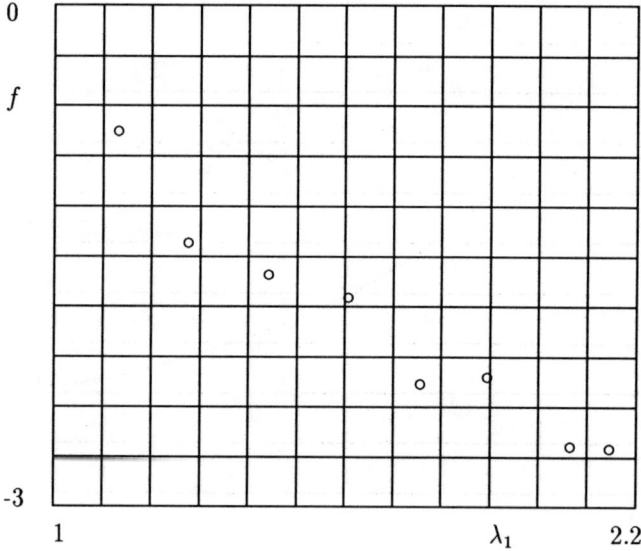

Fig. 1.10. Function f in the right-hand side of Eqn. (1.46) *versus* the extension ratio λ_1 for styrene butadiene rubber. Circles correspond to experimental data for $\lambda_2 = 1$ obtained by Glucklich & Landel (1977).

To verify hypotheses (i) – (iv), we find the derivatives of the strain energy density from Eq. (1.40).

Exercise 1.11. Derive the following formulas:

$$\frac{\partial W}{\partial I_1} = \frac{\lambda_1^2 \lambda_2^2}{2(\lambda_1^2 - \lambda_2^2)} \Big(\frac{\sigma_1 \lambda_1^2}{\lambda_1^4 \lambda_2^2 - 1} - \frac{\sigma_2 \lambda_2^2}{\lambda_1^2 \lambda_2^4 - 1} \Big),$$

$$\frac{\partial W}{\partial I_2} = \frac{\lambda_1^2 \lambda_2^2}{2(\lambda_1^2 - \lambda_2^2)} \Big(\frac{\sigma_2}{\lambda_1^2 \lambda_2^4 - 1} - \frac{\sigma_1}{\lambda_1^4 \lambda_2^2 - 1} \Big). \quad \square \qquad (1.42)$$

First, we suppose that function W depends on I_1 only, and $\partial W/\partial I_2 = 0$. In this case, Eqs. (1.42) imply that

$$\frac{\sigma_1}{\sigma_2} = \frac{\lambda_1^4 \lambda_2^2 - 1}{\lambda_1^2 \lambda_2^4 - 1}. \qquad (1.43)$$

140

The ratios f_1 in the left and f_2 in the right-hand side of Eq. (1.43) are plotted in Fig. 1.8 $versus$ λ_1. The results presented in Fig. 1.8 show that assumption (i) cannot be accepted because of essential differences in the behavior of functions f_1 and f_2. This implies that both the neo-Hookean model (1.31) and the Knowles model (1.33) should fail in describing biaxial tension.

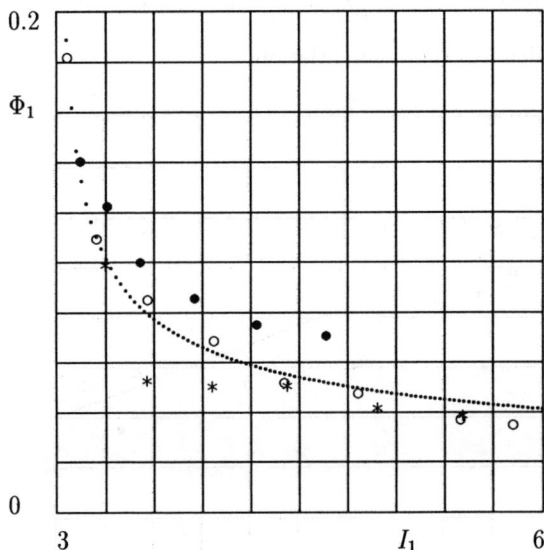

Fig. 1.11. Function Φ_1 $versus$ the first principal invariant I_1. Large circles and asterisks correspond to experimental data for styrene butadiene rubber obtained by Glucklich & Landel (1977) for different programs of tension. Small circles provide the approximation of experimental data by the power function (1.47) with $A_1 = 0.0634$ and $\alpha_1 = 0.389$.

Let us now consider hypothesis (ii) and suppose that function W depends on I_2 only. It follows from this assumption and Eqs. (1.42) that

$$\frac{\sigma_1}{\sigma_2} = \frac{\lambda_2^2(\lambda_1^4\lambda_2^2 - 1)}{\lambda_1^2(\lambda_1^2\lambda_2^4 - 1)}. \tag{1.44}$$

The ratios f_1 in the left and f_2 in the right-hand side of Eq. (1.44) are plotted in Fig. 1.9 $versus$ λ_1. These results demonstrate that assumption (ii) describes correctly a decrease in the ratio σ_1/σ_2 with the growth of λ_1. Nevertheless, this hypothesis leads to significant discrepancies between experimental data and their prediction, and it cannot be employed. This is a reason for the failure of the Mooney–Rivlin model (1.32), since this model with the parameters found in uniaxial experiments satisfies assumption (ii).

141

Let us now consider the self-similarity condition (iii). This condition together with Eqs. (1.42) implies that

$$\frac{W'}{I_2^\nu} = \frac{\lambda_1^2 \lambda_2^2}{2(\lambda_1^2 - \lambda_2^2)} \left(\frac{\sigma_1 \lambda_1^2}{\lambda_1^4 \lambda_2^2 - 1} - \frac{\sigma_2 \lambda_2^2}{\lambda_1^2 \lambda_2^4 - 1} \right),$$

$$\frac{\nu I_1 W_1'}{I_2^{\nu+1}} = \frac{\lambda_1^2 \lambda_2^2}{2(\lambda_1^2 - \lambda_2^2)} \left(\frac{\sigma_1}{\lambda_1^4 \lambda_2^2 - 1} - \frac{\sigma_2}{\lambda_1^2 \lambda_2^4 - 1} \right), \qquad (1.45)$$

where the prime denotes the differentiation.

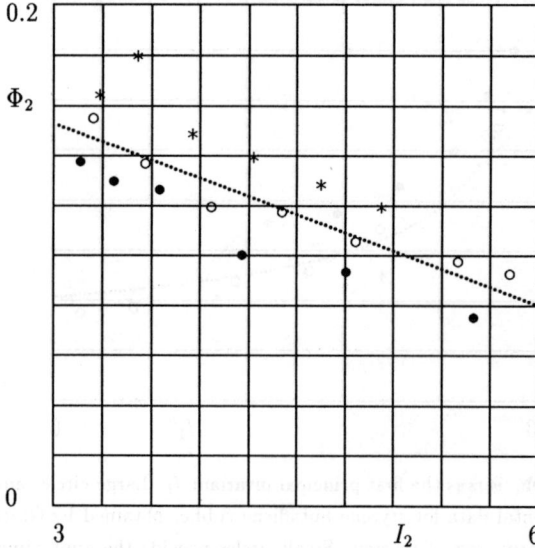

Fig. 1.12. Function Φ_2 *versus* the second principal invariant I_2. Large circles and asterisks correspond to experimental data for styrene butadiene rubber obtained by Glucklich & Landel (1977) for various programs of tension. Small circles provide the approximation of experimental data by the linear function (1.48) with $B_1 = 0.2259$ and $B_2 = 0.0243$.

Exercise 1.12. By dividing the first equation (1.45) by the other, derive the following formula:

$$\nu = \frac{\lambda_1^2 + \lambda_2^2 + \lambda_1^4 \lambda_2^4}{1 + \lambda_1^2 \lambda_2^4 + \lambda_1^4 \lambda_2^2} \frac{\sigma_1(\lambda_1^2 \lambda_2^4 - 1) - \sigma_2(\lambda_1^4 \lambda_2^2 - 1)}{\sigma_1 \lambda_1^2(\lambda_1^2 \lambda_2^4 - 1) - \sigma_2 \lambda_2^2(\lambda_1^4 \lambda_2^2 - 1)}. \qquad \square \qquad (1.46)$$

Function f in the right-hand side of Eq. (1.46) is plotted in Fig. 1.10 *versus* the extension ratio λ_1. This function is rather far from being constant, which means that equality (1.46) is not fulfilled. Therefore, neither can assumption (iii) describe biaxial deformations.

Finally, we consider hypothesis (iv). According to it, Eqs. (1.42) may be treated as two independent equations for functions $W_1(I_1)$ and $W_2(I_2)$. The corresponding dependencies are plotted in Figs. 1.11 and 1.12.

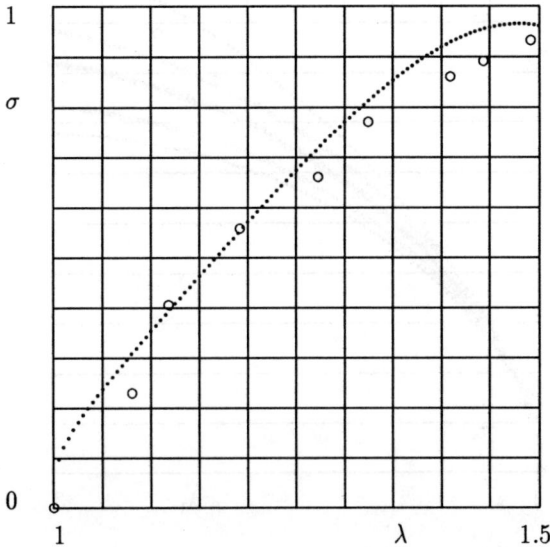

Fig. 1.13. Stresses $\sigma_1 = \sigma_2 = \sigma$ (MPa) *versus* the extension ratios $\lambda_1 = \lambda_2 = \lambda$ for styrene butadiene rubber. Large circles correspond to experimental data obtained by Glucklich & Landel (1977). Small circles provide prediction of experimental data by using model (1.41), (1.47), and (1.48).

Fig. 1.11 shows that function $\Phi_1 = \partial W/\partial I_1$ depends weakly on the second invariant I_2. Deviations from the approximating curve (small circles) which correspond to both errors in measurements and the effect of argument I_2, are less than 10 per cent (taking into account that I_2 changes in experiments from 3.04 to 5.81). This allows the dependence of Φ_1 on I_2 to be neglected, which is equivalent to assumption (iv).

Function Φ_1 may be approximated by the power function

$$\Phi_1 = \frac{A_1}{(I_1 - 3)^{\alpha_1}}. \tag{1.47}$$

Eq. (1.47) can be employed sufficiently far from the point $I_1 = 3$, which corresponds to the initial configuration. This means that the above approximation is valid only for large deformations and cannot be used at infinitesimal strains (in the latter case the derivative $\partial W_1/\partial I_1$ should tend to a finite constant as $I_1 \to 3$).

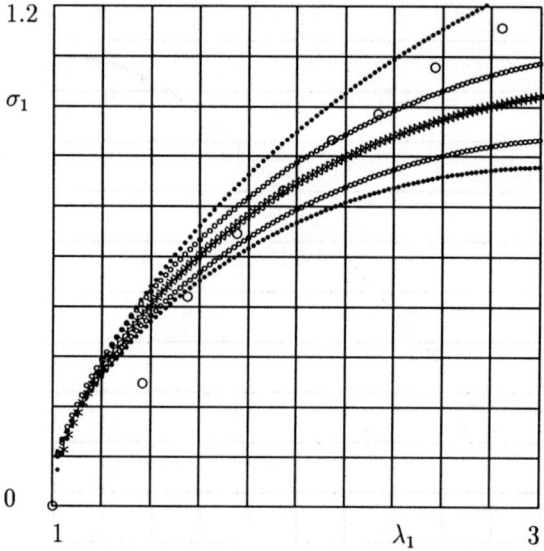

Fig. 1.14. The principal stress σ_1 (MPa) *versus* the extension ratio λ_1 for styrene butadiene rubber. Large circles correspond to experimental data obtained by Glucklich & Landel (1977). Asterisks provide the approximation of experimental data by using model (1.41), (1.47), and (1.48). Small circles demonstrate discrepancies from the predicted values caused by deviations of experimental data in Fig. 1.11 (filled circles) and Fig. 1.12 (unfilled circles).

Fig. 1.12 shows that function $\Phi_2 = \partial W/\partial I_2$ depends weakly on argument I_1, and this dependence may be neglected as well. For example, both errors of measurements and the dependence on the first invariant I_1 lead to deviations of Φ_2 from the approximating curve of about 10 per cent. Up to this level of accuracy, function Φ_2 may be approximated by the linear function

$$\Phi_2 = B_1 - B_2 I_2. \tag{1.48}$$

It is worth comparing the above results with the results obtained by Rivlin for rubber specimens, see Rivlin (1956). In those experiments function $\partial W_1/\partial I_1$ was found to be practically constant, while function $\partial W_2/\partial I_2$ decreased according to the power law.

The next step of our analysis consists in testing dependencies (1.47) and (1.48) for other types of loading, which have not been used in constructing the model. For this purpose, we employ biaxial tension of a sheet with equal

144

extension ratios $\lambda_1 = \lambda_2 = \lambda$ and uniaxial tension of a bar. The results for biaxial tension are plotted in Fig. 1.13. They demonstrate excellent agreement between experimental data and their theoretical prediction.

The results for uniaxial tension of a bar are plotted in Fig. 1.14. This figure demonstrates fair correspondence between experimental data and their prediction, except for a narrow region of relatively small deformations when $1 \leq \lambda_1 < 1.5$. To exclude discrepancies between experimental and theoretical results, a more accurate approximation of the derivative $\partial W_1 / \partial I_1$ is necessary. For example, instead of Eq. (1.47) we may employ the dependence

$$\frac{\partial W_1}{\partial I_1} = \frac{A_1}{A_2 + (I_1 - 3)^{\alpha_1}},$$

which leads to an increase in the number of material parameters to be found in experiments, cf. the Alexander model (3.2.111).

1.7. Concluding Remarks

In this Section we analyze affine transformations of a hyperelastic medium. First, we demonstrate that these transformations exhaust entirely the class of universal solutions for a compressible medium (the Ericksen theorem). Afterward, we concentrate on particular affine transformations, which correspond to dilatation of a medium, to uniaxial stretching of a bar, and to biaxial tension of a sheet.

It is shown that explicit solutions to these problems for compressible elastic materials can be derived for specific strain energy densities only. Even when such solutions exist, the sphere of their application may be extremely narrow. For example, for the Agarwal model, a physically realistic relationship between external pressure and the extension ratio is valid for elongations less than 12 per cent.

For incompressible hyperelastic solids, an explicit solution can be derived for any homogeneous deformation. In this case, an important problem arises of choosing an appropriate constitutive model which describes adequately experimental data. It is shown that uniaxial tension of a bar (which is treated as the standard test in linear elasticity) is not sufficient at finite deformations. On the one hand, different constitutive models with parameters found by fitting experimental data ensure the same level of accuracy. On the other hand, models which demonstrate fair prediction of experimental results for uniaxial tension fail to fit data for other types of loading.

Finally, a series of hypotheses is suggested which allow a correct constitutive model to be derived. By verifying these assumptions, we demonstrate a way to develop a model which provides excellent agreement with experimental data for various types of loads.

2. Simple Shear

This Section is concerned with a particular homogeneous quasi-static deformation of a hyperelastic medium occupying the whole space. Our objective is to derive the Lodge–Meissner formula which is fulfilled for an arbitrary strain energy density. As a consequence of the Lodge–Meissner formula we develop the Lodge equality and present some experimental data which confirm it.

Relations similar to the Lodge–Meissner formula play the key role in the nonlinear mechanics of continua. Owing to their independence of constitutive models, such formulas allow the basic assumptions of the theory (homogeneity, isotropy, etc.) to be verified.

2.1. Formulation of the Problem

Definition. *Simple shear* is a homogeneous deformation, which may be presented in appropriate Cartesian coordinates as

$$x^1 = X^1 + \kappa X^2, \qquad x^2 = X^2, \qquad x^3 = X^3. \tag{2.1}$$

Here $\{X^i\}$ and $\{x^i\}$ are Cartesian coordinates in the initial and actual configurations, respectively, κ is a coefficient of shear.

Plane $X^3 = $ const. is called *the plane of shear*, axis X^3 is called *the axis of shear*, plane $X^2 = $ const. is called *the shearing plane*, and angle $\alpha = \tan^{-1}\kappa$ is called *the angle of shear*.

Radius-vectors $\bar{r}_0$ and $\bar{r}$ in the initial and actual configurations are calculated as follows:

$$\bar{r}_0 = X^1 \bar{e}_1 + X^2 \bar{e}_2 + X^3 \bar{e}_3, \qquad \bar{r} = (X^1 + \kappa X^2)\bar{e}_1 + X^2 \bar{e}_2 + X^3 \bar{e}_3, \tag{2.2}$$

where $\bar{e}_i$ are unit vectors of the Cartesian coordinate frame.

Exercise 2.1. By differentiation of expressions (2.2), show that

$$\begin{aligned}
\bar{g}_{01} &= \bar{e}_1, & \bar{g}_{02} &= \bar{e}_2, & \bar{g}_{03} &= \bar{e}_3, \\
\bar{g}_0^1 &= \bar{e}_1, & \bar{g}_0^2 &= \bar{e}_2, & \bar{g}_0^3 &= \bar{e}_3, \\
\bar{g}_1 &= \bar{e}_1, & \bar{g}_2 &= \kappa \bar{e}_1 + \bar{e}_2, & \bar{g}_3 &= \bar{e}_3, \\
\bar{g}^1 &= \bar{e}_1 - \kappa \bar{e}_2, & \bar{g}^2 &= \bar{e}_2, & \bar{g}^3 &= \bar{e}_3. \quad \square
\end{aligned} \tag{2.3}$$

Exercise 2.2. By substituting expressions (2.3) into Eqs. (2.1.5) and (2.1.25), check that

$$\bar{\nabla}_0 \bar{r} = \bar{e}_1 \bar{e}_1 + \bar{e}_2 \bar{e}_2 + \bar{e}_3 \bar{e}_3 + \kappa \bar{e}_2 \bar{e}_1,$$

146

$$\hat{F} = (1 + \kappa^2)\bar{e}_1\bar{e}_1 + \kappa(\bar{e}_1\bar{e}_2 + \bar{e}_2\bar{e}_1) + \bar{e}_2\bar{e}_2 + \bar{e}_3\bar{e}_3,$$
$$\hat{F}^2 = [(1 + \kappa^2)^2 + \kappa^2]\bar{e}_1\bar{e}_1 + \kappa(2 + \kappa^2)(\bar{e}_1\bar{e}_2 + \bar{e}_2\bar{e}_1)$$
$$+(1 + \kappa^2)\bar{e}_2\bar{e}_2 + \bar{e}_3\bar{e}_3. \quad \square \qquad (2.4)$$

Eqs. (2.4) imply the following matrix presentation of the Finger tensor $\hat{F}$:

$$\hat{F} = \begin{bmatrix} 1 + \kappa^2 & \kappa & 0 \\ \kappa & 1 & 0 \\ 0 & 0 & 1 \end{bmatrix}.$$

Exercise 2.3. Derive the characteristic polynomial for the tensor $\hat{F}$

$$\det(\hat{F} - \lambda\hat{I}) = -\lambda^3 + (3 + \kappa^2)\lambda^2 - (3 + \kappa^2)\lambda + 1, \qquad (2.5)$$

where $\hat{I}$ is the unit tensor. $\square$

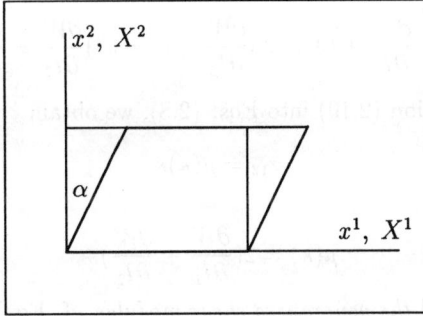

Fig. 2.1. Simple shear of an elastic medium.

Since the principal invariants of the Finger tensor coincide (up to their signs) with coefficients of the characteristic polynomial, Eq. (2.5) yields

$$I_1(\hat{F}) = 3 + \kappa^2, \quad I_2(\hat{F}) = 3 + \kappa^2, \quad I_3(\hat{F}) = 1. \qquad (2.6)$$

Definition. A deformation which preserves the volume element, i.e. which satisfies the condition $I_3(\hat{F}) = 1$, is called *isochoric*.

According to (2.6), simple shear is an example of isochoric deformations.

Exercise 2.4. Provide several examples of isochoric deformations. $\square$

Substitution of expressions (2.4) and (2.6) into Eqs. (3.2.18) implies that

$$\hat{\sigma} = \sigma_1\bar{e}_1\bar{e}_1 + \sigma_2\bar{e}_2\bar{e}_2 + \sigma_3\bar{e}_3\bar{e}_3 + \sigma_{12}(\bar{e}_1\bar{e}_2 + \bar{e}_2\bar{e}_1), \qquad (2.7)$$

147

where

$$\sigma_1 = 2\{\psi_0 + (1 + \kappa^2)\psi_1 + [(1 + \kappa^2)^2 + \kappa^2]\psi_2\},$$
$$\sigma_2 = 2[\psi_0 + \psi_1 + (1 + \kappa^2)\psi_2], \qquad \sigma_3 = 2(\psi_0 + \psi_1 + \psi_2),$$
$$\sigma_{12} = 2\kappa[\psi_1 + (2 + \kappa^2)\psi_2]. \tag{2.8}$$

Exercise 2.5. Derive formulas (2.8). □

The Cauchy stress tensor $\hat{\sigma}$ is independent of spatial coordinates. Thus, in the absence of body forces, the equilibrium equation is fulfilled identically for any κ value.

Substitution of expressions (2.6) into Eq. (3.2.12) yields

$$W = W(3 + \kappa^2, 3 + \kappa^2, 1) = \mathcal{W}(\kappa), \tag{2.9}$$

where $\mathcal{W}$ is a prescribed function.

Let us transform the fourth expression (2.8) with the use of Eqs. (2.6) and (3.2.19). We find

$$\psi_1 + (2 + \kappa^2)\psi_2 = \frac{\partial W}{\partial I_1} + (3 + \kappa^2)\frac{\partial W}{\partial I_2} - (2 + \kappa^2)\frac{\partial W}{\partial I_2} = \frac{\partial W}{\partial I_1} + \frac{\partial W}{\partial I_2}. \tag{2.10}$$

Substituting expression (2.10) into Eqs. (2.8), we obtain

$$\sigma_{12} = \mu(\kappa)\kappa, \tag{2.11}$$

where

$$\mu(\kappa) = 2\left(\frac{\partial W}{\partial I_1} + \frac{\partial W}{\partial I_2}\right). \tag{2.12}$$

Parameter μ is called *the generalized shear modulus*, cf. Eq. (3.2.28). Eq. (2.11) may be treated as an extension to finite strains of the well-known relationship in the linear theory of elasticity, $\sigma_{12} = \mu\alpha$, between the shear stress σ_{12} and the shear angle α. Bearing in mind that

$$\frac{d\mathcal{W}}{d\kappa} = 2\kappa\left(\frac{\partial W}{\partial I_1} + \frac{\partial W}{\partial I_2}\right), \tag{2.13}$$

we can also present the shear stress σ_{12} as follows:

$$\sigma_{12} = \frac{d\mathcal{W}}{d\kappa}(\kappa). \tag{2.14}$$

Exercise 2.6. Derive Eq. (2.14). □

Let us now transform the expression $\psi_0 + \psi_1 + (1 + \kappa^2)\psi_2$. By using Eqs. (2.6) and (3.2.19) we obtain

$$\psi_0 + \psi_1 + (1 + \kappa^2)\psi_2 = \frac{1}{2}\Sigma, \tag{2.15}$$

148

where

$$\Sigma(\kappa) = 2\left(\frac{\partial W}{\partial I_1} + 2\frac{\partial W}{\partial I_2} + \frac{\partial W}{\partial I_3}\right). \qquad (2.16)$$

Exercise 2.7. By substitution of Eqs. (2.10) and (2.15) into Eqs. (2.8), derive the equalities

$$\sigma_1 = \Sigma(\kappa) + \mu(\kappa)\kappa^2, \qquad \sigma_2 = \Sigma(\kappa), \qquad \sigma_3 = \Sigma(\kappa) + 2\kappa^2\frac{\partial W}{\partial I_2}. \qquad \square \qquad (2.17)$$

Eqs. (2.11) and (2.17) imply that

$$\Delta\sigma = \kappa\sigma_{12}, \qquad (2.18)$$

where $\Delta\sigma = \sigma_1 - \sigma_2$ is the difference of normal stresses. Eq. (2.18) should be fulfilled in any isotropic hyperelastic material. Since this equation is independent of the concrete form of strain energy density W, it may be employed to test experimentally the characteristic features of hyperelastic media with finite strains.

Eq. (2.18) is called *the Lodge–Meissner formula*, see e.g. Lodge & Meissner (1972), where Eq. (2.18) was applied to study polymeric liquids.

Substitution of expression (2.11) into Eq. (2.18) implies that

$$\Delta\sigma = \frac{\sigma_{12}^2}{\mu}. \qquad (2.19)$$

For a constant modulus μ, Eq. (2.19) means that the difference of normal stresses is proportional to the square of the shear stress. This equality is called *the Lodge formula*, see Lodge (1964).

According to Eq. (2.12), parameter μ is constant provided the strain energy density depends linearly on the first and second principal invariants (for example, for the neo-Hookean and Mooney–Rivlin media). Moreover, this parameter may also be treated as constant either when the nonlinear dependence of W on the first and second invariants of $\hat{F}$ is feeble, or when the shear angle α is relatively small. Thus, the sphere of applications of the Lodge formula is rather wide. This is confirmed by experimental data plotted in Fig. 2.2. These data show that in the "log–log" coordinates, experimental points are extremely close to the straight lines with the slope equal to 2, which corresponds to the theoretical prediction based on formula (2.19).

It follows from Eq. (2.18) that for a non-zero shear stress $\sigma_{12} \neq 0$, normal stresses σ_1 and σ_2 cannot vanish simultaneously. This means that simple shear cannot be produced by applying only shear stresses, while normal stresses are also necessary. This phenomenon (existence of normal stresses in simple shear) is called *the Weissenberg effect*. A number of experimental data confirming the Weissenberg effect are presented by Bird et al. (1977), Lodge (1964), and Vinogradov & Malkin (1980).

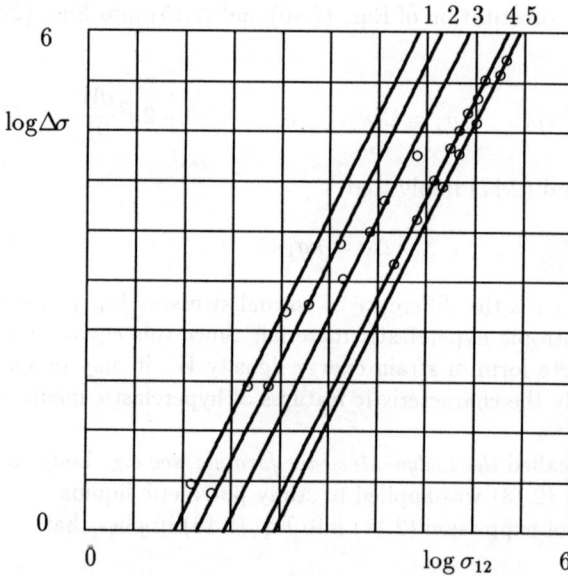

Fig. 2.2. $\log \Delta\sigma$ (Pa) *versus* $\log \sigma_{12}$ (Pa). Circles correspond to experimental data for butyl rubber solution in cetane obtained by Malkin (1995). Straight lines provide their least square approximation by the function $\log \Delta\sigma = 2\log \sigma_{12} + a$, where a is found by fitting experimental data. Different curves correspond to different concentrations c of solutions. Curve 1: $c = 9\%$, curve 2: $c = 20\%$, curve 3: $c = 51\%$, curve 4: $c = 85\%$, curve 5: $c = 100\%$.

According to Eqs. (3.2.23), (2.7), and (2.17), up to the second order terms compared with κ, the Cauchy stress tensor $\hat{\sigma}$ at finite strains coincides with the stress tensor in the linear theory of elasticity

$$\hat{\sigma} = \sigma_{12}(\bar{e}_1\bar{e}_2 + \bar{e}_2\bar{e}_1). \tag{2.20}$$

It is quite natural to analyze motions where Eq. (2.20) is fulfilled not approximately, but exactly, for an arbitrary κ value.

Definition. A homogeneous deformation satisfying condition (2.20) is called *pure shear.*

Pure shear is a significantly more complicated deformation compared with simple shear. We do not intend to provide its detailed analysis and leave this problem to the reader as an exercise referring to Lurie (1990).

Problem 2.1. Calculate the displacement field for pure shear of the Blatz–Ko material (3.2.66). □

150

Rajagopal and Wineman (1987) and Wineman & Gandhi (1984) considered the following homogeneous deformation:

$$x^1 = \lambda_1 X^1 + \kappa \lambda_2 X^2, \qquad x^2 = \lambda_2 X^2, \qquad x^3 = \lambda_3 X^3. \qquad (2.21)$$

Eq. (2.21) is turned into (2.1) when $\lambda_1 = \lambda_2 = \lambda_3 = 1$.

Problem 2.2. Show that Eq. (2.21) implies the universal relation

$$\sigma_{11} - \sigma_{22} = \frac{\lambda_1^2 + \lambda_2^2(\kappa^2 - 1)}{\kappa \lambda_2^2} \sigma_{12}. \qquad \square \qquad (2.22)$$

Exercise 2.8. Show that for $\lambda_1 = \lambda_2$, formula (2.22) is reduced to the Lodge–Meissner equation (2.18). $\square$

2.2. Concluding Remarks

In this Section we consider quasi-static shear deformation of a hyperelastic medium. We derive the Lodge–Meissner formula, which connects the first difference of normal stresses with the shear stress, and demonstrate the Weissenberg effect: the presence of normal stresses in simple shear.

3. Torsion of a Circular Cylinder

This Section is concerned with torsion of a circular hyperelastic cylinder under the action of compressive forces and torques applied to its ends. Torsion of a circular cylinder is one of the classical problems in finite elasticity, see Rivlin (1949). We provide an explicit solution to this problem and discuss the Poynting effect.

3.1. Formulation of the Problem

Let us consider a circular elastic cylinder with length l and radius a made of an incompressible hyperelastic material. Deformation of the cylinder from its natural (stress-free) state occurs under the action of surface loads applied to the cylinder's edges. The lateral surface of the cylinder is traction-free. Body forces are absent.

The surface traction at the edges is statically equivalent to a force $\bar{P}$ and a *torque* $\bar{M}$, which are directed along the longitudinal axis of the cylinder.

Combined *torsion* and tension of a circular cylinder is described by the formulas

$$r = cR, \qquad \phi = \Phi + \alpha \lambda Z, \qquad z = \lambda Z. \qquad (3.1)$$

Here $\{R, \Phi, Z\}$ and $\{r, \phi, z\}$ are cylindrical coordinates in the initial and actual configurations with unit vectors $\bar{e}_R$, $\bar{e}_\Phi$, $\bar{e}_Z$ and $\bar{e}_r$, $\bar{e}_\phi$, $\bar{e}_z$, respectively, α is *twist angle* per unit length, λ and c are constants to be found.

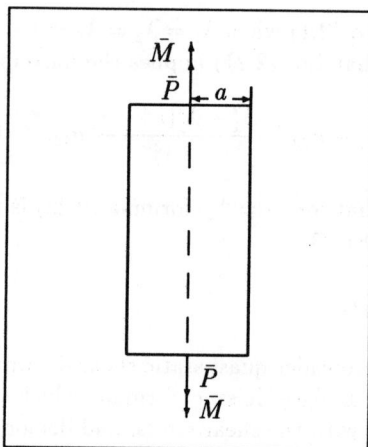

Fig. 3.1. Combined torsion and tension of a circular cylinder.

It follows from Eq. (3.1) that radius-vectors $\bar{r}_0$ and $\bar{r}$ of a point with Lagrangian coordinates $\{R, \Phi, Z\}$ in the initial and actual configurations equal

$$\bar{r}_0 = R\bar{e}_R + Z\bar{e}_Z, \qquad \bar{r} = cR\bar{e}_r + \lambda Z\bar{e}_Z.$$

Differentiating these equalities with the use of Eqs. (1.1.1) we find tangent vectors in the initial and actual configurations

$$\bar{g}_{0\,1} = \bar{e}_R, \qquad \bar{g}_{0\,2} = R\bar{e}_\Phi, \qquad \bar{g}_{0\,3} = \bar{e}_Z,$$
$$\bar{g}_1 = c\bar{e}_r, \qquad \bar{g}_2 = r\bar{e}_\phi, \qquad \bar{g}_3 = \lambda(\alpha r\bar{e}_\phi + \bar{e}_Z). \tag{3.2}$$

Exercise 3.1. Show that dual vectors in the initial and actual configurations are calculated as follows:

$$\bar{g}_0^1 = \bar{e}_R, \qquad \bar{g}_0^2 = \frac{1}{R}\bar{e}_\Phi, \qquad \bar{g}_0^3 = \bar{e}_Z,$$
$$\bar{g}^1 = \frac{1}{c}\bar{e}_r, \qquad \bar{g}^2 = \frac{1}{r}\bar{e}_\phi - \alpha\bar{e}_Z, \qquad \bar{g}^3 = \frac{1}{\lambda}\bar{e}_Z. \qquad \Box \tag{3.3}$$

Eqs. (3.2) and (3.3) together with Eq. (2.1.5) imply that

$$\bar{\nabla}_0\bar{r} = c(\bar{e}_R\bar{e}_r + \bar{e}_\Phi\bar{e}_\phi) + \lambda\bar{e}_Z(\alpha r\bar{e}_\phi + \bar{e}_Z). \tag{3.4}$$

152

Multiplying Eq. (3.4) by the transpose formula

$$\bar{\nabla}_0 \bar{r}^T = c(\bar{e}_r \bar{e}_R + \bar{e}_\phi \bar{e}_\Phi) + \lambda(\alpha r \bar{e}_\phi + \bar{e}_Z)\bar{e}_Z,$$

we find with the use of Eq. (2.1.25)

$$\hat{F} = c^2 \bar{e}_r \bar{e}_r + (c^2 + \alpha^2 \lambda^2 r^2)\bar{e}_\phi \bar{e}_\phi + \lambda^2 \bar{e}_Z \bar{e}_Z + \alpha \lambda^2 r(\bar{e}_\phi \bar{e}_Z + \bar{e}_Z \bar{e}_\phi). \qquad (3.5)$$

Exercise 3.2. Check that

$$\hat{F}^2 = c^4 \bar{e}_r \bar{e}_r + [(c^2 + \alpha^2 \lambda^2 r^2)^2 + \alpha^2 \lambda^4 r^2]\bar{e}_\phi \bar{e}_\phi$$

$$+\lambda^4(1 + \alpha^2 r^2)\bar{e}_Z \bar{e}_Z + \alpha \lambda^2 r(\lambda^2 + c^2 + \alpha^2 \lambda^2 r^2)(\bar{e}_\phi \bar{e}_Z + \bar{e}_Z \bar{e}_\phi). \quad \Box \qquad (3.6)$$

In the cylindrical coordinates $\{r, \phi, Z\}$, tensor $\hat{F}$ can be presented in the matrix form as

$$\hat{F} = \begin{bmatrix} c^2 & 0 & 0 \\ 0 & c^2 + \alpha^2 \lambda^2 r^2 & \alpha \lambda^2 r \\ 0 & \alpha \lambda^2 r & \lambda^2 \end{bmatrix}.$$

Let $\hat{I}$ be the unit matrix, and κ be a scalar variable. Calculating determinant of the matrix $(\hat{F} - \kappa \hat{I})$, we obtain

$$\det(\hat{F} - \kappa \hat{I}) = -\kappa^3 + (2c^2 + \lambda^2 + \alpha^2 \lambda^2 r^2)\kappa^2$$

$$-c^2(c^2 + 2\lambda^2 + \alpha^2 \lambda^2 r^2)\kappa + c^4 \lambda^2. \qquad (3.7)$$

Since the principal invariants of the Finger tensor $\hat{F}$ coincide (up to their signs) with coefficients of the characteristic equation, Eq. (3.7) yields

$$I_1(\hat{F}) = 2c^2 + \lambda^2 + \alpha^2 \lambda^2 r^2, \qquad I_2(\hat{F}) = c^2(c^2 + 2\lambda^2 + \alpha^2 \lambda^2 r^2),$$
$$I_3(\hat{F}) = c^4 \lambda^2. \qquad (3.8)$$

Eqs. (3.8) together with the incompressibility condition $I_3(\hat{F}) = 1$ imply that

$$c = \lambda^{-1/2}. \qquad (3.9)$$

Substitution of expression (3.9) into Eqs. (3.5), (3.6), and (3.8) yields

$$\hat{F} = \frac{1}{\lambda}\bar{e}_r \bar{e}_r + (\frac{1}{\lambda} + \alpha^2 \lambda^2 r^2)\bar{e}_\phi \bar{e}_\phi + \lambda^2 \bar{e}_Z \bar{e}_Z + \alpha \lambda^2 r(\bar{e}_\phi \bar{e}_Z + \bar{e}_Z \bar{e}_\phi),$$

$$\hat{F}^2 = \frac{1}{\lambda^2}\bar{e}_r \bar{e}_r + [(\frac{1}{\lambda} + \alpha^2 \lambda^2 r^2)^2 + \alpha^2 \lambda^4 r^2]\bar{e}_\phi \bar{e}_\phi$$

$$+\lambda^4(1 + \alpha^2 r^2)\bar{e}_Z \bar{e}_Z + \alpha \lambda^2 r(\frac{1}{\lambda} + \lambda^2 + \alpha^2 \lambda^2 r^2)(\bar{e}_\phi \bar{e}_Z + \bar{e}_Z \bar{e}_\phi),$$

$$I_1(\hat{F}) = \frac{2}{\lambda} + \lambda^2 + \alpha^2 \lambda^2 r^2, \qquad I_2(\hat{F}) = \frac{1}{\lambda^2} + 2\lambda + \alpha^2 \lambda r^2. \qquad (3.10)$$

153

Exercise 3.3. Derive formulas (3.10). □

Exercise 3.4. By substitution of expressions (3.10) into Eq. (3.2.99), show that

$$\hat{\sigma} = \sigma_{rr}\bar{e}_r\bar{e}_r + \sigma_{\phi\phi}\bar{e}_\phi\bar{e}_\phi + \sigma_{ZZ}\bar{e}_Z\bar{e}_Z + \sigma_{\phi Z}(\bar{e}_\phi\bar{e}_Z + \bar{e}_Z\bar{e}_\phi). \tag{3.11}$$

Here

$$\sigma_{rr} = -p + \frac{2}{\lambda}(\psi_1 + \frac{1}{\lambda}\psi_2),$$

$$\sigma_{\phi\phi} = -p + 2\{(\frac{1}{\lambda} + \alpha^2\lambda^2 r^2)\psi_1 + [(\frac{1}{\lambda} + \alpha^2\lambda^2 r^2)^2 + \alpha^2\lambda^4 r^2]\psi_2\},$$

$$\sigma_{ZZ} = -p + 2\lambda^2[\psi_1 + \lambda^2(1 + \alpha^2 r^2)\psi_2],$$

$$\sigma_{\phi Z} = 2\alpha\lambda^2 r[\psi_1 + (\frac{1}{\lambda} + \lambda^2 + \alpha^2\lambda^2 r^2)\psi_2], \tag{3.12}$$

where functions ψ_1 and ψ_2 are determined by Eq. (3.2.98). □

Let us transform expression (3.12) for $\sigma_{\phi Z}$.

Exercise 3.5. By using Eqs. (3.2.98) and (3.10), check that

$$\psi_1 + (\frac{1}{\lambda} + \lambda^2 + \alpha^2\lambda^2 r^2)\psi_2 = \frac{1}{\lambda}(\lambda\frac{\partial W}{\partial I_1} + \frac{\partial W}{\partial I_2}). \quad\square \tag{3.13}$$

On the other hand, expressions (3.2.98) and (3.10) imply that

$$\frac{dW}{dr} = \frac{\partial W}{\partial I_1}\frac{dI_1}{dr} + \frac{\partial W}{\partial I_2}\frac{dI_2}{dr} = 2\alpha^2\lambda r(\lambda\frac{\partial W}{\partial I_1} + \frac{\partial W}{\partial I_2}). \tag{3.14}$$

Eqs. (3.12) – (3.14) yield

$$\sigma_{\phi Z} = \frac{1}{\alpha}\frac{dW}{dr}. \tag{3.15}$$

It follows from Eqs. (3.12) that

$$\sigma_{\phi\phi} = \sigma_{rr} + 2\alpha^2\lambda^2 r^2[\psi_1 + (\frac{2}{\lambda} + \lambda^2 + \alpha^2\lambda^2 r^2)\psi_2]. \tag{3.16}$$

Equalities (3.10), (3.16), and (3.2.98) imply that

$$\sigma_{\phi\phi} = \sigma_{rr} + 2\alpha^2\lambda^2 r^2(\psi_1 + I_1\psi_2) = \sigma_{rr} + 2\alpha^2\lambda^2 r^2\frac{\partial W}{\partial I_1}. \tag{3.17}$$

We now subtract the first equation (3.12) from the third and find

$$\sigma_{ZZ} = \sigma_{rr} + 2\{(\lambda^2 - \frac{1}{\lambda})\psi_1 + [\lambda^4(1 + \alpha^2 r^2) - \frac{1}{\lambda^2}]\psi_2\}. \tag{3.18}$$

It follows from Eqs. (3.2.98), (3.10), and (3.14) that

$$\psi_1 = \frac{\partial W}{\partial I_1} + I_1\frac{\partial W}{\partial I_2} = \frac{1}{2\alpha^2\lambda^2 r}\frac{dW}{dr} + (\frac{1}{\lambda} + \lambda^2 + \alpha^2\lambda^2 r^2)\frac{\partial W}{\partial I_2}.$$

Exercise 3.6. By substituting this expression and Eq. (3.2.98) into Eq. (3.18), show that

$$\sigma_{ZZ} = \sigma_{rr} + \frac{\lambda^3 - 1}{\alpha^2 \lambda^3 r} \frac{dW}{dr} - 2\alpha^2 \lambda r^2 \frac{\partial W}{\partial I_2}. \qquad \square \qquad (3.19)$$

Exercise 3.7. Show that the only equilibrium equation (2.2.29) is written as

$$\frac{d\sigma_{rr}}{dr} + \frac{1}{r}(\sigma_{rr} - \sigma_{\phi\phi}) = 0, \qquad (3.20)$$

provided pressure p depends on r only. $\square$

On the traction-free lateral surface of the cylinder we have

$$\sigma_{rr}|_{R=a} = 0. \qquad (3.21)$$

Integration of Eq. (3.20) from r to ca with the use of (3.21) yields

$$\sigma_{rr}(r) = \int_r^{ca} \frac{\sigma_{rr} - \sigma_{\phi\phi}}{r_1} dr_1.$$

Exercise 3.8. By substituting expression (3.17) into this equality and using Eq. (3.9), show that

$$\sigma_{rr}(r) = -2\alpha^2 \lambda^2 \int_r^{\frac{a}{\sqrt{\lambda}}} \frac{\partial W}{\partial I_1} r_1 dr_1. \qquad \square \qquad (3.22)$$

We write the boundary conditions at the cylinder's ends in the integral form

$$\bar{P} = \int_S \bar{n} \cdot \hat{\sigma} dS, \qquad \bar{M} = \int_S \bar{r} \times (\bar{n} \cdot \hat{\sigma}) dS, \qquad (3.23)$$

where S is the cylinder cross-section, dS is the surface element, and $\bar{n} = \pm \bar{e}_Z$ is the outward normal vector in the actual configuration.

Exercise 3.9. By employing Eqs. (3.11) and (3.23), show that at the end with $\bar{n} = +\bar{e}_Z$:

$$\bar{n} \cdot \hat{\sigma} = \sigma_{ZZ} \bar{e}_Z + \sigma_{\phi Z} \bar{e}_\phi, \qquad \bar{r} \times (\bar{n} \cdot \hat{\sigma}) = -z\sigma_{\phi Z} \bar{e}_r - r\sigma_{ZZ} \bar{e}_\phi + r\sigma_{\phi Z} \bar{e}_Z. \qquad \square \quad (3.24)$$

Exercise 3.10. Prove that for any function f independent of the polar angle ϕ

$$\int_S f\bar{e}_r dS = 0, \qquad \int_S f\bar{e}_\phi dS = 0. \qquad \square \qquad (3.25)$$

By using Eqs. (3.24) and (3.25), we can present conditions (3.23) as follows:

$$P = \int_S \sigma_{ZZ} dS = \int_0^{2\pi} \int_0^{ca} \sigma_{ZZ} r \, dr \, d\phi,$$

$$M = \int_S \sigma_{\phi Z} r \, dS = \int_0^{2\pi} \int_0^{ca} \sigma_{\phi Z} r^2 \, dr \, d\phi.$$

155

Since components of the Cauchy stress tensor are independent of ϕ, the integrals over ϕ can be calculated explicitly. As a result, we obtain

$$P = 2\pi \int_0^{ca} \sigma_{ZZ} r \, dr, \qquad M = 2\pi \int_0^{ca} \sigma_{\phi Z} r^2 \, dr. \qquad (3.26)$$

Our objective now is to express the integrals in the right-hand side of Eq. (3.26) in terms of strain energy density W. To calculate the integral $\int_0^{ca} \sigma_{ZZ} r \, dr$, we, first, substitute Eq. (3.19) into this expression

$$\int_0^{ca} \sigma_{ZZ} r \, dr = \int_0^{ca} \sigma_{rr} r \, dr + \frac{\lambda^3 - 1}{\alpha^2 \lambda^3} \int_0^{ca} \frac{dW}{dr} dr - 2\alpha^2 \lambda \int_0^{ca} \frac{\partial W}{\partial I_2} r^3 \, dr.$$

Afterward, we substitute expression (3.22) into the first integral, calculate the second integral in the right-hand side, and find

$$\int_0^{ca} \sigma_{ZZ} r \, dr = \frac{\lambda^3 - 1}{\alpha^2 \lambda^3}(W|_{R=a} - W|_{R=0})$$
$$-2\alpha^2 \lambda^2 [\int_0^{\frac{a}{\sqrt{\lambda}}} r \, dr \int_r^{\frac{a}{\sqrt{\lambda}}} \frac{\partial W}{\partial I_1} r_1 \, dr_1 + \frac{1}{\lambda} \int_0^{\frac{a}{\sqrt{\lambda}}} \frac{\partial W}{\partial I_2} r^3 \, dr]. \qquad (3.27)$$

Changing the order of integration in the first integral in the right-hand side of Eqn. (3.27) implies that

$$\int_0^{\frac{a}{\sqrt{\lambda}}} r \, dr \int_r^{\frac{a}{\sqrt{\lambda}}} \frac{\partial W}{\partial I_1} r_1 \, dr_1 = \int_0^{\frac{a}{\sqrt{\lambda}}} \frac{\partial W}{\partial I_1} r \, dr \int_0^r r_1 \, dr_1 = \frac{1}{2} \int_0^{\frac{a}{\sqrt{\lambda}}} \frac{\partial W}{\partial I_1} r^3 \, dr.$$

Finally, substitution of this expression into Eq. (3.27) yields

$$\int_0^{ca} \sigma_{ZZ} r \, dr = \frac{\lambda^3 - 1}{\alpha^2 \lambda^3}(W|_{R=a} - W|_{R=0}) - \alpha^2 \lambda^2 \int_0^{\frac{a}{\sqrt{\lambda}}} (\frac{\partial W}{\partial I_1} + \frac{2}{\lambda} \frac{\partial W}{\partial I_2}) r^3 \, dr. \qquad (3.28)$$

In order to calculate the integral $\int_0^{ca} \sigma_{\phi Z} r^2 \, dr$, we employ Eqs. (3.9), (3.15) and obtain

$$\int_0^{ca} \sigma_{\phi Z} r^2 \, dr = \frac{1}{\alpha} \int_0^{\frac{a}{\sqrt{\lambda}}} \frac{dW}{dr} r^2 \, dr = \frac{a^2}{\alpha \lambda} W|_{R=a} - \frac{2}{\alpha} \int_0^{\frac{a}{\sqrt{\lambda}}} W r \, dr. \qquad (3.29)$$

Substitution of expressions (3.28) and (3.29) into Eqs. (3.26) yields

$$M = \frac{2\pi}{\alpha}(\frac{a^2}{\lambda} W|_{R=a} - 2 \int_0^{\frac{a}{\sqrt{\lambda}}} W r \, dr),$$
$$P = 2\pi [\frac{\lambda^3 - 1}{\alpha^2 \lambda^3}(W|_{R=a} - W|_{R=0}) - \alpha^2 \lambda^2 \int_0^{\frac{a}{\sqrt{\lambda}}} (\frac{\partial W}{\partial I_1} + \frac{2}{\lambda} \frac{\partial W}{\partial I_2}) r^3 \, dr]. \qquad (3.30)$$

For a given strain energy density W and given loads P and M, Eqs. (3.30) may be treated as nonlinear algebraic equations for twist α per unit undeformed length and the extension ratio λ.

Exercise 3.11. Check that for the neo-Hookean solid (3.2.103), Eqs. (3.30) are written as

$$M = \frac{1}{2}\pi\mu a^4\alpha, \qquad P = \pi\mu a^2\left(\frac{\lambda^3 - 1}{\lambda^2} - \frac{\alpha^2 a^2}{4}\right). \qquad \square \qquad (3.31)$$

It follows from the first equation (3.31) that for a neo-Hookean cylinder, the relationship between torque M and twist α per unit length is linear both for finite and infinitesimal strains.

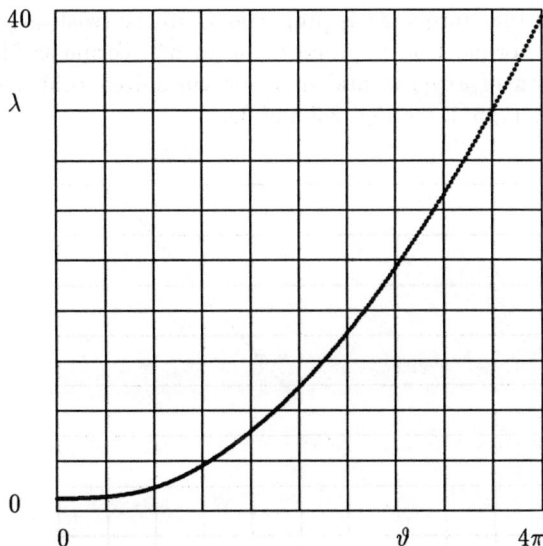

Fig. 3.2. The extension ratio λ *versus* twist angle ϑ for a circular cylinder made of neo-Hookean material.

The second equation (3.31) implies that in the absence of axial load, i.e. for $P = 0$, the extension ratio λ satisfies the equation

$$\lambda - \frac{1}{\lambda^2} = \frac{1}{4}\left(\frac{a}{l}\right)^2\vartheta^2, \qquad (3.32)$$

where $\vartheta = \alpha l$ is the twist angle.

Exercise 3.12. Show that $\lambda > 1$ for any $\vartheta \neq 0$. The latter means that for any twist ϑ, we observe elongation of the cylinder, while λ grows with an increase in the twist angle ϑ. $\square$

This phenomenon (elongation of an elastic cylinder under torsion) was discovered by Poynting in 1909 in experiments on torsion of steel wires, and it is called *the Poynting effect*.

Dependence λ *versus* ϑ is plotted in Fig. 3.2 for a "thick" cylinder, when its radius a coincides with its length l. Experiments on torsion of butyl rubber samples with the ratio $a/l = 1$ were carried out by Rivlin & Saunders (1951) and Hausler & Sayir (1995).

When the axial elongation is absent, $\lambda = 1$, torque M and normal force P are calculated as follows:

$$M = \frac{1}{2}\pi\mu a^4\alpha, \qquad P = \frac{1}{4}\pi\mu a^4\alpha^2. \tag{3.33}$$

Eqs. (3.33) imply that torque M is proportional to the twist angle α per unit length, while the normal load is proportional to α^2. Formulas (3.33) provide excellent prediction of experimental data for vulcanized rubber obtained by Rivlin & Saunders (1951), see Fig. 3.3 and 3.4.

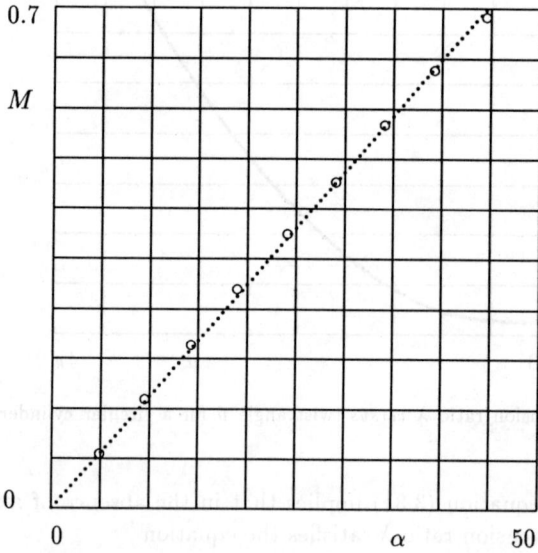

Fig. 3.3. Torque M (N·m) *versus* twist angle α per unit length (rad/m). Large circles correspond to experimental data obtained by Rivlin & Saunders (1951) for a vulcanized rubber cylinder with $a = l = 0.025$ (m). Small circles provide their approximation by the linear function $M = 0.0157\alpha$.

Exercise 3.13. Derive formulas for axial elongation and twist angle in a circular cylinder made of the Mooney–Rivlin material (3.2.104). □

Exercise 3.14. Derive relationships similar to Eqs. (3.31) for a circluar cylinder made of the Knowles material (3.2.115). □

Torsion of a circular cylinder made of a compressible hyperelastic material is an essentially more sophisticated problem. We do not intend to analyze this issue here. Nevertheless, we formulate two basic topics for investigation, and leave them to the reader as exercises with appropriate references.

Problem 3.1. Derive governing equation for pure torsion (without tension) of a compressible elastic cylinder. Show that for the Blatz–Ko material (3.2.66), equations for twist angle α and axial elongation λ are independent. For a detailed analysis of this problem, see Green & Adkins (1970). $\square$

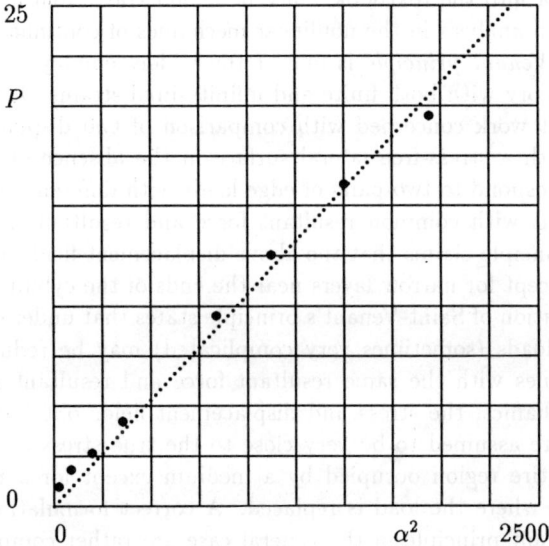

Fig. 3.4. Normal load P (N) *versus* square of the twist angle α^2 per unit length (rad/m)^2. Large circles correspond to experimental data obtained by Rivlin & Saunders (1951) for a vulcanized rubber cylinder with $a = l = 0.025$ (m). Small circles provide their approximation by the quadratic function $P = 0.0106\alpha^2$.

Problem 3.2. Torsion of a compressible elastic cylinder is not a universal solution. Derive conditions on a strain energy density W which permit such a deformation. For a detailed analysis of this issue, see Polignone & Horgan (1991). $\square$

3.2. Concluding Remarks

In this Section we provide an explicit solution to the problem of torsion for a circular cylinder and demonstrate the Poynting effect: axial elongation of a cylinder under the action of torques applied to its ends. The formulas derived show excellent agreement with experimental data for vulcanized rubber.

4. Saint-Venant's Principle

This Section is concerned with anti-plane deformation of a rod. Our objective is to formulate and prove the Saint-Venant principle (Proposition 4.1). The reader interested in mechanical phenomena exclusively, may omit the proof without any detriment for understanding the material below. Nevertheless, we include the proof into the textbook, since it demonstrates mathematical tools typical of modern analysis in the nonlinear mechanics of continua.

The Saint-Venant principle is one of the widely employed assertions in the elasticity theory with both finite and infinitesimal strains. It goes back to the Saint-Venant work concerned with comparison of two displacement fields in a cylinder with a stress-free lateral surface in the absence of body forces. These fields correspond to two pairs of edge loads with different detailed stress distributions, but with common resultant force and resultant moment. The Saint-Venant principle claims that the above displacement fields are very close to each other except for narrow layers near the ends of the cylinder.

A generalization of Saint-Venant's principle states that under some natural conditions real loads (sometimes very complicated) may be reduced to sufficiently simple ones with the same resultant force and resultant moment. In engineering mechanics, the stress and displacement fields obtained after such a replacement are assumed to be very close to the true stresses and displacements in the entire region occupied by a medium except for a narrow layer near the surface where the load is replaced. A correct formulation and proof of the Saint-Venant principle in the general case are rather complicated. We confine ourselves to the particular case when anti-plane deformation occurs in an elastic incompressible body in the form of a rectilinear rod. Body forces are neglected, and a dead surface traction $\bar{b}$ is assumed to be directed in parallel to the rod's longitudinal axis. Since anti-plane deformation cannot occur in an arbitrary hyperelastic medium, specific potential energy W is assumed to have the Knowles form (3.2.114). The exposition follows Horgan & Payne (1990).

4.1. Anti-plane Deformation

Let us consider an elastic body which occupies in the initial configuration a cylindrical domain $\{(x^1, x^2) \in \mathcal{S}, \ -\infty < x^3 < \infty\}$, where $\{x^i\}$ are Cartesian coordinates with unit vectors $\bar{e}_i$, $\mathcal{S} = \{0 \leq x^1 \leq l, \ 0 \leq x^2 \leq h\}$ is a rectangular cross-section with length l and width h.

Let $\bar{u} = u_1 \bar{e}_1 + u_2 \bar{e}_2 + u_3 \bar{e}_3$ be a displacement vector.

Definition. A deformation is called *anti-plane* if

$$u_1 = 0, \qquad u_2 = 0, \qquad u_3 = u(x^1, x^2). \tag{4.1}$$

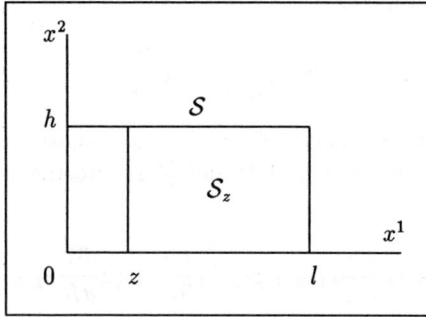

Fig. 4.1. The bar cross-section.

It follows from Eq. (4.1) that

$$\bar{r} = \bar{r}_0 + u\bar{e}_3, \tag{4.2}$$

where $\bar{r}_0$ and $\bar{r}$ are the radius-vectors in the initial and actual configurations, respectively. Differentiation of Eq. (4.2) implies that

$$\bar{\nabla}_0\bar{r} = \hat{I} + \frac{\partial u}{\partial x^1}\bar{e}_1\bar{e}_3 + \frac{\partial u}{\partial x^2}\bar{e}_2\bar{e}_3. \tag{4.3}$$

Exercise 4.1. Check that

$$\bar{\nabla}_0\bar{r}^{-1} = \hat{I} - \frac{\partial u}{\partial x^1}\bar{e}_1\bar{e}_3 - \frac{\partial u}{\partial x^2}\bar{e}_2\bar{e}_3. \qquad \square \tag{4.4}$$

Exercise 4.2. By using Eqs. (4.3) and (2.1.25) show that the Finger tensor $\hat{F}$ is written as

$$\hat{F} = \bar{e}_1\bar{e}_1 + \bar{e}_2\bar{e}_2 + [1 + (\frac{\partial u}{\partial x^1})^2 + (\frac{\partial u}{\partial x^2})^2]\bar{e}_3\bar{e}_3$$
$$+ \frac{\partial u}{\partial x^1}(\bar{e}_1\bar{e}_3 + \bar{e}_3\bar{e}_1) + \frac{\partial u}{\partial x^2}(\bar{e}_2\bar{e}_3 + \bar{e}_3\bar{e}_2). \qquad \square \tag{4.5}$$

Tensor $\hat{F}$ can be presented in the matrix form

$$\hat{F} = \begin{bmatrix} 1 & 0 & \frac{\partial u}{\partial x^1} \\ 0 & 1 & \frac{\partial u}{\partial x^2} \\ \frac{\partial u}{\partial x^1} & \frac{\partial u}{\partial x^2} & 1 + (\frac{\partial u}{\partial x^1})^2 + (\frac{\partial u}{\partial x^2})^2 \end{bmatrix}.$$

161

Exercise 4.3. Derive the following characteristic equation for the tensor $\hat{F}$:

$$\lambda^3 - [3 + (\frac{\partial u}{\partial x^1})^2 + (\frac{\partial u}{\partial x^2})^2]\lambda^2 + [3 + (\frac{\partial u}{\partial x^1})^2 + (\frac{\partial u}{\partial x^2})^2]\lambda - 1 = 0. \qquad \Box \quad (4.6)$$

It follows from (4.6) that

$$I_1(\hat{F}) = I_2(\hat{F}) = 3 + (\frac{\partial u}{\partial x^1})^2 + (\frac{\partial u}{\partial x^2})^2, \qquad I_3(\hat{F}) = 1. \qquad (4.7)$$

Eqs. (4.7) imply that anti-plane deformation is isochoric.

Substitution of expressions (4.3) and (4.4) into the constitutive equation (3.2.116) yields

$$\hat{P} = (2\frac{dW}{dI_1} - p)\hat{I} + (2\frac{dW}{dI_1}\bar{e}_1\bar{e}_3 + p\bar{e}_3\bar{e}_1)\frac{\partial u}{\partial x^1} + (2\frac{dW}{dI_1}\bar{e}_2\bar{e}_3 + p\bar{e}_3\bar{e}_2)\frac{\partial u}{\partial x^2}. \quad (4.8)$$

Exercise 4.4. Derive Eq. (4.8). $\Box$

In the absence of body forces, Eq. (4.8) together with the equilibrium equation (2.2.45) implies that

$$\frac{\partial}{\partial x^1}(2\frac{dW}{dI_1} - p) + \frac{\partial}{\partial x^3}(p\frac{\partial u}{\partial x^1}) = 0,$$

$$\frac{\partial}{\partial x^2}(2\frac{dW}{dI_1} - p) + \frac{\partial}{\partial x^3}(p\frac{\partial u}{\partial x^2}) = 0,$$

$$2[\frac{\partial}{\partial x^1}(\frac{dW}{dI_1}\frac{\partial u}{\partial x^1}) + \frac{\partial}{\partial x^2}(\frac{dW}{dI_1}\frac{\partial u}{\partial x^2})] + \frac{\partial}{\partial x^3}(2\frac{dW}{dI_1} - p) = 0.$$

Since u and I_1 are independent of x^3, these equalities can be simplified as follows:

$$\frac{\partial}{\partial x^1}(2\frac{dW}{dI_1} - p) + \frac{\partial u}{\partial x^1}\frac{\partial p}{\partial x^3} = 0,$$

$$\frac{\partial}{\partial x^2}(2\frac{dW}{dI_1} - p) + \frac{\partial u}{\partial x^2}\frac{\partial p}{\partial x^3} = 0,$$

$$2[\frac{\partial}{\partial x^1}(\frac{dW}{dI_1}\frac{\partial u}{\partial x^1}) + \frac{\partial}{\partial x^2}(\frac{dW}{dI_1}\frac{\partial u}{\partial x^2})] = \frac{\partial p}{\partial x^3}. \qquad (4.9)$$

The left-hand side of the latter equation (4.9) depends on x^1 and x^2 only. Therefore,

$$\frac{\partial p}{\partial x^3} = \Theta(x^1, x^2),$$

where Θ is a function to be found. Integration of this relation yields

$$p = \Theta(x^1, x^2)x^3 + \vartheta(x^1, x^2), \qquad (4.10)$$

where ϑ is an unknown function. Substitution of expression (4.10) into the first two equalities (4.9) yields

$$\frac{\partial}{\partial x^1}(2\frac{dW}{dI_1} - \Theta x^3 - \vartheta) + \Theta\frac{\partial u}{\partial x^1} = 0,$$

$$\frac{\partial}{\partial x^2}(2\frac{dW}{dI_1} - \Theta x^3 - \vartheta) + \Theta\frac{\partial u}{\partial x^2} = 0. \tag{4.11}$$

Since Θx^3 is the only term in Eq. (4.11) depending on x^3, these equations imply that

$$\frac{\partial\Theta}{\partial x^1} = 0, \qquad \frac{\partial\Theta}{\partial x^2} = 0.$$

This means that

$$\Theta = c, \tag{4.12}$$

where c is a constant. It follows from Eqs. (4.11) and (4.12) that

$$\frac{\partial}{\partial x^1}(2\frac{dW}{dI_1} - \vartheta + cu) = 0, \qquad \frac{\partial}{\partial x^2}(2\frac{dW}{dI_1} - \vartheta + cu) = 0. \tag{4.13}$$

Since the function in the brackets in Eqs. (4.13) depends on x^1 and x^2, Eqs. (4.13) yield

$$\vartheta = 2\frac{dW}{dI_1} + cu + c_1,$$

where c_1 is a constant. Substitution of this expression and Eq. (4.12) into Eq. (4.10) implies that

$$p = 2\frac{dW}{dI_1} + c(u + x^3) + c_1. \tag{4.14}$$

Eq. (4.14) determines pressure p up to an unknown function $u(x^1, x^2)$. To find this function we substitute expression (4.14) into the latter equation (4.9) and obtain

$$2[\frac{\partial}{\partial x^1}(\frac{dW}{dI_1}\frac{\partial u}{\partial x^1}) + \frac{\partial}{\partial x^2}(\frac{dW}{dI_1}\frac{\partial u}{\partial x^2})] = c. \tag{4.15}$$

Eq. (4.15) is a nonlinear *elliptic* partial differential equation for the unknown function u.

Let us now derive boundary conditions for this equation. Since the domain occupied by the rod is infinite along the x^3 direction, boundary conditions are prescribed on the lateral surface of the cylinder only. Denote by $\bar{n}_0 = n_{x^1}\bar{e}_1 + n_{x^2}\bar{e}_2$ the unit normal to the lateral surface in the initial configuration. It follows from Eq. (4.8) that

$$\bar{n}_0 \cdot \hat{P} = n_{x^1}(2\frac{dW}{dI_1} - p)\bar{e}_1 + n_{x^2}(2\frac{dW}{dI_1} - p)\bar{e}_2 + 2\frac{dW}{dI_1}(n_{x^1}\frac{\partial u}{\partial x^1} + n_{x^2}\frac{\partial u}{\partial x^2})\bar{e}_3. \tag{4.16}$$

163

For dead forces, see Eq. (3.3.6), boundary condition (2.2.42) may be written as

$$\bar{n}_0 \cdot \hat{P} \,|_{\partial S} = \bar{b}_0, \tag{4.17}$$

where $\bar{b}_0 = b(x^1, x^2)\bar{e}_3$ for anti-plane deformation. Substitution of expression (4.16) into Eq. (4.17) implies that for any $(x^1, x^2) \in \partial S$

$$p = 2\frac{dW}{dI_1}, \tag{4.18}$$

$$2\frac{dW}{dI_1}\left(n_{x^1}\frac{\partial u}{\partial x^1} + n_{x^2}\frac{\partial u}{\partial x^2}\right) = b. \tag{4.19}$$

It follows from Eqs. (4.14) and (4.18) that $c = c_1 = 0$. Finally, by substituting this expression into Eq. (4.15) we obtain the nonlinear partial differential equation

$$\frac{\partial}{\partial x^1}\left(\frac{dW}{dI_1}\frac{\partial u}{\partial x^1}\right) + \frac{\partial}{\partial x^2}\left(\frac{dW}{dI_1}\frac{\partial u}{\partial x^2}\right) = 0 \tag{4.20}$$

with boundary condition (4.19).

For example, for the neo-Hookean material (3.2.103), Eqs. (4.19) and (4.20) are transformed into *the Laplace equation*

$$\frac{\partial^2 u}{\partial (x^1)^2} + \frac{\partial^2 u}{\partial (x^1)^2} = 0 \tag{4.21}$$

with *the von-Neumann boundary condition*

$$\frac{\partial u}{\partial n_0}\,|_{\partial S} = \frac{b}{\mu}. \tag{4.22}$$

Any solution of Eq. (4.21) is *a harmonic function*. That is why neo-Hookean material is called *harmonic* as well.

In the general case, we rewrite Eq. (4.20) with the use of Eq. (4.7) as

$$\bar{\nabla}' \cdot [\phi(|\bar{\nabla}'u|)\bar{\nabla}'u] = 0, \tag{4.23}$$

where $\bar{\nabla}'$ is a "plane" Hamilton operator

$$\bar{\nabla}'u = \bar{e}_1\frac{\partial u}{\partial x^1} + \bar{e}_2\frac{\partial u}{\partial x^2},$$

and

$$\phi(|\bar{\nabla}'u|) = 2\frac{dW}{dI_1}\Big|_{I_1=3+|\bar{\nabla}'u|^2}.$$

Exercise 4.5. Show that for the power-law medium (3.2.115)

$$\phi(|\bar{\nabla}u|) = \mu(1 + \frac{b}{n}|\bar{\nabla}u|^2)^{n-1}. \quad \square \qquad (4.24)$$

The long sides of the rectangular $\mathcal{S}$ are assumed to be stress-free, while the side $x^1 = 0$ is subjected to a prescribed surface traction $b(x^2)$ with an average shear load

$$\tilde{b} = \frac{1}{h}\int_0^h b(x^2)dx^2, \qquad (4.25)$$

and the side $x^1 = l$ is subjected to the uniformly distributed shear traction $\tilde{b}$ with the opposite direction. These conditions together with Eq. (4.19) imply

$$\frac{\partial u}{\partial x^2}\mid_{x^2=0} = 0, \qquad \frac{\partial u}{\partial x^2}\mid_{x^2=h} = 0,$$

$$\phi(|\bar{\nabla}'u|)\frac{\partial u}{\partial x^1}\mid_{x^1=0} = b(x^2), \qquad \phi(|\bar{\nabla}'u|)\frac{\partial u}{\partial x^1}\mid_{x^1=l} = \tilde{b}. \qquad (4.26)$$

Below we will discuss specific features of boundary problem (4.23) and (4.26). For simplicity, the prime will be omitted.

4.2. The Saint-Venant Principle

Let us denote by v a solution to boundary problem (4.23) and (4.26) for a constant b, i.e. when $b = \tilde{b}$.

Exercise 4.6. Check that function v has the form $v(x^1) = \kappa x^1$, and, therefore,

$$|\bar{\nabla}v| = |\kappa|, \qquad (4.27)$$

where constant κ satisfies the algebraic equation

$$\phi(|\kappa|)\kappa = \tilde{b}. \quad \square \qquad (4.28)$$

We assume that Eq. (4.28) has a unique solution.

The Saint-Venant principle states that the effect of the shape of function b weakens with the growth of distance from the side $x^1 = 0$. To estimate this effect, various norms may be used. We employ *the energy norm* and calculate the difference between solutions u and v as

$$E(z) = \int_{S_z} \phi(|\bar{\nabla}u|)|\bar{\nabla}w|^2 dx^1 dx^2, \qquad (4.29)$$

where

$$w(x^1, x^2) = u(x^1, x^2) - v(x^1) \qquad (4.30)$$

and $S_z = \{z \leq x^1 \leq l,\ 0 \leq x^2 \leq h\}$. Another approach based on point-wise estimations of the difference $u - v$ was suggested by Horgan & Knowles (1981).

Let us impose the following restrictions on function $\phi(y)$:

(i) there exists a positive constant ϕ_0 such that for any $y \geq 0$

$$\phi(y) \geq \phi_0; \tag{4.31}$$

(ii) there exists a positive constant $C < \kappa^{-1}$ such that for any $y_1,\ y_2 \geq 0$:

$$|\ \phi(y_2) - \phi(y_1)\ | \leq C|y_2 - y_1|\phi(y_1). \tag{4.32}$$

Exercise 4.7. Check that the power-law material (3.2.115) with $n \geq 1$ satisfies conditions (4.31) and (4.32). $\square$

Proposition 4.1 (Saint-Venant's principle). Let conditions (4.31) and (4.32) be fulfilled. Then there exist positive constants N and ν such that

$$E(z) \leq N \exp(-\nu z). \quad \square \tag{4.33}$$

Since $E(z) \geq \phi_0 \int_{S_z} |\bar{\nabla}(u-v)|^2 dx^1 dx^2$, inequality (4.33) means that the energy norm of the difference between two solutions corresponding to two different distributions of surface loads decays exponentially with the growth of distance from the boundary.

4.3. The Cauchy and Wirtinger Inequalities

The proof of the Saint-Venant principle is rather cumbersome due to a number of inequalities similar to each other. All of them are based on the following two assertions.

Proposition 4.2 (The Cauchy inequality). For any two sufficiently smooth vector functions $\bar{f}_1$ and $\bar{f}_2$ and for any bounded connected domain Ω with a smooth boundary Γ

$$|\int_\Omega \bar{f}_1 \cdot \bar{f}_2 dV| \leq (\int_\Omega \bar{f}_1 \cdot \bar{f}_1 dV)^{\frac{1}{2}}(\int_\Omega \bar{f}_2 \cdot \bar{f}_2 dV)^{\frac{1}{2}},$$

$$|\int_\Gamma \bar{f}_1 \cdot \bar{f}_2 dS| \leq (\int_\Gamma \bar{f}_1 \cdot \bar{f}_1 dS)^{\frac{1}{2}}(\int_\Gamma \bar{f}_2 \cdot \bar{f}_2 dS)^{\frac{1}{2}}, \tag{4.34}$$

where dV is the volume element, and dS is the surface element. $\square$

Proposition 4.3 (The Wirtinger inequality). For any function f of several real variables with the zero mean value, the integral of the square of this function is less than or equal to the integral of the square of the gradient of function f with a coefficient c depending on the domain of integration only

$$\int_\Omega f^2 dV \leq c \int_\Omega |\bar{\nabla}f|^2 dV. \quad \square \tag{4.35}$$

Exercise 4.8. Show that the Wirtinger inequality follows from the Cauchy inequality. □

Proof. We provide a sketch of proof of the Wirtinger inequality for a function $f(t)$ of one variable and a finite interval $\Omega = [0, T]$. A detailed proof is left to the reader as an exercise.

Our objective is to show that for any smooth function $f(t)$

$$\int_0^T f^2(t)dt \leq \frac{T^2}{4\pi^2} \int_0^T (\frac{df}{dt}(t))^2 dt \qquad (4.36)$$

provided $\int_0^T f(t)dt = 0$.

To derive Eq. (4.36), it suffices to expand function $f(t)$ into *the Fourier series*

$$f(t) = \frac{1}{2}a_0 + \sum_{n=1}^{\infty}(a_n \cos \frac{2\pi nt}{T} + b_n \sin \frac{2\pi nt}{T}),$$

$$\frac{df}{dt}(t) = \frac{2\pi}{T} \sum_{n=1}^{\infty} n(-a_n \sin \frac{2\pi nt}{T} + b_n \cos \frac{2\pi nt}{T})$$

with $a_0 = 0$ and to employ *the Parseval theorem*. □

Exercise 4.9. Derive the Wirtinger-type inequalities

(i) for any continuously differentiable function $f(t)$ with $f(0) = 0$

$$\int_0^T f^2(t)dt \leq \frac{4T^2}{\pi^2} \int_0^T (\frac{df}{dt}(t))^2 dt,$$

(ii) for any continuously differentiable function $f(t)$ with $f(0) = f(T) = 0$

$$\int_0^T f^2(t)dt \leq \frac{T^2}{\pi^2} \int_0^T (\frac{df}{dt}(t))^2 dt.$$

4.4. Proof of the Saint-Venant Principle

The proof is divided into nine steps.

Step 1. Let us transform expression (4.29) as follows:

$$E(z) = \int_{S_z} \phi(|\bar{\nabla}u|)\bar{\nabla}w \cdot \bar{\nabla}w dS$$

$$= \int_{S_z} \{\bar{\nabla} \cdot [\phi(|\bar{\nabla}u|)w\bar{\nabla}w] - w\bar{\nabla} \cdot [\phi(|\bar{\nabla}u|)\bar{\nabla}w]\}dS,$$

where $dS = dx^1 dx^2$. With the use of Eq. (4.30) this equality can be written as

$$E(z) = \int_{S_z} \{\bar{\nabla} \cdot [\phi(|\bar{\nabla}u|)w\bar{\nabla}w]$$

$$- w\bar{\nabla} \cdot [\phi(|\bar{\nabla}u|)\bar{\nabla}u] + w\bar{\nabla} \cdot [\phi(|\bar{\nabla}u|)\bar{\nabla}v]\}dS.$$

It follows from Eq. (4.23) that $\bar{\nabla} \cdot [\phi(|\bar{\nabla}u|)\bar{\nabla}u] = 0$ and $\bar{\nabla} \cdot [\phi(|\bar{\nabla}v|)\bar{\nabla}v] = 0$. Therefore, the second term under the integral sign can be replaced by $w\bar{\nabla} \cdot [\phi(|\bar{\nabla}v|)\bar{\nabla}v]$. As a result, we find

$$E(z) = \int_{S_z} \{\bar{\nabla} \cdot [\phi(|\bar{\nabla}u|)w\bar{\nabla}w] + w\bar{\nabla} \cdot [(\phi(|\bar{\nabla}u|) - \phi(|\bar{\nabla}v|))\bar{\nabla}v]\}dS.$$

Introduce w under the gradient sign in the second term in the right-hand side of this formula

$$E(z) = \int_{S_z} \{\bar{\nabla} \cdot [\phi(|\bar{\nabla}u|)w\bar{\nabla}w] + \bar{\nabla} \cdot [(\phi(|\bar{\nabla}u|) - \phi(|\bar{\nabla}v|))w\bar{\nabla}v]$$
$$-[\phi(|\bar{\nabla}u|) - \phi(|\bar{\nabla}v|)]\bar{\nabla}w \cdot \bar{\nabla}v\}dS.$$

Transformation of the first two terms with the use of (4.30) yields

$$E(z) = \int_{S_z} \bar{\nabla} \cdot [(\phi(|\bar{\nabla}u|)\bar{\nabla}u - \phi(|\bar{\nabla}v|)\bar{\nabla}v)w]dS + J_1(z),$$

where

$$J_1(z) = \int_{S_z} [\phi(|\bar{\nabla}v|) - \phi(|\bar{\nabla}u|)]\bar{\nabla}w \cdot \bar{\nabla}vdS. \qquad (4.37)$$

Finally, we employ the Stokes formula and boundary conditions (4.26)

$$E(z) = \int_{\partial S_z} \bar{n} \cdot [(\phi(|\bar{\nabla}u|)\bar{\nabla}u - \phi(|\bar{\nabla}v|)\bar{\nabla}v)w]ds + J_1(z)$$
$$= J_1(z) + J_2(z). \qquad (4.38)$$

Here ds is the arc element and

$$J_2(z) = \int_0^h [\phi(|\bar{\nabla}v|)\frac{\partial v}{\partial x^1} - \phi(|\bar{\nabla}u|)\frac{\partial u}{\partial x^1}]w \mid_{x^1=z} dx^2. \qquad (4.39)$$

Step 2. Let us estimate functional $J_1(z)$ by employing Eqs. (4.30) and (4.32)

$$|J_1(z)| \leq C \int_{S_z} |\bar{\nabla}u - \bar{\nabla}v|\phi(|\bar{\nabla}u|)|\bar{\nabla}w||\bar{\nabla}v|dS$$
$$= C \int_{S_z} |\bar{\nabla}v|\phi(|\bar{\nabla}u|)|\bar{\nabla}w|^2dS.$$

This inequality together with Eqs. (4.27) and (4.29) implies that

$$|J_1(z)| \leq C\kappa E(z). \qquad (4.40)$$

Substitution of (4.40) into Eq. (4.38) yields

$$(1 - C\kappa)E(z) \leq J_2(z). \qquad (4.41)$$

Step 3. We now transform the functional $J_2(z)$. For this purpose we introduce the average quantities

$$\phi^\circ(z) = \int_0^h \phi(|\bar\nabla u|)|_{x^1=z} dx^2, \quad w^\circ(z) = \frac{1}{\phi^\circ(z)} \int_0^h [\phi(|\bar\nabla u|)w]_{x^1=z} dx^2. \quad (4.42)$$

Let us integrate Eq. (4.23) over $\mathcal{S}_z$. By using the Stokes formula, boundary conditions (4.26) and Eq. (4.25) we find

$$0 = \int_{\mathcal{S}_z} \bar\nabla \cdot [\phi(|\bar\nabla u|)\bar\nabla u] dS = \int_{\partial \mathcal{S}_z} \bar n \cdot [\phi(|\bar\nabla u|)\bar\nabla u] ds$$

$$= \int_0^h \phi(|\bar\nabla u|)\frac{\partial u}{\partial x^1} |_{x^1=l} dx^2 - \int_0^h \phi(|\bar\nabla u|)\frac{\partial u}{\partial x^1} |_{x^1=z} dx^2$$

$$= \tilde b h - \int_0^h \phi(|\bar\nabla u|)\frac{\partial u}{\partial x^1} |_{x^1=z} dx^2.$$

Therefore,

$$\int_0^h \phi(|\bar\nabla u|)\frac{\partial u}{\partial x^1} |_{x^1=z} dx^2 = \tilde b h. \quad (4.43)$$

By using similar reasoning, we obtain

$$\int_0^h \phi(|\bar\nabla v|)\frac{\partial v}{\partial x^1} |_{x^1=z} dx^2 = \tilde b h. \quad (4.44)$$

Eqs. (4.43) and (4.44) imply that

$$\int_0^h [\phi(|\bar\nabla v|)\frac{\partial v}{\partial x^1} - \phi(|\bar\nabla u|)\frac{\partial u}{\partial x^1}]_{x^1=z} dx^2 = 0. \quad (4.45)$$

It follows from Eqs. (4.39) and (4.45) that

$$J_2(z) = \int_0^h [\phi(|\bar\nabla v|)\frac{\partial v}{\partial x^1} - \phi(|\bar\nabla u|)\frac{\partial u}{\partial x^1}](w - w^\circ) |_{x^1=z} dx^2$$

$$+ w^\circ \int_0^h [\phi(|\bar\nabla v|)\frac{\partial v}{\partial x^1} - \phi(|\bar\nabla u|)\frac{\partial u}{\partial x^1}]_{x^1=z} dx^2$$

$$= \int_0^h [\phi(|\bar\nabla v|)\frac{\partial v}{\partial x^1} - \phi(|\bar\nabla u|)\frac{\partial u}{\partial x^1}]\tilde w |_{x^1=z} dx^2, \quad (4.46)$$

where $\tilde w = w - w^\circ$. By replacing u by $v + w$ in Eq. (4.46), we obtain

$$J_2(z) = \int_0^h [\phi(|\bar\nabla v|)\frac{\partial v}{\partial x^1} - \phi(|\bar\nabla u|)\frac{\partial v}{\partial x^1} - \phi(|\bar\nabla u|)\frac{\partial w}{\partial x^1}]\tilde w |_{x^1=z} dx^2$$

$$= J_3(z) + J_4(z), \quad (4.47)$$

where

$$J_3(z) = -\int_0^h \phi(|\bar\nabla u|)\frac{\partial w}{\partial x^1}\tilde w |_{x^1=z} dx^2,$$

$$J_4(z) = \int_0^h [\phi(|\bar\nabla v|) - \phi(|\bar\nabla u|)]\frac{\partial v}{\partial x^1}\tilde w |_{x^1=z} dx^2. \quad (4.48)$$

Step 4. Let us estimate integrals $J_3(z)$ and $J_4(z)$ by employing the Cauchy inequality, expression (4.27) and assumption (4.32)

$$|J_3(z)| \leq [\int_0^h \phi(\bar{\nabla}u)(\frac{\partial w}{\partial x^1})^2 \mid_{x^1=z} dx^2]^{\frac{1}{2}} [\int_0^h \phi(\bar{\nabla}u)\tilde{w}^2 \mid_{x^1=z} dx^2]^{\frac{1}{2}}$$

$$\leq K(z)[\int_0^h \phi(\bar{\nabla}u)|\bar{\nabla}w|^2 \mid_{x^1=z} dx^2]^{\frac{1}{2}},$$

$$|J_4(z)| \leq \int_0^h |\phi(\bar{\nabla}v) - \phi(\bar{\nabla}u)||\frac{\partial v}{\partial x^1}||\tilde{w}| \mid_{x^1=z} dx^2$$

$$\leq C\kappa \int_0^h \phi(\bar{\nabla}u)|\bar{\nabla}v - \bar{\nabla}u||\tilde{w}| \mid_{x^1=z} dx^2$$

$$= C\kappa \int_0^h \phi(\bar{\nabla}u)|\bar{\nabla}w||\tilde{w}| \mid_{x^1=z} dx^2$$

$$\leq C\kappa K(z)[\int_0^h \phi(\bar{\nabla}u)|\bar{\nabla}w|^2 \mid_{x^1=z} dx^2]^{\frac{1}{2}}, \qquad (4.49)$$

where

$$K^2(z) = \int_0^h \phi(\bar{\nabla}u)\tilde{w}^2 \mid_{x^1=z} dx^2. \qquad (4.50)$$

Substitution of expression (4.48) into Eq. (4.47) with the use of (4.49) yields

$$|J_2(z)| \leq (1 + C\kappa)K(z)[\int_0^h \phi(\bar{\nabla}u)|\bar{\nabla}w|^2 \mid_{x^1=z} dx^2]^{\frac{1}{2}}. \qquad (4.51)$$

Step 5. Our objective now is to estimate $K(z)$. For this purpose, we use Eq. (4.42) and find

$$\int_0^h \phi(\bar{\nabla}u)\tilde{w} \mid_{x^1=z} dx^2 = 0.$$

By introducing the new variable

$$y(z) = \int_0^{x^2} \phi(\bar{\nabla}u) \mid_{x^1=z} d\eta,$$

we rewrite this equality as follows:

$$\int_0^{\phi^\circ(z)} \tilde{w} \mid_{x^1=z} dy = 0. \qquad (4.52)$$

Eq. (4.52) implies that the mean value of function $\tilde{w}$ equals zero and the Wirtinger inequality can be applied. As a result, we find

$$K^2(z) \leq (\frac{\phi^\circ(z)}{2\pi})^2 \int_0^{\phi^\circ(z)} (\frac{\partial \tilde{w}}{\partial y})^2 \mid_{x^1=z} dy$$

$$= (\frac{\phi^\circ(z)}{2\pi})^2 \int_0^{\phi^\circ(z)} (\frac{\partial \tilde{w}}{\partial x^2})^2 (\frac{dy}{dx^2})^{-2} \mid_{x^1=z} dy$$

$$= (\frac{\phi^\circ(z)}{2\pi})^2 \int_0^h \frac{1}{\phi(\bar{\nabla}u)} (\frac{\partial \tilde{w}}{\partial x^2})^2 \mid_{x^1=z} dx^2. \qquad (4.53)$$

The expressions under the integral sign are transformed with the use of assumption (4.31) as follows:

$$\frac{1}{\phi(\bar{\nabla}u)} \le \frac{\phi(\bar{\nabla}u)}{\phi_0^2}, \qquad (\frac{\partial\tilde{w}}{\partial x^2})^2 = (\frac{\partial w}{\partial x^2})^2 \le |\bar{\nabla}w|^2.$$

Substitution of these expressions into (4.53) yields

$$K(z) \le \frac{\phi^\diamond(z)}{2\pi\phi_0}[\int_0^h \phi(\bar{\nabla}u)|\bar{\nabla}w|^2 \,|_{x^1=z} \, dx^2]^{\frac{1}{2}}. \tag{4.54}$$

Step 6. We now derive a differential inequality for the function $E(z)$. First, we substitute expression (4.54) into (4.51)

$$|J_2(z)| \le \frac{(1+C\kappa)\phi^\diamond(z)}{2\pi\phi_0} \int_0^h \phi(\bar{\nabla}u)|\bar{\nabla}w|^2 \,|_{x^1=z} \, dx^2.$$

Afterward, the expression obtained is substituted into inequality (4.41)

$$\int_0^h \phi(\bar{\nabla}u)|\bar{\nabla}w|^2 \,|_{x^1=z} \, dx^2 \ge \frac{2\pi\phi_0(1-C\kappa)}{\psi^\diamond(z)(1+C\kappa)}E(z). \tag{4.55}$$

It follows from (4.29) that

$$E(z) = \int_z^l d\zeta \int_0^h \phi(\bar{\nabla}u)|\bar{\nabla}w|^2 \,|_{x^1=\zeta} \, dx^2.$$

Differentiation of this equality with respect to z yields

$$\frac{dE}{dz}(z) = -\int_0^h \phi(\bar{\nabla}u)|\bar{\nabla}w|^2 \,|_{x^1=z} \, dx^2. \tag{4.56}$$

Eqs. (4.55) and (4.56) imply that

$$\frac{dE}{dz}(z) \le -\frac{\alpha}{\phi^\diamond(z)}E(z) \tag{4.57}$$

with $\alpha = 2\pi\phi_0(1-C\kappa)(1+C\kappa)^{-1}$.

Step 7. We integrate inequality (4.57) and derive an exponential estimate for the function $E(z)$. Integration of Eq. (4.57) with the initial condition $E(0) = E_0$ implies that

$$\ln\frac{E(z)}{E_0} \le -\alpha\int_0^z \frac{d\zeta}{\phi^\diamond(\zeta)}.$$

Therefore,

$$E(z) \le E_0\exp[-\alpha\int_0^z \frac{d\zeta}{\phi^\diamond(\zeta)}]. \tag{4.58}$$

171

Applying the Cauchy inequality we find

$$z^2 = \left(\int_0^z d\zeta \right)^2 = \left(\int_0^z \sqrt{\phi^\circ(\zeta)} \frac{1}{\sqrt{\phi^\circ(\zeta)}} d\zeta \right)^2 \leq \int_0^z \phi^\circ(\zeta) d\zeta \int_0^z \frac{d\zeta}{\phi^\circ(\zeta)}.$$

Thus,

$$\int_0^z \frac{d\zeta}{\phi^\circ(\zeta)} \geq \frac{z^2}{\int_0^z \phi^\circ(\zeta) d\zeta}.$$

This inequality together with Eq. (4.58) implies that

$$E(z) \leq E_0 \exp[-\frac{\alpha z^2}{\Phi(z)}], \tag{4.59}$$

where, according to Eq. (4.42),

$$\Phi(z) = \int_0^z \phi^\circ(\zeta) d\zeta = \int_0^z \int_0^h \phi(\bar{\nabla} u) dx^1 dx^2. \tag{4.60}$$

Step 8. We now estimate function $\Phi(z)$ from above. First, we employ notation (4.30) and assumption (4.32), which lead to the inequality

$$\begin{aligned}
\phi(|\bar{\nabla} u|) &= \phi(|\bar{\nabla} v|) + \phi(|\bar{\nabla} u|) - \phi(|\bar{\nabla} v|) \\
&\leq \phi(|\bar{\nabla} v|) + |\phi(|\bar{\nabla} u|) - \phi(|\bar{\nabla} v|)| \\
&\leq \phi(|\bar{\nabla} v|) + C|\bar{\nabla} u - \bar{\nabla} v|\phi(|\bar{\nabla} v|) \\
&= \phi(|\kappa|)[1 + C|\bar{\nabla} w|].
\end{aligned} \tag{4.61}$$

Eqs. (4.60) and (4.61) imply that

$$\Phi(z) \leq \phi(|\kappa|)[hz + C \int_0^z \int_0^h |\bar{\nabla} w| dx^1 dx^2]. \tag{4.62}$$

To estimate the integral we use the Cauchy inequality together with Eqs. (4.29) and (4.31)

$$\begin{aligned}
\int_0^z \int_0^h |\bar{\nabla} w| dx^1 dx^2 &\leq [\int_0^z \int_0^h dx^1 dx^2]^{\frac{1}{2}} [\int_0^z \int_0^h |\bar{\nabla} w|^2 dx^1 dx^2]^{\frac{1}{2}} \\
&\leq \sqrt{hz} [\int_0^z \int_0^h \frac{\phi(|\bar{\nabla} u|)}{\phi_0} |\bar{\nabla} w|^2 dx^1 dx^2]^{\frac{1}{2}} \leq \sqrt{\frac{hz[E_0 - E(z)]}{\phi_0}} \\
&\leq \sqrt{\frac{hzE_0}{\phi_0}} \leq hz + \frac{E_0}{4\phi_0}.
\end{aligned}$$

This inequality and Eq. (4.62) yield

$$\Phi(z) \leq \phi(|\kappa|)[(1 + C)hz + \frac{CE_0}{4\phi_0}]. \tag{4.63}$$

172

Step 9. To complete the proof, it suffices to substitute expression (4.63) into Eq. (4.59)

$$E(z) \le E_0 \exp\{-\frac{4\alpha\phi_0 z^2}{\phi(|\kappa|)[4(1+C)\phi_0 hz + CE_0]}\}. \tag{4.64}$$

For any positive x we have $1 - x^2 \le 1$. Since $1 - x^2 = (1-x)(1+x)$, the latter inequality implies that $(1+x)^{-1} \ge 1 - x$. Therefore, for any positive a and b

$$\frac{x^2}{ax+b} = \frac{x}{a}(1 + \frac{b}{ax})^{-1} \ge \frac{x}{a}(1 - \frac{b}{ax}) = \frac{x}{a} - \frac{b}{a^2}.$$

Applying this inequality to Eq. (4.64) we obtain Eq. (4.33) with

$$N = E_0 \exp\frac{\alpha C E_0}{4\phi_0 \phi(|\kappa|)(1+C)^2 h^2}, \qquad \nu = \frac{\alpha}{(1+C)h\phi(|\kappa|)}.$$

The proof is complete. $\square$

4.5. Concluding Remarks

In this Section we formulate and prove the Saint-Venant principle, which allows complicated external loads to be replaced by simple loads with the same resultant force and resultant moment. This principle serves as the main tool in applied problems when it is necessary to derive simple explicit solutions to engineering problems.

5. One-dimensional Motions of a Hyperelastic Medium

This Section is concerned with elongation and shear motions of a hyperelastic medium. In subsections 1 and 2 we derive nonlinear hyperbolic equations describing these motions. Subsections 3 and 4 deal with characteristic features of quasi-linear partial differential equations. We introduce the characteristic surfaces and the Riemann invariants, and derive nonlinear differential equations for the derivatives of the Riemann invariants along characteristics. In subsection 5, two assertions are proved regarding the intervals of existence for solutions to quadratic differential equations. Finally, in subsection 6, we estimate the critical time before blow-up of solutions to quasi-linear systems, and apply the obtained formula to elongation and shear motions. The exposition of subsections 4 and 5 follows Lax (1964).

5.1. Elongation Motion

Let us consider a viscoelastic medium which occupies the whole space. Denote by $\{X^i\}$ and $\{x^i\}$ Cartesian coordinates in the initial and actual configurations, and by $\bar{e}_i$ their unit vectors. We are concerned with *elongation*

173

motions which are described by the equations

$$x^1 = X^1 + u(t, X^1), \qquad x^2 = X^2, \qquad x^3 = X^3, \tag{5.1}$$

where $u(t, X^1)$ is a function to be found. The radius-vectors of a point with Cartesian coordinates $\{X^i\}$ in the initial and actual configurations equal

$$\bar{r}_0 = X^1 \bar{e}_1 + X^2 \bar{e}_2 + X^3 \bar{e}_3, \qquad \bar{r} = [X^1 + u(t, X^1)]\bar{e}_1 + X^2 \bar{e}_2 + X^3 \bar{e}_3. \tag{5.2}$$

Differentiating equalities (5.2) we obtain tangent vectors

$$\bar{g}_{0\,1} = \bar{e}_1, \qquad \bar{g}_{0\,2} = \bar{e}_2, \qquad \bar{g}_{0\,3} = \bar{e}_3,$$
$$\bar{g}_1 = (1 + \kappa)\bar{e}_1, \qquad \bar{g}_2 = \bar{e}_2, \qquad \bar{g}_3 = \bar{e}_3, \tag{5.3}$$

where

$$\kappa(t, X^1) = \frac{\partial u}{\partial X^1}(t, X^1). \tag{5.4}$$

Exercise 5.1. Derive the following formulas for dual vectors:

$$\bar{g}_0^1 = \bar{e}_1, \qquad \bar{g}_0^2 = \bar{e}_2, \qquad \bar{g}_0^3 = \bar{e}_3,$$
$$\bar{g}^1 = \frac{1}{1 + \kappa}\bar{e}_1, \qquad \bar{g}^2 = \bar{e}_2, \qquad \bar{g}^3 = \bar{e}_3. \quad \Box \tag{5.5}$$

Substitution of Eqs. (5.3) and (5.5) into Eq. (2.1.5) implies that

$$\bar{\nabla}_0 \bar{r} = (1 + \kappa)\bar{e}_1 \bar{e}_1 + \bar{e}_2 \bar{e}_2 + \bar{e}_3 \bar{e}_3. \tag{5.6}$$

It follows from Eq. (5.6) and formula (2.1.25) that

$$\hat{F} = (1 + \kappa)^2 \bar{e}_1 \bar{e}_1 + \bar{e}_2 \bar{e}_2 + \bar{e}_3 \bar{e}_3,$$
$$\hat{F}^2 = (1 + \kappa)^4 \bar{e}_1 \bar{e}_1 + \bar{e}_2 \bar{e}_2 + \bar{e}_3 \bar{e}_3. \tag{5.7}$$

Let us calculate the characteristic polynomial for tensor $\hat{F}$

$$\det(\hat{F} - \lambda \hat{I}) = \det \begin{bmatrix} (1+\kappa)^2 - \lambda & 0 & 0 \\ 0 & 1 - \lambda & 0 \\ 0 & 0 & 1 - \lambda \end{bmatrix}$$
$$= -\lambda^3 + [2 + (1 + \kappa)^2]\lambda^2 - [1 + 2(1 + \kappa)^2]\lambda + (1 + \kappa)^2,$$

where $\hat{I}$ is the unit tensor. Since the principal invariants coincide (up to their signs) with coefficients of this polynomial, we find

$$I_1(\hat{F}) = 2 + (1 + \kappa)^2, \qquad I_2(\hat{F}) = 1 + 2(1 + \kappa)^2, \qquad I_3(\hat{F}) = (1 + \kappa)^2. \tag{5.8}$$

174

Substitution of expressions (5.7) and (5.8) into Eqs. (3.2.18) and (3.2.19) implies that

$$\hat{\sigma} = \sigma_{11}\bar{e}_1\bar{e}_1 + \sigma_{22}\bar{e}_2\bar{e}_2 + \sigma_{33}\bar{e}_3\bar{e}_3. \tag{5.9}$$

Here

$$\sigma_{11} = 2(1+\kappa)\Psi_0, \qquad \sigma_{22} = \sigma_{33} = \frac{2}{1+\kappa}[\Psi_1 + (1+\kappa)^2\Psi_2], \tag{5.10}$$

where

$$\Psi_0 = (\frac{\partial}{\partial I_1} + 2\frac{\partial}{\partial I_2} + \frac{\partial}{\partial I_3})W, \qquad \Psi_1 = (\frac{\partial}{\partial I_1} + \frac{\partial}{\partial I_2})W,$$

$$\Psi_2 = (\frac{\partial}{\partial I_2} + \frac{\partial}{\partial I_3})W. \tag{5.11}$$

The displacement vector equals $\bar{u} = u\bar{e}_1$. Differentiation of this expression yields the acceleration vector

$$\bar{a} = \frac{\partial^2 u}{\partial t^2}\bar{e}_1. \tag{5.12}$$

Substituting expression (5.8) into mass conservation law (2.2.10), we obtain

$$\rho = \frac{\rho_0}{1+\kappa}, \tag{5.13}$$

where ρ_0 and ρ are mass densities in the initial and actual configurations.

Exercise 5.2. By using Eqs. (5.5) and (5.9), show that

$$\bar{\nabla} \cdot \hat{\sigma} = \frac{1}{1+\kappa}\frac{\partial \sigma_{11}}{\partial X^1}\bar{e}_1. \qquad \square \tag{5.14}$$

External body and surface forces are assumed to vanish. Substitution of expressions (5.12) – (5.14) into the motion equation $\rho\bar{a} = \bar{\nabla} \cdot \hat{\sigma}$ yields

$$\rho_0\frac{\partial^2 u}{\partial t^2} = 2\frac{\partial}{\partial X^1}[(1+\kappa)\Psi_0]. \tag{5.15}$$

It follows from Eq. (5.8) that function W depends on κ only:

$$W(I_1, I_2, I_3) = W(2 + (1+\kappa)^2, 1 + 2(1+\kappa)^2, (1+\kappa)^2) = \mathcal{W}(\kappa). \tag{5.16}$$

Exercise 5.3. Prove that

$$\mathcal{W}' = 2(1+\kappa)\Psi_0, \tag{5.17}$$

where the prime denotes the differentiation. $\square$

175

Substitution of expression (5.17) into Eq. (5.15) implies that

$$\frac{\partial^2 u}{\partial t^2} = \frac{\partial}{\partial x} H\left(\frac{\partial u}{\partial x}\right), \tag{5.18}$$

where

$$H(\kappa) = \frac{1}{\rho_0} W'(\kappa), \tag{5.19}$$

and we set $x = X^1$ for simplicity.

Exercise 5.4. Derive the following equalities:

$$\begin{aligned} \mathcal{W}'' &= 2\mathbf{M}W + 2(1+\kappa)^2 \mathbf{M}^2 W, \\ \mathcal{W}''' &= 12(1+\kappa)\mathbf{M}^2 W + 8(1+\kappa)^3 \mathbf{M}^3 W, \end{aligned} \tag{5.20}$$

where operator $\mathbf{M}$ is defined by the formula

$$\mathbf{M} = \frac{\partial}{\partial I_1} + 2\frac{\partial}{\partial I_2} + \frac{\partial}{\partial I_3}. \qquad \square$$

Exercise 5.5. Check that Eq. (3.2.23) implies that $\mathbf{M}W(3,3,1) = 0$, provided the initial configuration is stress-free. $\square$

Exercise 5.6. Show that

$$\mathcal{W}''(0) = 4\mathbf{M}^2 W(3,3,1), \qquad \mathcal{W}'''(0) = 12\mathbf{M}^2 W(3,3,1) + 8\mathbf{M}^3 W(3,3,1). \qquad \square \tag{5.21}$$

We return to the analysis of Eq. (5.18) in subsection 3. Our objective now is to derive a similar governing equation for shear motions.

5.2. Shear Motion

In Section 2, quasi-static shear deformation was studied for a hyperelastic medium with a uniform angle of shear. In this subsection we generalize the above results to motions when the angle of shear is assumed to be a function of time and a spatial variable. We are concerned with motions which are described by the following equations:

$$x^1 = X^1 + u(t, X^2), \qquad x^2 = X^2, \qquad x^3 = X^3, \tag{5.22}$$

where $u(t, X^2)$ is a function to be found. The radius-vector of a point with Cartesian coordinates $\{X^i\}$ in the actual configuration equals

$$\bar{r} = [X^1 + u(t, X^2)]\bar{e}_1 + X^2 \bar{e}_2 + X^3 \bar{e}_3. \tag{5.23}$$

Differentiating Eq. (5.23) we obtain tangent vectors

$$\bar{g}_1 = \bar{e}_1, \qquad \bar{g}_2 = \kappa\bar{e}_1 + \bar{e}_2, \qquad \bar{g}_3 = \bar{e}_3, \tag{5.24}$$

where

$$\kappa(t, X^2) = \frac{\partial u}{\partial X^2}(t, X^2).$$ (5.25)

Exercise 5.7. Derive the following formulas for dual vectors:

$$\bar{g}^1 = \bar{e}_1 - \kappa\bar{e}_2, \qquad \bar{g}^2 = \bar{e}_2, \qquad \bar{g}^3 = \bar{e}_3. \qquad \Box$$ (5.26)

Substitution of Eqs. (5.5) and (5.24) into Eq. (2.1.5) implies that

$$\bar{\nabla}_0 \bar{r} = \bar{e}_1\bar{e}_1 + \bar{e}_2\bar{e}_2 + \bar{e}_3\bar{e}_3 + \kappa\bar{e}_2\bar{e}_1.$$ (5.27)

It follows from Eqs. (5.27) and (2.1.25) that

$$\begin{aligned}
\hat{F} &= (1 + \kappa^2)\bar{e}_1\bar{e}_1 + \bar{e}_2\bar{e}_2 + \bar{e}_3\bar{e}_3 + \kappa(\bar{e}_1\bar{e}_2 + \bar{e}_2\bar{e}_1), \\
\hat{F}^2 &= [(1 + \kappa^2)^2 + \kappa^2]\bar{e}_1\bar{e}_1 + (1 + \kappa^2)\bar{e}_2\bar{e}_2 + \bar{e}_3\bar{e}_3 \\
&\quad + \kappa(2 + \kappa^2)(\bar{e}_1\bar{e}_2 + \bar{e}_2\bar{e}_1).
\end{aligned}$$ (5.28)

The Finger tensor $\hat{F}$ can be presented in the matrix form

$$\hat{F} = \begin{bmatrix} 1 + \kappa^2 & \kappa & 0 \\ \kappa & 1 & 0 \\ 0 & 0 & 1 \end{bmatrix}.$$

Calculation of the characteristic polynomial for this matrix implies that

$$\det(\hat{F} - \lambda\hat{I}) = -\lambda^3 + (3 + \kappa^2)\lambda^2 - (3 + \kappa^2)\lambda + 1.$$ (5.29)

Since the principal invariants of $\hat{F}$ coincide (up to their signs) with coefficients of the characteristic polynomial, we find

$$I_1(\hat{F}) = 3 + \kappa^2, \qquad I_2(\hat{F}) = 3 + \kappa^2, \qquad I_3(\hat{F}) = 1.$$ (5.30)

Eqs. (5.30) demonstrate that any shear motion is isochoric, i.e. it preserves the volume element.

Substitution of expressions (5.28) and (5.30) into the constitutive equations (3.2.18) and (3.2.19) yields

$$\hat{\sigma} = \sigma_{11}\bar{e}_1\bar{e}_1 + \sigma_{22}\bar{e}_2\bar{e}_2 + \sigma_{33}\bar{e}_3\bar{e}_3 + \sigma_{12}(\bar{e}_1\bar{e}_2 + \bar{e}_2\bar{e}_1).$$ (5.31)

Here

$$\begin{aligned}
\sigma_{11} &= 2(\Psi_0 + \kappa^2\Psi_1), & \sigma_{22} &= 2\Psi_0, \\
\sigma_{33} &= 2(\Psi_0 + \kappa^2\frac{\partial W}{\partial I_2}), & \sigma_{12} &= 2\kappa\Psi_1.
\end{aligned}$$ (5.32)

177

Differentiation of the displacement vector $\bar{u} = u\bar{e}_1$ with respect to time implies formula (5.12). Since shear motion is isochoric, we have

$$\rho(t) = \rho_0. \tag{5.33}$$

Exercise 5.8. By using Eqs. (5.26), and (5.31), derive the formula

$$\bar{\nabla} \cdot \hat{\sigma} = \frac{\partial \sigma_{12}}{\partial X^2}\bar{e}_1 + \frac{\partial \sigma_{22}}{\partial X^2}\bar{e}_2. \quad \square \tag{5.34}$$

Substitution of expressions (5.12), (5.31), (5.33), and (5.34) into the motion equation

$$\rho\bar{a} = \bar{\nabla} \cdot \hat{\sigma} + \rho\bar{B}$$

yields

$$\rho_0\frac{\partial^2 u}{\partial t^2} = \frac{\partial \sigma_{12}}{\partial X^2} + \rho_0 B_1, \qquad 0 = \frac{\partial \sigma_{22}}{\partial X^2} + \rho_0 B_2, \qquad 0 = \rho_0 B_3. \tag{5.35}$$

Here $\bar{B} = B_1\bar{e}_1 + B_2\bar{e}_2 + B_3\bar{e}_3$ is a body force.

It follows from Eqs. (5.32) and (5.35) that for shear motion

$$B_2 = -\frac{2}{\rho_0}\frac{\partial \Psi_0}{\partial X^2}, \qquad B_3 = 0. \tag{5.36}$$

We assume additionally that component B_1 vanishes. Then the only governing equation is

$$\rho_0\frac{\partial^2 u}{\partial t^2} = 2\frac{\partial(\Psi_1\kappa)}{\partial X^2}. \tag{5.37}$$

It follows from Eq. (5.30) that strain energy density W depends on κ only:

$$W(I_1, I_2, I_3) = W(3 + \kappa^2, 3 + \kappa^2, 1) = \mathcal{W}(\kappa). \tag{5.38}$$

Exercise 5.9. Show that

$$\mathcal{W}'(\kappa) = 2\Psi_1\kappa. \quad \square \tag{5.39}$$

Substitution of expression (5.39) into Eq. (5.37) implies Eq. (5.18), where we set $x = X^2$ for simplicity. This means that elongation motion and shear motion of an elastic medium is described by the same nonlinear hyperbolic equation (5.18).

Exercise 5.10. By using Eqs. (5.30) and (5.38), show that

$$\mathcal{W}'' = 2NW + 4\kappa^2 N^2 W, \qquad \mathcal{W}''' = 12\kappa N^2 W + 8\kappa^3 N^3 W, \tag{5.40}$$

178

where operator N has the form

$$N = \frac{\partial}{\partial I_1} + \frac{\partial}{\partial I_2}. \qquad \square$$

Exercise 5.11. Check that

$$\mathcal{W}''(0) = 2NW(3,3,1), \qquad \mathcal{W}'''(0) = 0. \qquad \square \tag{5.41}$$

5.3. Transformation of the Governing Equation

The objective of this subsection is to transform Eq. (5.18) into the canonical form.

Let us calculate the derivative in the right-hand side of Eq. (5.18). As a result, we obtain

$$\frac{\partial^2 u}{\partial t^2} = H'(\frac{\partial u}{\partial x})\frac{\partial^2 u}{\partial x^2}. \tag{5.42}$$

If the derivative $\partial u/\partial x$ is sufficiently small, we can linearize Eq. (5.42) and derive *the linear wave equation*

$$\frac{\partial^2 u}{\partial t^2} = H'(0)\frac{\partial^2 u}{\partial x^2}. \tag{5.43}$$

Exercise 5.12. Check that for arbitrary smooth functions $f_1(t)$ and $f_2(t)$, the function

$$u(t,x) = f_1(x - ct) + f_2(x + ct) \tag{5.44}$$

with

$$c = \sqrt{H'(0)} \tag{5.45}$$

is a solution of Eq. (5.43). $\square$

Eq. (5.44) determines two *waves* which move in positive and negative x directions with velocity c. The constant c is called *the wave speed* for *linear waves*. Functions $f_i(t)$ determine *stationary waves* which preserve their shape in time.

Introduce the notation

$$v_1 = \frac{\partial u}{\partial t}, \qquad v_2 = \frac{\partial u}{\partial x}. \tag{5.46}$$

Substitution of expressions (5.46) into Eq. (5.18) implies that

$$\frac{\partial v_1}{\partial t} = \frac{\partial}{\partial x}H(v_2). \tag{5.47}$$

179

It follows from Eqs. (5.46) that

$$\frac{\partial v_2}{\partial t} = \frac{\partial v_1}{\partial x}. \tag{5.48}$$

Eqs. (5.47) and (5.48) are treated as a system of nonlinear partial differential equations for the description of one-dimensional waves. This system is a particular case of a system of two first-order *quasi-linear partial differential equations*

$$\frac{\partial v_1}{\partial t} + a_{11}\frac{\partial v_1}{\partial x} + a_{12}\frac{\partial v_2}{\partial x} = 0,$$
$$\frac{\partial v_2}{\partial t} + a_{21}\frac{\partial v_1}{\partial x} + a_{22}\frac{\partial v_2}{\partial x} = 0 \tag{5.49}$$

with

$$a_{11} = 0, \qquad a_{12} = -H'(v_2), \qquad a_{21} = -1, \qquad a_{22} = 0. \tag{5.50}$$

Let us calculate the characteristic polynomial for matrix

$$\hat{A} = \begin{bmatrix} a_{11} & a_{12} \\ a_{21} & a_{22} \end{bmatrix}. \tag{5.51}$$

Exercise 5.13. By using Eqs. (5.50) and (5.51), show that

$$\det(\hat{A} - \lambda\hat{I}) = \lambda^2 - H'(\kappa) = \lambda^2 - \frac{1}{\rho_0}\mathcal{W}''(\kappa). \qquad \square \tag{5.52}$$

We confine ourselves to materials which satisfy the inequality

$$\mathcal{W}''(\kappa) > 0 \tag{5.53}$$

for any $\kappa \in (-\infty, \infty)$. In this case, matrix $\hat{A}$ has real and distinct eigenvalues λ_1 and λ_2. The latter means that system (5.49) is hyperbolic. A precise definition of hyperbolicity and some features of hyperbolic quasi-linear systems are discussed in subsection 4.

Exercise 5.14. Show that for the Agarwal medium (3.2.53) with parameters μ and k, inequality (5.53) implies $\mu > 0$ for shear motion and

$$\mu[1 + \frac{k}{(1+\kappa)^{2(1+k)}}] > 0$$

for elongation motion. $\square$

5.4. Quasi-linear Hyperbolic Systems

In this subsection we discuss some properties of quasi-linear homogeneous systems in the form

$$\frac{\partial \bar{v}}{\partial t} + \hat{A}(t, x, \bar{v}) \cdot \frac{\partial \bar{v}}{\partial x} = 0. \tag{5.54}$$

180

Here $\hat{A}$ is a prescribed matrix function which is assumed to be sufficiently smooth, and $\bar{v} = \begin{bmatrix} v_1 \\ v_2 \end{bmatrix}$ is an unknown vector function.

Definition. A surface $\Phi(t, x) = 0$ is called *characteristic*, if the normal derivative of $\bar{v}$ on this surface cannot be expressed in terms of the tangential derivatives by using Eq. (5.54).

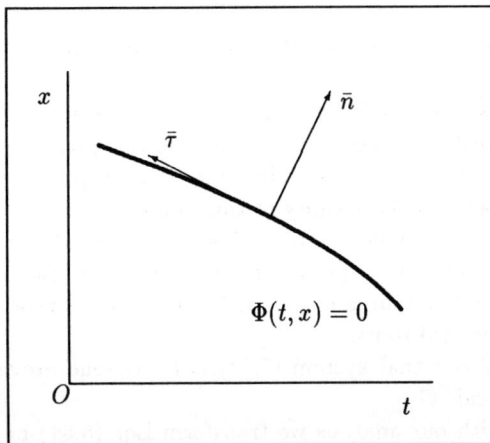

Fig. 5.1. Characteristic surface for a hyperbolic system.

According to this definition, the Cauchy problem for Eq. (5.54) cannot be solved (formally), provided the initial data are prescribed on a characteristic surface.

In order to derive an equation for the characteristic surface, we calculate the normal $\partial \bar{v}/\partial n$ and tangential $\partial \bar{v}/\partial \tau$ derivatives. By using Fig. 5.1, we obtain

$$\bar{n} = \begin{bmatrix} \frac{\partial \Phi}{\partial t} \\ \frac{\partial \Phi}{\partial x} \end{bmatrix}, \qquad \bar{\tau} = \begin{bmatrix} -\frac{\partial \Phi}{\partial x} \\ \frac{\partial \Phi}{\partial t} \end{bmatrix}. \tag{5.55}$$

Eqs. (5.54) and (5.55) imply that

$$\frac{\partial \bar{v}}{\partial n} = \frac{\partial \bar{v}}{\partial t}\frac{\partial \Phi}{\partial t} + \frac{\partial \bar{v}}{\partial x}\frac{\partial \Phi}{\partial x} = -\hat{A} \cdot \frac{\partial \bar{v}}{\partial x}\frac{\partial \Phi}{\partial t} + \frac{\partial \bar{v}}{\partial x}\frac{\partial \Phi}{\partial x}$$
$$= (\frac{\partial \Phi}{\partial x}\hat{I} - \frac{\partial \Phi}{\partial t}\hat{A}) \cdot \frac{\partial \bar{v}}{\partial x},$$

$$\frac{\partial \bar{v}}{\partial \tau} = -\frac{\partial \bar{v}}{\partial t}\frac{\partial \Phi}{\partial x} + \frac{\partial \bar{v}}{\partial x}\frac{\partial \Phi}{\partial t} = \hat{A} \cdot \frac{\partial \bar{v}}{\partial x}\frac{\partial \Phi}{\partial x} + \frac{\partial \bar{v}}{\partial x}\frac{\partial \Phi}{\partial t}$$
$$= (\frac{\partial \Phi}{\partial t}\hat{I} + \frac{\partial \Phi}{\partial x}\hat{A}) \cdot \frac{\partial \bar{v}}{\partial x}. \tag{5.56}$$

181

It follows from Eqs. (5.56) that a surface $\Phi(t, x) = 0$ is characteristic if and only if

$$\det(\frac{\partial \Phi}{\partial t} \hat{I} + \frac{\partial \Phi}{\partial x} \hat{A}) = 0. \tag{5.57}$$

By assuming $\frac{\partial \Phi}{\partial x} \neq 0$, and setting $\lambda = -\frac{\partial \Phi}{\partial t} / \frac{\partial \Phi}{\partial x}$, we can present Eq. (5.57) as

$$\det(\hat{A} - \lambda \hat{I}) = 0. \tag{5.58}$$

Systems (5.54) are classified (locally) according to the number of their characteristic surfaces.

Definition. A system (5.54) is called *hyperbolic* at a point (t, x) if it has two different characteristic surfaces in the vicinity of this point.

Definition. A system (5.54) is called *elliptic* at a point (t, x) if it has no characteristic surfaces in the vicinity of this point.

According to Eq. (5.58), system (5.54) is hyperbolic at a point (t, x) if for any $\bar{v}$ the characteristic equation for matrix $\hat{A}$ has two real and distinct eigenvalues, and this system is elliptic if for any $\bar{v}$ the characteristic equation for matrix $\hat{A}$ has no real roots.

Exercise 5.15. Prove that system (5.54) is hyperbolic provided matrix $\hat{A}$ is real and symmetrical. $\square$

To proceed with our analysis we transform Eq. (5.54) by introducing the Riemann invariants.

Let $\bar{l}^{(1)} = \begin{bmatrix} l_1^{(1)} \\ l_2^{(1)} \end{bmatrix}$ be a left eigenvector of matrix $\hat{A}$ corresponding to an eigenvalue λ_1. This means that $\bar{l}^{(1)}$ is a solution of the algebraic equation

$$\bar{l}^{(1)} \cdot \hat{A} = \lambda_1 \bar{l}^{(1)}. \tag{5.59}$$

Since $\hat{A}$ depends on t, x, and $\bar{v}$, functions λ_1 and $\bar{l}^{(1)}$ depend on the same arguments as well.

We assume that there are a function $\theta_1 \neq 0$ (an integrating factor) and a function $R_1(\bar{v})$ such that

$$\theta_1 \bar{l}^{(1)} = \frac{\partial R_1}{\partial \bar{v}}. \tag{5.60}$$

Vector equation (5.60) is equivalent to the following two scalar equations:

$$\theta_1 l_1^{(1)} = \frac{\partial R_1}{\partial v_1}, \qquad \theta_1 l_2^{(1)} = \frac{\partial R_1}{\partial v_2}. \tag{5.61}$$

We multiply the first equation (5.61) by dv_1, the other equation by dv_2 and add together. As a result, we obtain

$$dR_1 = \theta_1(l_1^{(1)} dv_1 + l_2^{(1)} dv_2). \tag{5.62}$$

182

We now differentiate the first equation (5.61) with respect to v_2, the other equation with respect to v_1 and subtract the second equality from the first. After simple algebra, we find

$$l_2^{(1)}\frac{\partial\theta_1}{\partial v_1} - l_1^{(1)}\frac{\partial\theta_1}{\partial v_2} = \theta_1\left(\frac{\partial l_1^{(1)}}{\partial v_2} - \frac{\partial l_2^{(1)}}{\partial v_1}\right). \qquad (5.63)$$

Exercise 5.16. Derive Eqs. (5.62) and (5.63). $\square$

For any fixed t and x, Eq. (5.63) is a linear partial differential equation for function θ_1. Therefore, its solution exists and can be found, e.g. by using *the method of characteristics*. Afterward, this solution θ_1 is substituted into Eq. (5.62) to find function R_1. This means that for any eigenvector $\bar{l}^{(1)}$, appropriate functions θ_1 and R_1 can be obtained (explicitly or implicitly).

Exercise 5.17. Show that for a linear system, when matrix $\hat{A}$ is independent of $\bar{v}$, we can set

$$\theta_1 = 1, \qquad R_1(\bar{v}) = \bar{l}^{(1)} \cdot \bar{v}. \qquad \square$$

Let us now multiply Eq. (5.54) by $\theta_1\bar{l}^{(1)}$ and use equality (5.59). As a result, we find

$$\theta_1\bar{l}^{(1)} \cdot \left(\frac{\partial\bar{v}}{\partial t} + \lambda_1\frac{\partial\bar{v}}{\partial x}\right) = 0.$$

This equality together with Eq. (5.60) implies that

$$\frac{\partial R_1}{\partial\bar{v}} \cdot \frac{\partial\bar{v}}{\partial t} + \lambda_1\frac{\partial R_1}{\partial\bar{v}} \cdot \frac{\partial\bar{v}}{\partial x} = 0,$$

which means that

$$\frac{\partial R_1}{\partial s_1} = 0, \qquad (5.64)$$

where

$$\frac{\partial}{\partial s_1} = \frac{\partial}{\partial t} + \lambda_1\frac{\partial}{\partial x}. \qquad (5.65)$$

Repeating the same calculations for the other eigenvalue λ_2 with the corresponding left eigenvector $\bar{l}^{(2)}$ and functions θ_2 and R_2, we obtain

$$\frac{\partial R_2}{\partial s_2} = 0, \qquad (5.66)$$

where

$$\frac{\partial}{\partial s_2} = \frac{\partial}{\partial t} + \lambda_2\frac{\partial}{\partial x}. \qquad (5.67)$$

Definition. Functions R_i which satisfy conservation laws (5.64) and (5.66) are called *the Riemann invariants*.

Let us calculate the determinant

$$|J| = \det \begin{bmatrix} \frac{\partial R_1}{\partial v_1} & \frac{\partial R_1}{\partial v_2} \\ \frac{\partial R_2}{\partial v_1} & \frac{\partial R_2}{\partial v_2} \end{bmatrix}.$$

According to Eq. (5.61) we have

$$|J| = \det \begin{bmatrix} \theta_1 l_1^{(1)} & \theta_1 l_2^{(1)} \\ \theta_2 l_1^{(2)} & \theta_2 l_2^{(2)} \end{bmatrix} = \theta_1 \theta_2 (l_1^{(1)} l_2^{(2)} - l_1^{(2)} l_2^{(1)}). \tag{5.68}$$

Functions θ_i do not vanish, and the expression in the brackets is not equal to zero. (If this expression vanishes, then vectors $\bar{l}^{(1)}$ and $\bar{l}^{(2)}$ become proportional, which contradicts the hyperbolicity condition.) Therefore, $|J| \neq 0$, and we may treat functions R_1 and R_2 as new variables (instead of v_1 and v_2) and write

$$\bar{v} = \bar{v}(t, x, R_1, R_2). \tag{5.69}$$

In particular, substituting expression (5.69) into the formula for $\lambda_i(t, x, \bar{v})$, we obtain

$$\lambda_i = \lambda_i(t, x, R_1, R_2).$$

For the further analysis, we should introduce additional hypotheses. Namely, we assume that matrix $\hat{A}$ in system (5.54) is independent of t and x. In this case, its eigenvalues are independent of these variables as well, and we can write

$$\lambda_i = \lambda_i(R_1, R_2). \tag{5.70}$$

Let us differentiate Eq. (5.64) with respect to x. By using Eqs. (5.65) and (5.70) we find

$$\frac{\partial^2 R_1}{\partial t \partial x} + \lambda_1 \frac{\partial^2 R_1}{\partial x^2} + \frac{\partial R_1}{\partial x}\left(\frac{\partial \lambda_1}{\partial R_1} \frac{\partial R_1}{\partial x} + \frac{\partial \lambda_1}{\partial R_2} \frac{\partial R_2}{\partial x} \right) = 0. \tag{5.71}$$

It follows from Eqs. (5.65) – (5.67) that

$$0 = \frac{\partial R_2}{\partial s_2} = \frac{\partial R_2}{\partial t} + \lambda_2 \frac{\partial R_2}{\partial x} = \frac{\partial R_2}{\partial t} + \lambda_1 \frac{\partial R_2}{\partial x} + (\lambda_2 - \lambda_1) \frac{\partial R_2}{\partial x}$$

$$= \frac{\partial R_2}{\partial s_1} + (\lambda_2 - \lambda_1) \frac{\partial R_2}{\partial x}.$$

Thus,

$$\frac{\partial R_2}{\partial x} = \frac{1}{\lambda_1 - \lambda_2} \frac{\partial R_2}{\partial s_1}. \tag{5.72}$$

Substitution of expression (5.72) into Eq. (5.71) with the use of Eq. (5.65) implies that

$$\frac{\partial}{\partial s_1}\left(\frac{\partial R_1}{\partial x} \right) + \frac{\partial \lambda_1}{\partial R_1}\left(\frac{\partial R_1}{\partial x} \right)^2 + \frac{1}{\lambda_1 - \lambda_2} \frac{\partial \lambda_1}{\partial R_2} \frac{\partial R_1}{\partial x} \frac{\partial R_2}{\partial s_1} = 0.$$

184

Letting $T_i = \partial R_i / \partial x$, we can rewrite this equality as follows:

$$\frac{\partial T_1}{\partial s_1} + \frac{\partial \lambda_1}{\partial R_1} T_1^2 + \frac{1}{\lambda_1 - \lambda_2} \frac{\partial \lambda_1}{\partial R_2} \frac{\partial R_2}{\partial s_1} T_1 = 0. \tag{5.73}$$

We introduce a new function $h_1(R_1, R_2)$ such that

$$\frac{\partial h_1}{\partial R_2} = \frac{1}{\lambda_1 - \lambda_2} \frac{\partial \lambda_1}{\partial R_2}. \tag{5.74}$$

It follows from Eqs. (5.64) and (5.74) that

$$\frac{\partial h_1}{\partial s_1} = \frac{\partial h_1}{\partial R_1} \frac{\partial R_1}{\partial s_1} + \frac{\partial h_1}{\partial R_2} \frac{\partial R_2}{\partial s_1} = \frac{1}{\lambda_1 - \lambda_2} \frac{\partial \lambda_1}{\partial R_2} \frac{\partial R_2}{\partial s_1}. \tag{5.75}$$

Substitution of expression (5.75) into Eq. (5.73) yields

$$\frac{\partial T_1}{\partial s_1} + \frac{\partial h_1}{\partial s_1} T_1 + \frac{\partial \lambda_1}{\partial R_1} T_1^2 = 0. \tag{5.76}$$

We multiply Eq. (5.76) by $\exp(h_1)$ and set $Z_1 = T_1 \exp(h_1)$. Using the identity

$$\frac{\partial Z_1}{\partial s_1} = \left(\frac{\partial T_1}{\partial s_1} + T_1 \frac{\partial h_1}{\partial s_1} \right) \exp(h_1),$$

we find

$$\frac{\partial Z_1}{\partial s_1} = -\frac{\partial \lambda_1}{\partial R_1} \exp(-h_1) Z_1^2. \tag{5.77}$$

Exercise 5.18. Let a function $h_2(R_1, R_2)$ satisfy the equality

$$\frac{\partial h_2}{\partial R_1} = -\frac{1}{\lambda_1 - \lambda_2} \frac{\partial \lambda_2}{\partial R_1}. \tag{5.78}$$

Check that function $Z_2 = T_2 \exp(h_2)$ is a solution of the differential equation

$$\frac{\partial Z_2}{\partial s_2} = -\frac{\partial \lambda_2}{\partial R_2} \exp(-h_2) Z_2^2. \quad \square \tag{5.79}$$

For given functions $\lambda_i(R_1, R_2)$, Eqs. (5.77) and (5.79) are nonlinear ordinary differential equations of the first order. The behavior of their solutions is discussed in the next subsection.

5.5. Some Features of Solutions of Nonlinear Differential Equations of the First Order

This subsection is concerned with estimates of solutions of the first-order differential equation with a time-depending coefficient β

$$\frac{dz}{dt} = \beta(t) z^2 \quad (0 \le t \le T), \tag{5.80}$$

subject to the initial condition

$$z(0) = z_0. \tag{5.81}$$

Proposition 5.1. Suppose that
(i) function $\beta(t)$ is continuous,
(ii) there is a $\beta_0 > 0$ such that for any $t \in [0, T]$

$$\beta(t) \geq \beta_0, \tag{5.82}$$

and (iii) $z_0 > 0$. Then

$$T \leq \frac{1}{\beta_0 z_0}. \tag{5.83}$$

Proof. Rewrite Eq. (5.80) as follows:

$$\frac{dz}{z^2} = \beta(t)dt.$$

Integration of this equation with the initial condition (5.81) implies that

$$\frac{1}{z_0} - \frac{1}{z} = \int_0^t \beta(s)ds \geq \beta_0 t. \tag{5.84}$$

Therefore,

$$z(t) \geq \frac{z_0}{1 - \beta_0 z_0 t}. \tag{5.85}$$

The right-hand side of Eq. (5.85) tends to infinity as $t \to (\beta_0 z_0)^{-1}$. Thus, the solution $z(t)$ cannot exist beyond this time. $\square$

Proposition 5.2. Suppose that function $\beta(t)$ is continuous and satisfies the inequality

$$|\beta(t)| \leq B < \infty. \tag{5.86}$$

Then problem (5.80) and (5.81) has a solution on the interval $[0, (Bz_0)^{-1}]$.

Proof. It follows from Eq. (5.84) that for any $t \in [0, (Bz_0)^{-1}]$

$$z(t) = \frac{z_0}{1 - z_0 \int_0^t \beta(s)ds} \leq \frac{z_0}{1 - Bz_0} < \infty. \quad \square$$

5.6. Discussion of the Obtained Results

Let us apply the above assertions to Eqs. (5.77) and (5.79). First, we assume that for any R_1 and R_2

$$\frac{\partial \lambda_i}{\partial R_i} < 0 \quad (i = 1, 2), \tag{5.87}$$

186

and there are positive constants $b_1^{(i)}$ and $b_2^{(i)}$ such that

$$b_1^{(i)} \leq |\frac{\partial \lambda_i}{\partial R_i} \exp(-h_i)| \leq b_2^{(i)}. \tag{5.88}$$

Introduce the notation

$$\zeta_i = \max_{R_1, R_2} Z_i^0, \tag{5.89}$$

where $Z_i^0 = Z_i|_{s_1=0,\, s_2=0}$. We suppose that $\zeta_i > 0$.

According to Propositions 5.1 and 5.2, the critical time T_{cr} before blow-up of solutions of Eqs. (5.77) and (5.79) with the initial conditions Z_i^0 is estimated as

$$\min_i \frac{1}{b_2^{(i)} \zeta_i} \leq T_{cr} \leq \min_i \frac{1}{b_1^{(i)} \zeta_i}. \tag{5.90}$$

Exercise 5.19. Derive this inequality. $\square$

For arbitrary initial conditions, bounds (5.90) are far from being sharp. Nevertheless, there is an important case when the above technique provides an asymptotically correct estimate. Let us assume that the initial conditions $R_i|_{t=0}$ are close to constants whereas their derivatives $T_i|_{t=0}$ can be arbitrarily large. The function $\alpha \sin(\omega x)$ with a small amplitude α and an arbitrary frequency ω provides an example of functions with such a property.

Functions R_i preserve their values along the characteristics. Therefore, for any point (t, x)

$$r_i - \eta \leq R_i(t, x) \leq r_i + \eta, \tag{5.91}$$

provided that

$$r_i - \eta \leq R_i(0, x) \leq r_i + \eta. \tag{5.92}$$

Here r_i is the average value of $R_i(0, x)$, and η is a small parameter.

We suppose that functions $\lambda_i(R_1, R_2)$ are continuously differentiable and their derivatives are bounded. It follows from Eq. (5.91) that we can write

$$\frac{\partial \lambda_i}{\partial R_i}(r_1, r_2) - O(\eta) \leq \frac{\partial \lambda_i}{\partial R_i}(t, x) \leq \frac{\partial \lambda_i}{\partial R_i}(r_1, r_2) + O(\eta), \tag{5.93}$$

where $0 < \lim_{\eta \to 0} O(\eta)/\eta < \infty$.

Functions $h_i(R_1, R_2)$ are determined by differential equations (5.74) and (5.78) up to their initial values. We set

$$h_1|_{R_2=r_2} = 0, \qquad h_2|_{R_1=r_1} = 0. \tag{5.94}$$

As a result, we find that for any point (t, x)

$$-O(\eta) \leq h_i(R_1, R_2) \leq O(\eta). \tag{5.95}$$

187

Eqs. (5.88), (5.93), and (5.95) imply that up to the terms of the first order compared with η, we can substitute

$$b_i^{(1)} = b_i^{(2)} = |\frac{\partial \lambda_i}{\partial R_i}(r_1, r_2)|$$

into Eq. (5.90), instead of their precise values. Furthermore, by using inequality (5.95), we can replace Z_i^0 in Eq. (5.89) by the initial values of functions T_i. As a result, we obtain $\zeta_i = \max_x |T_i(0, x)|$, which implies that

$$T_{cr} \leq \min_{i=1,2}[-\frac{\partial \lambda_i}{\partial R_i}(r_1, r_2) \max_x |\frac{\partial R_i}{\partial x}(0, x)|]^{-1}. \tag{5.96}$$

Let us now return to elongation and shear motions of a hyperelastic medium, and assume that initial velocities vanish, and initial displacement are described by a sinusoidal function

$$u(0, x) = \eta \sin(\frac{2\pi x}{L}), \qquad \frac{\partial u}{\partial t}(0, x) = 0. \tag{5.97}$$

Here η is an amplitude of initial perturbations, and L is *a wavelength*.

Further analysis is left to the reader as a series of exercises.

Exercise 5.20. Check that initial conditions (5.97) can be written as

$$v_1(0, x) = 0, \qquad v_2(0, x) = \frac{2\pi\eta}{L}\cos(\frac{2\pi x}{L}). \quad \square \tag{5.98}$$

Exercise 5.21. By using Eq. (5.52) show that

$$\lambda_1 = \sqrt{H'(v_2)}, \qquad \lambda_2 = -\sqrt{H'(v_2)}. \quad \square \tag{5.99}$$

Exercise 5.22. Derive the following expressions for the eigenvectors $\bar{l}^{(i)}$:

$$\bar{l}^{(1)} = \begin{bmatrix} 1 \\ -\sqrt{H(v_2)} \end{bmatrix}, \qquad \bar{l}^{(2)} = \begin{bmatrix} 1 \\ \sqrt{H(v_2)} \end{bmatrix}. \quad \square \tag{5.100}$$

Exercise 5.23. Show that $\theta_1 = 1$ is a solution of Eq. (5.63). Prove that in this case Eq. (5.62) yields

$$R_1(v_1, v_2) = v_1 - \int_0^{v_2} \sqrt{H'(v)}dv. \quad \square \tag{5.101}$$

Exercise 5.24. Show that $\theta_2 = 1$ and

$$R_2(v_1, v_2) = v_1 + \int_0^{v_2} \sqrt{H'(v)}dv. \quad \square \tag{5.102}$$

Exercise 5.25. Check that Eqs. (5.101) and (5.102) imply that

$$v_1 = \frac{R_2 + R_1}{2}, \qquad \int_0^{v_2} \sqrt{H'(v)}\, dv = \frac{R_2 - R_1}{2}. \qquad (5.103)$$

In order to calculate the derivatives $\partial \lambda_i / \partial R_i$ we employ the formula

$$\frac{\partial \lambda_i}{\partial R_i} = \frac{\partial \lambda_i}{\partial v_1} \frac{\partial v_1}{\partial R_i} + \frac{\partial \lambda_i}{\partial v_2} \frac{\partial v_2}{\partial R_i}. \qquad (5.104)$$

Exercise 5.26. By using Eq. (5.99), show that

$$\frac{\partial \lambda_1}{\partial v_1} = 0, \qquad \frac{\partial \lambda_1}{\partial v_2} = \frac{H''(v_2)}{2\sqrt{H'(v_2)}}, \qquad \frac{\partial \lambda_2}{\partial v_1} = 0, \qquad \frac{\partial \lambda_2}{\partial v_2} = -\frac{H''(v_2)}{2\sqrt{H'(v_2)}}. \qquad \square$$

$$(5.105)$$

Exercise 5.27. By employing Eq. (5.99), derive the following formulas:

$$\frac{\partial v_1}{\partial R_1} = \frac{1}{2}, \qquad \frac{\partial v_1}{\partial R_2} = \frac{1}{2}, \qquad \frac{\partial v_2}{\partial R_1} = -\frac{1}{2\sqrt{H'(v_2)}}, \qquad \frac{\partial v_2}{\partial R_2} = \frac{1}{2\sqrt{H'(v_2)}}. \qquad \square$$

$$(5.106)$$

Exercise 5.28. Check that Eqs. (5.104) – (5.106) imply that

$$\frac{\partial \lambda_1}{\partial R_1} = \frac{\partial \lambda_2}{\partial R_2} = -\frac{H''(v_2)}{4 H'(v_2)}. \qquad \square \qquad (5.107)$$

Exercise 5.29. By using Eqs. (5.101) and (5.102), show that

$$\frac{\partial R_1}{\partial x} = \frac{\partial v_1}{\partial x} - \sqrt{H'(v_2)}\frac{\partial v_2}{\partial x}, \qquad \frac{\partial R_2}{\partial x} = \frac{\partial v_1}{\partial x} + \sqrt{H'(v_2)}\frac{\partial v_2}{\partial x}. \qquad \square \qquad (5.108)$$

We now substitute expressions (5.98) into Eqs. (5.107) and (5.108). Since η is small, we can replace argument v_2 of functions H' and H'' by zero. Finally, substituting expressions (5.107) and (5.108) into inequality (5.96) we obtain

$$T_{\mathrm{cr}} \leq \frac{L^2 \sqrt{H'(0)}}{\pi^2 \eta H''(0)}. \qquad (5.109)$$

Exercise 5.30. Derive formula (5.109). $\square$

Dividing the wavelength L by the speed c of linear waves, we may introduce *the period of oscillations* T. According to Eq. (5.45),

$$T = \frac{L}{c} = \frac{L}{\sqrt{H'(0)}}.$$

Substitution of this expression into Eq. (5.109) implies that

$$\frac{T_{\mathrm{cr}}}{T} = \frac{1}{\pi^2} \frac{L}{\eta} \frac{H'(0)}{H''(0)}. \qquad (5.110)$$

It follows from Eq. (5.110) that the critical number of oscillations $N_{\mathrm{cr}} = T_{\mathrm{cr}}/T$ is proportional to the dimensionless wavelength L/η and is determined by the only material parameter

$$\chi = \frac{H'(0)}{H''(0)} = \frac{W''(0)}{W'''(0)}. \qquad (5.111)$$

It is of special interest to calculate the χ value for elongation and shear of an elastic medium. According to Eq. (5.21), for elongation motion

$$\chi = \frac{1}{3 + 2\nu}, \qquad \nu = \frac{M^3 W(3,3,1)}{M^2 W(3,3,1)}. \qquad (5.112)$$

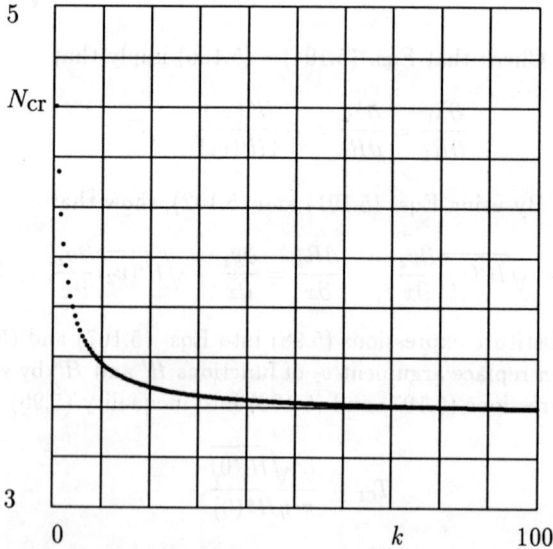

Fig. 5.2. The critical number of oscillations N_{cr} for elongation motion of the Agarwal medium *versus* the material parameter k.

It follows from Eq. (5.41) that for shear motion $\chi = \infty$. This means that there is a fundamental difference between elongation and shear motions with finite strains. Shear motion with sufficiently smooth initial conditions

190

exists at any moment of time, whereas elongation motion exists within a limited interval of time which is determined by dimensionless wavelengths L/η of initial perturbations, and by the material parameter ν.

As an example, let us consider the Agarwal medium (3.2.53).

Exercise 5.31. Show that for the Agarwal medium

$$\nu = -\frac{1}{k+2}, \tag{5.113}$$

where k is the material parameter. □

The dependence of the critical number of oscillations N_{cr} on the material parameter k is plotted in Fig 5.2. Calculations are carried out for the Agarwal medium and for initial perturbations with $L/\eta = 100$. The results show that the critical number of oscillations is not very large (3 to 5), it decreases with the growth of k and tends to its limiting value 3.4 as $k \to \infty$.

5.7. Concluding Remarks

In this Section we consider two one-dimensional motions of a hyperelastic medium: elongation and shear. We derive hyperbolic equations which describe these motions and analyze the characteristic features of nonlinear hyperbolic systems. In particular, we develop estimates for the time intervals where solutions to these equations exist. It is shown that for shear motions with sufficiently smooth initial conditions, solutions exist on any time interval. For elongation motions, the critical time is finite and it is characterized by the wavelengths of initial perturbations and a material parameter. This parameter is determined by the third derivatives of strain energy density and has no analogues in linear elasticity.

6. Nonlinear Oscillations of an Elastic Rod

This Section is concerned with small nonlinear oscillations of an elastic rod. The problem is characterized by the essential nonlinearity of deformations and the presence of two small geometrical parameters (the ratio of the rod deflection to its thickness and the ratio of the rod thickness to its length). We derive a governing equation for the transversal oscillations of the rod and demonstrate that under some conditions this equation can be reduced to the modified Korteweg – de Vries equation. We discuss several properties of this equation, and, in particular, derive N-solitons solutions.

The theory of bending for elastic rods with finite strains was developed in the 60s and 70s, see e.g. Antman (1972, 1973). We do not intend to discuss this theory in detail, and confine ourselves to the simplest version corresponding to plane bending. Our introduction to the theory of nonlinear wave equations follows (partially) Ablowitz & Segur (1981) and Whitham (1974).

6.1. Formulation of the Problem and Governing Equations

Let us consider *plane bending* of a thin elastic rod with rectangular cross-section $\mathcal{S}$. Let l be length, b width, h thickness of the rod. The rod is located along X^1 axis, see Fig. 6.1, and is under tensile forces P applied to its ends. Forces P produce an axial stress $\sigma_{11}^0 = P/S$, where $S = bh$ is the cross-section area.

At instant $t = 0$, the rod begins to deform in plane (X^1, X^2). Denote by $\bar{w}(t, X)$ vector of additional displacements caused by the rod's bending. We assume that the longitudinal displacement w_1 (the displacement in X^1-direction) is significantly less than *the deflection w_2* (the displacement in X^2-direction), and we can neglect w_1 compared with $w_2 = w$.

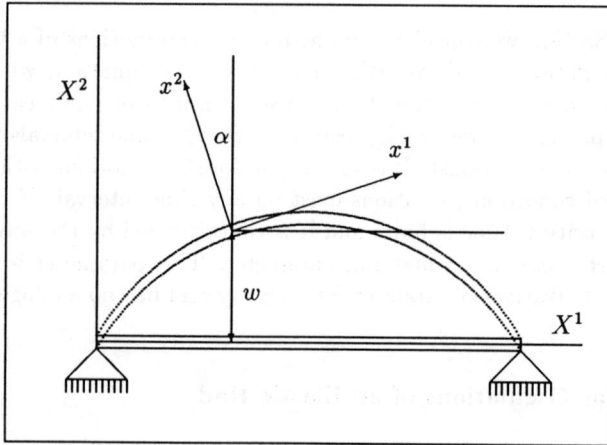

Fig. 6.1. Plane bending of a rod.

In addition to the Lagrangian coordinates $\{X^1, X^2\}$ we introduce Cartesian coordinates $\{x^1, x^2\}$ in the actual configuration. Axis x^1 is directed along the longitudinal axis of the rod, whereas axis x^2 is perpendicular to x^1. The angle between axes X^1 and x^1 is denoted by α,

$$\tan \alpha = \frac{\partial w}{\partial X^1}. \tag{6.1}$$

The arc element equals dX^1 in the initial configuration and ds in the actual configuration. It follows from Fig 6.1 that

$$ds = \sqrt{1 + (\frac{\partial w}{\partial X^1})^2} dX^1 = \frac{dX^1}{\cos \alpha}. \tag{6.2}$$

It is worth noting the well-known formulas for the curvature κ of the deformed longitudinal axis

$$\kappa = -[1 + (\frac{\partial w}{\partial X^1})^2]^{-\frac{3}{2}} \frac{\partial^2 w}{\partial (X^1)^2}, \tag{6.3}$$

and for the corresponding radius of curvature R

$$R = \frac{1}{\kappa}. \tag{6.4}$$

The main assumptions of the technical theory of bending were formulated by Kirchhoff, and they are called *the Kirchhoff–Love hypotheses*, see e.g. Novozhilov (1953). According to *the kinematic hypothesis*, all the strains should be neglected, except for the axial strain $\epsilon_{11} = \epsilon$. For points of the longitudinal axis we set $\epsilon = \epsilon_0$, where

$$\epsilon_0 = \frac{ds - dx}{dx} = \sqrt{1 + (\frac{\partial w}{\partial X^1})^2} - 1. \tag{6.5}$$

For a point located at distance z from the longitudinal axis, the axial deformation is calculated as

$$\epsilon = \epsilon_0 - z\kappa. \tag{6.6}$$

Exercise 6.1. Explain the physical meaning of Eq. (6.6). □

It follows from the Kirchhoff hypothesis that volume deformation φ, see Eq. (2.1.75), equals ϵ. This assertion together with mass conservation law (2.2.10) yields

$$\rho ds = \rho_0 dX^1, \tag{6.7}$$

where ρ_0 and ρ are material densities in the initial and actual configurations, respectively.

Let us write the motion equations for an element of the rod. For this purpose we consider a small element in the actual configuration and denote by $\bar{N}$, $\bar{Q}$, and M *the axial* force, *the shear force*, and *the bending moment* applied to some cross-section. It is of interest to compare these quantities with others introduced in Section 2.2 for the description of surface traction. We should note two main differences:

(i) the theory of bending for *thin-walled structural members* (bars, plates, and shells) postulates the key role of bending moments which are absent in the classical theory of elasticity. Formally, analogs of bending moments may be introduced as additional characteristics of stresses (together with the standard tensors of stresses). This leads to the so-called *Cosserat continua* and *micropolar media*. These theories allow several physical phenomena to be explained adequately, for example, in the theory of cracks. Regretfully, on the one hand, they lead to significant complications of formulas in finite elasticity, and on the other hand, their experimental verification is far from being completed. This forces us not to include the theories of micropolar media in the textbook referring to Green & Nahgdi (1995), where a detailed list of bibliography is presented;

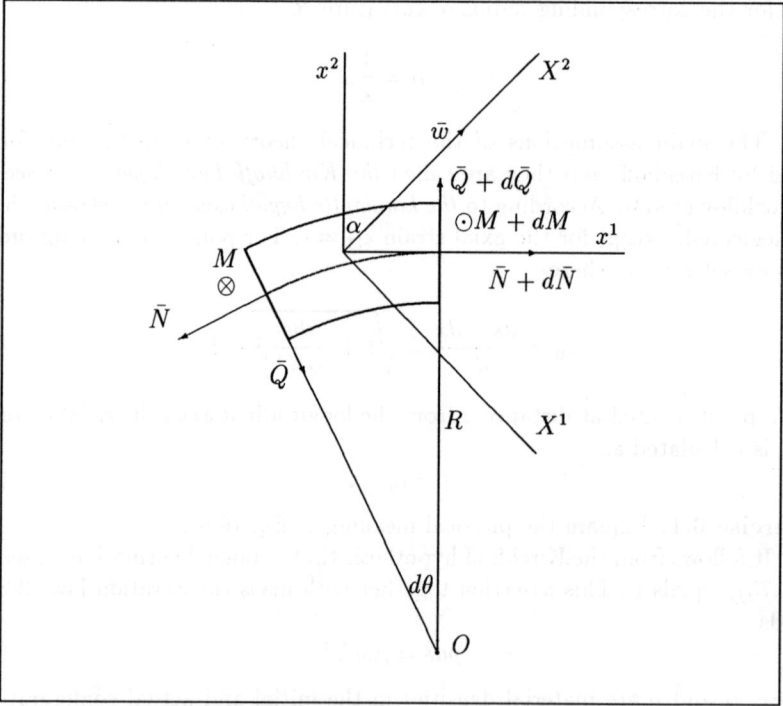

Fig. 6.2. Forces applied to an element of a rod.

(ii) in the classical theory of elasticity both normal and transverse loads are expressed in terms of the strain tensor with the use of constitutive equations. In contrast, in the technical theory of bending, only normal force $\bar{N}$ satisfies the constitutive equation, whereas the shear force is assumed to be arbitrary (*the dynamic hypothesis* by Kirchhoff). The transverse force $\bar{Q}$ is found from the motion equation which expresses the principle of angular momentum.

194

Let us consider an element of the rod and derive the motion equations for this element, see Fig. 6.2. We suppose that forces $\bar{N}$, $\bar{Q}$ and bending moment M are applied to the left edge of the element, whereas forces $\bar{N} + d\bar{N}$, $\bar{Q} + d\bar{Q}$ and bending moment $M + dM$ are applied to its right edge. Directions of the forces and moments are demonstrated in Fig. 6.2. As usual, signs $\odot$ and $\otimes$ mean that a vector is directed to and from us, respectively.

The principle of linear momentum (2.2.17) written in projection on X^2 axis yields

$$\rho ds \frac{\partial^2 w}{\partial t^2} = [(N + dN)\sin\alpha + (Q + dQ)\cos\alpha]$$
$$- [N\sin(\alpha + d\theta) + Q\cos(\alpha + d\theta)]. \tag{6.8}$$

Exercise 6.2. Deduce from Eqs. (6.7) and (6.8) the following formula:

$$\rho_0 dX^1 \frac{\partial^2 w}{\partial t^2} = (dN\sin\alpha - N\cos\alpha d\theta) + (dQ\cos\alpha + Q\sin\alpha d\theta). \qquad \square$$

Since $d\theta = -d\alpha$, this equality implies that

$$\rho_0 \frac{\partial^2 w}{\partial t^2} = \frac{\partial}{\partial X^1}(N\sin\alpha + Q\cos\alpha). \tag{6.9}$$

Let us now write the principle of angular momentum (2.2.24) with respect to the center of the right edge of the deformed element. Our purpose is to derive a differential equation for the bending moment, and we neglect terms of the second order compared with ds. In particular, we neglect the momentum of the inertia force, since this force is proportional to ds, and it should be multiplied by its arm which is proportional to ds as well. Referring to Fig. 6.2, we obtain

$$(M + dM) - M + (Q\cos d\theta + N\sin d\theta)R\sin d\theta$$
$$+ (Q\sin d\theta - N\cos d\theta)(R - R\cos d\theta) = 0.$$

Exercise 6.3. Check that this equation can be presented in the form

$$\frac{\partial M}{\partial \theta} = -RQ. \qquad \square \tag{6.10}$$

Finally, taking into account that $Rd\theta = ds$ and using Eq. (6.2), we find

$$Q = -\cos\alpha \frac{\partial M}{\partial X^1}. \tag{6.11}$$

Substitution of expression (6.11) into Eq. (6.9) yields

$$\rho_0 \frac{\partial^2 w}{\partial t^2} = \frac{\partial}{\partial X^1}(N\sin\alpha - \cos^2\alpha \frac{\partial M}{\partial X^1}). \tag{6.12}$$

To proceed with our analysis, it suffices to derive equalities for axial force N and bending moment M. We encounter here two opportunities: either to postulate dependencies

$$N = N(\epsilon, \kappa), \qquad M = M(\epsilon, \kappa) \tag{6.13}$$

as constitutive equations for a hyperelastic rod, see e.g. Antman (1972, 1973), or to derive these relationships by using additional hypotheses regarding the stress-strain distribution across the cross-section, see Novozhilov (1959). We choose the latter method and assume the simplest (linear) dependence of stresses on the strain ϵ. We suppose that only the axial component $\sigma_{11} = \sigma$ of the Cauchy stress tensor does not vanish, and that the Hooke law is valid

$$\sigma = E\epsilon, \tag{6.14}$$

where E is *Young's modulus*, and ϵ is determined by Eqs. (6.5) and (6.6). We substitute expression (6.14) into the formulas for the axial force and bending moment

$$N = P + \int_S \sigma \, dS = P + b \int_{-\frac{h}{2}}^{\frac{h}{2}} \sigma \, dz,$$

$$M = -\int_S \sigma z \, dS = -b \int_{-\frac{h}{2}}^{\frac{h}{2}} \sigma z \, dz.$$

Exercise 6.4. By using Eqs. (6.3) and (6.5), derive the following equalities:

$$N = P + Ebh\epsilon_0 = P + Ebh\{[1 + (\frac{\partial w}{\partial X^1})^2]^{\frac{1}{2}} - 1\},$$

$$M = -\frac{Ebh^3}{12}\kappa = \frac{Ebh^3}{12}[1 + (\frac{\partial w}{\partial X^1})^2]^{-\frac{3}{2}}\frac{\partial^2 w}{\partial (X^1)^2}. \qquad \square \tag{6.15}$$

Exercise 6.5. By using Eq. (6.1), show that

$$\sin \alpha = [1 + (\frac{\partial w}{\partial X^1})^2]^{-\frac{1}{2}}\frac{\partial w}{\partial X^1}, \qquad \cos \alpha = [1 + (\frac{\partial w}{\partial X^1})^2]^{-1}. \qquad \square \tag{6.16}$$

Eqs. (6.12), (6.15), and (6.16) imply the following nonlinear partial differential equation for the deflection w:

$$\rho_0 \frac{\partial^2 w}{\partial t^2} = \frac{\partial}{\partial X^1}\{[P + Ebh((1 + (\frac{\partial w}{\partial X^1})^2)^{\frac{1}{2}} - 1)][1 + (\frac{\partial w}{\partial X^1})^2]^{-\frac{1}{2}}\frac{\partial w}{\partial X^1}$$

$$- \frac{Ebh^3}{12}\frac{\partial}{\partial X^1}[1 + (\frac{\partial w}{\partial X^1})^2]^{-\frac{3}{2}}\frac{\partial^2 w}{\partial (X^1)^2}\}. \tag{6.17}$$

6.2. The Korteweg – de Vries Equations

Introduce dimensionless variables

$$w = w_0 u, \qquad X^1 = lx, \qquad t = T\tau,$$

$$T = \sqrt{\frac{\rho_0 l^2}{Ebh}}; \qquad p = \frac{P}{Ebh}, \qquad \mu_1 = (\frac{w_0}{l})^2, \qquad \mu_2 = \frac{1}{12}(\frac{h}{l})^2. \qquad (6.18)$$

Substitution of expressions (6.18) into Eq. (6.17) yields

$$\frac{\partial^2 u}{\partial \tau^2} = \frac{\partial}{\partial x}\{[p + ((1 + \mu_1(\frac{\partial u}{\partial x})^2)^{\frac{1}{2}} - 1)][1 + \mu_1(\frac{\partial u}{\partial x})^2]^{-\frac{1}{2}}\frac{\partial u}{\partial x}$$
$$- \mu_2 \frac{\partial}{\partial x}[1 + \mu_1(\frac{\partial u}{\partial x})^2]^{-\frac{3}{2}}\frac{\partial^2 u}{\partial x^2}\}. \qquad (6.19)$$

Let us differentiate Eq. (6.19) with respect to x and introduce the notation $U = \partial u/\partial x$. We find

$$\frac{\partial^2 U}{\partial \tau^2} = \frac{\partial^2}{\partial x^2}\{[p + ((1 + \mu_1 U^2)^{\frac{1}{2}} - 1)](1 + \mu_1 U^2)^{-\frac{1}{2}}U$$
$$- \mu_2 \frac{\partial}{\partial x}(1 + \mu_1 U^2)^{-\frac{3}{2}}\frac{\partial U}{\partial x}\}. \qquad (6.20)$$

Finally, we expand functions in the right-hand side of Eq. (6.20) into series in μ_1 and μ_2 and neglect terms of the second order compared with the small parameters. As a result, we obtain the equation

$$\frac{\partial^2 U}{\partial \tau^2} = \frac{\partial^2}{\partial x^2}(pU + \frac{1-p}{2}\mu_1 U^3 - \mu_2\frac{\partial^2 U}{\partial x^2}). \qquad (6.21)$$

As common practice in the theory of the Korteweg-de Vries equations, we concentrate on the case when parameters μ_1 and μ_2 coincide: $\mu_1 = \mu_2 = \mu$.

First, let us consider the case of "small" tensile loads $p < 1$. Introducing the new variables

$$U = 2\sqrt{\frac{p}{1-p}}U_*, \qquad \tau = \frac{\tau_*}{p}, \qquad x = \frac{x_*}{\sqrt{p}}, \qquad (6.22)$$

we rewrite Eq. (6.21) as follows:

$$\frac{\partial^2 U_*}{\partial \tau_*^2} = \frac{\partial^2}{\partial x_*^2}(U_* + 2\mu U_*^3 - \mu\frac{\partial^2 U_*}{\partial x_*^2}). \qquad (6.23)$$

We seek solutions of Eq. (6.23) in the form

$$U_* = U_*(\xi, \eta) \tag{6.24}$$

where

$$\xi = \tau_* - x_*, \qquad \eta = \frac{\mu}{2}\tau_*.$$

Exercise 6.6. Check that

$$\frac{\partial^2 U_*}{\partial \tau_*^2} = \frac{\partial^2 U_*}{\partial \xi^2} + \mu \frac{\partial^2 U_*}{\partial \xi \partial \eta} + \frac{\mu^2}{4} \frac{\partial^2 U_*}{\partial \eta^2}, \qquad \frac{\partial^2 U_*}{\partial x_*^2} = \frac{\partial^2 U_*}{\partial \xi^2}. \qquad \square \tag{6.25}$$

Substitution of expressions (6.25) into Eq. (6.23) implies that

$$\frac{\partial^2 U_*}{\partial \xi \partial \eta} + \frac{\mu}{4} \frac{\partial^2 U_*}{\partial \eta^2} = \frac{\partial^2}{\partial \xi^2}\left(2U_*^3 - \frac{\partial^2 U_*}{\partial \xi^2}\right). \tag{6.26}$$

We expand a solution of Eq. (6.26) in a series in the small parameter

$$U_* = U_0(\xi, \eta) + \mu U_1(\xi, \eta) + \dots, \tag{6.27}$$

substitute expression (6.27) into Eq. (6.26) and equate terms independent of the small parameter. As a result, we find

$$\frac{\partial^2 U_0}{\partial \xi \partial \eta} - \frac{\partial^2}{\partial \xi^2}\left(2U_0^3 - \frac{\partial^2 U_0}{\partial \xi^2}\right) = 0. \tag{6.28}$$

Eq. (6.28) can be integrated by ξ. Neglecting the constant of integration (which means that function U_0 and its derivatives decay at infinity), we obtain

$$\frac{\partial U_0}{\partial \eta} - 6U_0^2 \frac{\partial U_0}{\partial \xi} + \frac{\partial^3 U_0}{\partial \xi^3} = 0. \tag{6.29}$$

Eq. (6.29) is called *the modified Korteweg – de Vries equation*. Owing to the specific choice of new variables (6.24), this equation is derived as the zero approximation (with respect to the small parameter μ) of Eq. (6.23).

It is worth noting two non-consistent moments in the above procedure, which are typical of the asymptotic analysis for *singularly perturbed* differential equations. First, we could neglect the terms proportional to the small parameter μ in Eq. (6.23). This would lead us to the linear wave equation, which would be valid for "small" ξ only (since we neglected the highest derivative of U_*). Thus, Eq. (6.26) holds for relatively large ξ values (of the order of μ^{-1}). Second, we could preserve the term proportional to μ in Eq. (6.26). Since this term contained the highest derivative, its effect would be observed for "large" η values only. Therefore, Eq. (6.29) holds for relatively small η values (of the order of

μ^{-1}). For finite X^1 values, the above estimates imply that Eq. (6.29) is valid for the dimensionless times τ_* of the order between μ^{-1} and μ^{-2}.

Solutions of Eq. (6.29) are closely connected with solutions of *the Korteweg – de Vries equation*

$$\frac{\partial U}{\partial \eta} + 6U\frac{\partial U}{\partial \xi} + \frac{\partial^3 U}{\partial \xi^3} = 0. \tag{6.30}$$

Eq. (6.30) was introduced by Korteweg and de Vries in 1895 for the description of *long waves* in a canal. A connection between Eqs. (6.29) and (6.30) is determined by the following assertion, see Miura (1968).

Proposition 6.1 (The Miura theorem). Let $U_0(\xi,\eta)$ be a solution of Eq. (6.29). Then the function

$$U = -\frac{\partial U_0}{\partial \xi} - U_0^2$$

is a solution of Eq. (6.30). $\square$

Exercise 6.7. Prove this assertion by direct calculations. $\square$

We introduced solutions of Eq. (6.29) which correspond to waves moving in X^1 direction. By using a similar procedure, we may consider waves moving in the opposite direction by looking for solutions $U_0(\xi,\eta)$ with

$$\xi = \tau_* + x_*, \qquad \eta = \frac{\mu}{2}\tau_*.$$

Exercise 6.8. Show that these solutions satisfy Eq. (6.29) as well. $\square$

Let us now return to the case of "large" tensile loads $p > 1$. We introduce the new variables

$$U = 2\sqrt{\frac{p}{p-1}}U_*, \qquad \tau = \frac{\tau_*}{p}, \qquad x = \frac{x_*}{\sqrt{p}}, \tag{6.31}$$

we rewrite Eq. (6.21) as follows:

$$\frac{\partial^2 U_*}{\partial \tau_*^2} = \frac{\partial^2}{\partial x_*^2}(U_* - 2\mu U_*^3 - \mu\frac{\partial^2 U_*}{\partial x_*^2}).$$

Repeating the above calculations, we arrive at the modified Korteweg – de Vries equation in the form

$$\frac{\partial U_0}{\partial \eta} + 6U_0^2\frac{\partial U_0}{\partial \xi} + \frac{\partial^3 U_0}{\partial \xi^3} = 0. \tag{6.32}$$

Exercise 6.9. Derive Eq. (6.32). $\square$

Comparison of Eqs. (6.29) and (6.32) implies that transverse oscillations of a rod under the action of small and large tensile loads are described by similar nonlinear equations which differ from each other by the signs of nonlinear terms.

6.3. Particular Solutions of the Modified Korteweg – de Vries Equation

The objective of this subsection is to present an algorithm to derive particular solutions to nonlinear partial differential equation (6.32). Evidently, there is no regular procedure to develop particular solutions to an arbitrary nonlinear equation. Nevertheless, for several specific equations of the Korteweg – de Vries type, such an algorithm exists. This algorithm was proposed by Hirota (1971), and it is based on a specific change of variables. For definiteness, we describe *Hirota's method* for Eq. (6.32), but it can be applied to Eq. (6.29) as well.

First, we introduce differential operators $\mathbf{D}_\xi$ and $\mathbf{D}_\eta$. For any smooth functions $f_1(\xi, \eta)$ and $f_2(\xi, \eta)$ and any positive integers m and n we set

$$\mathbf{D}_\xi^m \mathbf{D}_\eta^n f_1 \cdot f_2 = (\frac{\partial}{\partial \xi} - \frac{\partial}{\partial \xi_1})^m (\frac{\partial}{\partial \eta} - \frac{\partial}{\partial \eta_1})^n f_1(\xi, \eta) f_2(\xi_1, \eta_1)|_{\xi_1 = \xi, \ \eta_1 = \eta}. \quad (6.33)$$

Exercise 6.10. Check the following properties of operators $\mathbf{D}_\xi$ and $\mathbf{D}_\eta$:

$$\mathbf{D}_\xi^m f_1 \cdot 1 = \frac{\partial^m f_1}{\partial \xi^m}, \qquad \mathbf{D}_\xi f_1 \cdot f_2 = (-1)^m \mathbf{D}_\xi^m f_2 \cdot f_1,$$

$$\mathbf{D}_\xi^m \mathbf{D}_\eta^n \exp(k_1 \xi - \omega_1 \eta) \cdot \exp(k_2 \xi - \omega_2 \eta)$$
$$= (k_1 - k_2)^m (\omega_2 - \omega_1)^n \exp[(k_1 + k_2)\xi - (\omega_1 + \omega_2)\eta]. \qquad \Box \qquad (6.34)$$

We present a solution U_0 of Eq. (6.32) in the form

$$U_0 = \frac{\Psi(\xi, \eta)}{\Phi(\xi, \eta)}, \qquad (6.35)$$

where Ψ and Φ are functions to be found, and substitute expression (6.35) into Eq. (6.32). After cumbersome calculations, which are left to the reader as an exercise, we arrive at the equation

$$(\mathbf{D}_\eta + \mathbf{D}_\xi^3)\Psi \cdot \Phi + \frac{3}{\Phi^2}(\mathbf{D}_\xi \Psi \cdot \Phi)(\mathbf{D}_\xi^2 \Phi \cdot \Phi - 2\Psi^2) = 0. \qquad (6.36)$$

Exercise 6.11. Derive Eq. (6.36). $\Box$

Since no restrictions are imposed on Φ and Ψ, we assume that the terms in the brackets vanish. This leads to a system of partial differential equations for functions Φ and Ψ

$$(\mathbf{D}_\eta + \mathbf{D}_\xi^3)\Psi \cdot \Phi = 0, \qquad \mathbf{D}_\xi^2 \Phi \cdot \Phi - 2\Psi^2 = 0. \qquad (6.37)$$

This procedure is acceptable, because we look for particular (not general) solutions of Eq. (6.32).

Equations (6.37) do not contain any small parameter. Nevertheless, in order to solve them formally, we introduce a small parameter β and seek solutions in the form of series in the small parameter

$$\Phi = 1 + \beta^2 \phi_2 + \beta^4 \phi_4 + \dots, \qquad \Psi = \beta \psi_1 + \beta^3 \psi_3 + \dots. \qquad (6.38)$$

We substitute expansions (6.38) into Eq. (6.37) and equate terms of the same order of magnitude. As a result, we obtain a system of equations for coefficients ϕ_i and ψ_i. Afterward, we set $\beta = 1$ and find solutions of Eqs. (6.37) as formal series. Convergence of the obtained series is not discussed here.

The above procedure is employed rather frequently in problems of mathematical physics. As common practice, it is extremely difficult to analyze the convergence of series (6.38) for $\beta = 1$, except for the case when only finite numbers of terms occur in series. For nonlinear oscillations of an elastic rod, only a finite number of terms N can be taken into account which determines the so-called N-solitons solutions of the modified Korteweg – de Vries equation.

Substitution of expressions (6.38) into Eqs. (6.37) implies that

$$(\mathbf{D}_\eta + \mathbf{D}_\xi^3)\psi_1 \cdot 1 = 0,$$
$$(\mathbf{D}_\eta + \mathbf{D}_\xi^3)\psi_1 \cdot \phi_2 + (\mathbf{D}_\eta + \mathbf{D}_\xi^3)\psi_3 \cdot 1 = 0,$$
$$\mathbf{D}_\xi^2 1 \cdot 1 = 0,$$
$$\mathbf{D}_\xi^2 \phi_2 \cdot 1 + \mathbf{D}_\xi^2 1 \cdot \phi_2 = 2\psi_1^2,$$
$$\mathbf{D}_\xi^2 \phi_2 \cdot \phi_2 + \mathbf{D}_\xi^2 \phi_4 \cdot 1 + \mathbf{D}_\xi^2 1 \cdot \phi_4 = 4\psi_1 \psi_3. \qquad (6.39)$$

By using Eq. (6.34), we can rewrite the first and the fourth equations (6.39) as follows:

$$\left(\frac{\partial}{\partial \eta} + \frac{\partial^3}{\partial \xi^3}\right)\psi_1 = 0, \qquad \frac{\partial^2 \phi_2}{\partial \xi^2} = \psi_1^2. \qquad (6.40)$$

The third equation is satisfied identically. The second and the fifth equations (6.39) imply that

$$\left(\frac{\partial}{\partial \eta} + \frac{\partial^3}{\partial \xi^3}\right)\psi_3 = -(\mathbf{D}_\eta + \mathbf{D}_\xi^3)\psi_1 \cdot \phi_2,$$

$$2\frac{\partial^2 \phi_4}{\partial \xi^2} = 4\psi_1 \psi_3 - \mathbf{D}_\xi^2 \phi_2 \cdot \phi_2. \qquad (6.41)$$

We look for a solution of the first equation (6.40) in the exponential form

$$\psi_1 = \exp(k_1 \xi - \omega_1 \eta + \lambda_1), \qquad (6.42)$$

where k_1, ω_1, and λ_1 are constant.

Exercise 6.12. Check that function (6.42) satisfies Eq. (6.40) for $\omega_1 = k_1^3$. $\square$

Substituting the function

$$\psi_1 = \exp(k_1\xi - k_1^3\eta + \lambda_1) \tag{6.43}$$

into the second equation (6.40), we find

$$\frac{\partial^2 \phi_2}{\partial \xi^2} = \exp[2(k_1\xi - k_1^3\eta + \lambda_1)]. \tag{6.44}$$

Exercise 6.13. Check that the function

$$\phi_2 = \frac{1}{4k_1^2} \exp[2(k_1\xi - k_1^3\eta + \lambda_1)] \tag{6.45}$$

is a solution of Eq. (6.40). $\square$

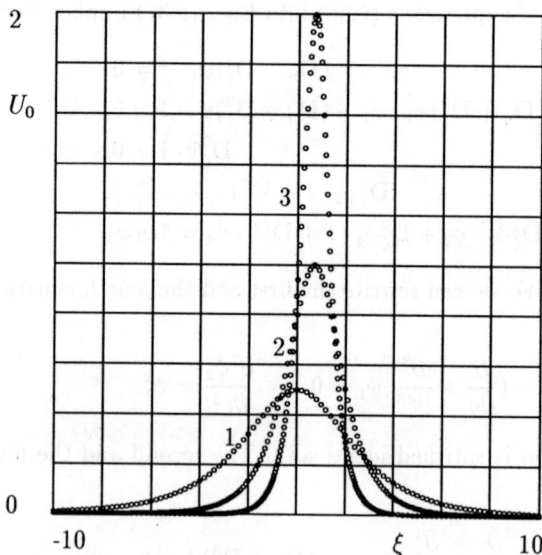

Fig. 6.3. One-soliton solution of the modified Korteweg – de Vries equation for $\lambda_1 = 0$ and $\eta = 0$. Curve 1: $k_1 = 0.5$, curve 2: $k_1 = 1.0$; curve 3: $k_1 = 2.0$.

Exercise 6.14. By direct calculations with the use of Eqs. (6.34), (6.43), and (6.45), show that

$$\mathbf{D}_\xi^2 \phi_2 \cdot \phi_2 = 0, \qquad (\mathbf{D}_\eta + \mathbf{D}_\xi^3)\psi_1 \cdot \phi_2 = 0. \qquad \square \tag{6.46}$$

Substitution of expressions (6.46) into Eqs. (6.41) implies that

$$\left(\frac{\partial}{\partial \eta} + \frac{\partial^3}{\partial \xi^3}\right)\psi_3 = 0, \qquad \frac{\partial^2 \phi_4}{\partial \xi^2} = 0,$$

and we can set $\psi_3 = 0$ and $\phi_4 = 0$.

Exercise 6.15. By employing these equalities together with Eqs. (6.37) and (6.38), prove that we may set $\psi_{2n+1} = 0$ and $\phi_{2n} = 0$ for any integer $n \geq 2$. $\square$

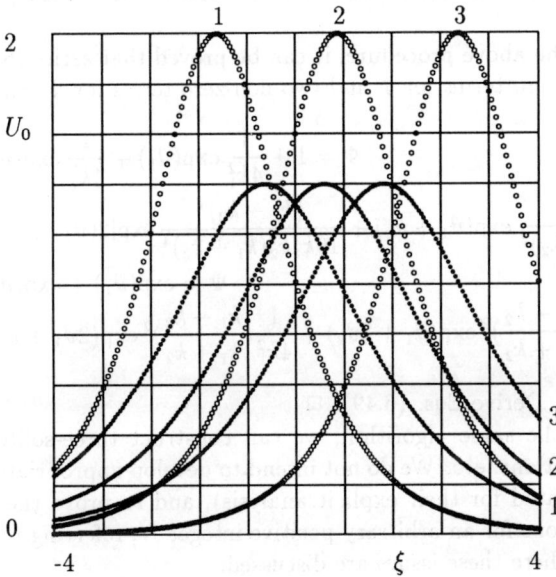

Fig. 6.4. One-soliton solution of the modified Korteweg – de Vries equation for $\lambda_1 = 0$. Curve 1: $\eta = -2$, curve 2: $\eta = 0$; curve 3: $\eta = 2$. Filled circles correspond to $k_1 = 0.5$, unfilled circles correspond to $k_1 = 1$.

Using the above assertion and setting $\beta = 1$ in Eqs. (6.38), we find that functions

$$\Phi = 1 + \phi_2 = 1 + \frac{1}{4k_1^2}\exp[2(k_1\xi - k_1^3\eta + \lambda_1)], \qquad \Psi = \psi_1 = \exp(k_1\xi - k_1^3\eta + \lambda_1)$$

provide exact solutions of Eqs. (6.37). Substitution of these expressions into Eq. (6.35) implies the following solution of the modified Korteweg – de Vries equation:

$$U_0(\xi,\eta) = \frac{\exp(k_1\xi - k_1^3\eta + \lambda_1)}{1 + (2k_1)^{-2}\exp[2(k_1\xi - k_1^3\eta + \lambda_1)]}. \tag{6.47}$$

203

Function U_0 is determined by two arbitrary parameters k_1 and λ_1. This function is called *one-soliton solution* of Eq. (6.32).

In order to derive *a two-soliton solution*, it suffices to take two exponential terms in expression (6.42) for function ψ_1, i.e. to replace expression (6.42) by the formula

$$\psi_1 = \exp(\vartheta_1) + \exp(\vartheta_2). \tag{6.48}$$

Here $\vartheta_i = k_i\xi - k_i^3\eta + \lambda_i$, where k_i and λ_i are arbitrary constants.

Exercise 6.16. Check that expression (6.48) satisfies the first equation (6.40). $\square$

By using the above procedure, it can be proved that series (6.38) contains only three non-zero terms for Φ and two non-zero terms for Ψ, and

$$\Phi = 1 + \frac{1}{4k_1^2}\exp(\vartheta_1) + \frac{1}{4k_2^2}\exp(\vartheta_2)$$

$$+\frac{2}{(k_1+k_2)^2}\exp(\vartheta_1+\vartheta_2) + \frac{(k_1-k_2)^4}{16k_1^2k_2^2(k_1+k_2)^4}\exp[2(\vartheta_1+\vartheta_2)],$$

$$\Psi = \exp(\vartheta_1) + \exp(\vartheta_2)$$

$$+\frac{1}{4k_2^2}(\frac{k_1-k_2}{k_1+k_2})^2\exp(\vartheta_1+2\vartheta_2) + \frac{1}{4k_1^2}(\frac{k_1-k_2}{k_1+k_2})^2\exp(2\vartheta_1+\vartheta_2). \tag{6.49}$$

Exercise 6.17. Derive Eqs. (6.49). $\square$

By using the same algorithm, we can construct three-soliton solutions, four-soliton solutions, etc. We do not intend to develop appropriate expressions (rather complicated for their explicit analysis), and to prove the existence of N-soliton solutions for an arbitrary positive integer N, referring to Ablowitz & Segur (1981), where these issues are discussed.

Our objective now is to analyze some interesting features of the so-called *elastic interaction* of solitons by using Eqs. (6.47) and (6.49).

First, let us consider the one-soliton solution (6.47). The graph of function $U_0(\xi,\eta)$ for $\eta = 0$ is plotted in Fig. 6.3. The curve U_0 *versus* ξ vanishes for practically any ξ except for a narrow region near $\xi = 0$. In this region, function U_0 grows rapidly in ξ, reaches its maximum, and decreases tending to zero. The length of the region and the maximum of U_0 depend essentially on k_1. With the growth of k_1, the length of the region decreases, and the maximum increases. The point where function U_0 reaches its maximum, also depends on k_1, but very weakly.

The one-soliton wave moves along ξ axis with the growth of "time" η without any change of its shape, see Fig. 6.4. The speed of this motion (wave speed) depends significantly on its amplitude. "Small" waves move rather slowly, whereas "large" waves move rapidly. This implies that for a system of solitons, "large" solitons should overtake "small" ones.

An interaction of solitons is plotted in Fig. 6.5 by using the two-soliton solution (6.49). Fig. 6.5(A) demonstrates two solitons at the initial moment $\eta = 0$, the large soliton being on the left of the small one. With the growth of time, the large soliton gains on the small one, see Fig 7.5(B), and overtakes it, see Fig. 6.5(C). The phenomenon of interest is the absence of any change in the shape of solitons.

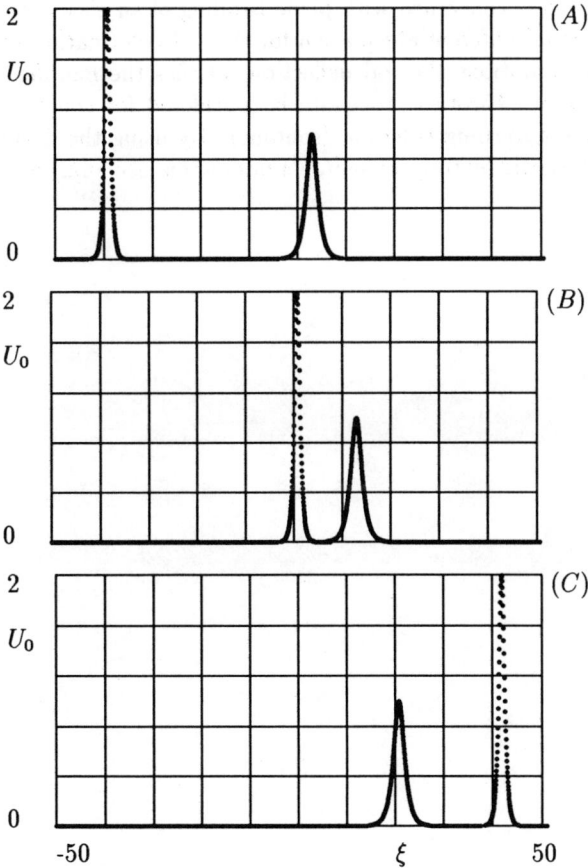

Fig. 6.5. Two-soliton solution of the modified Korteweg – de Vries equation for $k_1 = 1$, $k_2 = 2$, $\lambda_1 = 0$, $\lambda_2 = 80$. Figure A: $\eta = 0$, figure B: $\eta = 10$; figure C: $\eta = 20$.

An interaction of waves without changes of their shape is called *elastic interaction*. This phenomenon is typical of linear waves, but soliton solutions show that the elastic interaction occurs in nonlinear waves as well.

To the best of our knowledge, there is no experimental confirmation for the elastic interaction of nonlinear waves in an elastic rod. However, experimental data in the fluid mechanics demonstrate the above discussed features of solitons, see e.g. Hammack & Segur (1974).

6.4. Concluding Remarks

This Section is concerned with plane bending of an elastic rod. We derive a nonlinear partial differential equation for the rod deformation and show that under some assumptions the rod deflection satisfies the modified Korteweg – de Vries equation. Hirota's procedure is considered for constructing explicit solutions of the governing equation (solitons). By using the derived formulas, several characteristic features of soliton's interaction are analyzed.

Chapter 5

Variational principles in elasticity

This Chapter is concerned with variational principles in finite elastostatics. In Section 1, we discuss the Lagrange principle and derive the equilibrium equation and the boundary condition in stresses as necessary conditions of minimum for the total energy. We show that the strong ellipticity conditions provide sufficient conditions of the minimum, and establish infinitesimal stability conditions for a hyperelastic medium. Section 2 deals with phase transitions in hyperelastic solids. By using the generalized Lagrange principle, a specific condition on the interface is derived, which is formulated in terms of the Eshelby tensor. For elastic liquids, the interface condition obtained is equivalent to the equality of chemical potentials.

1. The Lagrange Variational Principle

In this Section, the Lagrange principle in finite elastostatics is formulated and discussed. It is shown that the equilibrium equation and the boundary condition in stresses for a hyperelastic medium are developed as necessary conditions of minimum for the total energy. A sufficient condition of minimum is provided by the infinitesimal stability condition. By introducing the Korn constant, we derive a simple inequality which ensures the infinitesimal stability at small deformations.

1.1. Formulation of the Lagrange Principle

Let us consider a hyperelastic medium which occupies a bounded, connected domain Ω_0 with a smooth boundary Γ_0 in the initial (stress-free) configuration. Points of the medium refer to Lagrangian coordinates $\xi = \{\xi^i\}$. Under the action of a body force $\bar{B}$ and a surface traction $\bar{b}$, the medium deforms and transforms into the actual configuration, where it occupies a domain Ω with a boundary Γ. The surface forces are prescribed on a part $\Gamma_0^{(\sigma)}$ of boundary Γ_0. The other part of the boundary $\Gamma_0^{(u)} = \Gamma_0 \setminus \Gamma_0^{(\sigma)}$ is clamped. We confine ourselves to isotropic hyperelastic media under the action of dead body forces and dead surface tractions, see Eq. (3.3.6).

Denote by $\bar{u}(\xi)$ a displacement vector, by

$$W^\dagger = \int_{\Omega_0} W \, dV_0 \tag{1.1}$$

the total strain energy, and by

$$A^\dagger = \int_{\Omega} \rho \bar{B} \cdot \bar{u} \, dV + \int_{\Gamma^{(\sigma)}} \bar{b} \cdot \bar{u} \, dS \tag{1.2}$$

the work of external loads. Here dV_0 and dV are the volume elements, dS_0 and dS are the surface elements, and ρ_0 and ρ are mass densities in the initial and actual configuration, respectively.

For dead loads, Eq. (1.2) can be presented in the form

$$A^\dagger = \int_{\Omega_0} \rho_0 \bar{B}_0 \cdot \bar{u} \, dV_0 + \int_{\Gamma_0^{(\sigma)}} \bar{b}_0 \cdot \bar{u} \, dS_0, \tag{1.3}$$

where $\bar{B}_0(\xi)$ and $\bar{b}_0(\xi)$ are given functions.

Definition. Functional $\Phi = W^\dagger - A^\dagger$ is called *total energy.*

Definition. A set Υ of continuously differentiable displacement fields $\bar{u}$ subject to the boundary condition

$$\bar{u} \big|_{\Gamma_0^{(u)}} = 0 \tag{1.4}$$

is the set of admissible displacement fields.

Proposition 1.1 (Lagrange's principle). For dead loads, the real displacement field $\bar{u}$ (i.e. the displacement field which occurs under the action of external loads) minimizes the total energy Φ on the set Υ of admissible displacements. $\square$

For isothermal loading the total energy Φ coincides (up to a constant) with the free energy of an elastic medium. In this case *the Lagrange principle* is equivalent to *the principle of minimum free energy,* see e.g. Landau & Lifshitz (1969).

1.2. Perturbations of Displacement Vectors and Deformation Tensors

To write the Lagrange principle in the differential form, we should fix a displacement field $\bar{u}$ which minimizes the functional Φ, consider its perturbation

$\bar{u}^* = \bar{u} + \delta\bar{u}$, where $\delta\bar{u}$ is a sufficiently small admissible displacement field, and calculate the difference between Φ values corresponding to the displacement fields $\bar{u}^*$ and $\bar{u}$. For this purpose, we derive some formulas for perturbations of the deformation tensors and their invariants.

The radius-vector $\bar{r}^*$ of a point ξ in the actual perturbed configuration equals

$$\bar{r}^*(t, \xi) = \bar{r}(t, \xi) + \delta\bar{u}(t, \xi) = \bar{r}_0(\xi) + \bar{u}(t, \xi) + \delta\bar{u}(t, \xi), \qquad (1.5)$$

where $\bar{r}_0(\xi)$ is the radius-vector in the initial configuration, and $\bar{r}(t, \xi)$ is the radius-vector in the non-perturbed actual configuration. Differentiation of Eq. (1.5) with respect to ξ^i yields

$$\bar{g}_i^* = \frac{\partial\bar{r}^*}{\partial\xi^i} = \frac{\partial\bar{r}}{\partial\xi^i} + \frac{\partial\delta\bar{u}}{\partial\xi^i} = \bar{g}_i + \frac{\partial\delta\bar{u}}{\partial\xi^i}. \qquad (1.6)$$

It follows from Eq. (1.3.4) that

$$\bar{\nabla}\delta\bar{u} = \bar{g}^j \frac{\partial\delta\bar{u}}{\partial\xi^j}.$$

By multiplying this equality by $\bar{g}_i$, we find

$$\frac{\partial\delta\bar{u}}{\partial\xi^j} = \bar{g}_i \cdot \bar{\nabla}\delta\bar{u}. \qquad (1.7)$$

Substitution of expression (1.7) into Eq. (1.6) implies that

$$\bar{g}_i^* = \bar{g}_i \cdot (\hat{I} + \bar{\nabla}\delta\bar{u}) = (\hat{I} + \bar{\nabla}\delta\bar{u}^T) \cdot \bar{g}_i. \qquad (1.8)$$

In order to derive expressions for dual vectors $\bar{g}^{*i}$, we present the unit tensor $\hat{I}$ in the form $\hat{I} = \bar{g}^{*i}\bar{g}_i^*$. Substitution of expression (1.8) into this equality yields

$$\hat{I} = \bar{g}^{*i}\bar{g}_i + \bar{g}^{*i}\bar{g}_i \cdot \bar{\nabla}\delta\bar{u}.$$

It follows from this formula and Eq. (1.5) that up to the second order terms compared with $\delta\bar{u}$

$$\hat{I} = \bar{g}^{*i}\bar{g}_i + \bar{g}^{*i}\bar{g}_i^* \cdot \bar{\nabla}\delta\bar{u} = \bar{g}^{*i}\bar{g}_i + \bar{\nabla}\delta\bar{u}.$$

Therefore,

$$\bar{g}^{*j}\bar{g}_j = \hat{I} - \bar{\nabla}\delta\bar{u}. \qquad (1.9)$$

By multiplying Eq. (1.9) by $\bar{g}^i$, we obtain

$$\bar{g}^{*i} = (\hat{I} - \bar{\nabla}\delta\bar{u}) \cdot \bar{g}^i = \bar{g}^i \cdot (\hat{I} - \bar{\nabla}\delta\bar{u}^T). \qquad (1.10)$$

Using Eq. (1.10), we find the nabla-operator in the actual perturbed configuration

$$\bar{\nabla}^* = \bar{g}^{*i}\frac{\partial}{\partial\xi^i} = (\bar{g}^i - \bar{\nabla}\delta\bar{u}\cdot\bar{g}^i)\frac{\partial}{\partial\xi^i} = \bar{\nabla} - \bar{\nabla}\delta\bar{u}\cdot\bar{\nabla}. \tag{1.11}$$

It follows from Eqs. (2.1.5) and (1.8) that

$$\bar{\nabla}_0\bar{r}^* = \bar{g}_0^i\bar{g}_i^* = \bar{g}_0^i\bar{g}_i\cdot(\hat{I} + \bar{\nabla}\delta\bar{u}) = \bar{\nabla}_0\bar{r}\cdot(\hat{I} + \bar{\nabla}\delta\bar{u}). \tag{1.12}$$

Obviously,

$$\bar{\nabla}_0\bar{r}^{*T} = (\hat{I} + \bar{\nabla}\delta\bar{u}^T)\cdot\bar{\nabla}_0\bar{r}^T. \tag{1.13}$$

Exercise 1.1. By employing Eq. (1.11), check that

$$\bar{\nabla}^*\bar{r}_0 = (\hat{I} - \bar{\nabla}\delta\bar{u})\cdot\bar{\nabla}\bar{r}_0, \qquad \bar{\nabla}^*\bar{r}_0^T = \bar{\nabla}\bar{r}_0^T\cdot(\hat{I} - \bar{\nabla}\delta\bar{u}^T). \qquad \square$$

Let us calculate perturbations of the Finger deformation tensor caused by a perturbation $\delta\bar{u}$ of the displacement field $\bar{u}$. Unlike formula (1.10), we preserve the second order terms and neglect the third order terms compared with $|\delta\bar{u}|$. It follows from Eqs. (2.1.25), (1.12), and (1.13) that

$$
\begin{aligned}
\hat{F}^* &= \bar{\nabla}_0\bar{r}^{*T}\cdot\bar{\nabla}_0\bar{r}^* = (\hat{I} + \bar{\nabla}\delta\bar{u}^T)\cdot\bar{\nabla}_0\bar{r}^T\cdot\bar{\nabla}_0\bar{r}\cdot(\hat{I} + \bar{\nabla}\delta\bar{u}) \\
&= \hat{F} + \bar{\nabla}\delta\bar{u}^T\cdot\hat{F} + \hat{F}\cdot\bar{\nabla}\delta\bar{u} + \bar{\nabla}\delta\bar{u}^T\cdot\hat{F}\cdot\bar{\nabla}\delta\bar{u}.
\end{aligned} \tag{1.14}
$$

This equality implies that up to the second order terms with respect to $|\delta\bar{u}|$ we obtain

$$\hat{F}^* = \hat{F} + \bar{\nabla}\delta\bar{u}^T\cdot\hat{F} + \hat{F}\cdot\bar{\nabla}\delta\bar{u}. \tag{1.15}$$

Exercise 1.2. By multiplying Eq. (1.14) by itself, show that

$$
\begin{aligned}
\hat{F}^{*2} &= \hat{F}^2 + \bar{\nabla}\delta\bar{u}^T\cdot\hat{F}^2 + \hat{F}^2\cdot\bar{\nabla}\delta\bar{u} + 2\hat{F}\cdot\hat{\epsilon}(\delta\bar{u})\cdot\hat{F} \\
&\quad + 2[\bar{\nabla}\delta\bar{u}^T\cdot\hat{F}\cdot\hat{\epsilon}(\delta\bar{u})\cdot\hat{F} + \hat{F}\cdot\hat{\epsilon}(\delta\bar{u})\cdot\hat{F}\cdot\bar{\nabla}\delta\bar{u}] \\
&\quad + \bar{\nabla}\delta\bar{u}^T\cdot\hat{F}^2\cdot\bar{\nabla}\delta\bar{u} + \hat{F}\cdot\bar{\nabla}\delta\bar{u}\cdot\bar{\nabla}\delta\bar{u}^T\cdot\hat{F}, \\
\hat{F}^{*3} &= \hat{F}^3 + \bar{\nabla}\delta\bar{u}^T\cdot\hat{F}^3 + \hat{F}^3\cdot\bar{\nabla}\delta\bar{u} \\
&\quad + 2[\hat{F}^2\cdot\hat{\epsilon}(\delta\bar{u})\cdot\hat{F} + \hat{F}\cdot\hat{\epsilon}(\delta\bar{u})\cdot\hat{F}^2] \\
&\quad + [\bar{\nabla}\delta\bar{u}^T\cdot\hat{F}^2\cdot\hat{\epsilon}(\delta\bar{u})\cdot\hat{F} + \hat{F}^2\cdot\hat{\epsilon}(\delta\bar{u})\cdot\hat{F}\cdot\bar{\nabla}\delta\bar{u} \\
&\quad + \bar{\nabla}\delta\bar{u}^T\cdot\hat{F}\cdot\hat{\epsilon}(\delta\bar{u})\cdot\hat{F}^2 + \hat{F}\cdot\hat{\epsilon}(\delta\bar{u})\cdot\hat{F}^2\cdot\bar{\nabla}\delta\bar{u}] \\
&\quad + 4\hat{F}\cdot\hat{\epsilon}(\delta\bar{u})\cdot\hat{F}\cdot\hat{\epsilon}(\delta\bar{u})\cdot\hat{F} + \bar{\nabla}\delta\bar{u}^T\cdot\hat{F}^3\cdot\bar{\nabla}\delta\bar{u} \\
&\quad + \hat{F}^2\cdot\bar{\nabla}\delta\bar{u}\cdot\bar{\nabla}\delta\bar{u}^T\cdot\hat{F} + \hat{F}\cdot\bar{\nabla}\delta\bar{u}\cdot\bar{\nabla}\delta\bar{u}^T\cdot\hat{F}^2,
\end{aligned} \tag{1.16}
$$

where

$$\hat{\epsilon}(\delta\bar{u}) = \frac{1}{2}(\bar{\nabla}\delta\bar{u} + \bar{\nabla}\delta\bar{u}^T). \qquad \square \tag{1.17}$$

210

It follows from Eqs. (1.14) and (1.16) that

$$
\begin{aligned}
I_1(\hat{F}^*) &= I_1(\hat{F}) + 2\hat{F} : \hat{\epsilon}(\delta\bar{u}) + I_1(\bar{\nabla}\delta\bar{u}^T \cdot \hat{F} \cdot \bar{\nabla}\delta\bar{u}), \\
I_1(\hat{F}^{*2}) &= I_1(\hat{F}^2) + 4[\hat{F}^2 : \hat{\epsilon}(\delta\bar{u}) + (\hat{F} \cdot \hat{\epsilon}(\delta\bar{u})) : (\hat{F} \cdot \hat{\epsilon}(\delta\bar{u}))] \\
&\quad + 2I_1(\bar{\nabla}\delta\bar{u}^T \cdot \hat{F}^2 \cdot \bar{\nabla}\delta\bar{u}), \\
I_1(\hat{F}^{*3}) &= I_1(\hat{F}^3) + 6\hat{F}^3 : \hat{\epsilon}(\delta\bar{u}) + 12(\hat{F}^2 \cdot \hat{\epsilon}(\delta\bar{u})) : (\hat{F} \cdot \hat{\epsilon}(\delta\bar{u})) \\
&\quad + 3I_1(\bar{\nabla}\delta\bar{u}^T \cdot \hat{F}^3 \cdot \bar{\nabla}\delta\bar{u}).
\end{aligned}
$$

These equalities together with Eqs. (1.2.30) imply that

$$
\begin{aligned}
I_1(\hat{F}^*) &= I_1(\hat{F}) + 2\hat{F} : \hat{\epsilon}(\delta\bar{u}) + I_1(\bar{\nabla}\delta\bar{u}^T \cdot \hat{F} \cdot \bar{\nabla}\delta\bar{u}), \\
I_2(\hat{F}^*) &= I_2(\hat{F}) + 2(I_1(\hat{F})\hat{F} - \hat{F}^2) : \hat{\epsilon}(\delta\bar{u}) \\
&\quad + 2[(\hat{F} : \hat{\epsilon}(\delta\bar{u}))^2 - (\hat{F} \cdot \hat{\epsilon}(\delta\bar{u})) : (\hat{F} \cdot \hat{\epsilon}(\delta\bar{u}))] \\
&\quad + I_1(\bar{\nabla}\delta\bar{u}^T \cdot (I_1(\hat{F})\hat{F} - \hat{F}^2) \cdot \bar{\nabla}\delta\bar{u}), \\
I_3(\hat{F}^*) &= I_3(\hat{F}) + 2I_3(\hat{F})I_1(\hat{\epsilon}(\delta\bar{u})) \\
&\quad - 4[(\hat{F} : \hat{\epsilon}(\delta\bar{u}))(\hat{F}^2 : \hat{\epsilon}(\delta\bar{u})) - (\hat{F} \cdot \hat{\epsilon}(\delta\bar{u})) : (\hat{F}^2 \cdot \hat{\epsilon}(\delta\bar{u}))] \\
&\quad + 2I_1(\hat{F})[(\hat{F} : \hat{\epsilon}(\delta\bar{u}))^2 - (\hat{F} \cdot \hat{\epsilon}(\delta\bar{u})) : (\hat{F} \cdot \hat{\epsilon}(\delta\bar{u}))] \\
&\quad + I_3(\hat{F})I_1(\bar{\nabla}\delta\bar{u}^T \cdot \bar{\nabla}\delta\bar{u}). \qquad (1.18)
\end{aligned}
$$

Exercise 1.3. Check that Eqs. (1.18) turn into equalities

$$
\begin{aligned}
I_1(\hat{F}) &= 3 + 2I_1(\hat{\epsilon}_0(\bar{u})) + I_1(\bar{\nabla}\bar{u}^T \cdot \bar{\nabla}\bar{u}), \\
I_2(\hat{F}) &= 3 + 4I_1(\hat{\epsilon}_0(\bar{u})) + 2I_1^2(\hat{\epsilon}_0(\bar{u})) - 2I_1(\hat{\epsilon}_0^2(\bar{u})) + 2I_1(\bar{\nabla}\bar{u}^T \cdot \bar{\nabla}\bar{u}), \\
I_3(\hat{F}) &= 1 + 2I_1(\hat{\epsilon}_0(\bar{u})) + 2I_1^2(\hat{\epsilon}_0(\bar{u})) - 2I_1(\hat{\epsilon}_0^2(\bar{u})) + I_1(\bar{\nabla}\bar{u}^T \cdot \bar{\nabla}\bar{u}), \quad (1.19)
\end{aligned}
$$

provided the non-perturbed actual configuration coincides with the initial one.
□

1.3. Necessary Conditions of Minimum

In this subsection, we show that the Lagrange principle implies the equilibrium equation (2.2.21) and the boundary condition in stresses (2.2.20) as necessary conditions of minimum. For this purpose, we calculate the perturbation of Φ caused by a perturbation $\delta\bar{u}$ of a displacement field $\bar{u}$. Since strain energy density W of an isotropic hyperelastic medium depends on the principal invariants I_k of the Finger tensor $\hat{F}$, we may write

$$
W^* = W + \frac{\partial W}{\partial I_k}\delta I_k + \frac{1}{2}\frac{\partial^2 W}{\partial I_m \partial I_n}\delta I_m \delta I_n + \ldots, \qquad (1.20)
$$

where W and its derivatives are calculated in the non-perturbed actual configuration. Substitution of expressions (1.18) into Eq. (1.20) yields (up to the third order terms compared with $|\delta\bar{u}|$)

$$W^* = W + 2[(\frac{\partial W}{\partial I_1} + I_1\frac{\partial W}{\partial I_2})\hat{F} - \frac{\partial W}{\partial I_2}\hat{F}^2 + I_3\frac{\partial W}{\partial I_3}\hat{I}] : \hat{\epsilon}(\delta\bar{u})$$
$$+2(\frac{\partial W}{\partial I_2} + I_1\frac{\partial W}{\partial I_3})[(\hat{F} : \hat{\epsilon}(\delta\bar{u}))^2 - (\hat{F}\cdot\hat{\epsilon}(\delta\bar{u})) : (\hat{F}\cdot\hat{\epsilon}(\delta\bar{u}))]$$
$$-4\frac{\partial W}{\partial I_3}[(\hat{F} : \hat{\epsilon}(\delta\bar{u}))(\hat{F}^2 : \hat{\epsilon}(\delta\bar{u})) - (\hat{F}\cdot\hat{\epsilon}(\delta\bar{u})) : (\hat{F}^2\cdot\hat{\epsilon}(\delta\bar{u}))]$$
$$+I_1(\bar{\nabla}\delta\bar{u}^T\cdot[(\frac{\partial W}{\partial I_1} + I_1\frac{\partial W}{\partial I_2})\hat{F} - \frac{\partial W}{\partial I_2}\hat{F}^2 + I_3\frac{\partial W}{\partial I_3}\hat{I}]\cdot\bar{\nabla}\delta\bar{u})$$
$$+2[(\hat{F}:\hat{\epsilon}(\delta\bar{u}))(\frac{\partial}{\partial I_1}+I_1\frac{\partial}{\partial I_2}) - (\hat{F}^2:\hat{\epsilon}(\delta\bar{u}))\frac{\partial}{\partial I_2}+I_3(\hat{I}:\hat{\epsilon}(\delta\bar{u}))\frac{\partial}{\partial I_3}]^2 W. \quad (1.21)$$

Eq. (1.21) together with Eqs. (1.1) and (1.3) implies that

$$\Phi(\bar{u}^*) = \Phi(\bar{u}) + 2\int_{\Omega_0}[(\frac{\partial W}{\partial I_1}+I_1\frac{\partial W}{\partial I_2})\hat{F}- \frac{\partial W}{\partial I_2}\hat{F}^2+I_3\frac{\partial W}{\partial I_3}\hat{I}]:\hat{\epsilon}(\delta\bar{u})dV_0$$
$$- \int_{\Omega_0}\rho_0\bar{B}_0\cdot\delta\bar{u}dV_0 - \int_{\Gamma_0^{(\sigma)}}\bar{b}_0\cdot\delta\bar{u}dS_0 + N(\delta\bar{u}), \quad (1.22)$$

where

$$N(\delta\bar{u}) = \int_{\Omega_o}\{2(\frac{\partial W}{\partial I_2} + I_1\frac{\partial W}{\partial I_3})[(\hat{F} : \hat{\epsilon}(\delta\bar{u}))^2 - (\hat{F}\cdot\hat{\epsilon}(\delta\bar{u})) : (\hat{F}\cdot\hat{\epsilon}(\delta\bar{u}))]$$
$$-4\frac{\partial W}{\partial I_3}[(\hat{F} : \hat{\epsilon}(\delta\bar{u}))(\hat{F}^2 : \hat{\epsilon}(\delta\bar{u})) - (\hat{F}\cdot\hat{\epsilon}(\delta\bar{u})) : (\hat{F}^2\cdot\hat{\epsilon}(\delta\bar{u}))]$$
$$+2[(\hat{F} : \hat{\epsilon}(\delta\bar{u}))(\frac{\partial}{\partial I_1} + I_1\frac{\partial}{\partial I_2}) - (\hat{F}^2 : \hat{\epsilon}(\delta\bar{u}))\frac{\partial}{\partial I_2} + (\hat{I} : \hat{\epsilon}(\delta\bar{u}))I_3\frac{\partial}{\partial I_3}]^2 W$$
$$+I_1(\nabla\delta\bar{u}^T\cdot[(\frac{\partial W}{\partial I_1} + I_1\frac{\partial W}{\partial I_2})\hat{F} - \frac{\partial W}{\partial I_2}\hat{F}^2 + I_3\frac{\partial W}{\partial I_3}\hat{I}]\cdot\nabla\delta\bar{u}\}dV_0. \quad (1.23)$$

By employing Eqs. (3.2.18) and (3.2.19), we present Eq. (1.22) as

$$\Phi(\bar{u}^*) = \Phi(\bar{u}) + \int_{\Omega_0}\sqrt{I_3}\hat{\sigma} : \hat{\epsilon}(\delta\bar{u})dV_0$$
$$- \int_{\Omega_0}\rho_0\bar{B}_0\cdot\delta\bar{u}dV_0 - \int_{\Gamma_0^{(\sigma)}}\bar{b}_0\cdot\delta\bar{u}dS_0 + N(\delta\bar{u}).$$

Finally, replacing integrals over domains Ω_0 and Γ_0 by the integrals over appropriate domains in the actual configuration, we obtain

$$\Phi(\bar{u}^*) = \Phi(\bar{u}) + \int_{\Omega}\hat{\sigma}:\hat{\epsilon}(\delta\bar{u})dV - \int_{\Omega}\rho\bar{B}\cdot\delta\bar{u}dV - \int_{\Gamma^{(\sigma)}}\bar{b}\cdot\delta\bar{u}dS + N(\delta\bar{u}). \quad (1.24)$$

We transform the second term in the right-hand side of Eq. (1.24) by using formula (1.3.61) and boundary condition (1.4)

$$\int_\Omega \hat{\sigma} : \hat{\epsilon}(\delta\bar{u})dV = \int_\Gamma \bar{n}\cdot\hat{\sigma}\cdot\delta\bar{u}dS - \int_\Omega (\bar{\nabla}\cdot\hat{\sigma})\cdot\delta\bar{u}dV$$

$$= \int_{\Gamma^{(\sigma)}} \bar{n}\cdot\hat{\sigma}\cdot\delta\bar{u}dS - \int_\Omega (\bar{\nabla}\cdot\hat{\sigma})\cdot\delta\bar{u}dV, \qquad (1.25)$$

where $\bar{n}$ is the unit outward normal vector to surface Γ. Substitution of expression (1.25) into Eq. (1.24) yields

$$\Phi(\bar{u}^*)=\Phi(\bar{u})-\int_\Omega (\bar{\nabla}\cdot\hat{\sigma}+\rho\bar{B})\cdot\delta\bar{u}dV+\int_{\Gamma^{(\sigma)}}(\bar{n}\cdot\hat{\sigma}-\bar{b})\cdot\delta\bar{u}dS+N(\delta\bar{u}). \qquad (1.26)$$

The displacement field $\bar{u}$ minimizes (locally) functional $\Phi(u)$ if
(i) the linear terms compared with $\delta\bar{u}$ in Eq. (1.26) vanish,
(ii) the quadratic functional $N(\delta\bar{u})$ is positive definite.

Condition (i) is called *the Euler–Lagrange condition*, while condition (ii) is called *the Legendre–Hadamard condition*.

It follows from Eq. (1.26), that the Euler–Lagrange condition (i), i.e. the necessary condition of minimum, implies the equilibrium equation (2.2.21) in Ω and the boundary condition (2.2.20) on $\Gamma^{(\sigma)}$.

1.4. Sufficient Conditions of Minimum

In this subsection we discuss the sufficient condition of local minimum

$$N(\delta\bar{u}) > 0. \qquad (1.27)$$

By using Eqs. (2.1.75), (3.2.18), and (3.2.19), we write Eq. (1.23) in the form

$$2N(\delta\bar{u}) = \int_\Omega K(\hat{\epsilon}(\delta\bar{u}))dV - \mathcal{D}(\delta\bar{u}), \qquad (1.28)$$

where

$$K(\hat{\epsilon}) = \frac{4}{\sqrt{I_3}}\{(\frac{\partial W}{\partial I_2} + I_1\frac{\partial W}{\partial I_3})[(\hat{F}:\hat{\epsilon})^2 - (\hat{F}\cdot\hat{\epsilon}):(\hat{F}\cdot\hat{\epsilon})]$$
$$-2\frac{\partial W}{\partial I_3}[(\hat{F}:\hat{\epsilon})(\hat{F}^2:\hat{\epsilon}) - (\hat{F}\cdot\hat{\epsilon}):(\hat{F}^2\cdot\hat{\epsilon})]$$
$$+[(\hat{F}:\hat{\epsilon})(\frac{\partial}{\partial I_1} + I_1\frac{\partial}{\partial I_2}) - (\hat{F}^2:\hat{\epsilon})\frac{\partial}{\partial I_2} + (\hat{I}:\hat{\epsilon})I_3\frac{\partial}{\partial I_3}]^2 W\} \quad (1.29)$$

and

$$\mathcal{D}(\delta\bar{u}) = -\int_\Omega I_1(\bar{\nabla}\delta\bar{u}^T\cdot\hat{\sigma}\cdot\bar{\nabla}\delta\bar{u})dV. \qquad (1.30)$$

213

Exercise 1.4. Check that for small deformations for transition from the initial to the non-perturbed actual configuration, Eq. (1.29) can be written as

$$K(\hat{\epsilon}) = 4\{[(\frac{\partial}{\partial I_2} + \frac{\partial}{\partial I_3})W]_{I_1=3,\ I_2=3,\ I_3=1}[I_1^2(\hat{\epsilon}) - I_1(\hat{\epsilon}^2)]$$

$$+[(\frac{\partial}{\partial I_1} + 2\frac{\partial}{\partial I_2} + \frac{\partial}{\partial I_3})^2 W]_{I_1=3,\ I_2=3,\ I_3=1}I_1^2(\hat{\epsilon})\}. \quad \square \qquad (1.31)$$

Exercise 1.5. By using Eqs. (3.2.29) and (3.2.30), rewrite Eq. (1.31) in the form

$$K(\hat{\epsilon}) = \lambda I_1^2(\hat{\epsilon}) + 2\mu I_1(\hat{\epsilon}^2), \qquad (1.32)$$

where λ and μ are the Lame parameters. $\square$

Since the Lame parameters are positive, inequality

$$K(\hat{\epsilon}(\delta\bar{u})) > 0 \qquad (1.33)$$

is fulfilled for an arbitrary tensor $\hat{\epsilon}$ for small deformations from the initial to the non-perturbed actual configuration. In the general case, we should require Eq. (1.33) to be valid in order to ensure inequality (1.27) to be fulfilled for an arbitrary stress tensor $\hat{\sigma}$ in the non-perturbed state.

Let us confine ourselves to affine transformations from the non-perturbed to the perturbed actual configuration

$$\bar{r}^* = \bar{r} \cdot (\hat{I} + \bar{h}_1 \bar{h}_2), \qquad (1.34)$$

where $\bar{h}_1$ and $\bar{h}_2$ are arbitrary constant vectors. It follows from Eq. (1.34) that $\delta\bar{u} = \bar{r}^* - \bar{r} = \bar{r} \cdot \bar{h}_1 \bar{h}_2$, $\bar{\nabla}\delta\bar{u} = \bar{h}_1 \bar{h}_2$, and

$$\hat{\epsilon}(\delta\bar{u}) = \frac{1}{2}(\bar{h}_1\bar{h}_2 + \bar{h}_2\bar{h}_1).$$

Substitution of this expression into (1.33) yields

$$K(\frac{1}{2}(\bar{h}_1\bar{h}_2 + \bar{h}_2\bar{h}_1)) > 0. \qquad (1.35)$$

Eq. (1.35) is called *the Hadamard condition*.

Exercise 1.6. Show that the condition of strong ellipticity (1.5.43) for strain energy density W implies the Hadamard condition (1.35). $\square$

Let us assume that inequality (1.33) is valid for an arbitrary admissible displacement field $\delta\bar{u}$. Then the sufficient condition of minimum (1.27) is fulfilled if

$$\Lambda = \sup_{\bar{w}} \frac{|\mathcal{D}(\bar{w})|}{\int_\Omega K(\hat{\epsilon}(\bar{w}))dV} < 1 \qquad (1.36)$$

for any admissible displacement field $\bar{w}$.

We do not intend to discuss the stability theory for elastic and viscoelastic media, referring to the recent book by Drozdov & Kolmanovskii (1994). However, it is worth noting that Eq. (1.36) coincides with the stability condition for an elastic medium under dead loads $\bar{B}$ and $\bar{b}$. This allows inequality (1.36), which guarantees the positivity of the second variation of the total energy, to be referred to as *infinitesimal stability condition*, see Truesdell & Noll (1965).

Exercise 1.7. Show that for small deformations at transition from the initial to the non-perturbed actual configuration, inequality (1.36) is valid if

$$\Lambda = \sup_{\bar{w}} \frac{\int_{\Omega_0} |I_1(\bar{\nabla}_0 \bar{w}^T \cdot \hat{\sigma} \cdot \bar{\nabla}_0 \bar{w})| dV_0}{\int_{\Omega_0} [\lambda I_1^2(\epsilon_0(\bar{w})) + 2\mu I_1(\hat{\epsilon}_0^2(\bar{w}))] dV_0} < 1, \tag{1.37}$$

where $\bar{\nabla}_0$ is the Hamilton operator in the initial configuration, and $\hat{\epsilon}_0(\bar{w})$ is the first infinitesimal strain tensor corresponding to the displacement vector $\bar{w}$. $\square$

1.5. Infinitesimal Stability Conditions

To derive explicit conditions of infinitesimal stability, we should maximize the functional in the left-hand side of Eq. (1.37). The latter problem is too complicated to be solved analytically or numerically. Therefore, additional simplifications of Eq. (1.37) are necessary. For this purpose, we replace the optimization problem (1.37) by appropriate estimates which impose rough (but explicit) restrictions on the non-perturbed stress tensor $\hat{\sigma}$.

Let us estimate the integrand in the numerator of Eq. (1.37)

$$|I_1(\bar{\nabla}\bar{w}^T \cdot \hat{\sigma} \cdot \bar{\nabla}\bar{w})| \leq |\sigma^{ij}\nabla_i w^k \nabla_j w_k| \leq \sigma^0 \nabla^i w^k \nabla_i w_k = \sigma^0 \bar{\nabla}\bar{w} : \bar{\nabla}\bar{w}^T,$$

where σ^0 is the maximal absolute value of *the principal stresses* (the eigenvalues of the Cauchy stress tensor $\hat{\sigma}$). Using Eq. (1.3.12), we obtain

$$|I_1(\bar{\nabla}\bar{w}^T \cdot \hat{\sigma} \cdot \bar{\nabla}\bar{w})| \leq \sigma^0 \bar{\nabla}\bar{w} : \bar{\nabla}\bar{w}^T = \sigma^0[\hat{\epsilon}(\bar{w}) : \hat{\epsilon}(\bar{w}) + \hat{\omega}(\bar{w}) : \hat{\omega}^T(\bar{w})], \tag{1.38}$$

where $\hat{\omega}(\bar{w}) = \frac{1}{2}(\bar{\nabla}_0 \bar{w}^T - \bar{\nabla}_0 \bar{w})$ is the spin tensor corresponding to the displacement vector $\bar{w}$.

The expression in the denominator is estimated as follows:

$$\lambda I_1^2(\epsilon_0(\bar{w})) + 2\mu I_1(\hat{\epsilon}_0^2(\bar{w})) \geq 2\mu I_1(\hat{\epsilon}_0^2(\bar{w})). \tag{1.39}$$

Substituting expressions (1.38) and (1.39) into Eq. (1.37), we obtain

$$\Lambda \leq \frac{\max_{\xi \in \Omega_0} \sigma^0(\xi)}{2\mu} [1 + \sup_{\bar{w}} \frac{\int_{\Omega_0} \hat{\omega}(\bar{w}) : \hat{\omega}^T(\bar{w}) dV_0}{\int_{\Omega_0} \hat{\epsilon}(\bar{w}) : \hat{\epsilon}(\bar{w}) dV_0}]. \tag{1.40}$$

215

Proposition 1.2 (The Korn inequality). There exists a positive constant C such that for any continuously differentiable function $\bar{w}$ satisfying boundary condition (1.4)

$$\int_{\Omega_0} \hat{\omega}(\bar{w}) : \hat{\omega}^T(\bar{w})dV_0 \leq C \int_{\Omega_0} \hat{\epsilon}(\bar{w}) : \hat{\epsilon}(\bar{w})dV_0. \qquad \square \qquad (1.41)$$

We do not intend to prove the Korn inequality, referring to Duvaut & Lions (1976), where a detailed proof is provided.

The minimal constant C which ensures Eq. (1.41) is called *the Korn constant*:

$$C = \sup_{\bar{w}} \frac{\int_{\Omega_0} \hat{\omega} : \hat{\omega}^T dV_0}{\int_{\Omega_0} \hat{\epsilon} : \hat{\epsilon}dV_0}. \qquad (1.42)$$

The constant C is a geometrical characteristic of domain Ω_0.

It follows from Eqs. (1.40) and (1.42) that

$$\Lambda \leq \frac{\max_{\xi \in \Omega_0} \sigma^0(\xi)}{2\mu}(C + 1). \qquad (1.43)$$

Therefore, inequality (1.37) is valid if

$$\max_{\xi \in \Omega_0} \sigma^0(\xi) < \frac{2\mu}{C + 1}. \qquad (1.44)$$

Eq. (1.44) provides a stability condition which is sufficiently rough, but easy to test. It is of interest to compare Eq. (1.44) with the condition of the material failure

$$\max_{\xi} \sigma^0(\xi) < \sigma^*, \qquad (1.45)$$

where σ^* is the ultimate stress, see Section 3.1. Since for a number of elastic materials $\sigma_* \ll \mu$, Eqs. (1.44) and (1.45) imply that the loss of stability precedes the material failure when a domain Ω_0 has a large Korn's constant $C \gg 1$.

2. Phase Transitions in Elastic Media

In this Section we study isothermal phase transitions in elastic solids with finite strains. Phase transformations in solids are in the focus of attention in the mechanics of continua owing to their physical applications (*twinning* in crystals, *austenite-martensite transformations* in metals, load-induced transformations in *shape memory alloys*, etc.), as well as due to a new mathematical apparatus developed for these problems with unknown boundaries. By using the generalized Lagrange principle, we derive the standard equilibrium equations and boundary conditions as necessary conditions of minimum for the total energy, as well as an additional condition (equality of the Eshelby tensors) which allows the interface position to be found.

2.1. Formulation of the Problem

Let us consider an elastic medium which occupies a bounded connected domain Ω_0 with a smooth boundary Γ_0. Points of Ω_0 refer to Lagrangian coordinates $\xi = \{\xi^i\}$. The medium can be in two solid phases. Particular results for phase transitions in elastic liquids are obtained as consequences of the general approach.

Denote by $\Omega_{0\,1}$ and $\Omega_{0\,2}$ domains in the initial configuration occupied by the substance in phases 1 and 2, respectively. For simplicity we suppose that domain $\Omega_{0\,2}$ is connected and is located inside domain $\Omega_{0\,1}$. Domains $\Omega_{0\,1}$ and $\Omega_{0\,2}$ are divided by a smooth *interface* γ_0 which does not cross Γ_0.

At instant $t = 0$, a body force $\bar{B}$ and a surface traction $\bar{b}$ are applied to the medium. The force $\bar{b}$ is given on a part $\Gamma_0^{(\sigma)}$ of boundary Γ_0, the other part of the boundary $\Gamma_0^{(u)}$ is clamped. Under the action of external loads, the medium deforms and a part of the material is transformed from phase 1 into phase 2. In the actual configuration, the medium occupies domains Ω_1 and Ω_2, divided by an interface γ, which does not intersect boundary Γ.

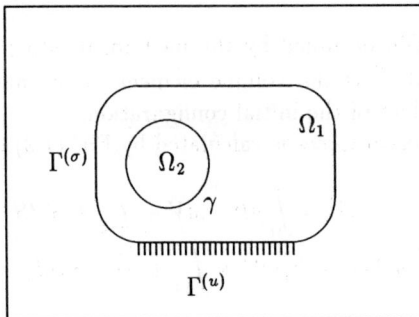

Fig. 2.1. Domain Ω occupied by an elastic medium in phases 1 and 2.

Our objective is to determine stresses and displacements in the domain $\Omega = \Omega_1 \bigcup \Omega_2$, as well as the interface position γ for given external forces.

2.2. Initial, Natural, and Actual Configurations

In "classical" problems in finite elasticity, two basic configurations are employed: *the initial*, where the displacement vector $\bar{u}$ is defined, and *the actual*, occupied by a medium under the action of applied loads. The third (*natural*) configuration, where a body's portion is in the stress-free state, is not used, since the initial configuration is assumed to be stress-free, i.e. to coincide with the natural configuration.

In the theory of phase transitions, we should distinguish initial and natural configurations. The initial configuration is introduced to provide a kinematic picture of deformations by using the displacement vector. Thus, practically any configuration may be treated as initial.

The natural configuration is characterized uniquely by the material properties: strain energy density W is determined as a function of the deformation gradient from the natural to the actual configuration. Denote by $\bar{r}_0$, $\bar{r}_0^{(k)}$, and $\bar{r}$ radius-vectors of a point ξ in the initial, natural in kth phase, and actual configurations. Let $\bar{\nabla}_0^{(k)}\bar{r}$ be the deformation gradient for transition from the natural configuration of kth phase to the actual configuration. Strain energy density of a material in kth phase is assumed to depend on $\bar{\nabla}_0^{(k)}\bar{r}$:

$$W^{(k)} = W^{(k)}(\bar{\nabla}_0^{(k)}\bar{r}).$$

Therefore, the total strain energy equals

$$W^\dagger = \int_{\Omega_0^{(1)}} W^{(1)}(\bar{\nabla}_0^{(1)}\bar{r})dV_0^{(1)} + \int_{\Omega_0^{(2)}} W^{(2)}(\bar{\nabla}_0^{(2)}\bar{r})dV_0^{(2)}, \qquad (2.1)$$

where $\Omega_0^{(k)}$ is a domain occupied by the medium in kth phase in the natural configuration, and $dV_0^{(k)}$ is the volume element in the natural configuration. Eq. (2.1) is independent of the initial configuration.

The work of external forces is calculated by Eq. (1.2)

$$A^\dagger = \int_\Omega \rho \bar{B} \cdot \bar{u}dV + \int_{\Gamma^{(\sigma)}} \bar{b} \cdot \bar{u}dS$$

$$= \int_\Omega \rho \bar{B} \cdot (\bar{r} - \bar{r}_0)dV + \int_{\Gamma^{(\sigma)}} \bar{b} \cdot (\bar{r} - \bar{r}_0)dS, \qquad (2.2)$$

where dV and dS are the volume and surface elements in the actual configuration, and $\bar{u} = \bar{r} - \bar{r}_0$ is the displacement vector from the initial to the actual configuration. Formally, Eq. (2.2) depends on the initial configuration through vector $\bar{r}_0$. On the other hand, this dependence has no mechanical meaning since only small perturbations of $A^\dagger$ are used, which, in turn, are independent of the initial configuration.

2.3. Basic Assumptions

Solid-solid phase transitions are studied under the following assumptions:

218

(i) body and surface forces are applied so slowly that dynamical effects can be neglected;

(ii) body force $\bar{B}$ and surface traction $\bar{b}$ are dead;

(iii) the temperature is fixed and coincides with the temperature of phase transition in the absence of stresses;

(iv) *the first order* phase transitions occur in the medium, which are accompanied by an additional (*phase*) deformation. This deformation is defined as a deformation which is necessary to transform the natural configuration for phase 1 into the natural configuration for phase 2;

(v) the natural configurations for phases 1 and 2 are given *a priori*, and are independent of the loading process. This assumption implies that phase transitions in elastic media are independent of the loading history. An alternative hypothesis implies a prescribed connection between the actual configuration of a material portion in the first phase immediately before phase transformation and its natural configuration in the second phase. Such a hypothesis allows the so-called *phase hysteresis* to be described, see Arutyunyan & Drozdov (1992) for details;

(vi) the material behavior in phases 1 and 2 is described by different strain energy densities $W^{(k)}$ and, as a consequence, by different constitutive equations;

(vii) the displacement field $\bar{u}$ is continuously differentiable in domain Ω_0. This hypothesis describes the so-called *coherent* phase transitions. It is worth noting that this definition coincides with the classical definition of coherent phase transitions. According to the standard definition, at coherent phase transitions the deformation gradient is continuous in $\Omega_{0\,k}$, and it may have finite jumps of discontinuity on the interface γ_0, see e.g. Gurtin (1993) for details. According to our approach, only the displacement field $\bar{u}$ is continuously differentiable, while the deformation gradients $\bar{\nabla}_0^{(k)}\bar{r}$ may have finite jumps of discontinuity on the interface owing to hypothesis (iv);

(viii) strain energy density of interface γ between two bulk phases is neglected. This hypothesis is introduced to avoid some technical difficulties caused by calculating differentials of surface integrals. A theory accounting for the interface energy is derived by Roitburd (1974) at infinitesimal strains and by Gurtin & Struthers (1990) at finite strains.

2.4. *The Generalized Lagrange Principle*

In the problem of phase transitions a new "variable" arises: the interface position. The governing equations developed in Section 3.3 are not sufficient to determine the interface location γ, and an additional equation is necessary.

To derive this equation we employ the Lagrange principle. This principle is treated here on a large scale, since we assume both the displacements and the interface position to be varied, see Ericksen (1991) for a discussion of this issue. For simplicity we confine ourselves only to necessary conditions of minimum (the Euler–Lagrange conditions), and calculate all the quantities up to the second order terms compared with small perturbations.

Proposition 2.1 (The generalized Lagrange principle). The real interface γ_0 and the real displacement field $\bar{u}$ which occur in an elastic medium minimize the total energy

$$\Phi = W^\dagger - A^\dagger \qquad (2.3)$$

on the set of arbitrary smooth interfaces which do not intersect a boundary Γ_0 and the set Υ of smooth displacement fields satisfying boundary condition (1.4). □

According to this assertion, for any admissible interface γ_0^* and any admissible displacement field $\bar{u}^*$ we have

$$\Phi(\bar{u}^*, \gamma_0^*) \geq \Phi(\bar{u}, \gamma_0). \qquad (2.4)$$

Eq. (2.4) is not convenient for applications, since no technique exists to calculate variations of the functional Φ on a set of smooth surfaces. Therefore, this problem should be replaced by another, where the calculations can be carried out. Instead of varying the interface position, it is convenient to vary the displacement fields which describe transitions from the initial configuration to the natural configurations in phases 1 and 2.

Let us fix an interface γ_0 and present transformation of the initial configuration into the natural configuration for kth phase as follows:

$$\bar{r}_0^{(k)} = \bar{r}_0 + \bar{w}^{(k)}, \qquad (2.5)$$

where $\bar{w}^{(k)}$ are displacement vectors. The displacement field

$$\bar{w} = \begin{cases} \bar{w}^{(1)}(\xi) & \xi \in \Omega_{0\,1}, \\ \bar{w}^{(2)}(\xi) & \xi \in \Omega_{0\,2} \end{cases}$$

is continuous in Ω_0 and is continuously differentiable in $\Omega_{0\,k}$. We assume additionally that

$$\bar{w}\,|_{\Gamma_0} = 0. \qquad (2.6)$$

Denote by $\bar{w}^*$ another displacement field satisfying the same conditions. We replace Eq. (2.4) by the problem of seeking displacement fields $\bar{u}$, $\bar{w}$, and interface γ_0 such that

$$\Phi(\bar{u}^*, \bar{w}^*) \geq \Phi(\bar{u}, \bar{w}) \qquad (2.7)$$

for any admissible displacement fields $\bar{u}^*$ and $\bar{w}^*$.

2.5. Necessary Conditions of Minimum

To derive necessary conditions for problem (2.7), we introduce perturbations of the displacement fields $\bar{w}^{(k)}$ in the form

$$\bar{r}_0^{(k)*} = \bar{r}_0^{(k)} + \delta\bar{w}^{(k)}, \tag{2.8}$$

where $\delta\bar{w}^{(k)}(\xi)$ are sufficiently small displacement fields. Differentiation of Eq. (2.8) yields

$$\bar{\nabla}_0^{(k)}\bar{r}_0^{(k)*} = \hat{I} + \bar{\nabla}_0^{(k)}\delta\bar{w}^{(k)}. \tag{2.9}$$

By repeating the same procedure which was used to derive Eqs. (1.8) and (1.10), we obtain

$$\bar{g}_{0\,i}^{(k)*} = \bar{g}_{0\,i}^{(k)} \cdot (\hat{I} + \bar{\nabla}_0^{(k)}\delta\bar{w}^{(k)}), \qquad \bar{g}_0^{(k)*\,i} = (\hat{I} - \bar{\nabla}_0^{(k)}\delta\bar{w}^{(k)}) \cdot \bar{g}_0^{(k)\,i}. \tag{2.10}$$

It follows from Eqs. (1.18) and (2.1.75) that up to the second order terms compared with $|\delta\bar{w}^{(k)}|$

$$\delta(\frac{dV_0^{(k)}}{dV_0}) = \delta\sqrt{I_3(\hat{F}_0^{(k)})} = \frac{1}{2\sqrt{I_3(\hat{F}_0^{(k)})}}\delta\hat{F}_0^{(k)}$$

$$= \sqrt{I_3(\hat{F}_0^{(k)})}\bar{\nabla}_0^{(k)} \cdot \delta\bar{w}^{(k)} = \frac{dV_0^{(k)}}{dV_0}\bar{\nabla}_0^{(k)} \cdot \delta\bar{w}^{(k)}, \tag{2.11}$$

where $\hat{F}_0^{(k)}$ is the Finger tensor for transition from the initial to the natural configuration for kth phase. Eq. (2.11) implies that

$$\delta(dV_0^{(k)}) = dV_0^{(k)}\bar{\nabla}_0^{(k)} \cdot \delta\bar{w}^{(k)}. \tag{2.12}$$

Transition from the initial to the non-perturbed actual configuration is described by the formula

$$\bar{r} = \bar{r}_0 + \bar{u}. \tag{2.13}$$

We seek its perturbation in the form

$$\bar{r}^* = \bar{r} + \delta\bar{u}, \tag{2.14}$$

where $\delta\bar{u}$ is a small perturbation of the displacement vector. Differentiation of Eq. (2.14) with respect to ξ^i yields

$$\frac{\partial\bar{r}^*}{\partial\xi^i} = \frac{\partial\bar{r}}{\partial\xi^i} + \frac{\partial\delta\bar{u}}{\partial\xi^i} = \bar{g}_i + \bar{g}_i \cdot \bar{\nabla}\delta\bar{u} = \bar{g}_i \cdot (\hat{I} + \bar{\nabla}\delta\bar{u}). \tag{2.15}$$

It follows from Eqs. (2.10) and (2.15) that with the requested level of accuracy

$$\bar{\nabla}_0^{(k)*}\bar{r}^* = \bar{g}_0^{(k)*\,i}\frac{\partial \bar{r}^*}{\partial \xi^i} = (\hat{I} - \bar{\nabla}_0^{(k)}\delta\bar{w}^{(k)}) \cdot \bar{g}_0^{(k)i}\bar{g}_i(\hat{I} + \bar{\nabla}\delta\bar{u})$$
$$= \bar{\nabla}_0^{(k)}\bar{r} - \bar{\nabla}_0^{(k)}\delta\bar{w}^{(k)} \cdot \bar{\nabla}_0^{(k)}\bar{r} + \bar{\nabla}_0^{(k)}\bar{r} \cdot \bar{\nabla}\delta\bar{u}.$$

Therefore,

$$\delta(\bar{\nabla}_0^{(k)}\bar{r}) = -\bar{\nabla}_0^{(k)}\delta\bar{w}^{(k)} \cdot \bar{\nabla}_0^{(k)}\bar{r} + \bar{\nabla}_0^{(k)}\bar{r} \cdot \bar{\nabla}\delta\bar{u}. \tag{2.16}$$

Let us calculate perturbation of the work $A^\dagger$. According to Eq. (3.3.6), there are such functions $\bar{B}_0^{(k)}$ and $\bar{b}_0$ that

$$A^\dagger = \sum_{k=1}^{2}\int_{\Omega_0^{(k)}} \rho_0^{(k)}\bar{B}_0^{(k)} \cdot (\bar{r} - \bar{r}_0)dV_0^{(k)} + \int_{\Gamma_0^{(\sigma)}} \bar{b}_0 \cdot (\bar{r} - \bar{r}_0)dS_0, \tag{2.17}$$

where $\rho_0^{(k)}$ is mass density in the natural configuration for kth phase, dS_0 is the surface element in the initial configuration. It follows from Eq. (2.17) that

$$\delta A^\dagger = \sum_{k=1}^{2}\int_{\Omega_0^{(k)}} \rho_0^{(k)}\bar{B}_0^{(k)} \cdot \delta\bar{r}dV_0^{(k)} + \int_{\Gamma_0^{(\sigma)}} \bar{b}_0 \cdot \delta\bar{r}dS_0. \tag{2.18}$$

The radius-vector $\bar{r}$ of a point ξ in the actual configuration is determined by its radius-vector $\bar{r}_0$ in the initial configuration, which, in turn, is an image of its radius-vector $\bar{r}_0^{(k)}$ in the natural configuration. Thus, to calculate $\delta\bar{r}$ we should use the formula

$$\delta\bar{r} = \bar{r}^*(\bar{r}_0(\bar{r}_0^{(k)*})) - \bar{r}(\bar{r}_0(\bar{r}_0^{(k)})).$$

It follows from this equality and Eqs. (2.5), (2.8), (2.13), and (2.14) that up to the second order terms compared with additional displacements

$$\delta\bar{r} = \bar{r}(\bar{r}_0(\bar{r}_0^{(k)*})) + \delta\bar{u}(\bar{r}_0(\bar{r}_0^{(k)})) - \bar{r}(\bar{r}_0(\bar{r}_0^{(k)}))$$
$$= \bar{r}(\bar{r}_0^{(k)} - \bar{w}^{(k)} - \delta\bar{w}^{(k)}) + \delta\bar{u}(\bar{r}_0^{(k)} - \bar{w}^{(k)}) - \bar{r}(\bar{r}_0^{(k)} - \bar{w}^{(k)})$$
$$= \delta\bar{u} - \bar{\nabla}_0^{(k)}\bar{r}^T \cdot \delta\bar{w}^{(k)}. \tag{2.19}$$

Substitution of expression (2.19) into the first term in the right-hand side of Eq. (2.18) with the use of Eq. (3.3.6) yields

$$\sum_{k=1}^{2}\int_{\Omega_0^{(k)}} \rho_0^{(k)}\bar{B}_0^{(k)} \cdot \delta\bar{r}dV_0^{(k)}$$
$$= \sum_{k=1}^{2}\int_{\Omega_0^{(k)}} \rho_0^{(k)}\bar{B}_0^{(k)} \cdot \delta\bar{u}dV_0^{(k)} - \sum_{k=1}^{2}\int_{\Omega_0^{(k)}} \rho_0^{(k)}\bar{B}_0^{(k)} \cdot \bar{\nabla}_0^{(k)}\bar{r}^T \cdot \delta\bar{w}^{(k)}dV_0^{(k)}$$
$$= \int_{\Omega} \rho\bar{B} \cdot \delta\bar{u}dV - \sum_{k=1}^{2}\int_{\Omega_0^{(k)}} \rho_0^{(k)}\bar{B}_0^{(k)} \cdot \bar{\nabla}_0^{(k)}\bar{r}^T \cdot \delta\bar{w}^{(k)}dV_0^{(k)}. \tag{2.20}$$

222

It follows from Eqs. (2.6) and (2.19) that $\delta\bar{r}|_{\Gamma_0} = \delta\bar{u}$. Bearing in mind this equality, we transform the second term in the right-hand side of Eq. (2.18) as follows:

$$\int_{\Gamma_0^{(\sigma)}} \bar{b}_0 \cdot \delta\bar{r}dS_0 = \int_{\Gamma_0^{(\sigma)}} \bar{b}_0 \cdot \delta\bar{u}dS_0 = \int_{\Gamma^{(\sigma)}} \bar{b} \cdot \delta\bar{u}dS. \tag{2.21}$$

Substitution of Eqs. (2.20) and (2.21) into Eq. (2.18) implies that

$$\delta A^\dagger = \int_\Omega \rho\bar{B} \cdot \delta\bar{u}dV + \int_{\Gamma^{(\sigma)}} \bar{b} \cdot \delta\bar{u}dS - J, \tag{2.22}$$

where

$$J = \sum_{k=1}^{2} \int_{\Omega_0^{(k)}} \rho_0^{(k)} \bar{B}_0^{(k)} \cdot \bar{\nabla}_0^{(k)} \bar{r}^T \cdot \delta\bar{w}^{(k)} dV_0^{(k)}. \tag{2.23}$$

Let us calculate perturbations of the total strain energy $W^\dagger$ caused by perturbations of displacement vectors $\delta\bar{u}$ and $\delta\bar{w}^{(k)}$. It follows from Eq. (2.1) that up to the second order terms compared with additional displacements

$$\begin{aligned}
\delta W^\dagger &= \sum_{k=1}^{2} \int_{\Omega_0^{(k)}} (W^{(k)} + \delta W^{(k)})(dV_0^{(k)} + \delta dV_0^{(k)}) - \sum_{k=1}^{2} \int_{\Omega_0^{(k)}} W^{(k)} dV_0^{(k)} \\
&= \sum_{k=1}^{2} \int_{\Omega_0^{(k)}} (\delta W^{(k)} + W^{(k)} \frac{\delta dV_0^{(k)}}{dV_0^{(k)}}) dV_0^{(k)}. \tag{2.24}
\end{aligned}$$

Eqs. (1.5.23) and (2.16) imply that

$$\delta W^{(k)} = W_{\bar{\nabla}_0^{(k)}\bar{r}} : \bar{\nabla}_0^{(k)} \bar{r}^T = W_{\bar{\nabla}_0^{(k)}\bar{r}} : (\bar{\nabla}\delta\bar{u}^T \cdot \bar{\nabla}_0^{(k)} \bar{r}^T - \bar{\nabla}_0^{(k)} \bar{r}^T \cdot \bar{\nabla}_0^{(k)} \delta\bar{w}^{(k)\,T}).$$

Substitution of this expression and Eq. (2.12) into Eq. (2.24) yields

$$\delta W^\dagger = J_1 - J_2 + J_3, \tag{2.25}$$

where

$$\begin{aligned}
J_1 &= \sum_{k=1}^{2} \int_{\Omega_0^{(k)}} W_{\bar{\nabla}_0^{(k)}\bar{r}}^{(k)} : (\bar{\nabla}\delta\bar{u}^T \cdot \bar{\nabla}_0^{(k)} \bar{r}^T) dV_0^{(k)}, \\
J_2 &= \sum_{k=1}^{2} \int_{\Omega_0^{(k)}} W_{\bar{\nabla}_0^{(k)}\bar{r}}^{(k)} : (\bar{\nabla}_0^{(k)} \bar{r}^T \cdot \bar{\nabla}_0^{(k)} \delta\bar{w}^{(k)\,T}) dV_0^{(k)}, \\
J_3 &= \sum_{k=1}^{2} \int_{\Omega_0^{(k)}} W^{(k)} \bar{\nabla}_0^{(k)} \cdot \delta\bar{w}^{(k)} dV_0^{(k)}. \tag{2.26}
\end{aligned}$$

Finally, it follows from Eqs. (2.3), (2.22), and (2.25) that

$$\delta\Phi = J_1 - J_2 + J_3 + J - \int_\Omega \rho\bar{B} \cdot \delta\bar{u}dV - \int_{\Gamma^{(\sigma)}} \bar{b} \cdot \delta\bar{u}dS. \tag{2.27}$$

223

Let us transform functional J_1. Using Eqs. (2.2.38) and (3.2.11) we obtain

$$W_{\bar{\nabla}_0^{(k)}\bar{r}}^{(k)} : (\bar{\nabla}\delta\bar{u}^T \cdot \bar{\nabla}_0^{(k)}\bar{r}^T) = \hat{P}^{(k)} : (\bar{\nabla}\delta\bar{u}^T \cdot \bar{\nabla}_0^{(k)}\bar{r}^T)$$

$$= I_1(\hat{P}^{(k)} \cdot \bar{\nabla}\delta\bar{u}^T \cdot \bar{\nabla}_0^{(k)}\bar{r}^T) = I_1(\bar{\nabla}_0^{(k)}\bar{r}^T \cdot \hat{P}^{(k)} \cdot \bar{\nabla}\delta\bar{u}^T)$$

$$= I_1(\sqrt{\frac{g}{g_0^{(k)}}}\hat{\sigma}^{(k)} \cdot \bar{\nabla}\delta\bar{u}^T) = \sqrt{\frac{g}{g_0^{(k)}}}\hat{\sigma}^{(k)} : \bar{\nabla}\delta\bar{u}^T, \tag{2.28}$$

where $\hat{P}^{(k)} = W_{\bar{\nabla}_0^{(k)}\bar{r}}^{(k)}$ is the Piola stress tensor, and $\hat{\sigma}^{(k)}$ is the Cauchy stress tensor. It follows from Eqs. (1.3.15), (1.3.59), (2.1.75), (2.28), and boundary condition (1.4) that

$$J_1 = \sum_{k=1}^{2} \int_{\Omega_0^{(k)}} \sqrt{\frac{g}{g_0^{(k)}}}\hat{\sigma}^{(k)} : \bar{\nabla}\delta\bar{u}^T dV_0^{(k)} = \sum_{k=1}^{2} \int_{\Omega_k} \hat{\sigma}^{(k)} : \bar{\nabla}\delta\bar{u}^T dV$$

$$= \int_\Gamma \bar{n}_1 \cdot \hat{\sigma}^{(1)} \cdot \delta\bar{u} dS + \int_\gamma \bar{n}_1 \cdot (\hat{\sigma}^{(1)} - \hat{\sigma}^{(2)}) \cdot \delta\bar{u} dS - \sum_{k=1}^{2} \int_{\Omega_k} (\bar{\nabla} \cdot \hat{\sigma}^{(k)}) \cdot \delta\bar{u} dV$$

$$= \int_{\Gamma(\sigma)} \bar{n}_1 \cdot \hat{\sigma}^{(1)} \cdot \delta\bar{u} dS + \int_\gamma \bar{n}_1 \cdot (\hat{\sigma}^{(1)} - \hat{\sigma}^{(2)}) \cdot \delta\bar{u} dS - \sum_{k=1}^{2} \int_{\Omega_k} (\bar{\nabla} \cdot \hat{\sigma}^{(k)}) \cdot \delta\bar{u} dV, \tag{2.29}$$

where $\bar{n}_1$ is the normal vector in the actual configuration outward with respect to the domain Ω_1.

We now transform functional J_2 by using Eq. (1.3.15) and the Stokes formula (1.3.59)

$$J_2 = \sum_{k=1}^{2} \int_{\Omega_0^{(k)}} \hat{P}^{(k)} \cdot \bar{\nabla}_0^{(k)}\bar{r}^T : \bar{\nabla}_0^{(k)}\delta\bar{w}^{(k)\,T} dV_0^{(k)}$$

$$= \int_{\Gamma_0^{(1)}} \bar{n}_0^{(1)} \cdot \hat{P}^{(1)} \cdot \bar{\nabla}_0^{(1)}\bar{r}^T \cdot \delta\bar{w}^{(1)} dS_0^{(1)}$$

$$+ \int_{\gamma^{(1)}} \bar{n}_0^{(1)} \cdot \hat{P}^{(1)} \cdot \bar{\nabla}_0^{(1)}\bar{r}^T \cdot \delta\bar{w}^{(1)} dS_0^{(1)}$$

$$+ \int_{\gamma_0^{(2)}} \bar{n}_0^{(2)} \cdot \hat{P}^{(2)} \cdot \bar{\nabla}_0^{(2)}\bar{r}^T \cdot \delta\bar{w}^{(2)} dS_0^{(2)}$$

$$- \sum_{k=1}^{2} \int_{\Omega_0^{(k)}} [\bar{\nabla}_0^{(k)} \cdot (\hat{P}^{(k)} \cdot \bar{\nabla}_0^{(k)}\bar{r}^T)] \cdot \delta\bar{w}^{(k)} dV_0^{(k)}. \tag{2.30}$$

Here $\gamma_0^{(k)}$ is the image of the interface γ at transition from the initial configuration to the natural configuration for kth phase, $\bar{n}_0^{(k)}$ is the unit outward normal to $\gamma_0^{(k)}$, $dS_0^{(k)}$ is the surface element in the natural configuration for kth phase. According to Eq. (2.6), the first term in the right-hand side of Eq. (2.30) vanishes. The second and third terms can be transformed by using Eq. (2.1.83) as

224

follows:

$$\int_{\gamma^{(1)}} \bar{n}_0^{(1)} \cdot \hat{P}^{(1)} \cdot \bar{\nabla}_0^{(1)} \bar{r}^T \cdot \delta \bar{w}^{(1)} dS_0^{(1)} + \int_{\gamma^{(2)}} \bar{n}_0^{(2)} \cdot \hat{P}^{(2)} \cdot \bar{\nabla}_0^{(2)} \bar{r}^T \cdot \delta \bar{w}^{(2)} dS_0^{(2)}$$

$$= \int_{\gamma_0} \bar{n}_{0\,1} \cdot [\sqrt{\frac{g_0^{(1)}}{g_0}} \bar{\nabla}_0^{(1)} \bar{r}_0^T \cdot \hat{P}^{(1)} \cdot \bar{\nabla}_0^{(1)} \bar{r}^T - \sqrt{\frac{g_0^{(2)}}{g_0}} \bar{\nabla}_0^{(2)} \bar{r}_0^T \cdot \hat{P}^{(2)} \cdot \bar{\nabla}_0^{(2)} \bar{r}^T] \cdot \delta \bar{w} dS_0,$$

where $\delta \bar{w} = \delta \bar{w}^{(1)} = \delta \bar{w}^{(2)}$ on γ_0 and $\bar{n}_{0\,1}$ is the unit normal to γ_0 outward with respect to $\Omega_{0\,1}$.

Exercise 2.1. By using Eqs. (1.2.8), (1.3.17), and (1.3.54), show that

$$\begin{aligned}
\bar{\nabla}_0^{(k)} \cdot (\hat{P}^{(k)} \cdot \bar{\nabla}_0^{(k)} \bar{r}^T) &= \bar{\nabla}_0^{(k)} \bar{r} \cdot (\bar{\nabla}_0^{(k)} \cdot \hat{P}^{(k)}) + \hat{P}^{(k)\,T} : \bar{\nabla}_0^{(k)} (\bar{\nabla}_0^{(k)} \bar{r}^T) \\
&= \bar{\nabla}_0^{(k)} \bar{r} \cdot (\bar{\nabla}_0^{(k)} \cdot \hat{P}^{(k)}) + \bar{\nabla}_0^{(k)} (\bar{\nabla}_0^{(k)} \bar{r}^T) : \hat{P}^{(k)} \\
&= (\bar{\nabla}_0^{(k)} \cdot \hat{P}^{(k)}) \cdot \bar{\nabla}_0^{(k)} \bar{r}^T + \bar{\nabla}_0^{(k)} (\bar{\nabla}_0^{(k)} \bar{r}^T) : \hat{P}^{(k)}. \qquad \square
\end{aligned}$$

Substitution of these expressions into Eq. (2.30) implies that

$$J_2 = \int_{\gamma_0} \bar{n}_{0\,1} \cdot [\sqrt{\frac{g_0^{(1)}}{g_0}} \bar{\nabla}_0^{(1)} \bar{r}_0^T \cdot \hat{P}^{(1)} \cdot \bar{\nabla}_0^{(1)} \bar{r}^T$$

$$- \sqrt{\frac{g_0^{(2)}}{g_0}} \bar{\nabla}_0^{(2)} \bar{r}_0^T \cdot \hat{P}^{(2)} \cdot \bar{\nabla}_0^{(2)} \bar{r}^T] \cdot \delta \bar{w} dS_0$$

$$- \sum_{k=1}^{2} \int_{\Omega_0^{(k)}} [(\bar{\nabla}_0^{(k)} \cdot \hat{P}^{(k)}) \cdot \bar{\nabla}_0^{(k)} \bar{r}^T + \bar{\nabla}_0^{(k)} (\bar{\nabla}_0^{(k)} \bar{r}^T) : \hat{P}^{(k)}] \cdot \delta \bar{w}^{(k)} dV_0^{(k)}. \quad (2.31)$$

To transform functional J_3 we rewrite it as follows:

$$J_3 = \sum_{k=1}^{2} \int_{\Omega_0^{(k)}} [\bar{\nabla}_0^{(k)} \cdot (W^{(k)} \delta \bar{w}^{(k)}) - \bar{\nabla}_0^{(k)} W^{(k)} \cdot \delta \bar{w}^{(k)}] dV_0^{(k)}. \quad (2.32)$$

By using Eqs. (1.5.27) and (3.2.11), we find

$$\bar{\nabla}_0^{(k)} W^{(k)} = \bar{\nabla}_0^{(k)} (\bar{\nabla}_0^{(k)} \bar{r}^T) : W_{\bar{\nabla}_0^{(k)} \bar{r}}^{(k)} = \bar{\nabla}_0^{(k)} (\bar{\nabla}_0^{(k)} \bar{r}^T) : \hat{P}^{(k)}.$$

Substitution of this expression into Eq. (2.32) with the use of the Stokes formula (1.3.57) yields

$$\begin{aligned}
J_3 &= \int_{\Gamma_0^{(1)}} W^{(1)} \bar{n}_0^{(1)} \cdot \delta \bar{w}^{(1)} dS_0^{(1)} + \int_{\gamma_0^{(1)}} W^{(1)} \bar{n}_0^{(1)} \cdot \delta \bar{w}^{(1)} dS_0^{(1)} \\
&+ \int_{\gamma_0^{(2)}} W^{(2)} \bar{n}_0^{(2)} \cdot \delta \bar{w}^{(2)} dS_0^{(2)} - \sum_{k=1}^{2} \int_{\Omega_0^{(k)}} \bar{\nabla}_0^{(k)} (\bar{\nabla}_0^{(k)} \bar{r}^T) : \hat{P}^{(k)} \cdot \delta \bar{w}^{(k)} dV_0^{(k)}.
\end{aligned}$$

225

It follows from Eq. (2.6) that the first term in the right-hand side of this equality equals zero. Transforming the second and third terms with the use of Eq. (2.1.83), we obtain

$$J_3 = \int_{\gamma_0} \bar{n}_{0\,1} \cdot [\sqrt{\frac{g_0^{(1)}}{g_0}} \bar{\nabla}_0^{(1)} \bar{r}_0^T W^{(1)} - \sqrt{\frac{g_0^{(2)}}{g_0}} \bar{\nabla}_0^{(2)} \bar{r}_0^T W^{(2)}] \cdot \delta \bar{w} dS_0$$

$$- \sum_{k=1}^{2} \int_{\Omega_0^{(k)}} \bar{\nabla}_0^{(k)} (\bar{\nabla}_0^{(k)} \bar{r}^T) : \hat{P}^{(k)} \cdot \delta \bar{w}^{(k)} dV_0^{(k)}. \tag{2.33}$$

Substitution of expressions (2.23), (2.29), (2.31), and (2.33) into Eq. (2.27) implies that

$$\delta \Phi = - \sum_{k=1}^{2} \int_{\Omega_k} (\bar{\nabla} \cdot \hat{\sigma}^{(k)} + \rho \bar{B}) \cdot \delta \bar{u} dV + \int_{\gamma_0} \bar{n}_{0\,1} \cdot (\hat{E}^{(1)} - \hat{E}^{(2)}) \cdot \delta \bar{w} dS_0$$

$$+ \int_{\Gamma^{(\sigma)}} (\bar{n}_1 \cdot \hat{\sigma}^{(1)} - \bar{b}) \cdot \delta \bar{u} dS + \int_{\gamma} \bar{n}_1 \cdot (\hat{\sigma}^{(1)} - \hat{\sigma}^{(2)}) \cdot \delta \bar{u} dS$$

$$+ \sum_{k=1}^{2} \int_{\Omega_0^{(k)}} (\bar{\nabla}_0^{(k)} \cdot \hat{P}^{(k)} + \rho_0^{(k)} \bar{B}_0^{(k)}) \cdot \bar{\nabla}_0^{(k)} \bar{r}^T \cdot \delta \bar{w}^{(k)} dV_0^{(k)}, \tag{2.34}$$

where

$$\hat{E}^{(k)} = \sqrt{\frac{g_0^{(k)}}{g_0}} \bar{\nabla}_0^{(k)} \bar{r}_0^T \cdot (W^{(k)} \hat{I} - \hat{P}^{(k)} \cdot \bar{\nabla}_0^{(k)} \bar{r}^T). \tag{2.35}$$

Tensor $\hat{E}^{(k)}$ is called *the Eshelby tensor* (or *tensor of the chemical potential*) for kth phase, see Eshelby (1975).

For arbitrary displacement fields $\delta \bar{u}$ and $\delta \bar{w}^{(k)}$, Eqs. (2.7) and (2.35) imply:
– the equilibrium equations for the Cauchy stress tensor $\hat{\sigma}^{(k)}$ in domains Ω_k

$$\bar{\nabla} \cdot \hat{\sigma}^{(k)} + \rho \bar{B} = 0,$$

– the equilibrium equations for the Piola stress tensor $\hat{\sigma}^{(k)}$ in domains $\Omega_0^{(k)}$

$$\bar{\nabla}_0^{(k)} \cdot \hat{P}^{(k)} + \rho_0^{(k)} \bar{B}_0^{(k)} = 0,$$

– boundary conditions in stresses on the surface $\Gamma^{(\sigma)}$

$$\bar{n}_1 \cdot \hat{\sigma}^{(1)} = \bar{b},$$

– the continuity condition for surface tractions on the interface γ

$$\bar{n}_1 \cdot \hat{\sigma}^{(1)} = \bar{n}_1 \cdot \hat{\sigma}^{(2)},$$

– the continuity conditions for the Eshelby tensors on the interface γ_0

$$\bar{n}_{0\,1} \cdot \hat{E}^{(1)} = \bar{n}_{0\,1} \cdot \hat{E}^{(2)}. \tag{2.36}$$

Eq. (2.36) is the only new governing equation which allows the interface γ to be found at phase transitions. It is called *the Maxwell rule*, see Fonseca (1989), Gurtin (1983), and James (1981).

For phase transitions in elastic liquids, see Eq. (3.2.49), we may simplify Eq. (2.36), since natural configurations for different phases may be chosen coinciding with the initial configuration. In this case, Eq. (2.35) implies that

$$\hat{E}^{(k)} = W^{(k)} \hat{I} - \hat{P}^{(k)} \cdot \bar{\nabla}_0^{(k)} \bar{r}^T. \tag{2.37}$$

It follows from Eqs. (2.1.9), (2.2.10), (2.2.37), and (3.2.51) that

$$\hat{P}^{(k)} = -p^{(k)} \sqrt{\frac{g}{g_0}} (\bar{\nabla}\bar{r}_0^{(k)})^T = -\sqrt{I_3} p^{(k)} (\bar{\nabla}_0^{(k)} \bar{r}^T)^{-1} = -\frac{p^{(k)} \rho_0}{\rho} (\bar{\nabla}_0^{(k)} \bar{r}^T)^{-1}, \tag{2.38}$$

where

$$p^{(k)} = \frac{\rho^2}{\rho_0} \frac{\partial W^{(k)}}{\partial \rho}.$$

Substitution of expression (2.38) into Eqs. (2.36) and (2.37) yields

$$\bar{n}_{0\,1} [(W^{(1)} + \frac{p^{(1)} \rho_0}{\rho}) - (W^{(2)} + \frac{p^{(2)} \rho_0}{\rho})] = 0.$$

This equality holds if and only if

$$W^{(1)} + \frac{p^{(1)} \rho_0}{\rho} = W^{(2)} + \frac{p^{(2)} \rho_0}{\rho} \tag{2.39}$$

on interface γ_0. Condition (2.39) means the equality of the specific chemical potentials of two liquids at the interface, see Landau & Lifshitz (1969).

Chapter 6

Constitutive models in finite viscoelasticity

This Chapter is concerned with constitutive models in finite viscoelasticity. In Section 1, we discuss basic physical phenomena typical of viscoelastic media and provide some experimental data. Section 2 deals with differential constitutive models, while in Section 3 we consider integral constitutive models. Finally, in Section 4 we discuss some features of creep kernels and relaxation kernels.

1. Creep and Relaxation

This Section provides an introduction to the theory of viscoelasticity. We demonstrate phenomena of creep, relaxation, and recovery, define the viscoelastic behavior of materials, and discuss the effect of temperature on the material response.

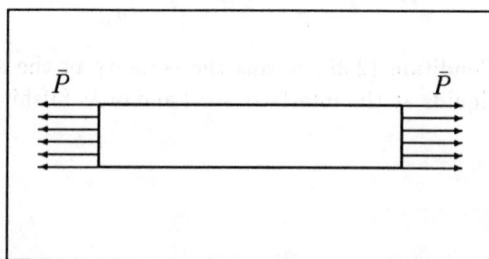

Fig. 1.1. A specimen under uniaxial load.

The viscoelastic behavior is typical of a number of materials which are extremely important for applications: polymers and plastics, see Aklonis et al.

(1972), Bicerano (1993), Ferry (1980), Struik (1978), and Ward (1971), composites, see Alberola et al. (1995), metals and alloys at elevated temperature, see Skrzypek (1993), Szczepinski (1990), and van Dam & Mischgofsky (1982), concrete, see Bazant (1988) and Neville et al. (1983), soils, see Adeyeri et al. (1970), road construction materials, see Maccarrone & Tiu (1988) and Vinogradov et al. (1977), building materials, see Papo (1988), biological tissues, see Deligianni et al. (1994), food-stuffs, see Robert & Sherman (1988) and Struik (1980), etc.

1.1. The Creep Phenomenon

Let a rod-shaped specimen be in its natural (stress-free) state. At instant $t = 0$, tensile forces P are applied to its ends and provide the stress $\sigma_0 = P/S$, where S is the cross-section area, see Fig 1.1.

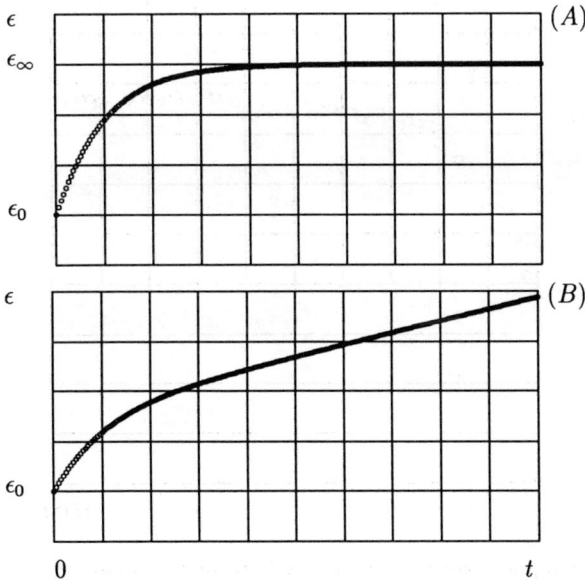

Fig. 1.2. Typical creep curves for inelastic solids (A) and inelastic liquids (B).

Immediately after application of tensile forces, an instant strain ϵ_0, see Eq. (3.1.1), arises in the specimen. For a purely elastic material, strain $\epsilon(t)$ at any instant $t \geq 0$ coincides with the initial strain ϵ_0. For an inelastic material, an additional strain $\epsilon(t) > \epsilon_0$ can be observed which increases monotonically in time: $\dot{\epsilon}(t) \geq 0$, where the superscript dot denotes the differentiation with respect to time. This phenomenon (growth of strains under a load uniform in time) is called *creep*.

The creep process is characterized either by *the creep strain* ϵ_c which is defined as the difference between the strain ϵ and the initial (elastic) strain ϵ_0

$$\epsilon_c(t) = \epsilon(t) - \epsilon_0, \qquad (1.1)$$

or by *the creep compliance* J which equals the ratio of the creep strain to the applied stress σ_0

$$J(t) = \frac{\epsilon_c(t)}{\sigma_0} = \frac{\epsilon(t) - \epsilon_0}{\sigma_0}.$$

The creep curve $\epsilon = \epsilon(t)$ is plotted in Fig. 1.4, while the creep compliance curve $J = J(t)$ is presented in Fig. 1.6.

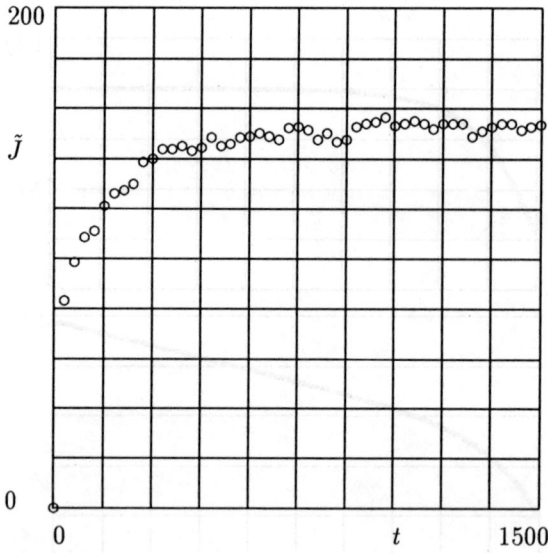

Fig. 1.3. The creep proper compliance $\tilde{J}$ (MPa^{-1}) *versus* time t (sec). Circles correspond to experimental data obtained by Gramespacher & Meissner (1995) for polystyrene melt.

The growth of strains in time is a characteristic feature of both inelastic liquids and inelastic solids. To distinguish these two substances we presume that for solid media $\epsilon(t)$ tends to a finite limit ϵ_∞ and its derivative $\dot{\epsilon}(t)$ tends to zero as $t \to \infty$, whereas for liquid media the rate of strains $\dot{\epsilon}(t)$ tends to a finite limit $\dot{\epsilon}_\infty$ which corresponds to a Newtonian flow. The standard dependencies $\epsilon(t)$ are plotted in Fig. 1.2.

For a Newtonian liquid under stress σ_0 uniform in time, the strain is proportional to time

$$\epsilon_n(t) = \frac{\sigma_0}{\eta} t, \tag{1.2}$$

where η is called *Newtonian viscosity*, and $\epsilon_n(t)$ is called *Newtonian strain*. Bearing in mind Eq. (1.1), it is convenient to extend the strain ϵ into the sum

$$\epsilon(t) = \epsilon_0 + \tilde{\epsilon}_c(t) + \epsilon_n(t), \tag{1.3}$$

where ϵ_0 is an instantaneous elastic strain, and $\tilde{\epsilon}_c(t) = \epsilon_c(t) - \epsilon_n(t)$ is *the creep proper strain*.

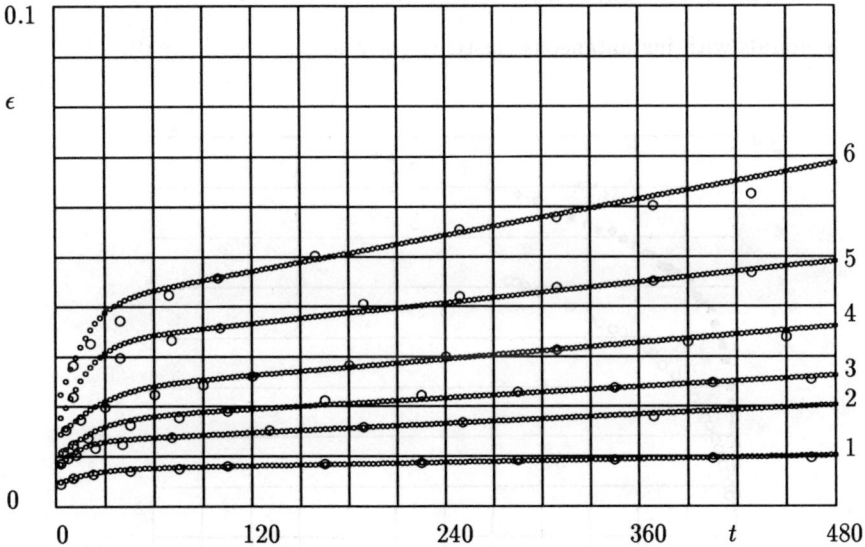

Fig. 1.4. Strain ϵ *versus* time t (min) for a polypropylene fiber. Large circles correspond to experimental data obtained by Barenblatt et al. (1974). Curve 1: $\sigma_0 = 7.46$ (MPa); curve 2: $\sigma_0 = 11.28$ (MPa); curve 3: $\sigma_0 = 13.54$ (MPa); curve 4: $\sigma_0 = 15.01$ (MPa); curve 5: $\sigma_0 = 16.82$ (MPa); curve 6: $\sigma_0 = 18.05$ (MPa).

Eq. (1.2) is valid for both solid and liquid media. The only difference is that $\eta = \infty$ for solids, and $\eta < \infty$ for liquids. By using Eq. (1.3), we may state that the creep proper strain tends to some finite limiting value for liquids as well as for solids, see Fig. 1.3, where *the creep proper compliance* $\tilde{J} = \tilde{\epsilon}_c/\sigma_0$ is plotted *versus* time t for a polymer melt.

The division of media into solids and liquids is rather conditional, and it depends essentially on temperature. For example, metals at room temperature

231

are elastic solids, whereas at elevated temperatures they demonstrate the creep typical of liquids, see Szczepinski (1990).

The strain ϵ_0 characterizes an instant response of material. In applications, it is convenient to distinguish media with instant response: $\epsilon_0 \neq 0$, and without instant response: $\epsilon_0 = 0$, see e.g. Giesekus (1995).

According to Giesekus' classification, four different kinds of materials may be distinguished:

- solids with instantaneous viscosity ($\epsilon_0 = 0$, $\epsilon_\infty < \infty$, $\dot{\epsilon}_\infty = 0$);

- solids with instantaneous elasticity ($\epsilon_0 \neq 0$, $\epsilon_\infty < \infty$, $\dot{\epsilon}_\infty = 0$);

- liquids with instantaneous viscosity ($\epsilon_0 = 0$, $\epsilon_\infty = \infty$, $\dot{\epsilon}_\infty \neq 0$);

- liquids with instantaneous elasticity ($\epsilon_0 \neq 0$, $\epsilon_\infty = \infty$, $\dot{\epsilon}_\infty \neq 0$).

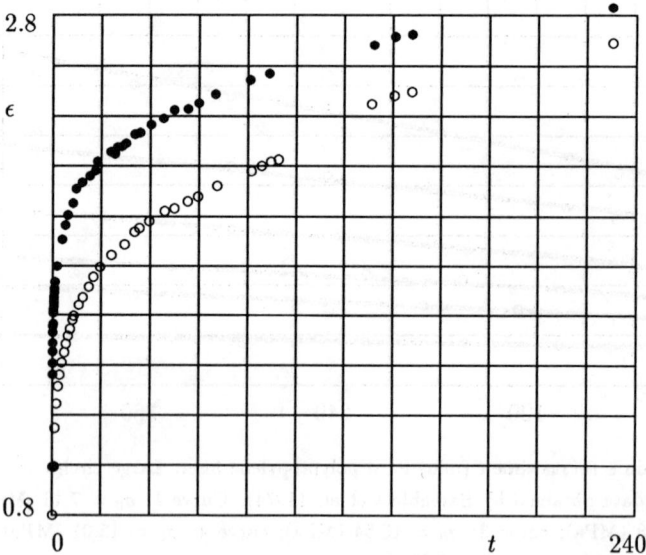

Fig. 1.5. Strain ϵ (per cent) *versus* time t (10^3 hours). Circles correspond to experimental data obtained by Findley (1987). Filled circles correspond to polyethylene under stress $\sigma = 1.55$ (MPa), unfilled circles correspond to PVC under stress $\sigma = 27.59$ (MPa).

The mechanical behavior of several polymers differs from the standard creep behavior in both solids and liquids: strain $\epsilon(t)$ increases monotonically and does not tend to any finite limit, while the rate of strain $\dot{\epsilon}(t)$ tends to zero with the growth of time.

We should be extremely careful regarding the asymptotical behavior of experimental data, since any data are available on a finite interval of time only. Nevertheless, when we employ data either obtained on, or extrapolated to extremely large intervals of time (compared with the lifetime of engineering structures), our assertion regarding the asymptotic behavior may seem plausible.

For example, Fig. 1.5 presents experimental data obtained during the 26-years interval of observations. In Fig. 1.6 experimental data are extrapolated (by using *the time-temperature superposition principle*) to the interval of about 350 years. The data presented in these figures show the absence of the limiting strain ϵ_∞.

The above figures demonstrate "pure" creep without the influence of other physical phenomena. Among "attendant" phenomena, *plasticity* and *damage* should be mentioned.

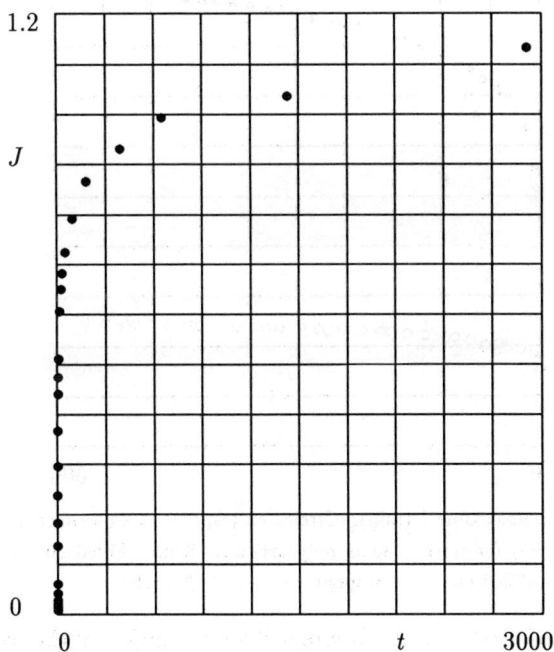

Fig. 1.6. Creep compliance J (MPa^{-1}) *versus* time t (10^3 hours). Circles correspond to experimental data obtained by Struik (1987) for unfilled polyurethane rubber quenched from 60°C to -55°C.

A typical behavior of a *viscoelastoplastic medium* is presented in Fig. 1.7. For small stresses, the material demonstrates a creep curve typical of solids:

233

the strain $\epsilon(t)$ tends to some limiting value ϵ_∞ with the growth of time. For relatively large stresses, we observe a creep curve typical of liquids: the strain $\epsilon(t)$ tends to infinity, while the rate of strain $\dot\epsilon(t)$ tends to some limiting value $\dot\epsilon_\infty \neq 0$.

The standard creep curves for very high intensities of stresses (83% of the ultimate stress for failure) are plotted in Fig. 1.8. Three regions may be distinguished where different physical processes are important. In region 1, for relatively small times t, the material behavior is similar to the behavior of solids. In region 2, a Newtonian flow can be observed typical of liquids. Finally, in region 3, strains grow rapidly (practically, as an exponential function of time) which is the characteristic feature of *the material damage*, cf. Section 3.1.

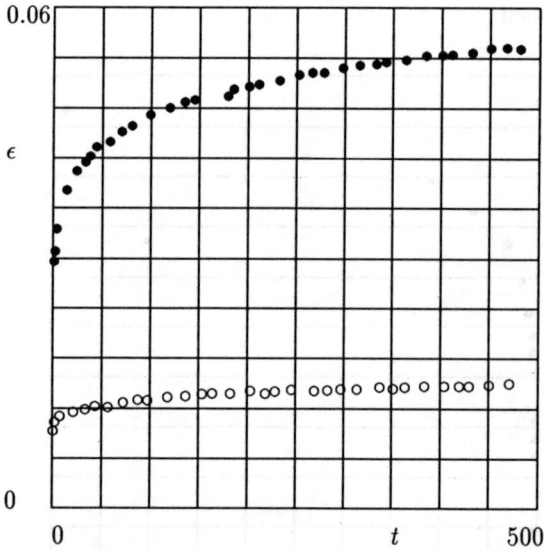

Fig. 1.7. Strain ϵ *versus* time t (min). Circles correspond to experimental data obtained by Wing et al. (1995) for microcellular polycarbonate foam. Filled circles correspond to $\sigma_0 = 27.58$ (MPa), unfilled circles correspond to $\sigma_0 = 13.79$ (MPa).

Phenomena of plasticity and failure will be the subject of the second volume of this textbook. Our objective here is to emphasize that these phenomena accompanying the material creep are observed in experiments.

1.2. The Relaxation Phenomenon

Let us now consider another type of loading, when at instant $t = 0$ the specimen is stretched up to a length l_1, and this new length remains fixed at

234

$t \geq 0$. The latter means that the strain $\epsilon(t)$ is also fixed and it remains equal to the initial strain ϵ_0.

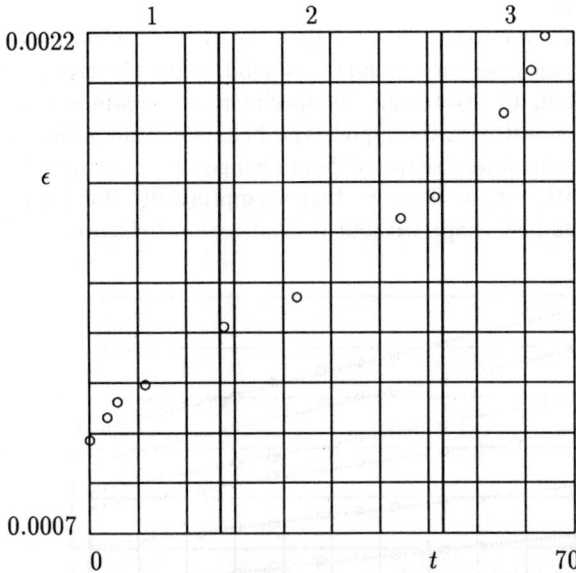

Fig. 1.8. Axial strain ϵ *versus* time t (min) for a concrete specimen. Circles correspond to experimental data obtained by Zhaoxia (1994) for axial load equal to 83% of the ultimate strength for compression.

For a purely elastic material, the stress σ_0 arises in the specimen at the initial instant and remains unchanged during the loading.

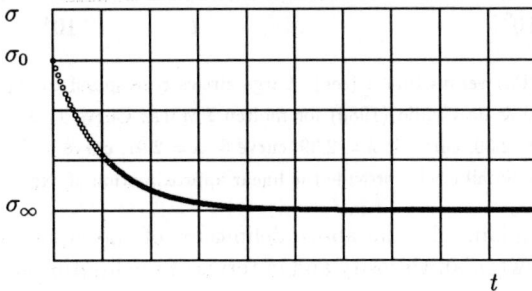

Fig. 1.9. A typical relaxation curve.

For an inelastic material, the stress σ equals σ_0 at the initial moment $t = 0$, and decreases monotonically in time for $t \geq 0$ tending to some limiting value σ_∞

235

as $t \to \infty$. This phenomenon (decrease of the stress in time for a fixed strain) is called *the stress relaxation*. A typical dependence of the stress on time is plotted in Fig. 1.9.

We distinguish two cases: when the limiting stress σ_∞ is positive, and when $\sigma_\infty = 0$. The former case corresponds to a solid, while the latter corresponds to a liquid. Indeed, by stretching the specimen, we create a new shape for it. A medium demonstrates the liquid-type behavior if no additional stresses arise when we provide a new shape without changes in its volume. For uniaxial loading, the condition $\sigma_\infty = 0$ means that asymptotically (for large times) the medium accepts its new shape without resistance, i.e. behaves as a liquid.

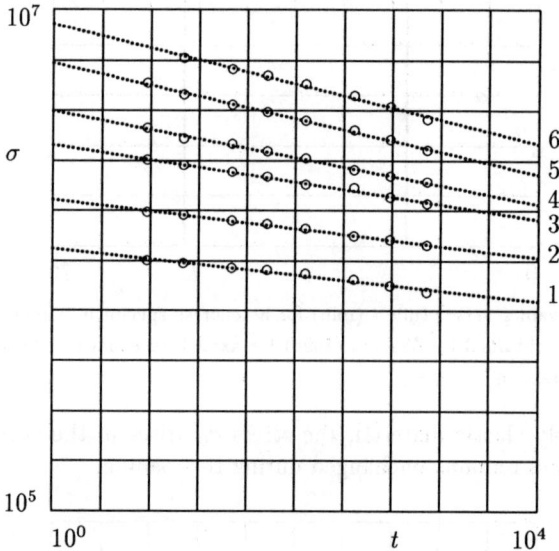

Fig. 1.10. Stress σ (Pa) *versus* time t (sec). Large circles correspond to experimental data obtained by Marrucci & de Cindio (1980) for molten PMMA. Curve 1: $\lambda = 1.32$, curve 2: $\lambda = 1.59$, curve 3: $\lambda = 2.00$, curve 4: $\lambda = 2.39$, curve 5: $\lambda = 2.91$, curve 6: $\lambda = 3.64$, where λ is the extension ratio. Small circles provide the linear approximation of experimental data.

Exercise 1.1. Explain why the above definitions of the liquid-type behavior (in terms of the Newtonian viscosity and in terms of the limiting stress) coincide with each other. □

As common practice, *the relaxation curve* $\sigma = \sigma(t)$ is plotted in *bi-logarithmic* coordinates. Several curves obtained in experiments are plotted in Figs. 1.10 – 1.12. These figures demonstrate various dependencies $\sigma = \sigma(t)$ which occur in polymeric materials, both solid and liquid: a linear dependence, see Fig. 1.10, a concave curve, see Fig. 1.11, and a convex-concave curve, see Fig. 1.12.

To describe the material response we introduce *the stress relaxation modulus*

$$E(t) = \frac{\sigma(t)}{\epsilon_0}.$$

(1.4)

For a linear elastic material, its elastic modulus $E = \sigma/\epsilon_0$ is inverse to the compliance $J = \epsilon/\sigma_0$. In general, the stress relaxation modulus $E(t)$ and the creep compliance $J(t)$ are not inverse to each other.

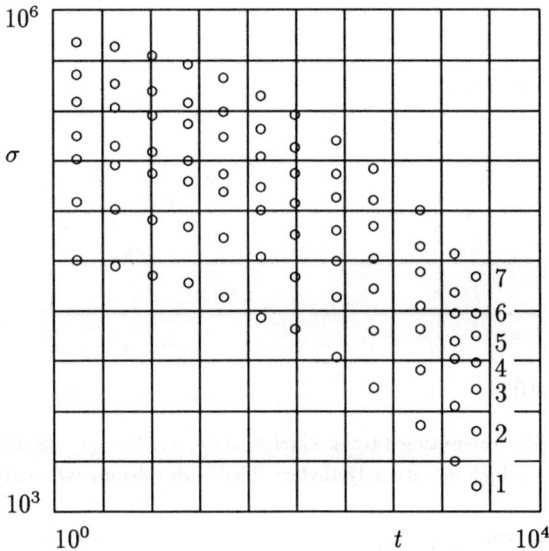

Fig. 1.11. Stress σ (Pa) *versus* time t (sec). Circles correspond to experimental data obtained by Titomanlio et al. (1980) for polyisobutylene. Curve 1: $\lambda = 1.07$, curve 2: $\lambda = 1.15$, curve 3: $\lambda = 1.26$, curve 4: $\lambda = 1.40$, curve 5: $\lambda = 1.75$, curve 6: $\lambda = 2.15$, curve 7: $\lambda = 3.42$, where λ is the extension ratio.

This part of the textbook is concerned with the study of viscoelastic media.
Definition. A material is called *viscoelastic*, if it demonstrates both creep and relaxation phenomena.

Like any definition, this definition is rather conditional. First, sometimes we refer to viscoelasticity when only one of these phenomena is taken into account. For example, some applied problems of interest are concerned with either creep or relaxation. In this case, engineering models are employed, see Section 2, which are too simple to describe these processes together, but present the

237

process under consideration only. Secondly, creep and relaxation are processes which correspond to loads uniform in time when either stress or strain remains constant. In applied problems, we encounter time-dependent loads. Under these loads, the material demonstrates such a behavior which is not relevant to "pure" creep or relaxation.

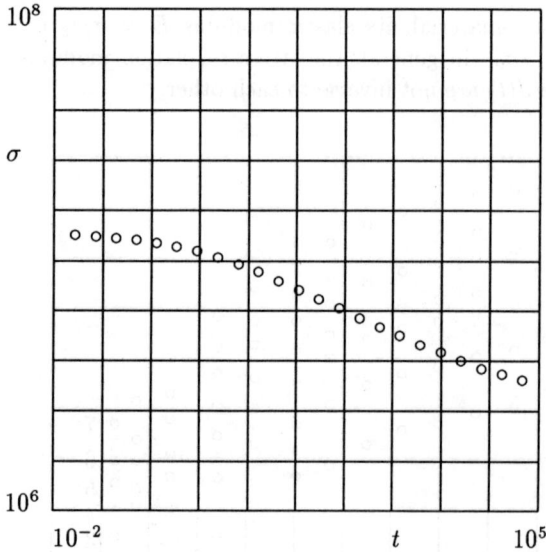

Fig. 1.12. Stress σ (Pa) *versus* time t (sec). Circles correspond to experimental data obtained by Ek et al. (1987) for high density polyethylene filled with calcium carbonate.

1.3. The Recovery Phenomenon

The creep and relaxation tests are the simplest experiments, when either stress or strain is a step-function

$$\sigma(t) = \sigma_0 \mathcal{H}(t), \qquad \epsilon(t) = \epsilon_0 \mathcal{H}(t),$$

where $\mathcal{H}(t)$ is *the Heaviside function*

$$\mathcal{H}(t) = \begin{cases} 1 & t \geq 0, \\ 0 & t < 0. \end{cases} \tag{1.5}$$

The above loadings cannot reveal all the characteristic features of the viscoelastic behavior, and more sophisticated programs are necessary. Several tests with time-dependent forces are employed, see e.g. Szczepinski (1990), but the most convenient for applications are experiments with piece-wise constant loads.

The program of loading in the form

$$\sigma(t) = \begin{cases} 0 & t < 0, \\ \sigma_0 & 0 \le t \le T, \\ 0 & T < t, \end{cases} \tag{1.6}$$

is used to analyze *the recovery phenomenon*. The functions $\sigma(t)$ and $\epsilon(t)$ corresponding to Eq. (1.6) are plotted in Fig. 1.13.

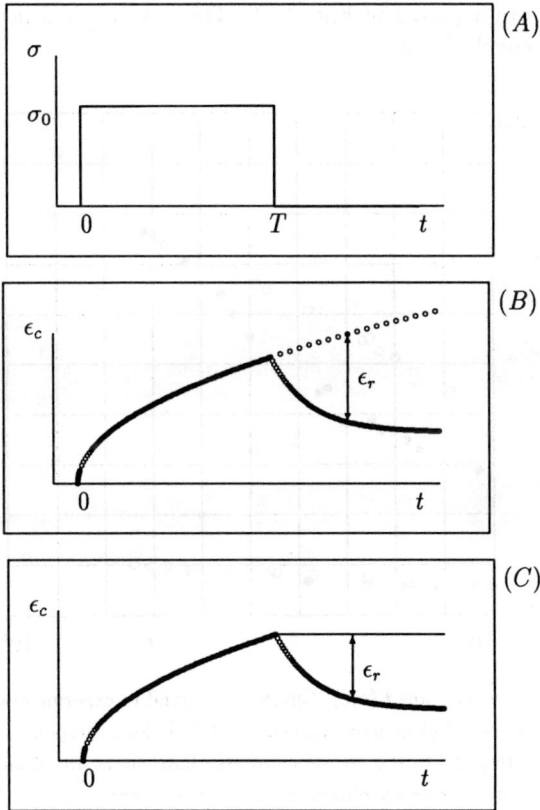

Fig. 1.13. A recovery program for a viscoelastic specimen.

The creep recovery is the difference between the creep strain under loading and the creep strain when external forces are removed. In bibliography we encounter two different definitions of the creep recovery.

239

According to Hadley & Ward (1965), Ward & Wolfe (1966), and Ward (1971), the creep recovery equals the difference between strains in two specimens: one of them remains under the action of longitudinal forces and the other is unloaded: ϵ_r in Fig. 1.13(B).

According to Schwartzl (1990) and Gramespacher & Meissner (1995), the creep recovery equals the difference between the strain in a specimen at some instant when tensile forces are removed and the strain at the current moment of time: ϵ_r in Fig. 1.13(C). We will use the latter definition of the recovery strain.

The typical dependencies of the total strain $\epsilon = \epsilon_0 + \epsilon_c$ on time t for creep and recovery are depicted in Fig. 1.14. The data plotted in this figure imply the following conclusions:

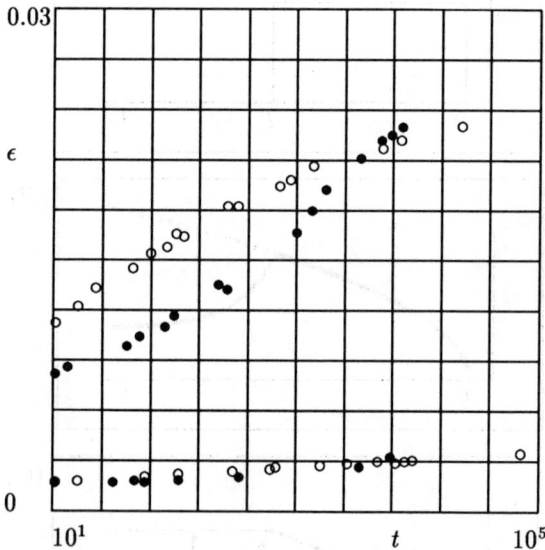

Fig. 1.14. Strain ϵ *versus* time t (sec). Circles correspond to experimental data presented by Ward (1971) for a polyethylene monofilament. Filled circles correspond to the creep process, unfilled circles correspond to the recovery process. Low curves are obtained for tensile load $P = 0.130$ (kg), upper curves are obtained for $P = 0.587$ (kg).

- for small stresses, creep and recovery curves practically coincide;

- for large stresses, the "instantaneous" or short-time recovery is always greater than the "instantaneous" creep;

- for any level of stresses, the creep and recovery strains increase in time monotonically.

240

The above assertions are not fulfilled for all the viscoelastic materials. For example, experimental data obtained by Gramespacher & Meissner (1995) demonstrate that the recovery strains in some polymeric blends are less than the creep strains at any instant, and that the recovery may be non-monotonic in time, see Fig. 1.15.

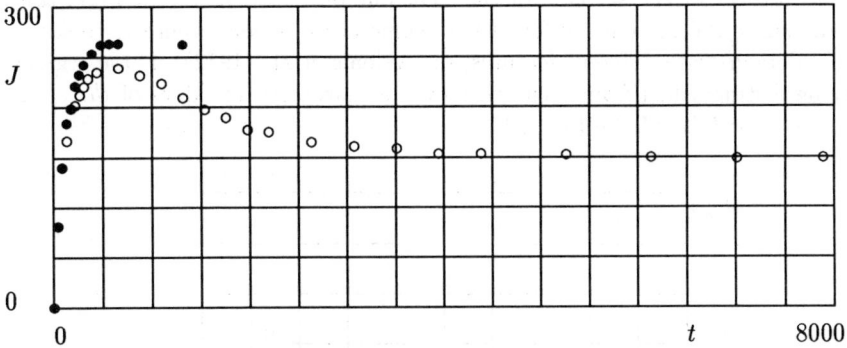

Fig. 1.15. Creep and recovery compliance J versus time t (sec). Circles correspond to experimental data obtained by Gramespacher & Meissner (1995) for a blend with 16 % of polystyrene in poly(methylmethacrylate). Filled circles correspond to the creep process, unfilled circles correspond to the recovery process.

1.4. Nonlinearity of the Material Response

A number of elastic materials are rather hard and brittle, and they demonstrate the linear response practically in the entire interval of strains before failure. The only exception from this rule of the thumb are rubbers and biological tissues which permit large deformations up to several hundred per cent.

For viscoelastic media, the opposite assertion is true: only a few materials demonstrate the linear response, whereas the majority of viscoelastic media are physically nonlinear.

For linear materials, the ratio $\epsilon_c(t)/\sigma_0$ which determines the creep compliance $J(t)$ should be independent of the applied stress σ_0. Experimental data for a wide range of polymeric materials show that this ratio depends essentially on the stress σ_0, even for relatively small deformations (less than $2-3$ per cent), see Fig. 1.16.

This leads us to the necessity to analyze the viscoelastic behavior in the framework of nonlinear mechanics of continua, accounting for both physical and geometrical nonlinearity.

1.5. Effect of Temperature on the Viscoelastic Behavior

This volume is concerned with isothermal processes, when the temperature remains constant. *The effect of temperature* on the mechanical behavior will be the subject of the second volume. For our purposes, it is now important to emphasize the presence of four basic temperature regions where viscoelastic media demonstrate different features, see Aklonis et al. (1972). These regions may be distinguished from each other by the characteristic values of *the shear relaxation modulus G*, see Fig. 1.17.

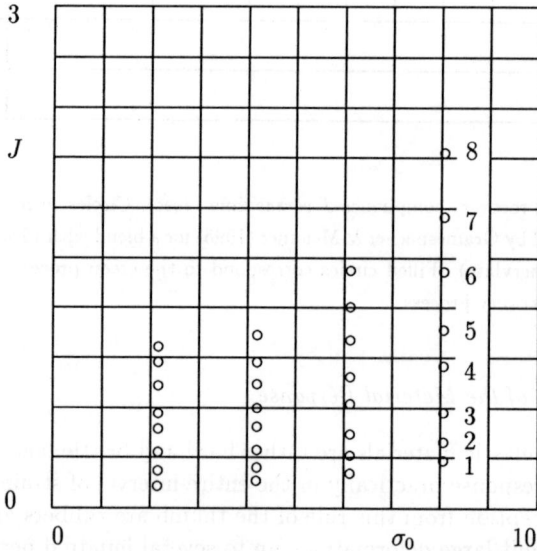

Fig. 1.16. The creep compliance J (GPa^{-1}) *versus* stress σ_0 (MPa). Circles correspond to experimental data obtained by Ward & Wolfe (1966) for polypropylene fibers. Curve 1: $t = 50$ (sec), curve 2: $t = 100$ (sec), curve 3: $t = 300$ (sec), curve 4: $t = 1000$ (sec), curve 5: $t = 2000$ (sec), curve 6: $t = 5000$ (sec), curve 7: $t = 10000$ (sec), curve 8: $t = 20000$ (sec).

In region 1, where temperature Θ is relatively low, the material is in the so-called *glassy state* characterized by a high level of resistance to external loads. The dependence of the relaxation modulus G on temperature Θ is extremely weak and may be neglected.

When the temperature increases and reaches *the glass transition temperature Θ_g*, the shear modulus decreases sharply (by several decades) and the

242

material is transformed into the so-called *rubbery state*, see region 3. This state is characterized by a relatively low resistance and a low hardness. In the rubbery state, the effect of temperature on the shear relaxation modulus is feeble and may be neglected as well.

In the rubber state, the material responses of *non-crosslinked* (liquid) and *crosslinked* (solid) polymers are identical. The difference between these two substances becomes evident when the temperature increases and reaches region 4. In this region, the relaxation modulus of a crosslinked polymer decreases slightly in temperature, but the mechanical behavior remains unchanged up to the temperature where chemical degradation begins. The shear relaxation modulus of a non-crosslinked polymer vanishes, and the material behavior becomes similar to the behavior of a viscoelastic liquid.

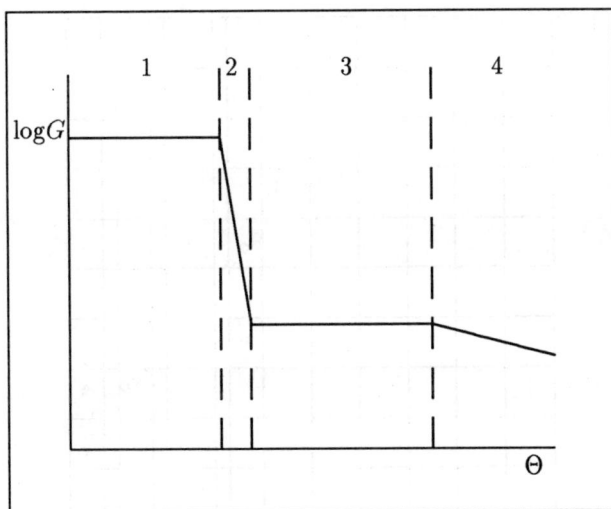

Fig. 1.17. Various regions of the viscoelastic behavior.

The stress relaxation modulus G is not the only parameter which characterizes these regions. Similar charges may be observed in *the coefficient of thermal expansion*, in *the heat capacity*, in *the yield stress*, etc., see e.g. Bicerano (1993), Zallen (1983).

The above scheme is valid for both *amorphous* and *semicrystalline* polymers, but the decrease in the relaxation modulus at the rubber-glass transition is stronger in amorphous polymers than in semicrystalline ones.

Fig. 1.17 demonstrates an idealized dependence of the relaxation modulus on temperature, while experimental data are plotted in Fig. 1.18. The curves in Figs. 1.17 and 1.18 are similar to each other, but Fig. 1.18 demonstrates additional features of the dependence $G = G(t)$: the so-called *secondary phase transitions* in the glassy region, see points A_1 and A_2 in Fig. 1.18. A detailed treatment of the glass transition is outside the scope of the present textbook.

1.6. The Aging Phenomenon

Aging is one of the interesting phenomena observed in viscoelastic media. Formally, aging is defined as any change of viscoelastic properties in time. As common practice, the material aging leads to an increase in elastic moduli and to a decrease in creep strains. It is convenient to distinguish *physical* (reversible) and *chemical* (irreversible) aging.

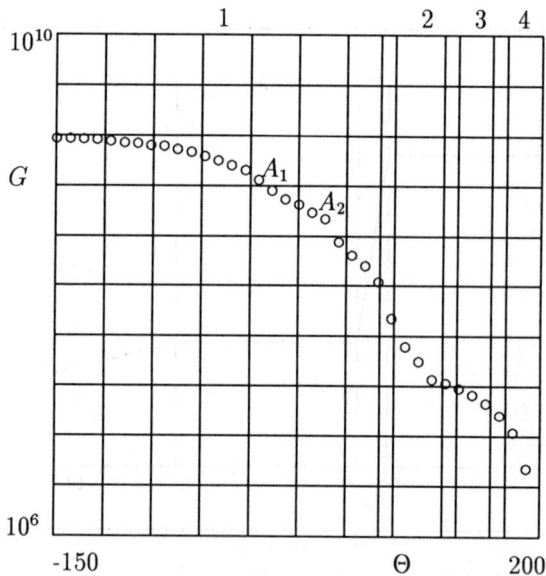

Fig. 1.18. The shear relaxation modulus G (Pa) *versus* temperature $\Theta°$C. Circles correspond to experimental data presented by Ward (1971) for polychlorotrifluoroethylene.

To demonstrate physical aging we heat a polymeric specimen until the glass transition temperature Θ_g and, afterwards, cool it to a temperature $\Theta < \Theta_g$. After waiting for a time $\rho > 0$ (the aging time), we stretch the specimen in a creep tester to measure the material creep compliance J. Typical experimental data are presented in Figs. 1.19 and 1.20.

For small intervals of aging, the creep strains are very large. With the growth of aging time ρ, the creep deformations decrease and tend to some limiting creep deformations of *the aged material* (i.e. the material in *the ther-modynamic equilibrium*).

The time necessary for the material aging depends drastically on the difference between the glass transition temperature Θ_g and the temperature of cooling Θ. According to Plazek et al. (1984), for the difference $\Theta_g - \Theta$ equal to 10°C the time of aging (the time necessary to reach the equilibrium thermodynamic state) is about a week, for the difference equal to 30°C the time of aging reaches a century, and for the difference equal to 50°C the time of aging is commensurable with the age of the universe.

The main feature of physical aging is its *reversibility*. Suppose that an aged material is heated once more until the glass transition temperature Θ_g and cooled according to the above program. Then its response at temperature Θ coincides with the mechanical response of a new material which underwent no preliminary treatment.

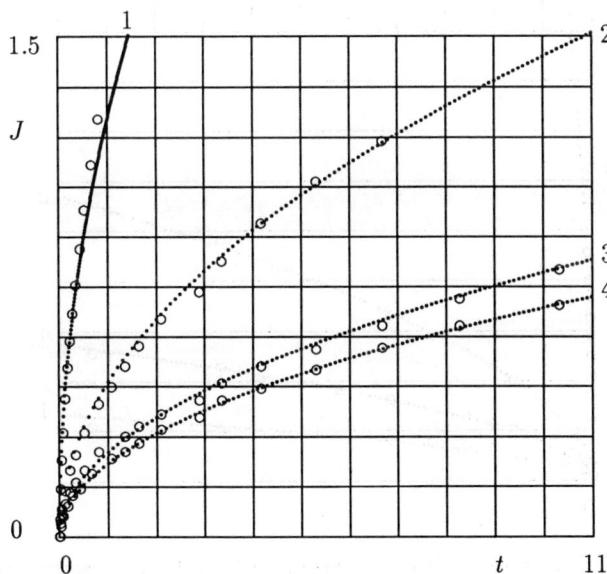

Fig. 1.19. Creep compliance J (GPa^{-1}) *versus* time t (min) for polystyrene quenched from the rubber-glass transition temperature $\Theta_g = 100$°C to temperature $\Theta = 95$°C. Large circles correspond to experimental data obtained by Plazek et al. (1984). Curve 1: $\rho = 5$ (min), curve 2: $\rho = 90$ (min), curve 3: $\rho = 1140$ (min), curve 4: $\rho = 2530$ (min).

Unlike physical aging which has a thermodynamic nature, chemical aging is caused by chemical reactions occurring in a viscoelastic medium. Concrete provides a typical example of a chemically aging material. Its mechanical characteristics change in time due to *the hydration process*, see Fig. 1.21.

245

Despite different natures, the physical and chemical processes of aging are extremely close to each other from the mechanical standpoint. In sequel, we will study the mechanical behavior of viscoelastic media subject to aging without distinguishing reversible and irreversible mechanisms of aging.

1.7. Concluding Remarks

In this Section we demonstrate the characteristic features of the viscoelastic behavior: creep, relaxation, recovery, and aging. A number of viscoelastic materials (above all, polymers) show a nonlinear response even at small strains, and an essentially nonlinear response at large deformations. The study of this response is the objective of this Chapter.

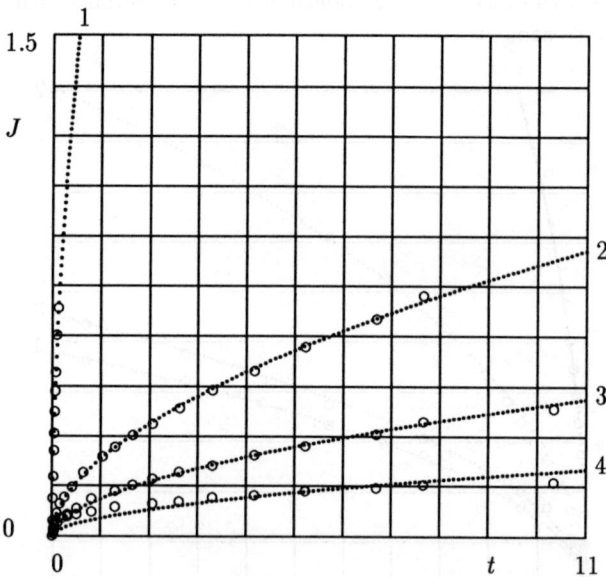

Fig. 1.20. Creep compliance J (GPa^{-1}) *versus* time t (min) for polystyrene quenched from the rubber-glass transition temperature $\Theta_g = 100°C$ to temperature $\Theta = 90°C$. Large circles correspond to experimental data obtained by Plazek et al. (1984). Curve 1: $\rho = 10$ (min), curve 2: $\rho = 56$ (min), curve 3: $\rho = 230$ (min), curve 4: $\rho = 7230$ (min).

We are concerned with the mechanical behavior of aging viscoelastic solids under isothermal loading. Owing to the energy dissipation in viscoelastic media, the temperature changes in time. However, since the material response depends weakly on temperature in both rubbery and glassy states, the influence of temperature may be neglected. More complicated mechanical problems accounting for the temperature effect on the material behavior (coupled problems in thermoviscoelasticity) will be considered in the second volume of the textbook.

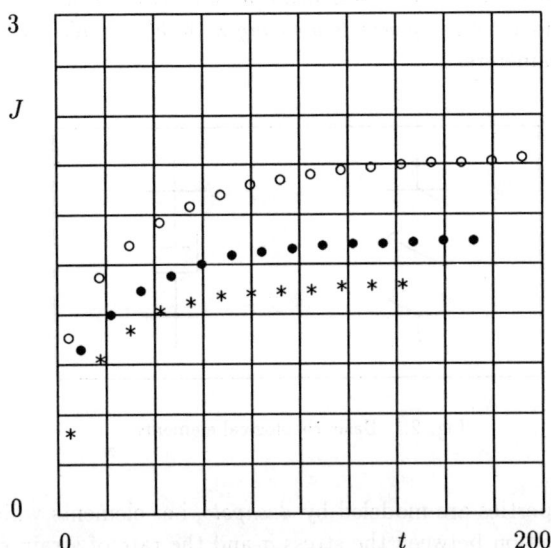

Fig. 1.21. Creep compliance $J \times 10^{-3}$ (GPa^{-1}) versus time t (days) for plain concrete. Circles and asterisks correspond to experimental data obtained by Ross (1958). Unfilled circles: $\rho = 8$ (days), filled circles: $\rho = 28$ (days), asterisks: $\rho = 60$ (days).

2. Differential Constitutive Models

In this Section we provide a brief survey of differential models in viscoelasticity, i.e. the constitutive models where the stress and the strain are connected by differential equations.

2.1. "Springs & Dashpots" Models

Viscoelasticity means that the material behavior demonstrates both elastic and viscous features. In rheology, elastic properties are described by springs, i.e. elements which provide a one-to-one connection between the stress σ and the strain ϵ. A linear *elastic spring* obeys *the Hooke law*

$$\sigma = E\epsilon, \tag{2.1}$$

247

where E is called Young's modulus. A nonlinear elastic spring satisfies the equality

$$\sigma = \Phi(\epsilon), \qquad (2.2)$$

where $\Phi(\epsilon)$ is a given, sufficiently smooth function. An example of the nonlinear response is provided by *the power-law spring* with $\Phi(\epsilon) = K\epsilon^n$, where K and n are material parameters.

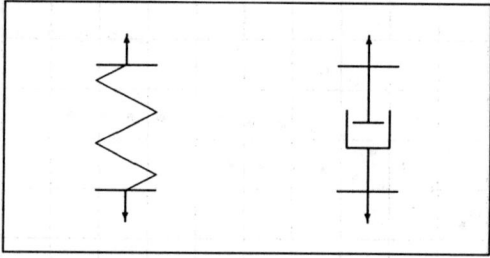

Fig. 2.1. Basic rheological elements.

Viscous properties are modeled by *dashpots*, i.e. elements which provide a one-to-one connection between the stress σ and the rate of strain $\dot{\epsilon}$. The linear dependence of the stress on the rate of strain

$$\sigma = \eta\dot{\epsilon} \qquad (2.3)$$

is called *the Newton law*, while the coefficient η is called Newtonian viscosity. A more sophisticated relation may be developed by assuming a nonlinear dependence of the stress σ on the rate of strain $\dot{\epsilon}$

$$\sigma = \Psi(\dot{\epsilon}), \qquad (2.4)$$

where $\Psi(\dot{\epsilon})$ is a given, sufficiently smooth function. An example of the nonlinear function $\Psi(\dot{\epsilon})$ is provided by *the Eyring dashpot* with the constitutive equation $\dot{\epsilon} = A\sinh(\alpha\sigma)$, where A and α are material parameters.

Basic rheological elements are plotted schematically in Fig. 2.1. To proceed with the construction of rheological models, it is natural to combine these elements in groups. Two opportunities arise in combining these elements: in series and in parallel. The former leads to *the Maxwell model*, see Fig. 2.2, whereas the latter leads to *the Kelvin–Voigt model*, see Fig. 2.3.

To generalize the obtained models, we combine the basic elements (springs and dashpots) as well as the Maxwell and Kelvin–Voigt elements in parallel and in series. For example, by choosing an arbitrary number of Maxwell models connected in parallel we obtain *the generalized Maxwell model*, or *the Maxwell–Weichert model*, see Fig. 2.4. Experimental data for this model are presented by Roylance & Wang (1978). By choosing an arbitrary number of the Kelvin–Voigt models connected in series, we arrive at *the generalized Voigt model*, see Fig. 2.5. For a discussion of this model and its accordance to experimental data see e.g. Smith (1973) and Giesekus (1995).

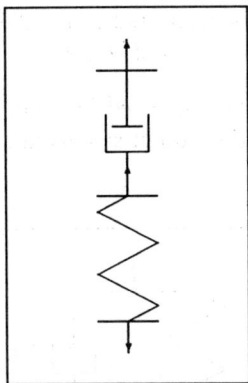

Fig. 2.2. The Maxwell model.

Taking the Maxwell model together with an elastic spring in parallel, we obtain *the standard viscoelastic solid (the Zener model)*, see Fig. 2.6.

The standard viscoelastic solid is one of the basic models for the theoretical analysis. This may be explained by the following:
(i) this model describes both creep and relaxation processes;
(ii) it is determined by the minimal number of material parameters: the material viscosity and the characteristic time of relaxation (retardation);
(iii) the standard viscoelastic solid is one of a few models which may be presented both in the differential and integral forms.

By adding to the Kelvin–Voigt model a spring and a dashpot in series, we arrive at *the Burgers model*, see Fig. 2.7.

By using this approach, an arbitrary number of constitutive models may be constructed which differ from each other by numbers of elements and their locations. Some of these models can be transformed into each other by replacing springs and dashpots, the others are independent.

Two basic presentations may be established for rheological models which are similar to the generalized Voigt model and to the Maxwell–Weichert model.

This issue is left to the reader as an exercise with reference to Giesekus (1994).

2.2. Differential Models with Small Strains

In this subsection we derive differential equations which are equivalent to rheological models introduced above. For this purpose we use the following rules:

- for a system of elements connected in parallel, their strains coincide and the total stress equals the sum of stresses in separate elements;

- for a system of elements connected in series, their stresses coincide and the total strain equals the sum of strains in separate elements.

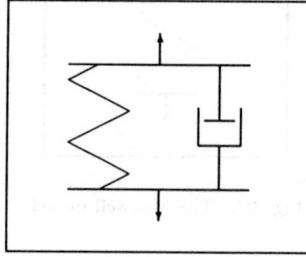

Fig. 2.3. The Kelvin–Voigt model.

Let us apply these assertions to the Maxwell solid presented in Fig. 2.2. We find

$$\epsilon = \epsilon_e + \epsilon_v, \tag{2.5}$$

where ϵ_e and ϵ_v are strains in the spring and in the dashpot. Differentiation of Eq. (2.5) with respect to time implies that

$$\dot{\epsilon} = \dot{\epsilon}_e + \dot{\epsilon}_v.$$

By assuming that the rheological elements are linear and using Eqs. (2.1) and (2.3), we obtain

$$\dot{\epsilon} = \frac{\dot{\sigma}}{E} + \frac{\sigma}{\eta}. \tag{2.6}$$

We now apply the above procedure to the Kelvin–Voigt model, see Fig. 2.3,

$$\sigma = \sigma_e + \sigma_v, \tag{2.7}$$

where σ_e and σ_v are stresses in the spring and in the dashpot, respectively. By assuming the linearity of the material response, we find

$$\sigma = E\epsilon + \eta\dot{\epsilon}. \qquad (2.8)$$

Let us employ the above procedure to develop the constitutive equation for the standard viscoelastic solid, see Fig. 2.6. For definiteness, we denote by E and η Young's modulus and Newtonian viscosity for the Maxwell element, and by E_1 the Young modulus of the additional spring. We have

$$\sigma = \sigma_e + \sigma_m \qquad (2.9)$$

where σ_e is the stress in the spring and σ_m is the stress in the Maxwell element.

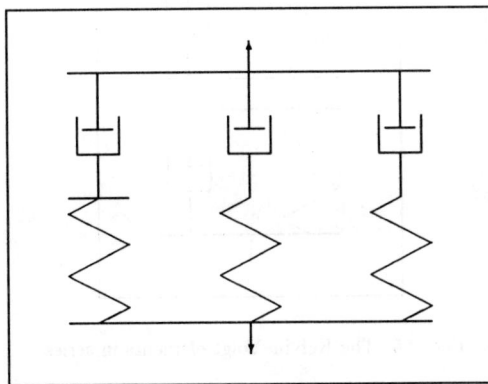

Fig. 2.4. The Maxwell–Weichert model with 3 elements.

Eqs. (2.1) and (2.9) yield

$$\sigma_m = \sigma - \sigma_e = \sigma - E_1\epsilon.$$

Exercise 2.1. By substituting this expression into Eq. (2.8), derive the following constitutive equation

$$\frac{\dot{\sigma}}{E} + \frac{\sigma}{\eta} = (1 + \frac{E_1}{E})\dot{\epsilon} + \frac{E_1}{\eta}\epsilon. \qquad \square \qquad (2.10)$$

Exercise 2.2. Derive the constitutive equation for the Burgers model

$$\ddot{\sigma} + (\frac{E_1}{\eta} + \frac{E}{\eta} + \frac{E_1}{\eta_1})\dot{\sigma} + \frac{EE_1}{\eta\eta_1}\sigma = E_1\ddot{\epsilon} + \frac{EE_1}{\eta}\dot{\epsilon}, \qquad (2.11)$$

251

where E and η are characteristics of the Kelvin–Voigt model, whereas E_1 and η_1 are parameters of the separate spring and dashpot. $\square$

It is of interest to analyze whether these rheological models can describe adequately the creep and relaxation processes.

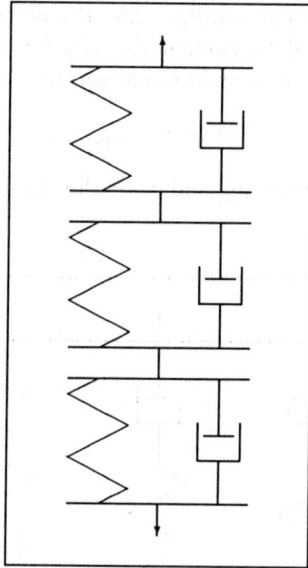

Fig. 2.5. The Kelvin–Voigt elements in series.

For the creep process, the stress $\sigma = \sigma_0$ is constant. Substituting this expression into Eqs. (2.6), (2.8), and (2.10) we find
– for the Maxwell model

$$\eta \dot{\epsilon} = \sigma_0,$$

– for the Kelvin–Voigt model

$$\eta \dot{\epsilon} + E\epsilon = \sigma_0,$$

– for the standard viscoelastic solid

$$\eta(1 + \frac{E_1}{E})\dot{\epsilon} + E_1\epsilon = \sigma_0.$$

It follows from these formulas that the Maxwell model cannot predict the material creep correctly (it predicts the Newtonian flow only), whereas the Kelvin–Voigt model is equivalent to the standard viscoelastic solid.

252

For the relaxation process, the strain $\epsilon = \epsilon_0$ is constant, and we obtain
– for the Maxwell model
$$\frac{\eta}{E}\dot{\sigma} + \sigma = 0,$$
– for the Kelvin–Voigt model
$$\sigma = E\epsilon_0,$$
– for the standard viscoelastic solid
$$\frac{\eta}{E}\dot{\sigma} + \sigma = E_1\epsilon_0.$$

These equalities imply that the Kelvin–Voigt model cannot predict the stress relaxation, whereas the Maxwell model and the standard viscoelastic solid lead to results similar to each other.

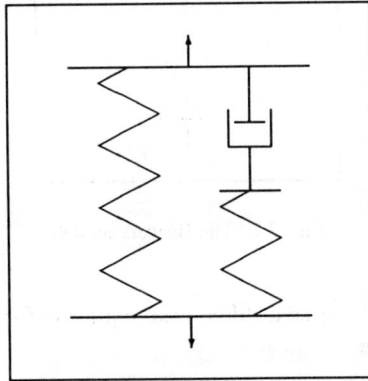

Fig. 2.6. The standard viscoelastic solid.

For a rheological model with an arbitrary number of springs and dashpots, the corresponding constitutive equation may be written in the form

$$A_0\sigma + A_1\frac{d\sigma}{dt} + \ldots + A_n\frac{d^n\sigma}{dt^n} = B_0\epsilon + B_1\frac{d\epsilon}{dt} + \ldots + B_m\frac{d^m\epsilon}{dt^m}, \qquad (2.12)$$

where m and n are positive integers, and A_k and B_l are material parameters.

For $m = n$, Eq. (2.12) may be written in the integral form. For simplicity, we prove this assertion for $m = n = 1$, i.e. for the standard viscoelastic solid. An extension of this statement to arbitrary integers $m = n$ is left to the reader as an exercise.

For $m = n = 1$, Eq. (2.12) implies that

$$A_1\dot{\sigma} + A_0\sigma = B_1\dot{\epsilon} + B_0\epsilon. \qquad (2.13)$$

253

For definiteness, we assume that at instant $t = 0$, the stress $\sigma(0)$ and the strain $\epsilon(0)$ vanish.

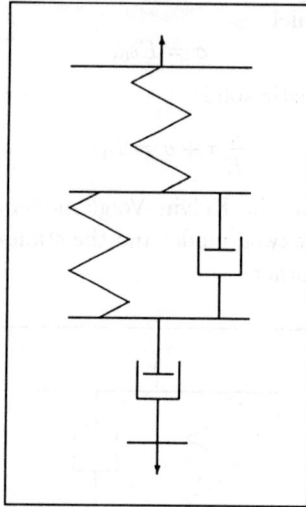

Fig. 2.7. The Burgers model.

Let us treat Eq. (2.13) as a differential equation for the stress $\sigma(t)$. Introducing the new variable

$$\sigma(t) = z(t) \exp(-\frac{A_0}{A_1}t), \qquad (2.14)$$

we rewrite Eq. (2.13) as follows:

$$\dot{z} = \frac{1}{A_1}(B_1\dot{\epsilon} + B_0\epsilon) \exp(\frac{A_0}{A_1}t).$$

Integration of this equation with the initial condition $z(0) = \sigma(0) = 0$ yields

$$z(t) = \frac{1}{A_1} \int_0^t [B_1\dot{\epsilon}(s) + B_0\epsilon(s)] \exp(\frac{A_0}{A_1}s)ds.$$

The first term in the integrand is transformed by integrating by parts with the use of the initial condition $\epsilon(0) = 0$

$$\int_0^t \dot{\epsilon}(s) \exp(\frac{A_0}{A_1}s)ds = \epsilon(t) \exp(\frac{A_0}{A_1}t) - \frac{A_0}{A_1} \int_0^t \epsilon(s) \exp(\frac{A_0}{A_1}s)ds.$$

Substitution of these expressions into Eq. (2.14) implies that

$$\sigma(t) = \frac{B_1}{A_1}\epsilon(t) + \frac{B_0}{A_1}(1 - \frac{A_0 B_1}{A_1 B_0}) \int_0^t \epsilon(s) \exp[-\frac{A_0}{A_1}(t - s)]ds. \qquad (2.15)$$

Exercise 2.3. Show that Eq. (2.15) is written in notation (2.10) as

$$\sigma(t) = (E + E_1)\{\epsilon(t) - \frac{E^2}{\eta(E + E_1)} \int_0^t \epsilon(s) \exp[-\frac{E}{\eta}(t - s)]ds\}. \qquad \square \quad (2.16)$$

By introducing the notation

$$E_0 = E + E_1, \qquad T = \frac{\eta}{E}, \qquad \chi = \frac{E}{E + E_1},$$

we find from Eq. (2.16) that

$$\sigma(t) = E_0[\epsilon(t) + \int_0^t \dot{Q}(t - s)\epsilon(s)ds], \qquad (2.17)$$

where

$$Q(t) = -\chi[1 - \exp(-\frac{t}{T})]. \qquad (2.18)$$

The constant E_0 is Young's modulus of the standard viscoelastic solid, and the function $Q(t)$ is its relaxation measure.

Eq. (2.17) may be treated as an integral relation between the stress $\sigma(t)$ at current instant t and the whole history of strains up to this moment of time. Such an equality is called *the Volterra equation* and it provides a typical example of integral constitutive models.

The above procedure shows that differential model (2.13) is equivalent to an integral model. In the general case, such an equivalence is absent. Some differential models (nonlinear and with time-varying coefficients) cannot be transformed into integral models. To the best of our knowledge, the problem of establishing a correspondence between differential and integral models in the nonlinear viscoelasticity is far from being complete.

Integral constitutive equations will be discussed in Section 3. We are concerned now with models which are "transient" between differential and integral models, and which are described by differential equations with the so-called fractional derivatives.

2.3. Fractional Differential Models with Small Strains

A fractional derivative is the Volterra integral operator with a kernel of a special form (Abel's kernel) which depends on a parameter α. When α is a positive integer, this operator is reduced to the standard operator of integration. For a negative integer α, this operator is equivalent to the operator of differentiation. For an arbitrary (non-integer) α value, this operator is called *fractional derivative*.

Differential models in linear viscoelasticity with fractional derivatives were suggested by Bagley & Torvik (1983, 1986), Friedrich (1991), Glockle & Nonnenmacher (1994). For a detailed exposition of the theory of fractional derivatives see Oldham & Spanier (1974).

Let $f(t)$ be a sufficiently smooth function which vanishes on the interval $(-\infty, 0]$ and which is integrable on $[0, T]$ for an arbitrary $T > 0$. The primitive $F_1(t)$ for the function $f(t)$ equals

$$F_1(t) = \int_0^t f(s)ds. \tag{2.19}$$

The primitive $F_2(t)$ for the function $F_1(t)$ is

$$F_2(t) = \int_0^t F_1(\tau)d\tau = \int_0^t d\tau \int_0^\tau f(s)ds.$$

By changing the order of integration, we obtain

$$F_2(t) = \int_0^t f(s)ds \int_s^t d\tau = \int_0^t (t-s)f(s)ds. \tag{2.20}$$

The primitive $F_3(t)$ for the function $F_2(t)$ is calculated as

$$F_3(t) = \int_0^t F_2(\tau)d\tau = \int_0^t d\tau \int_0^\tau (\tau-s)f(s)ds$$
$$= \int_0^t f(s)ds \int_s^t (\tau-s)d\tau = \frac{1}{2}\int_0^t (t-s)^2 f(s)ds. \tag{2.21}$$

Exercise 2.4. Derive the formula for the nth primitive of function $f(t)$:

$$F_n(t) = \frac{1}{(n-1)!} \int_0^t (t-s)^{n-1} f(s)ds, \tag{2.22}$$

where $n! = 1 \cdot 2 \cdot \ldots \cdot n$. $\square$

Introduce *the Euler gamma-function* $\Gamma(z)$ of a complex variable z by the formula

$$\Gamma(z) = \int_0^\infty t^{z-1} \exp(-t)dt. \tag{2.23}$$

For $\Re z > 0$ the integral in the right-hand side of Eq. (2.23) converges, for $\Re z < 0$ and $z \neq -1, -2, \ldots$ it is treated in the generalized sense.
Exercise 2.5. By integrating Eq. (2.23) by parts, show that for any z

$$\Gamma(1+z) = z\Gamma(z). \quad \square \tag{2.24}$$

Exercise 2.6. By using Eq. (2.24), check that for any positive integer n and any complex z

$$\Gamma(n + z) = z(1 + z)(2 + z)\ldots(n - 1 + z)\Gamma(z). \qquad \square \qquad (2.25)$$

Exercise 2.7. By employing Eq. (2.25) and the equality

$$\Gamma(1) = \int_0^\infty \exp(-t)dt = 1,$$

prove that for any positive integer n

$$\Gamma(n) = (n - 1)!. \qquad \square \qquad (2.26)$$

Substitution of expression (2.26) into Eq. (2.22) implies that

$$F_n(t) = \frac{1}{\Gamma(n)} \int_0^t (t - s)^{n-1} f(s)ds.$$

We introduce the new variable $\tau = t - s$, use the equality $f(t) = 0$ for $t < 0$, and find

$$F_n(t) = \frac{1}{\Gamma(n)} \int_0^t \tau^{n-1} f(t - \tau)d\tau = \frac{1}{\Gamma(n)} \int_0^\infty \tau^{n-1} f(t - \tau)d\tau. \qquad (2.27)$$

Our objective now is to define the function $F_\alpha(t)$ for an arbitrary real α. For this purpose, we introduce the function

$$\mathcal{J}_\alpha(t) = \frac{t^\alpha}{\Gamma(1 + \alpha)}, \qquad (2.28)$$

which is called *the Abel kernel.*

Exercise 2.8. By using Eqs. (2.26) and (2.28), show that for any real α and any positive integer n

$$\frac{d^n}{dt^n} \mathcal{J}_{\alpha+n}(t) = \mathcal{J}_\alpha(t). \qquad \square \qquad (2.29)$$

For $\alpha > 0$, the integral

$$I_\alpha(\phi) = \int_0^\infty \mathcal{J}_{\alpha-1}(s)\phi(s)ds \qquad (2.30)$$

converges for any continuous function $\phi(t)$ such that $\int_0^\infty |\phi(t)|dt < \infty$ and $\phi(0) = 0$. Therefore, we can define function $F_\alpha(t)$ for an arbitrary positive real α as follows:

$$F_\alpha(t) = \int_0^t \mathcal{J}_{\alpha-1}(s)f(t - s)ds. \qquad (2.31)$$

In order to define the functional $I_\alpha(\phi)$ for $\alpha < 0$ we impose additional restrictions on function $\phi(t)$. Namely, for $\alpha \in (-n, -n+1)$ we require the function $\phi(t)$ to have n continuous derivatives which vanish at $t = 0$. Afterward, we integrate Eq. (2.30) n times by parts and employ formula (2.29). As a result, we obtain

$$
\begin{aligned}
I_\alpha(\phi) &= \int_0^\infty \mathcal{J}_{\alpha-1}(s)\phi(s)ds = \int_0^\infty \frac{d\mathcal{J}_\alpha}{ds}(s)\phi(s)ds \\
&= \mathcal{J}_\alpha(s)\phi(s)\big|_0^\infty - \int_0^\infty \mathcal{J}_\alpha(s)\frac{d\phi}{ds}(s)ds = -\int_0^\infty \frac{d\mathcal{J}_{\alpha+1}}{ds}(s)\frac{d\phi}{ds}(s)ds \\
&= \ldots = (-1)^n \int_0^\infty \mathcal{J}_{\alpha+n-1}(s)\frac{d^n\phi}{ds^n}(s)ds. \quad (2.32)
\end{aligned}
$$

The integral in the right-hand side of Eq. (2.32) converges, and we define the functional $I_\alpha(\phi)$ for $\alpha \in (-n, -n+1)$ according to formula (2.32).

Returning to the function $f(t)$ which vanishes at $t = 0$ with its derivatives we set

$$
F_\alpha(t) = \int_0^\infty \mathcal{J}_{\alpha+n-1}(s)f^{(n)}(t-s)ds = \int_0^t \mathcal{J}_{\alpha+n-1}(s)f^{(n)}(t-s)ds, \quad (2.33)
$$

provided $\alpha \in (-n, -n+1)$. Here n is a positive integer and $f^{(n)}(t) = d^n f(t)/dt^n$.

The function $F_\alpha(t)$ is defined now for any real α except for non-positive integers. To define it on the whole real axis we use Eq. (2.32) and set

$$
F_{-n}(t) = \lim_{\alpha \to -n-0} F_\alpha(t).
$$

Substitution of expressions (2.28) and (2.33) into this equality implies that

$$
F_{-n}(t) = \lim_{\alpha \to -n-0} \int_0^t \mathcal{J}_{\alpha+n}(s)f^{(n+1)}(t-s)ds
$$

$$
= \lim_{\alpha \to -n-0} \int_0^t \frac{s^{\alpha+n}}{\Gamma(1+\alpha+n)}f^{(n+1)}(t-s)ds = \int_0^t f^{(n+1)}(t-s)ds = f^{(n)}(t). \quad (2.34)
$$

According to Eq. (2.34), for any positive integer n function $F_{-n}(t)$ determines the nth derivative of the function $f(t)$. We extend this concept to an arbitrary negative real α and define the fractional derivative of the order $\alpha \in [0, 1]$ as follows:

$$
\frac{d^\alpha f}{dt^\alpha}(t) = F_{-\alpha}(t) = \int_0^t \mathcal{J}_{-\alpha-1}(s)f(t-s)ds. \quad (2.35)
$$

Eq. (2.3) determines a Newtonian dashpot where the stress is proportional to the first derivative of the strain. A natural generalization of this rheological element is the so-called *fractional dashpot* with the constitutive equation

$$
\sigma = \eta \frac{d^\alpha \epsilon}{dt^\alpha}, \quad (2.36)
$$

which is characterized by two material parameters: $\alpha \in (0,1)$ and η. The limiting cases correspond to the ideal plasticity ($\alpha = 0$) and to the Newtonian liquid ($\alpha = 1$). For $\alpha \in (0,1)$, the fractional derivative leads to the non-exponential creep observed in experiments.

Exercise 2.9. Derive the following constitutive equation for the Maxwell model with the fractional viscosity:

$$\frac{d^\alpha \sigma}{dt^\alpha} + \frac{E}{\eta}\sigma = E\frac{d^\alpha \epsilon}{dt^\alpha}. \qquad \Box \qquad (2.37)$$

Exercise 2.10. Show that the Kelvin–Voigt model with the fractional viscosity satisfies the equation

$$\sigma = E\epsilon + \eta\frac{d^\alpha \epsilon}{dt^\alpha}. \qquad \Box \qquad (2.38)$$

Exercise 2.11. Derive the constitutive equation of the standard viscoelastic solid with the fractional viscosity

$$\frac{d^\alpha \sigma}{dt^\alpha} + \frac{E}{\eta}\sigma = (E + E_1)\frac{d^\alpha \epsilon}{dt^\alpha} + \frac{EE_1}{\eta}\epsilon. \qquad \Box \qquad (2.39)$$

The above relationships may be easily generalized by introducing derivatives of different fractional orders. For example, Glockle & Nonnenmacher (1994) suggested the following equation for the standard viscoelastic solid:

$$\frac{d^\alpha \sigma}{dt^\alpha} + \frac{1}{T_1}\sigma = E_1\frac{d^\beta \epsilon}{dt^\beta} + \frac{E_2}{T_2}\epsilon, \qquad (2.40)$$

where E_1, E_2, T_1, T_2, α, and β are material constants.

The above generalization of "springs & dashpots" models allows experimental data to be predicted adequately. However, even for the simplest programs of loading (creep, relaxation, recovery, etc.) the material response of model (2.40) is described in terms of special functions (either *the generalized Mittag–Leffler functions* or *the Wright functions*). The latter depreciates the value of fractional constitutive models for applications.

2.4. Nonlinear Constitutive Models

"Springs & dashpots" rheological models demonstrate linear or nonlinear behavior depending on the assumptions regarding the mechanical response of the basic elements. We should mention a number of nonlinear constitutive models which predict adequately the behavior of viscoelastic media. As common practice, the onset of these models has no conscious explanations, and it is difficult to establish rational reasons why one or another model is chosen to fit experimental data except the taste of the researcher.

We confine ourselves to a list of several models in nonlinear viscoelasticity with reference to Papo (1988), where the constitutive models were compared with experimental data for gypsum plaster paste. The models are classified formally according to the number of adjustable parameters.

Models with two material parameters:

- *the Eyring model*

$$\sigma = A \sinh^{-1}(\alpha \dot{\epsilon}), \tag{2.41}$$

where $A > 0$ and $\alpha > 0$;

- *the Ostwald–de Waele* (power-law) *model*

$$\sigma = A \dot{\epsilon}^{\alpha}, \tag{2.42}$$

where $A > 0$ and $\alpha \in (0,1)$;

Models with three material parameters:

- *the Briant model*

$$\sigma = A \dot{\epsilon}(1 + \frac{B}{\dot{\epsilon}})^{\alpha}, \tag{2.43}$$

where $\alpha > 0$, $A > 0$, and $B > 0$;

- *the Powell–Eyring model*

$$\sigma = A_{\infty} \dot{\epsilon} + (A_0 - A_{\infty}) \frac{\sinh^{-1}(\alpha \dot{\epsilon})}{\alpha}, \tag{2.44}$$

where $\alpha > 0$, $A_0 > 0$, and $A_{\infty} > 0$;

- *the Robertson–Stiff model*

$$\sigma = A(\dot{\epsilon} + a)^{\alpha}, \tag{2.45}$$

where $\alpha > 0$, $a > 0$, and $A > 0$;

- *the Shangraw–Grim–Mattocks model*

$$\sigma = A \dot{\epsilon} + B[1 - \exp(-\alpha \dot{\epsilon})], \tag{2.46}$$

where $\alpha > 0$, $A > 0$, and $B > 0$;

- *the Sisko model*

$$\sigma = A \dot{\epsilon} + B \dot{\epsilon}^{\alpha}, \tag{2.47}$$

where $\alpha > 0$, $A > 0$, and $B > 0$;

- *the Williamson model*

$$\sigma = A\dot{\epsilon} + B\frac{\dot{\epsilon}}{\dot{\epsilon} + a}, \qquad (2.48)$$

where $a > 0$, $A > 0$, and $B > 0$;

Models with four material parameters:

- *the Carreau model*

$$\sigma = A_\infty \dot{\epsilon} + (A_0 - A_\infty)\dot{\epsilon}(1 + a\dot{\epsilon})^{-\alpha}, \qquad (2.49)$$

where $\alpha \in (0, 1)$, $a > 0$, $A_0 > 0$, and $A_\infty > 0$.

This list does not appear to be exhaustive. Nevertheless, the following conclusions may be formulated:
(i) there is no rational procedure to construct differential constitutive models in nonlinear viscoelasticity;
(ii) the existing constitutive equations contain a small number of material parameters which may be found by fitting experimental data;
(iii) only a few constitutive equations suggested for uniaxial deformations may be extended to three-dimensional and finite deformations.

2.5. Differential Models with Finite Strains

The basic method of constructing differential models in finite viscoelasticity consists in replacing scalar stresses and strains in constitutive equations at infinitesimal strains by "finite" tensors of stresses and strains. For this purpose, the strain tensors and corotational derivatives of the stress tensors should be chosen correctly, since the material time derivatives are not objective. Even for models which contain the first temporal derivative only, several different versions of the same constitutive equation arise; every version corresponds to the specific corotational derivative. The number of modifications grows rapidly with an increase in the order of the derivatives employed. The choice of corotational derivatives is between equally acceptable alternatives, and it is a matter of taste and convenience of the researcher. Due to the nonlinearity of corotational derivatives, see Section 1.4, extensions of linear constitutive relations to large deformations become essentially nonlinear. To the best of our knowledge, no rational classification of constitutive equations in finite viscoelasticity exists even for "linear" models. Thus, we confine ourselves to several typical examples.

2.5.1. The Rivlin–Ericksen models

The Rivlin–Ericksen models may be treated as generalizations of the simplest version of Eq. (2.12) which does not contain temporal derivatives of stresses

$$\sigma = B_0\epsilon + B_1\frac{d\epsilon}{dt} + \ldots + B_n\frac{d^n\epsilon}{dt^n}.$$

Replacing the stress σ by the Cauchy stress tensor $\hat{\sigma}$, and the derivative $d^n\epsilon/dt^n$ by the Rivlin–Ericksen tensor $\hat{A}_n$ ($n = 1, 2, \ldots$), see Eq. (1.4.47), we obtain the Cauchy stress as a combinations (linear or nonlinear) of tensors $\hat{A}_n$.

The linear model with $n = 1$ is called Newtonian liquid

$$\hat{\sigma} = \eta\hat{A}_1 = 2\eta\hat{D}, \tag{2.50}$$

where the material parameter η is Newtonian viscosity. Traditionally, the constitutive equation of a Newtonian liquid contains the coefficient η for uniaxial deformations, see Eq. (2.3), and the coefficient 2η for three-dimensional deformations, see Eq. (2.50).

The second order model (with respect to $|\bar{\nabla}\bar{v}|$) obeys the equation

$$\hat{\sigma} = \eta\hat{A}_1 + c_{12}\hat{A}_1^2 + c_{21}\hat{A}_2, \tag{2.51}$$

where η and c_{ij} are material parameters, see Rivlin & Ericksen (1955) and Coleman & Noll (1960). Model (2.51) is widely used for the numerical analysis of laminar flows, but it predicts experimental data only qualitatively, see Astarita & Marrucci (1974).

2.5.2. The Kelvin–Voigt models

The Kelvin–Voigt models are based on the presentation of the Cauchy stress tensor $\hat{\sigma}$ as a sum of two tensors $\hat{\sigma}_e$ and $\hat{\sigma}_v$, see Eq. (2.7),

$$\hat{\sigma} = \hat{\sigma}_e + \hat{\sigma}_v. \tag{2.52}$$

Tensor $\hat{\sigma}_e$ determines the elastic response, while tensor $\hat{\sigma}_v$ characterizes additional stresses arisen due to the material viscosity. Model (2.52) treats a viscoelastic medium as a mixture of two continua: an elastic and a viscous, which cannot slide with respect to each other. These continua have the same strains, while the resulting stress equals the sum of stresses in the continua.

By assuming that the material is homogeneous, isotropic, and possesses a strain energy density W, we may write, see Eq. (3.2.18),

$$\hat{\sigma}_e = \frac{2}{\sqrt{I_3}}(\psi_0\hat{I} + \psi_1\hat{F} + \psi_2\hat{F}^2), \tag{2.53}$$

where $\hat{I}$ is the unit tensor, $\hat{F}$ is the Finger tensor for transition from the initial to the actual configuration, I_k ($k = 1, 2, 3$) are the principal invariants of the

Finger tensor, and functions ψ_i are expressed in terms of the strain energy density W by Eqs. (3.2.19). To describe the material viscosity, *the generalized Newton law* (2.50) is employed

$$\hat{\sigma}_v = 2\eta\hat{D}. \tag{2.54}$$

Eqs. (2.52) – (2.54) for incompressible media were used by Engler (1989) and Renardy et al. (1987) to analyze the existence and uniqueness of solutions to boundary value problems in finite viscoelasticity.

2.5.3. The Maxwell models

Unlike the Kelvin–Voigt model, two different approaches may be distinguished to constructing analogues of the Maxwell model (2.6) with large deformations.

The first approach is based on the "visual presentation" of the Maxwell model as a system containing an elastic and a viscous elements connected in series. To explain the mechanical meaning of these elements at finite strains, we introduce three basic configurations at the current instant: (i) the initial (reference) configuration where the medium is located before application of external loads, (ii) *the intermediate* (viscous) configuration which characterizes a purely viscous deformation of the medium, and (iii) the actual (viscoelastic) configuration which determines the current state of the medium. Transition from the initial to the intermediate configuration is described by the generalized Newton law (2.54)

$$\hat{\sigma}_v = 2\eta\hat{D}_v. \tag{2.55}$$

Transition from the intermediate configuration to the actual configuration is described by the constitutive equation of a hyperelastic medium (2.53)

$$\hat{\sigma} = \frac{2}{\sqrt{I_{3\,e}}}(\psi_0\hat{I} + \psi_1\hat{F}_e + \psi_2\hat{F}_e^2), \tag{2.56}$$

where the subscript indices e and v stand for transitions from the intermediate configuration to the actual configuration, and from the initial to the intermediate configuration, respectively.

Tensors $\hat{\sigma}$ and $\hat{\sigma}_v$ determine stresses in the actual and intermediate configurations. It follows from Eq. (2.56) that the stress tensor $\hat{\sigma}$ is symmetrical. To ensure the tensor $\hat{\sigma}_v$ to be symmetrical, which is necessary to satisfy Eq. (2.55), an additional assumption should be introduced.

In order to close the above system of constitutive equations for the Maxwell model we should add (i) a kinematic equation which expresses the deformation gradient for transition from the initial to the actual configuration in terms of the deformation gradients for transitions from the initial to the intermediate and from the intermediate to the actual configurations, and (ii) an equation which allows the stress tensors $\hat{\sigma}$ and $\hat{\sigma}_v$ to be expressed in terms of each other.

To derive the latter equation, we recall that $\hat{\sigma}$ is the Cauchy stress tensor in the actual configuration, and $\hat{\sigma}_v$ is the Cauchy stress tensor in the intermediate configuration, i.e. that these tensors determine the same stresses, but applied to surface elements in different configurations. Thus, we may use Eq. (2.2.37) and write

$$\hat{\sigma}_v = \sqrt{I_{3\,e}}\,\bar{\nabla}\bar{r}_e^T \cdot \hat{\sigma}, \tag{2.57}$$

where $\bar{\nabla}\bar{r}_e$ is the deformation gradient for transition from the intermediate to the actual configuration.

To develop a kinematic relation between the intermediate and actual configurations, we assume that the deformation gradient $\bar{\nabla}_e\bar{r}$ is symmetrical. This hypothesis is admissible, since Eq. (2.55) does not determine the intermediate configuration uniquely: for a given stress tensor $\hat{\sigma}_v$ we find the rate-of-strain tensor $\hat{D}_v$ only, while the vorticity tensor $\hat{Y}_v$ remains unknown. By using the polar decomposition formula $\bar{\nabla}_e\bar{r} = \hat{O}_e \cdot \hat{U}_e$, we obtain

$$\bar{\nabla}_e\bar{r} = \hat{U}_e, \qquad \hat{O}_e = \hat{I}, \tag{2.58}$$

where $\hat{O}_e$ and $\hat{U}_e$ are the rotation tensor and the right stretch tensor for transition from the intermediate to the actual configuration. Substitution of expression (2.58) into Eqs. (2.1.9) and (2.1.25) implies that $\bar{\nabla}\bar{r}_e = \hat{U}_e^{-1}$ and $\hat{F}_e = \hat{U}_e^2$. It follows from these expressions and Eqs. (2.56) – (2.58) that

$$\hat{\sigma}_v = 2(\psi_0\hat{U}_e^{-1} + \psi_1\hat{U}_e + \psi_2\hat{U}_e^3).$$

Therefore, the stress tensor $\hat{\sigma}_v$ is symmetrical, since all the tensors in the right-hand side of this equality are symmetrical.

The reader can easily see an analogy between the intermediate configuration in the Maxwell model and *the unloaded configuration* in finite elastoplasticity. Our assumption regarding the absence of rotation for transition from the intermediate to the actual configuration is similar to the neglecting of plastic spin. These issues will be discussed in detail in the second volume of the textbook.

The above model is close to the Leonov model, see Leonov (1976), where the Cauchy stress tensor $\hat{\sigma}$ is expressed by Eq. (2.56) in terms of some Finger tensor $\hat{F}$ which, in turn, satisfies a kinematic equation. This equation connects the Jaumann derivative of $\hat{F}$ and a tensor $\hat{D}_p$ which serves as a measure of *dissipative deformations*. The dissipative rate-of-strain tensor is proportional to the Cauchy stress tensor $\hat{\sigma}$, see Eq. (2.50), with a viscosity η depending on the principal invariants of $\hat{F}$. Leonov's model predicts correctly experimental data for some incompressible polymeric liquids, see Leonov et al. (1976). Nevertheless, it contains such a number of material functions to be found that it is too complicated to be employed in applications.

Another approach to constructing the Maxwell models with finite strains is based on replacing the strain ϵ and the stress σ in Eq. (2.6) by appropriate strain and stress tensors. By using the Jaumann derivative of the Cauchy stress tensor

$$\hat{\sigma}^\circ = \frac{\partial \hat{\sigma}}{\partial t} + \hat{\sigma} \cdot \hat{Y} - \hat{Y} \cdot \hat{\sigma},$$

we obtain the constitutive model

$$\hat{\sigma}^\circ + \frac{1}{T}\hat{\sigma} = 2\mu\hat{D}, \tag{2.59}$$

where we set $T = \eta/E$ and $\mu = E/2$. Eq. (2.59) predicts correctly flows of viscoelastic liquids, see Johnson & Segalman (1977) and Pearson & Middleman (1978).

By choosing the lower convected derivative of the stress tensor

$$\hat{\sigma}^\nabla = \frac{\partial \hat{\sigma}}{\partial t} + \hat{\sigma} \cdot \bar{\nabla}\bar{v}^T + \bar{\nabla}\bar{v} \cdot \hat{\sigma}$$

and the upper convected derivative

$$\hat{\sigma}^\Delta = \frac{\partial \hat{\sigma}}{\partial t} - \hat{\sigma} \cdot \bar{\nabla}\bar{v} - \bar{\nabla}\bar{v}^T \cdot \hat{\sigma},$$

we obtain two other Maxwell models

$$\hat{\sigma}^\nabla + \frac{1}{T}\hat{\sigma} = 2\mu\hat{D}, \tag{2.60}$$

and

$$\hat{\sigma}^\Delta + \frac{1}{T}\hat{\sigma} = 2\mu\hat{D}, \tag{2.61}$$

see Oldroyd (1958). Experimental data confirming constitutive equation (2.61) are presented by Pearson & Middleman (1978) and Janssen & Janssen - van Rosmalen (1979). Model (2.60) with coefficients T and μ depending on the second invariant of the rate-of-strain tensor $\hat{D}$ was studied by White & Metzner (1963) and Pearson & Middleman (1978).

Constitutive equation (2.61) may be easily generalized by replacing the Maxwell model by the Maxwell–Weichert model, see Fig. 2.4. As a result, we arrive at the constitutive relation

$$\hat{\sigma} = \sum_{n=1}^{N} \hat{\sigma}_n, \quad \hat{\sigma}_n^\Delta + \frac{1}{T_n}\hat{\sigma}_n = 2\mu_n\hat{D}, \tag{2.62}$$

265

which is characterized by a number N of parallel Maxwell elements, their relaxation times T_n, and their viscosities μ_n. Model (2.62) with coefficients T_n and μ_n depending on the principal invariants of the stress tensor $\hat{\sigma}$ was suggested and verified by La Mantia (1977) and La Mantia & Titomanlio (1979).

Astarita & Marrucci (1974) generalized constitutive equation (2.62) by replacing the upper corotational derivative $\hat{\sigma}^\triangle$ of the stress tensor by the general corotational derivative $\hat{\sigma}^\square$, see Eq. (1.4.46), and compared the obtained results with experimental data for polymeric melts.

As common practice, polymeric media with finite strains are incompressible. This means that their constitutive models should imply the rate-of-strain tensor with the zero first invariant for an arbitrary stress tensor.

Exercise 2.12. Show that for an arbitrary Cauchy stress tensor, the Maxwell models (2.59) – (2.62) may imply the rate-of-strain tensor with a non-zero first invariant. For a detailed discussion of this issue, see Astarita & Marrucci (1974). □

To ensure the incompressibility condition, we should present the Cauchy stress tensor $\hat{\sigma}$ in the form

$$\hat{\sigma} = -p\hat{I} + \hat{s}, \tag{2.63}$$

where p is a pressure, and $\hat{s}$ is the deviatoric part of the stress tensor, and assume that tensor $\hat{s}$ (instead of $\hat{\sigma}$) obeys the constitutive equations (2.59) – (2.62).

2.5.4. The standard viscoelastic solids

To extend the Zener model (2.13) to finite strains, we should replace the infinitesimal strain ϵ by a strain tensor at finite strain, the derivative of the infinitesimal strain tensor $\dot{\epsilon}$ by the rate-of-strain tensor $\hat{D}$, and the derivative of the stress tensor $\hat{\sigma}$ by an appropriate corotational derivative of the Cauchy stress tensor $\hat{\sigma}$. Evidently, this procedure is not determined uniquely, since we may employ various strain tensors and various corotational derivatives. For example, in *the Hausler–Sayir model*, see Hausler & Sayir (1995), the Finger strain tensor $\hat{E}_F$, see Eq. (2.1.36), and the upper convected derivative, see Eq. (1.4.44), of the stress tensor are used. As a result, we obtain

$$A_1 \hat{s}^\triangle + A_0 \hat{s} = B_1 \hat{D} + B_0 \hat{E}_F. \tag{2.64}$$

Model (2.63) and (2.64) describes adequately the viscoelastic behavior of butyl rubber. Experimental data show that the use of the low convected derivative $\hat{s}^\triangledown$ instead of $\hat{s}^\triangle$ leads to non-realistic results.

By adding higher order terms to the right-hand side of Eq. (2.64), we may construct more sophisticated differential models. For example, according to *the generalized Hausler–Sayir model*, see Hausler & Sayir (1995), the second order terms compared with the strain tensor and the rate-of-strain tensor are included into the constitutive equation

$$A_1 \hat{s}^\triangle + A_0 \hat{s} = B_1 \hat{D} + B_0 \hat{E}_F + C_{11} \hat{D}^2 + C_{12}(\hat{D} \cdot \hat{E}_F + \hat{E}_F \cdot \hat{D}) + C_{22} \hat{E}_F^2. \quad (2.65)$$

Here A_i, B_i, and C_{ij} are material parameters (which may depend on the principal invariants of the Finger tensor) found by fitting experimental data. Model (2.65) predicts correctly the viscoelastic behavior of carbon black reinforced rubber for deformations of up to 20 per cent.

3. Integral Constitutive Models

In this Section, some integral constitutive models are discussed for viscoelastic media with infinitesimal and finite strains. We begin our consideration with linear viscoelastic solids at infinitesimal strains. Afterward, we introduce a model of adaptive links which allows linear constitutive equations to be treated in terms of a system with a variable number of elastic springs (links). We formulate the Lagrange principle in finite viscoelasticity and (by using this variational principle) extend linear constitutive relations to nonlinear media with finite strains. Finally, we provide a brief survey of constitutive models for viscoelastic media with finite trains.

3.1. Linear Viscoelastic Media

In this subsection we formulate constitutive equations for uniaxial deformations of a viscoelastic specimen and discuss basic properties of creep and relaxation kernels.

3.1.1. The Boltzmann superposition principle

Let us consider a viscoelastic specimen in the form of a rectilinear rod. The specimen is in its natural (stress-free) state. At instant $\tau \geq 0$, unit tensile forces are applied to the ends of the rod. The longitudinal strain $e(t, \tau)$ at moment $t \geq \tau$ is presented as the sum

$$e(t, \tau) = e_0(\tau) + e_1(t, \tau), \quad (3.1)$$

where $e_0 = E^{-1}(\tau)$ is the instant strain, and $e_1 = C(t, \tau)$ is an additional (creep) strain caused by the material viscosity. $E(\tau)$ is called current elastic modulus, and $C(t, \tau)$ is called *creep measure*. Function $C(t, \tau)$ is assumed to be sufficiently smooth and to satisfy the condition

$$C(\tau, \tau) = 0. \quad (3.2)$$

We now suppose that at instant $t = 0$ a time-varying longitudinal load is applied to the specimen and denote by $\sigma(t)$ the longitudinal stress. Function $\sigma(t)$ is continuously differentiable and satisfies the equality

$$\sigma(0) = 0. \tag{3.3}$$

For linear viscoelastic media the following superposition principle is valid:

Superposition principle (Boltzmann). The strain $\epsilon(t)$ at instant t caused by a stress history $\{\sigma(\tau) \ (0 \leq \tau \leq t)\}$ equals the sum of the strains caused by elementary stresses:

$$\epsilon(t) = \int_0^t \frac{d\sigma(\tau)}{d\tau} e(t, \tau) d\tau. \qquad \square \tag{3.4}$$

For experimental verification of the Boltzmann principle, see Bugakov (1989) and Ferry (1980).

Integration of Eq. (3.4) by parts with the use of Eqs. (3.1) – (3.3) yields

$$\epsilon(t) = \sigma(\tau)e(t,\tau)|_{\tau=0}^{\tau=t} - \int_0^t \frac{\partial e}{\partial \tau}(t,\tau)\sigma(\tau)d\tau = \frac{\sigma(t)}{E(t)} + \int_0^t \frac{\partial \mathcal{Y}}{\partial \tau}(t,\tau)\sigma(\tau)d\tau, \tag{3.5}$$

where

$$\mathcal{Y}(t,\tau) = -[\frac{1}{E(\tau)} + C(t,\tau)]. \tag{3.6}$$

The function

$$K(t,\tau) = -E(t)\frac{\partial}{\partial \tau}[\frac{1}{E(\tau)} + C(t,\tau)] \tag{3.7}$$

is called *creep kernel*. By using Eq. (3.7), constitutive equation (3.5) can be presented in the form

$$\epsilon(t) = \frac{1}{E(t)}[\sigma(t) + \int_0^t K(t,\tau)\sigma(\tau)d\tau]. \tag{3.8}$$

The first term in the right-hand side of Eq. (3.8) determines the instant elastic deformation, while the other term determines the creep deformation. Eq. (3.8) describes the mechanical behavior of aging viscoelastic media with a time-dependent response.

The simplest constitutive model of aging media is *an aging elastic material* with $C(t,\tau) = 0$

$$\epsilon(t) = \frac{\sigma(t)}{E(t)} - \int_0^t \sigma(\tau)\frac{\partial}{\partial \tau}\frac{1}{E(\tau)}d\tau.$$

Differentiation of this equality implies that

$$\dot{\epsilon}(t) = \frac{\dot{\sigma}(t)}{E(t)}, \tag{3.9}$$

268

where the superscript dot denotes differentiation with respect to time. It is worth noting that Eq. (3.9) differs from the model

$$\epsilon(t) = \frac{\sigma(t)}{E(t)}, \tag{3.10}$$

which is employed in applications as well.

For a number of materials, the effect of aging is weak and it may be neglected. From the mathematical standpoint, this means that the current elastic modulus E may be treated as a constant, $E(t) = E$, while the creep kernel K may be considered as a function of the difference $t - \tau$ only, $K(t, \tau) = K(t - \tau)$. For a non-aging viscoelastic medium, constitutive equation (3.8) implies that

$$\epsilon(t) = \frac{1}{E}[\sigma(t) + \int_0^t K(t - \tau)\sigma(\tau)d\tau] = \frac{1}{E}[\sigma(t) + \int_0^t \dot{C}_0(t - \tau)\sigma(\tau)d\tau], \tag{3.11}$$

where

$$K(t) = \dot{C}_0(t), \qquad C_0(t) = EC(t). \tag{3.12}$$

Eqs. (3.7) and (3.11) which determine the strain ϵ as a function of the stress σ are called *creep equations*. For a given strain ϵ, they may be treated as linear integral equations for the stress σ. By solving these equations, we obtain *relaxation equations* which express the stress σ as a function of the strain ϵ. For an aging viscoelastic medium we find

$$\sigma(t) = E(t)\epsilon(t) - \int_0^t \frac{\partial \mathcal{X}}{\partial \tau}(t, \tau)\epsilon(\tau)d\tau. \tag{3.13}$$

The function $\mathcal{X}(t, \tau)$ is presented in the form

$$\mathcal{X}(t, \tau) = E(\tau) + Q(t, \tau), \tag{3.14}$$

where function $Q(t, \tau)$ is called *relaxation measure*. This function is assumed to satisfy the condition

$$Q(\tau, \tau) = 0. \tag{3.15}$$

Similar to Eq. (3.7), we introduce *the relaxation kernel*

$$R(t, \tau) = \frac{1}{E(t)} \frac{\partial}{\partial \tau}[E(\tau) + Q(t, \tau)]. \tag{3.16}$$

In the new notation, Eq. (3.13) is written as

$$\sigma(t) = E(t)[\epsilon(t) - \int_0^t R(t, \tau)\epsilon(\tau)d\tau]. \tag{3.17}$$

269

For non-aging viscoelastic media, Eq. (3.17) is simplified

$$\sigma(t) = E[\epsilon(t) - \int_0^t R(t-\tau)\epsilon(\tau)d\tau] = E[\epsilon(t) + \int_0^t \dot{Q}_0(t-\tau)\epsilon(\tau)d\tau], \quad (3.18)$$

where

$$R(t) = -\dot{Q}_0(t), \qquad Q_0(t) = \frac{Q(t)}{E}. \quad (3.19)$$

In applied studies, instead of the relaxation measure $Q_0(t)$, *the relaxation function $Q_*(t)$ is used.* The functions $Q_0(t)$ and $Q_*(t)$ are connected by the equality

$$Q_*(t) = 1 + Q_0(t). \quad (3.20)$$

Eqs. (3.15) and (3.20) imply that

$$Q_*(0) = 1. \quad (3.21)$$

By using Eq. (3.20) and integrating by parts, we transform Eq. (3.18) as follows:

$$\sigma(t) = E[\epsilon(t) + \int_0^t \dot{Q}_*(t-\tau)\epsilon(\tau)d\tau] = E[\epsilon(t) - Q_*(t-\tau)\epsilon(\tau)|_{\tau=0}^{\tau=t}$$

$$+ \int_0^t Q_*(t-\tau)\dot{\epsilon}(\tau)d\tau] = E\int_0^t Q_*(t-\tau)\dot{\epsilon}(\tau)d\tau. \quad (3.22)$$

Exercise 3.1. By carrying out similar transformations for Eq. (3.13), derive the formula

$$\sigma(t) = \int_0^t \mathcal{X}(t,\tau)\dot{\epsilon}(\tau)d\tau. \quad \Box \quad (3.23)$$

Eqs. (3.8) and (3.17) determine *a homogeneous viscoelastic medium* with the response independent of spatial coordinates. For *a non-homogeneous viscoelastic medium*, Young's modulus, as well as the creep and relaxation measures depend explicitly on the longitudinal coordinate x

$$\epsilon(t,x) = \frac{1}{E(t,x)}[\sigma(t,x) + \int_0^t K(t,\tau,x)\sigma(\tau,x)d\tau],$$

$$\sigma(t,x) = E(t,x)[\epsilon(t,x) - \int_0^t R(t,\tau,x)\epsilon(\tau,x)d\tau]. \quad (3.24)$$

Eqs. (3.24) characterize an arbitrary non-homogeneity. For *non-homogeneously aging media*, we assume that different points were manufactured at different instants which preceded the initial moment $t = 0$, see Arutyunyan et al. (1987). To describe the non–homogeneity, a piece-wise continuous and bounded function $\kappa(x)$ is introduced which equals the material age at point x at the initial moment $t = 0$. Since the material response is characterized by the internal time $t + \kappa(x)$, the constitutive equations of a non-homogeneously aging viscoelastic medium are written as

$$\epsilon(t,x) = \frac{1}{E(t+\kappa(x))}[\sigma(t,x) + \int_0^t K(t+\kappa(x),\tau+\kappa(x))\sigma(\tau,x)d\tau],$$

$$\sigma(t,x) = E(t+\kappa(x))[\epsilon(t,x) - \int_0^t R(t+\kappa(x),\tau+\kappa(x))\epsilon(\tau,x)d\tau]. \quad (3.25)$$

Three different approaches may be distinguished:
(i) $\kappa(x)$ is prescribed *a priori* to characterize the age distribution in a medium;
(ii) $\kappa(x)$ is a control function which is chosen to ensure optimal properties of a structure;
(iii) $\kappa(x)$ is a random variable which describes *stochastic aging* due to the *temperature*, see Lee & McKenna (1990b), *humidity*, see Knauss & Kenner (1980), *radiation*, see McHerron & Wilkes (1993), etc.

A specific spatial non-homogeneity is typical of *growing viscoelastic media*. The growth means an increase in mass owing to the material influx from the environment, see Section 7.6. For a growing viscoelastic medium, two specific functions are used in constitutive models. The first is the instant $\varpi(x)$ when a material portion at a point x is manufactured. The other is the first instant $\tau(x)$ when external forces are applied to this portion. The constitutive equations of a growing viscoelastic medium are written as

$$\epsilon(t,x) = \frac{1}{E(t-\varpi(x))}[\sigma(t,x) + \int_{\tau(x)}^t K(t-\varpi(x),\tau-\varpi(x))\sigma(\tau,x)d\tau],$$

$$\sigma(t,x) = E(t-\varpi(x))[\epsilon(t,x) - \int_{\tau(x)}^t R(t-\varpi(x),\tau-\varpi(x))\epsilon(\tau,x)d\tau]. \quad (3.26)$$

3.1.2. Connections between creep and relaxation kernels

Let us derive an integral equation which expresses creep and relaxation kernels in terms of each other. For this purpose, we substitute expression (3.5) into Eq. (3.13)

$$
\begin{aligned}
\sigma(t) &= E(t)[\frac{\sigma(t)}{E(t)} + \int_0^t \frac{\partial \mathcal{Y}}{\partial s}(t,s)\sigma(s)ds] \\
&\quad - \int_0^t \frac{\partial \mathcal{X}}{\partial s}(t,s)[\frac{\sigma(s)}{E(s)} + \int_0^s \frac{\partial \mathcal{Y}}{\partial \tau}(s,\tau)\sigma(\tau)d\tau]ds \\
&= \sigma(t) + \int_0^t [E(t)\frac{\partial \mathcal{Y}}{\partial s}(t,s) - \frac{1}{E(s)}\frac{\partial \mathcal{X}}{\partial s}(t,s)]\sigma(s)ds \\
&\quad - \int_0^t \frac{\partial \mathcal{X}}{\partial s}(t,s)ds \int_0^s \frac{\partial \mathcal{Y}}{\partial \tau}(s,\tau)\sigma(\tau)d\tau.
\end{aligned}
$$

271

Exercise 3.2. Check that this equality implies that

$$E(t)\frac{\partial \mathcal{Y}}{\partial s}(t,s) - \frac{1}{E(s)}\frac{\partial \mathcal{X}}{\partial s}(t,s) = \int_s^t \frac{\partial \mathcal{X}}{\partial \tau}(t,\tau)\frac{\partial \mathcal{Y}}{\partial s}(\tau,s)d\tau. \qquad \square \qquad (3.27)$$

Substitution of expressions (3.6) and (3.14) into Eq. (3.27) yields

$$E(t)\frac{\partial}{\partial s}[\frac{1}{E(s)}+C(t,s)] + \frac{1}{E(s)}\frac{\partial}{\partial s}[E(s)+Q(t,s)]$$

$$= \int_s^t \frac{\partial}{\partial \tau}[E(\tau)+Q(t,\tau)]\frac{\partial}{\partial s}[\frac{1}{E(s)}+C(\tau,s)]d\tau. \qquad (3.28)$$

By integrating Eq. (3.28) from T to t, we obtain

$$E(t)[(\frac{1}{E(t)}+C(t,t))-(\frac{1}{E(T)}+C(t,T))] + \int_T^t \frac{1}{E(s)}\frac{\partial}{\partial s}[E(s)+Q(t,s)]ds$$

$$= \int_T^t ds \int_s^t \frac{\partial}{\partial \tau}[E(\tau)+Q(t,\tau)]\frac{\partial}{\partial s}[\frac{1}{E(s)}+C(\tau,s)]d\tau. \qquad (3.29)$$

To transform the right-hand side of Eq. (3.29), we change the order of integration and find

$$\int_T^t ds \int_s^t \frac{\partial}{\partial \tau}[E(\tau)+Q(t,\tau)]\frac{\partial}{\partial s}[\frac{1}{E(s)}+C(\tau,s)]d\tau$$

$$= \int_T^t \frac{\partial}{\partial \tau}[E(\tau)+Q(t,\tau)]d\tau \int_T^\tau \frac{\partial}{\partial s}[\frac{1}{E(s)}+C(\tau,s)]ds.$$

Calculation of the integral with the use of Eq. (3.2) implies that

$$\int_T^t \frac{\partial}{\partial \tau}[E(\tau)+Q(t,\tau)]d\tau \int_T^\tau \frac{\partial}{\partial s}[\frac{1}{E(s)}+C(\tau,s)]ds$$

$$= \int_t^t \frac{\partial}{\partial \tau}[E(\tau)+Q(t,\tau)][(\frac{1}{E(\tau)}+C(\tau,\tau))-(\frac{1}{E(T)}+C(\tau,T))]d\tau$$

$$= \int_t^t \frac{\partial}{\partial \tau}[E(\tau)+Q(t,\tau)][\frac{1}{E(\tau)}-\frac{1}{E(T)}-C(\tau,T)]d\tau.$$

Substitution of this expression into Eq. (3.29) yields

$$1-E(t)[\frac{1}{E(T)}+C(t,T)] = -\int_T^t \frac{\partial}{\partial \tau}[E(\tau)+Q(t,\tau)][\frac{1}{E(T)}+C(\tau,T)]d\tau. \qquad (3.30)$$

We integrate the right-hand side of Eq. (3.30) by parts and use Eq. (3.2). As a result, we obtain

$$\int_T^t \frac{\partial}{\partial \tau}[E(\tau)+Q(t,\tau)][\frac{1}{E(T)}+C(\tau,T)]d\tau$$

272

$$= [E(t) + Q(t,t)][\frac{1}{E(T)} + C(t,T)] - [E(T) + Q(t,T)][\frac{1}{E(T)} + C(T,T)]$$

$$- \int_T^t [E(\tau) + Q(t,\tau)]\frac{\partial C}{\partial \tau}(\tau,T)d\tau$$

$$= E(t)[\frac{1}{E(T)} + C(t,T)] - 1 - \frac{Q(t,T)}{E(T)} - \int_T^t [E(\tau) + Q(t,\tau)]\frac{\partial C}{\partial \tau}(\tau,T)d\tau.$$

Substitution of this expression into Eq. (3.30) implies that

$$\frac{Q(t,s)}{E(s)} + \int_s^t [E(\tau) + Q(t,\tau)]\frac{\partial C}{\partial \tau}(\tau,s)d\tau = 0. \tag{3.31}$$

For a given creep measure $C(t,s)$ and a given Young modulus $E(t)$, Eq. (3.31) is a linear *Volterra equation* for relaxation measure $Q(t,s)$. By introducing the notation

$$M(t,s) = 1 + E(s)C(t,s), \tag{3.32}$$

we rewrite Eq. (3.31) in the form

$$Q(t,s) + \int_s^t \frac{\partial M}{\partial \tau}(\tau,s)Q(t,\tau)d\tau = - \int_s^t E(\tau)\frac{\partial M}{\partial \tau}(\tau,s)d\tau. \tag{3.33}$$

For non-aging media, it is convenient to return to Eq. (3.28) which is written as

$$\dot{C}_0(t-s) + \dot{Q}_0(t-s) = - \int_s^t \dot{Q}_0(t-\tau)\dot{C}_0(\tau-s)d\tau.$$

Introducing the new variables $t_1 = t - s$ and $\tau_1 = \tau - s$, we present this equation in the form

$$\dot{C}_0(t_1) + \dot{Q}_0(t_1) = - \int_0^{t_1} \dot{Q}_0(t_1-\tau_1)\dot{C}_0(\tau_1)d\tau_1.$$

Finally, by using Eqs. (3.12) and (3.19), we find

$$K(t) - R(t) = \int_0^t R(t-s)K(s)ds. \tag{3.34}$$

Eq. (3.34) expresses the creep and relaxation kernels in terms of each other. For a given function $R(t)$, Eq. (3.34) is transformed into a linear Volterra equation for function $K(t)$

$$K(t) - \int_0^t R(t-s)K(s)ds = R(t). \tag{3.35}$$

For a given function $K(t)$, Eq. (3.34) is a linear Volterra equation for function $R(t)$. Introducing the new variable under $\tau = t - s$, we obtain

$$R(t) + \int_0^t K(t-\tau)R(\tau)d\tau = K(t). \tag{3.36}$$

273

Let us introduce *the integral operators* **K** and **R** by the formulas

$$\mathbf{K}f = \int_0^t K(t-s)f(s)ds, \qquad \mathbf{R}f = \int_0^t R(t-s)f(s)ds, \qquad (3.37)$$

where $f(t)$ is a smooth function. Then Eqs. (3.35) and (3.36) are written as

$$(\mathbf{I}+\mathbf{K})R = K, \qquad (\mathbf{I}-\mathbf{R})K = R, \qquad (3.38)$$

where **I** is the unit operator.

3.1.3. Basic properties of the Volterra operators

Let us consider the linear integral equation

$$x(t) - \lambda \int_0^t V(t,s)x(s)ds = f(t). \qquad (3.39)$$

Here $f(t)$ is a given function, $x(t)$ is an unknown function, $V(t,s)$ is a sufficiently smooth kernel, and λ is a complex parameter. Eq. (3.39) is written in the operator form as

$$(\mathbf{I}-\lambda\mathbf{V})x = f,$$

where **V** is *the Volterra operator* with the kernel $V(t,s)$. For any sufficiently smooth function $f(t)$

$$\mathbf{V}f = \int_0^t V(t,s)f(s)ds.$$

Any solution of Eq. (3.39) is written as $x = \mathbf{P}(\lambda)f$, where

$$\mathbf{P}(\lambda) = (\mathbf{I}-\lambda\mathbf{V})^{-1} \qquad (3.40)$$

is called *resolvent operator*.

Exercise 3.3. Prove that for any complex λ, $\mathbf{P}(\lambda)$ is a linear Volterra operator.
□

Denote by $\lambda P(t,s,\lambda)$ the kernel of the resolvent operator $\mathbf{P}(\lambda)$

$$x(t) = f(t) + \lambda \int_0^t P(t,s,\lambda)f(s)ds. \qquad (3.41)$$

Function $P(t,s,\lambda)$ is called *resolvent kernel* for the kernel $V(t,s)$.
Exercise 3.4. Show that $P(t,s,\lambda)$ satisfies the linear integral equation

$$P(t,s,\lambda) - \lambda \int_s^t V(t,\tau)P(\tau,s,\lambda)d\tau = V(t,s). \qquad \square \qquad (3.42)$$

Introduce the notation $V_1(t, \tau) = V(t, \tau)$. By applying operator $\mathbf{V}$ to Eq. (3.39), we obtain

$$\mathbf{V}^2 f = \mathbf{V} \cdot \mathbf{V} f = \int_0^t V(t, s) ds \int_0^s V_1(s, \tau) f(\tau) d\tau.$$

Changing the order of integration implies that

$$\mathbf{V}^2 f = \int_0^t V_2(t, \tau) f(\tau) d\tau,$$

where

$$V_2(t, \tau) = \int_\tau^t V(t, s) V_1(s, \tau) ds.$$

Exercise 3.5. Check that for any positive integer n

$$\mathbf{V}^n f = \int_0^t V_n(t, \tau) f(\tau) d\tau \tag{3.43}$$

is a linear Volterra operator with the kernel

$$V_n(t, \tau) = \int_\tau^t V(t, s) V_{n-1}(s, \tau) ds. \qquad \square$$

Let $F(x)$ be a smooth function which is expanded in the Taylor series

$$F(x) = \sum_{n=0}^{\infty} a_n x^n, \tag{3.44}$$

converging at $|x| \leq \delta$ for a positive δ. We define *the function $F(\mathbf{V})$ of a Volterra operator* $\mathbf{V}$ as a Volterra operator which acts on a smooth function f according to the formula

$$F(\mathbf{V}) f = \sum_{n=0}^{\infty} a_n \mathbf{V}^n f. \tag{3.45}$$

Exercise 3.6. Prove that series (3.45) converges for any $t \geq 0$. $\square$

Substitution of expression (3.43) into Eq. (3.45) implies that $F(\mathbf{V}) = a_0 \mathbf{I} + \mathbf{\Phi}$, where $\mathbf{\Phi}$ is a Volterra operator with the kernel

$$\Phi(t, s) = \sum_{n=1}^{\infty} a_n V_n(t, s). \tag{3.46}$$

In particular, for $F(x) = (1 - \lambda x)^{-1}$ we find $a_n = \lambda^n$ $(n = 0, 1, \ldots)$. Therefore,

$$\mathbf{P}(\lambda) = \mathbf{I} + \sum_{n=1}^{\infty} \lambda^n \mathbf{V}^n. \tag{3.47}$$

It follows from Eqs. (3.41), (3.46), and (3.47) that

$$P(t, s, \lambda) = \sum_{n=1}^{\infty} \lambda^{n-1} V_n(t, s). \tag{3.48}$$

According to Eq. (3.45), if a function $F(x)$ satisfies some identity, then the operator $F(\mathbf{V})$ satisfies the same identity. This allows several important formulas to be derived.

Exercise 3.7. Check the equalities for the operator $\mathbf{\Pi}(\lambda) = \mathbf{V} \cdot (\mathbf{I} - \lambda \mathbf{V})^{-1}$

$$\mathbf{\Pi}(\lambda_1) \cdot \mathbf{\Pi}(\lambda_2) = \frac{1}{\lambda_1 - \lambda_2} [\mathbf{\Pi}(\lambda_1) - \mathbf{\Pi}(\lambda_2)],$$

$$\mathbf{I} + \lambda_1 \mathbf{\Pi}(\lambda_1 + \lambda_2) = [\mathbf{I} - \lambda_1 \mathbf{\Pi}(\lambda_2)]^{-1}, \qquad \frac{\partial \mathbf{\Pi}}{\partial \lambda}(\lambda) = \mathbf{\Pi}^2(\lambda). \qquad \Box \tag{3.49}$$

A Volterra operator $\mathbf{V}$ is called *convolutive*, if its kernel $V(t, s)$ depends on the difference $t - s$ only, $V = V(t - s)$.

Exercise 3.8. Prove that for any convolutive operator $\mathbf{V}$, its resolvent operator $\mathbf{P}(\lambda)$ is convolutive as well. $\Box$

Exercise 3.9. Show that for a convolutive operator $\mathbf{V}$, Eq. (3.42) is written as

$$P(t, \lambda) - \lambda \int_0^t V(t - s) P(s, \lambda) ds = V(t). \qquad \Box \tag{3.50}$$

To derive a connection between kernels $V(t)$ and $P(t, \lambda)$ of a convolutive operator $\mathbf{V}$ and its resolvent operator $\mathbf{P}(\lambda)$ we use *the Laplace transformation method*.

Definition. For any integrable function $f(t)$ of a real variable t, its *Laplace transform* $\tilde{f}(p)$ is a function

$$\tilde{f}(p) = \int_0^\infty f(t) \exp(-pt) dt \tag{3.51}$$

of a complex variable $p = \alpha + i\omega$, where $i = \sqrt{-1}$.

The inverse Laplace transform is determined by the formula

$$f(t) = \frac{1}{2\pi i} \int_{c-i\omega}^{c+i\omega} \tilde{f}(p) \exp(pt) dp. \tag{3.52}$$

We assume that there is a constant c_0 such that function $\tilde{f}(p)$ is analytical for $\Re p \geq c_0$, where $\Re$ stands for the real part of a complex number. In this case, c is an arbitrary real number, satisfying the inequality $c \geq c_0$.

To derive an expression for the function $\tilde{P}(p, \lambda)$ we apply the Laplace transform to Eq. (3.50) and use the following

Proposition 3.1. For any smooth functions $f(t)$ and $g(t)$

$$\int_0^\infty \exp(-pt)dt \int_0^t f(t-s)g(s)ds = \tilde{f}(p)\tilde{g}(p). \qquad \square \qquad (3.53)$$

Exercise 3.10. Prove this assertion. $\square$

As a result, we find

$$\tilde{P}(p,\lambda) - \lambda\tilde{P}(p,\lambda)\tilde{V}(p) = \tilde{V}(p),$$

which implies that

$$\tilde{P}(p,\lambda) = \frac{\tilde{V}(p)}{1 - \lambda\tilde{V}(p)}, \qquad \tilde{V}(p) = \frac{\tilde{P}(p,\lambda)}{1 + \lambda\tilde{P}(p,\lambda)}. \qquad (3.54)$$

In particular, by applying formulas (3.54) to Eqs. (3.35) and (3.36), we obtain

$$\tilde{R}(p) = \frac{\tilde{K}(p)}{1 + \tilde{K}(p)}, \qquad \tilde{K}(p) = \frac{\tilde{R}(p)}{1 - \tilde{R}(p)}. \qquad (3.55)$$

3.1.4. Three-dimensional constitutive equations

We may distinguish three basic approaches to constructing linear constitutive models for three-dimensional deformations. All of them are based on the hypothesis that constitutive relations for a viscoelastic medium may be derived from the constitutive equations for an elastic material by replacing elastic moduli by appropriate Volterra operators. For example, Eqs. (3.8) and (3.17) may be treated as extensions of *Hooke's law*, where the Young modulus E is replaced by *the relaxation operator* $E(\mathbf{I} - \mathbf{R})$, whereas E^{-1} is replaced by *the creep operator* $E^{-1}(\mathbf{I} + \mathbf{K})$.

Let us write the constitutive equations of an isotropic elastic medium

$$\hat{\sigma} = \frac{E}{1+\nu}\left(\hat{\epsilon} + \frac{\nu}{1-2\nu}\epsilon\hat{I}\right), \qquad \hat{\epsilon} = \frac{1}{E}[(1+\nu)\hat{\sigma} - \nu\sigma\hat{I}]. \qquad (3.56)$$

Here ν is *Poisson's ratio*, $\sigma = I_1(\hat{\sigma})$ and $\epsilon = I_1(\hat{\epsilon})$ are the first invariants of the stress and strain tensors. If the Young modulus is replaced by an appropriate integral operator and Poisson's ratio is assumed to be constant, we obtain the following constitutive equations of a linear isotropic viscoelastic medium:

$$\hat{\sigma} = \frac{E}{1+\nu}(\mathbf{I} - \mathbf{R})\left(\hat{\epsilon} + \frac{\nu}{1-2\nu}\epsilon\hat{I}\right), \qquad \hat{\epsilon} = \frac{1}{E}(\mathbf{I} + \mathbf{K})[(1+\nu)\hat{\sigma} - \nu\sigma\hat{I}], \quad (3.57)$$

where operators $\mathbf{K}$ and $\mathbf{R}$ are determined by Eqs. (3.37). Eqs. (3.57) allow explicit solutions to problems in linear viscoelasticity to be derived by using solutions for appropriate elastic problems. A method for constructing solutions to viscoelastic problems by using solutions to elastic problems is called *the correspondence principle* (the Volterra principle). We do not intend to discuss this method in detail referring to Tsien (1950).

Experimental studies show that Eqs. (3.57) have a narrow sphere of applications, since Poisson's ratio for a number of viscoelastic solids is not constant, see Bertilsson et al. (1993), Power & Caddell (1972), and Zhaoxia (1994).

Constitutive equations (3.56) may be written as

$$\sigma = 3\mathcal{K}\epsilon, \qquad \hat{s} = 2G\hat{e}. \tag{3.58}$$

Here $\hat{e}$, $\hat{s}$ are the deviatoric parts of the strain and stress tensors: $\hat{\sigma} = \frac{1}{3}\sigma\hat{I} + \hat{s}$, $\hat{\epsilon} = \frac{1}{3}\epsilon\hat{I} + \hat{e}$, and $\mathcal{K}$, G are *bulk* and *shear elastic moduli* which are connected with the Young modulus E and Poisson's ratio ν by the formulas

$$\mathcal{K} = \frac{E}{3(1 - 2\nu)}, \quad G = \frac{E}{2(1 + \nu)}, \quad E = \frac{9\mathcal{K}G}{3\mathcal{K} + G}, \quad \nu = \frac{3\mathcal{K} - 2G}{2(3\mathcal{K} + G)}. \tag{3.59}$$

By replacing constants $\mathcal{K}$ and G in Eqs. (3.58) by appropriate Volterra operators, we obtain the other version of constitutive equations for an isotropic linear viscoelastic material

$$\sigma = 3\mathcal{K}(\mathbf{I} - \mathbf{R}_b)\epsilon, \qquad \hat{s} = 2G(\mathbf{I} - \mathbf{R}_s)\hat{e}, \tag{3.60}$$

where $\mathbf{R}_b$ and $\mathbf{R}_s$ are *bulk* and *shear relaxation operators* with kernels $R_b(t, \tau)$ and $R_s(t, \tau)$. Since a number of viscoelastic materials demonstrate elastic bulk response, whereas their shear deformation is viscoelastic, Eqs. (3.60) may be simplified by setting $\mathbf{R}_b = 0$ and $\mathbf{R}_s = \mathbf{R}$

$$\sigma = 3\mathcal{K}\epsilon, \qquad \hat{s} = 2G(\mathbf{I} - \mathbf{R})\hat{e}. \tag{3.61}$$

It is worth noting a specific constitutive model, which combines "solid" and "liquid" shear responses. This model may be treated as a version of the Kelvin-Voigt model, see Fig. 2.3, which contains a viscoelastic spring with constitutive relations (3.61) located in parallel with a viscous dashpot with the constitutive equation (2.50). Assuming volume deformation to be purely elastic, we derive the constitutive equations in the form

$$\sigma = 3\mathcal{K}\epsilon, \qquad \hat{s} = 2G(\mathbf{I} - \mathbf{R})\hat{e} + 2\eta\frac{d\hat{e}}{dt}(t). \tag{3.62}$$

Eqs. (3.62) were used by Renardy et al. (1987) to study waves in a linear viscoelastic medium.

278

To derive the third version of constitutive equations for a viscoelastic medium, we rewrite Eqs. (3.56) as

$$\hat{\sigma} = \lambda \epsilon \hat{I} + 2\mu \hat{e},$$

(3.63)

where λ and μ are the Lame parameters, which are connected with Young's modulus and Poisson's ratio by the formulas

$$\lambda = \frac{E\nu}{(1-2\nu)(1+\nu)}, \quad \mu = \frac{E}{2(1+\nu)}, \quad E = \frac{(3\lambda+2\mu)\mu}{\lambda+\mu}, \quad \nu = \frac{\lambda}{2(\lambda+\mu)}.$$

(3.64)

By preserving a constant λ and replacing μ by an integral operator, we obtain from Eq. (3.63)

$$\hat{\sigma} = \lambda \epsilon \hat{I} + 2\mu(\mathbf{I} - \mathbf{R}_\mu)\hat{e},$$

(3.65)

where $\mathbf{R}_\mu$ is the relaxation operator with a kernel $R_\mu(t,\tau)$. Eq. (3.65) is widely employed in linear viscoelasticity, see Christensen (1982).

Eqs. (3.57), (3.60), and (3.65) may be extended to anisotropic viscoelastic materials. The stress-strain relation for an anisotropic, linear, viscoelastic medium is written as

$$\hat{\sigma}(t) = \hat{G}(t) : \hat{e}(t) - \int_0^t \hat{R}(t,\tau) : \hat{e}(\tau)d\tau,$$

where $\hat{G}(t)$ is a fourth-rank tensor of current elastic moduli and $\hat{R}(t,\tau)$ is a fourth-rank tensor of relaxation kernels, see Ward (1971).

3.2. A Model of Adaptive Links

In Section 2, we introduced "springs & dashpots" rheological models for the mechanical behavior of viscoelastic media. In this subsection, we demonstrate that the response of an aging viscoelastic material may be described by a model containing only elastic springs (without dashpots) provided the springs replace each other according to a given law. For this purpose, we transform Eq. (3.13) with the use of Eqs. (3.14) and (3.15)

$$
\begin{aligned}
\sigma(t) &= E(t)\epsilon(t) - \int_0^t \frac{\partial \mathcal{X}}{\partial \tau}(t,\tau)\epsilon(t)d\tau + \int_0^t \frac{\partial \mathcal{X}}{\partial \tau}(t,\tau)[\epsilon(t) - \epsilon(\tau)]d\tau \\
&= [E(t) - \mathcal{X}(t,t) + \mathcal{X}(t,0)]\epsilon(t) + \int_0^t \frac{\partial \mathcal{X}}{\partial \tau}(t,\tau)[\epsilon(t) - \epsilon(\tau)]d\tau \\
&= \mathcal{X}(t,0)\epsilon_*(t,0) + \int_0^t \frac{\partial \mathcal{X}}{\partial \tau}(t,\tau)\epsilon_*(t,\tau)d\tau,
\end{aligned}
$$

(3.66)

where $\epsilon_*(t,\tau) = \epsilon(t) - \epsilon(\tau)$ is the relative strain for transition from the actual configuration at instant τ to the actual configuration at instant t. Function $\epsilon_*(t,\tau)$ characterizes the specimen deformation in the interval $[\tau, t]$.

279

For definiteness, we suggest an interpretation of Eq. (3.66) for polymeric materials. However, the results developed are valid for an arbitrary viscoelastic medium as well.

Let us consider a system of parallel elastic springs (which model links between polymeric molecules). At instant $t = 0$, the system consists of N_0 springs in the natural (stress-free) state. Rigidity of any spring is $c = E(0)/N_0$. Under the action of external loads, the springs deform and their number changes in time. In the interval $[\tau, \tau + d\tau]$,

$$dN_0(\tau) = \frac{N_0}{E(0)} \frac{\partial \mathcal{X}}{\partial \tau}(\tau, \tau) d\tau$$

new springs merge with the system. These springs are connected in parallel to the initial springs. The initial length of springs which join the system at instant τ equals the length of the deformed system at the moment of joining. This means that the strain at instant t in springs which join the system at instant τ, equals $\epsilon_*(t, \tau)$.

Springs of the system may arise and collapse. The number of springs joining the system at instant τ and existing at instant $t \geq \tau$ equals

$$dN(t, \tau) = \frac{N_0}{E(0)} \frac{\partial \mathcal{X}}{\partial \tau}(t, \tau) d\tau.$$

The number of initial springs existing at instant $t \geq 0$ is

$$N(t, 0) = \frac{N_0}{E(0)} \mathcal{X}(t, 0).$$

To calculate the response of the system, stresses in all the springs should be added

$$\sigma(t) = \sigma_0(t) + \int_0^t d\sigma(t, \tau). \tag{3.67}$$

Here $\sigma_0(t)$ is the stress at instant t in the initial system, and $d\sigma(t, \tau)$ is the stress at instant t in springs which join the system at instant τ. According to Hooke's law

$$\sigma_0(t) = c\epsilon_*(t, 0)N(t, 0), \qquad d\sigma(t, \tau) = c\epsilon_*(t, \tau)dN(t, \tau).$$

Substitution of these expressions into Eq. (3.67) yields

$$\sigma(t) = c\epsilon_*(t, 0)N(t, 0) + \int_0^t c\epsilon_*(t, \tau)dN(t, \tau)$$
$$= \mathcal{X}(t, 0)\epsilon_*(t, 0) + \int_0^t \frac{\partial \mathcal{X}}{\partial \tau}(t, \tau)\epsilon_*(t, \tau)d\tau. \tag{3.68}$$

280

Expressions (3.66) and (3.68) coincide, which means that the behavior of the system of adaptive links coincides with the behavior of an aging linear viscoelastic medium. Therefore, the system of adaptive links may model the mechanical behavior of a viscoelastic specimen.

The reason for this assertion lies deeper than a simple coincidence of equations. According to Struik (1978), a polymeric material may be treated as a set of large molecules, linked to each other by chemical bonds. These molecules move relatively to each other (*the micro-Brownian motion*). When the displacements of portions connected by *chemical links* reach some ultimate value, the links fail, and molecules acquire "free edges" which are ready to create new links. These links emerge when appropriate "free edges" are located sufficiently close to each other due to random wandering of molecules. After their onset, new links oppose the displacements of molecules relatively to their positions at the moment when the links arise.

This scenario for the interaction of polymeric molecules coincides with the above scenario for the system of adaptive links if chemical links are treated as appropriate elastic springs. The function $\mathcal{X}(t, \tau)$ is an average (deterministic) characteristic of *random motion* of molecules at the micro-level. $\mathcal{X}(t, \tau)$ is proportional to the number of links, existing at moment t and arisen before instant τ. The derivative $\partial \mathcal{X}/\partial \tau\,(t, \tau)$ determines the rate of creation (at instant τ) of new links which have not collapsed before instant t.

To determine the potential energy of a system of parallel elastic springs, we should add together the energies of separate springs. The potential energy at instant t of springs joining the system at instant τ equals

$$\frac{1}{2}c\epsilon_*^2(t, \tau)dN(t, \tau) = \frac{1}{2}\frac{\partial \mathcal{X}}{\partial \tau}(t, \tau)\epsilon_*^2(t, \tau)d\tau.$$

The potential energy of the system (strain energy density of an aging viscoelastic medium under uniaxial stresses) is calculated as

$$W(t) = \frac{1}{2}[\mathcal{X}(t, 0)\epsilon_*^2(t, 0) + \int_0^t \frac{\partial \mathcal{X}}{\partial \tau}(t, \tau)\epsilon_*^2(t, \tau)d\tau]$$

$$= \frac{1}{2}[\mathcal{X}(t, 0)\epsilon^2(t) + \int_0^t \frac{\partial \mathcal{X}}{\partial \tau}(t, \tau)(\epsilon(t) - \epsilon(\tau))^2 d\tau]. \tag{3.69}$$

A similar expression was suggested by Dafermos (1970) as *the Lyapunov functional* for a viscoelastic medium.

It is quite natural to extend the model of adaptive links to large deformations of a viscoelastic medium, and to treat a nonlinear viscoelastic material as a system with a varying number of parallel nonlinearly elastic springs. The springs are assumed to be made of an isotropic hyperelastic material with a strain energy density W. Since the natural configuration of "newborn" springs

281

coincides with the actual configuration of the system at the moment of their birth, we may write

$$W = W(I_1(t, \tau), I_2(t, \tau), I_3(t, \tau)),$$

where $I_k(t, \tau)$ are the principal invariants of the relative Finger tensor for transition from the actual configuration at instant τ to the actual configuration at moment t.

Strain energy density of a viscoelastic medium equals the sum of energies for separate springs. To obtain strain energy density at instant t we should add together strain energy densities for all the springs existing at instant t. Similarly to Eq. (3.69), we obtain

$$W(t) = \mathcal{X}(t, 0)W(I_k(t, 0)) + \int_0^t \frac{\partial \mathcal{X}}{\partial \tau}(t, \tau)W(I_k(t, \tau))d\tau. \qquad (3.70)$$

Model (3.70) may be easily generalized by assuming that there are M kinds of springs, and the processes of creation and breakage for different kinds of springs are governed by different functions

$$\mathcal{X}_m(t, \tau) = \mu_m(\tau) + Q_m(t, \tau), \qquad (3.71)$$

where $\mu_m(t)$ are generalized elastic moduli, $Q_m(t, \tau)$ are relaxation measures $(m = 1, \ldots, M)$. This hypothesis together with Eq. (3.70) implies the following formula for strain energy density of an isotropic, aging, *hyper-viscoelastic medium* with finite strains:

$$W(t) = \sum_{m=1}^{M} [\mathcal{X}_m(t, 0)W_m(I_k(t, 0)) + \int_0^t \frac{\partial \mathcal{X}_m}{\partial \tau}(t, \tau)W_m(I_k(t, \tau))d\tau]. \qquad (3.72)$$

Formula (3.72) provides constitutive equations for an extensive class of viscoelastic media. By assuming elastic moduli $\mu_m(t)$ and relaxation measures $Q_m(t, \tau)$ to be independent of strains, we obtain constitutive equations of *operator-linear* viscoelastic materials. When elastic moduli and/or relaxation measures depend on the strain history, Eq. (3.72) is an *operator-nonlinear* constitutive equation.

A rheological model of elastic links between polymeric chains was suggested by Green and Tobolsky (1946). In that work, one-dimensional constitutive equations were proposed for elongation and shear of non-aging polymers with the exponential relaxation kernel.

Yamamoto (1956) generalized the Green–Tobolsky model and suggested a statistical theory for calculating relaxation kernels under some general assumptions regarding breakage of polymeric chains. As a result, integro-differential equations were derived for a chain-distribution function, a chain-reformation function, and a chain-breakage function. For a comprehensive exposition of statistical models in viscoelasticity, see e.g. Lodge (1989).

In our opinion, Yamamoto's approach is too cumbersome for applications, since (i) it requires the solution of a system of integro-differential equations for chain-distribution functions, and (ii) no experimental confirmation exists for relationships between chain-distribution and chain-reformation functions.

Constitutive relations of the Green–Tobolsky and Yamamoto models contain one relaxation function. On the other hand, the material behavior is described by at least two functions: relaxation kernels for compression and shear, cf. Eqs. (3.60). To refine these models, a version of the model of adaptive links was suggested by Drozdov (1993). According to Drozdov's model, several kinds of links are introduced which arise and collapse independently of each other. Any kind of links is characterized by its strain energy density (instead of the dissipation rate in the Yamamoto model) and by its relaxation measure. Kernel $\mathcal{X}(t,\tau)$ is considered as a function of two arguments (to account for the material aging) and it is treated as a quantity independent of strain energy density (unlike Yamamoto's model, where it is connected with the specific free energy by an integral equation). The use of kernel $\mathcal{X}(t,\tau)$ and strain energy density $W(I_1, I_2, I_3)$ as material functions seems preferable, since they can be easily determined in experiments.

3.3. The Lagrange Variational Principle

The Lagrange variational principle in finite elasticity was discussed in Section 5.1. We now extend the Lagrange principle to finite viscoelasticity preserving the notation of Section 5.1, and derive (by using this variational principle) constitutive equations for a viscoelastic medium with strain energy density (3.72).

Let us fix an instant $t \geq 0$ and a deformation history $\{\bar{u}(\tau, \xi) \ (0 \leq \tau < t)\}$. Denote by $\Upsilon(t)$ a set of admissible displacement fields, i.e. continuously differentiable displacement fields $\bar{u}(t, \xi)$ satisfying boundary condition (5.1.4). Let $\bar{u}^*(t, \xi)$ be an admissible displacement field, where the asterisk denotes an admissible quantity. The quantity which is observed in the deformation process is denoted by the same symbol without an asterisk.

A viscoelastic medium occupies a domain Ω_0 with a smooth boundary Γ_0 in the initial configuration. The total strain energy of the medium equals

$$W^\dagger(t) = \int_{\Omega_0} W(t) dV_0, \tag{3.73}$$

where dV_0 is the volume element in the initial configuration.

We confine ourselves to dead body forces and dead surface loads, see Eq. (3.3.6). The work of external forces on the displacement from the actual config-

uration at moment $t - 0$ to an admissible actual configuration at moment $t + 0$
equals

$$A^\dagger(t) = \int_{\Omega_0} \rho_0 \bar{B}_0(t) \cdot [\bar{u}^*(t) - \bar{u}_-(t)] dV_0 + \int_{\Gamma_0^{(\sigma)}} \bar{b}_0(t) \cdot [\bar{u}^*(t) - \bar{u}_-(t)] dS_0, \quad (3.74)$$

where the symbol f_- denotes the limit of a function $f(\tau)$ as $\tau \to t - 0$, dS_0 is
the surface element, and ρ_0 is mass density in the initial configuration.

The total energy of the "body & external forces" system equals

$$\Phi(t) = W^\dagger(t) - A^\dagger(t). \quad (3.75)$$

Proposition 3.2 (Lagrange's principle). For a given history of deformations
up to an instant $t \geq 0$, the real displacement field at moment t minimizes the
total energy $\Phi(t)$ on the set $\Upsilon(t)$ of admissible displacement fields. $\square$

The Lagrange principle in finite elasticity is widely accepted, while its
implementation for viscoelastic media was questionable for a long time. By
using the model of adaptive links, it is easy to explain why this variational
principle may be applied to quasi-static problems in viscoelasticity. Let us
introduce three characteristic times:
(i) the characteristic time for elastic deformations;
(ii) the characteristic time for the change of external loads;
(iii) the characteristic time for the stress relaxation (which coincides with the
characteristic time for links' replacing).

In the quasi-static theory, the first characteristic time is essentially less than
the others. This is equivalent to the assumption regarding instantaneous elastic
deformations. If the characteristic time for replacing links between polymeric
molecules exceeds essentially the characteristic time of external loads, then at
any instant in the time scale of external loads, a viscoelastic medium may be
treated as a system with a fixed number of elastic springs. The Lagrange principle is valid for this elastic medium, since viscous effects have the characteristic
time of the same order of magnitude as the characteristic time of relaxation.

As shown in Section 5.1, the Lagrange principle implies constitutive equations for a hyperelastic material. We now use this principle to derive constitutive
equations for an aging viscoelastic medium with strain energy density (3.72).
For this purpose, we calculate the increment $\delta\Phi(t)$ of the functional $\Phi(t)$ caused
by an admissible perturbation $\delta\bar{u}(t)$ of the displacement field $\bar{u}(t)$. Confining
ourselves to the necessary conditions of minimum, we calculate $\delta\Phi(t)$ up to the
first order terms compared with $|\delta\bar{u}|$. It follows from Eqs. (3.73) – (3.75) that

$$\delta\Phi(t) = \int_{\Omega_0} \delta W(t) dV_0 - \int_{\Omega_0} \rho_0 \bar{B}_0(t) \cdot \delta\bar{u}(t) dV_0 - \int_{\Gamma_0^{(\sigma)}} \bar{b}_0(t) \cdot \delta\bar{u}(t) dS_0. \quad (3.76)$$

To find $\delta W(t)$ we employ Eqs. (3.72), (5.1.18), and (5.1.20). Similar to Eq. (5.1.21), we obtain

$$\delta W(t) = 2 \sum_{m=1}^{M} \{ \mathcal{X}_m(t,0)\hat{\Theta}_m(t,0) + \int_0^t \frac{\partial \mathcal{X}_m}{\partial \tau}(t,\tau)\hat{\Theta}_m(t,\tau)d\tau \} : \delta\hat{\epsilon}(t). \quad (3.77)$$

Here $\delta\hat{\epsilon}(t) = \frac{1}{2}(\bar{\nabla}\delta\bar{u} + \bar{\nabla}\delta\bar{u}^T)$, $\bar{\nabla}$ is the Hamilton operator in the actual configuration at instant t, and

$$\hat{\Theta}_m(t,\tau) = [\frac{\partial W_m}{\partial I_1}(I_k(t,\tau)) + I_1(t,\tau)\frac{\partial W_m}{\partial I_2}(I_k(t,\tau))]\hat{F}_*(t,\tau)$$
$$-\frac{\partial W_m}{\partial I_2}(I_k(t,\tau))\hat{F}_*^2(t,\tau) + I_3(t,\tau)\frac{\partial W_m}{\partial I_3}(I_k(t,\tau))\hat{I}. \quad (3.78)$$

We introduce tensor $\hat{\sigma}(t)$ by the formula

$$\hat{\sigma}(t) = \frac{2}{\sqrt{I_3(t,0)}} \sum_{m=1}^{M} [\mathcal{X}_m(t,0)\hat{\Theta}_m(t,0) + \int_0^t \frac{\partial \mathcal{X}_m}{\partial \tau}(t,\tau)\hat{\Theta}_m(t,\tau)d\tau], \quad (3.79)$$

substitute expressions (3.77) and (3.79) into Eq. (3.76), and transform the integrals over Ω_0 and $\Gamma_0^{(\sigma)}$ with the use of mass conservation law (2.2.10). As a result, we obtain

$$\delta\Phi(t) = \int_{\Omega(t)} \hat{\sigma}(t):\delta\hat{\epsilon}(t)dV(t)$$
$$- \int_{\Omega(t)} \rho(t)\bar{B}(t)\cdot\delta\bar{u}(t)dV(t) - \int_{\Gamma^{(\sigma)}(t)} \bar{b}(t)\cdot\delta\bar{u}(t)dS(t), \quad (3.80)$$

where $dV(t)$ and $dS(t)$ are the volume and surface elements in the actual configuration at instant t, respectively. We substitute expression (3.80) into *the Euler–Lagrange condition* of minimum

$$\delta\Phi(\delta\bar{u}(t)) = 0,$$

transform the left-hand side by using the Stokes formula (1.3.61), and find

$$\int_{\Gamma^{(\sigma)}(t)} [\bar{n}(t)\cdot\hat{\sigma}(t) - \bar{b}(t)]\cdot\delta\bar{u}(t)dS(t)$$
$$- \int_{\Omega(t)} [\bar{\nabla}\cdot\hat{\sigma}(t) + \rho(t)\bar{B}(t)]\cdot\delta\bar{u}(t)dV(t) = 0,$$

where $\bar{n}(t)$ is the unit outward normal vector to $\Gamma(t)$. Since $\delta\bar{u}(t)$ is an arbitrary displacement field, this equality implies the equilibrium equation (2.2.21) in $\Omega(t)$ and boundary condition in stresses (2.2.20) on surface $\Gamma^{(\sigma)}(t)$ provided Eq.

285

(3.79) determines the Cauchy stress tensor. Thus, the Laplace principle implies formula (3.79) for the Cauchy stress tensor in the model of adaptive links.

Substitution of expression (3.78) into Eq. (3.79) yields the following constitutive equation of an aging viscoelastic medium:

$$
\hat{\sigma}(t) = \frac{2}{\sqrt{I_3(t,0)}} \sum_{m=1}^{M} \{ \mathcal{X}_m(t,0)[(\frac{\partial W_m}{\partial I_1}(I_k(t,0)) + I_1(t,0)\frac{\partial W_m}{\partial I_2}(I_k(t,0)))\hat{F}(t)
$$
$$
- \frac{\partial W_m}{\partial I_2}(I_k(t,0))\hat{F}^2(t) + I_3(t,0)\frac{\partial W_m}{\partial I_3}(I_k(t,0))\hat{I}]
$$
$$
+ \int_0^t \frac{\partial \mathcal{X}_m}{\partial \tau}[(\frac{\partial W_m}{\partial I_1}(I_k(t,\tau)) + I_1(t,\tau)\frac{\partial W_m}{\partial I_2}(I_k(t,\tau)))\hat{F}_*(t,\tau)
$$
$$
- \frac{\partial W_m}{\partial I_2}(I_k(t,\tau))\hat{F}_*^2(t,\tau) + I_3(t,\tau)\frac{\partial W_m}{\partial I_3}(I_k(t,\tau))\hat{I}]d\tau \}. \quad (3.81)
$$

For $M = 1$, $\mathcal{X}_1 = \mathcal{X}$, and $W_1 = W$, Eq. (3.81) is presented as

$$
\hat{\sigma}(t) = \frac{2}{\sqrt{I_3(t,0)}} \{ \mathcal{X}(t,0)[(\frac{\partial W}{\partial I_1}(I_k(t,0)) + I_1(t,0)\frac{\partial W}{\partial I_2}(I_k(t,0)))\hat{F}(t)
$$
$$
- \frac{\partial W}{\partial I_2}(I_k(t,0))\hat{F}^2(t) + I_3(t,0)\frac{\partial W}{\partial I_3}(I_k(t,0))\hat{I}]
$$
$$
+ \int_0^t \frac{\partial \mathcal{X}}{\partial \tau}[(\frac{\partial W}{\partial I_1}(I_k(t,\tau)) + I_1(t,\tau)\frac{\partial W}{\partial I_2}(I_k(t,\tau)))\hat{F}_*(t,\tau)
$$
$$
- \frac{\partial W}{\partial I_2}(I_k(t,\tau))\hat{F}_*^2(t,\tau) + I_3(t,\tau)\frac{\partial W}{\partial I_3}(I_k(t,\tau))\hat{I}]d\tau \}. \quad (3.82)
$$

For $\mu(t) = \mu = $ const., and $Q(t,\tau) = 0$, Eq. (3.82) implies the Finger presentation (3.2.18) and (3.2.19) for the Cauchy stress tensor in a hyperelastic material.

For an incompressible viscoelastic medium, the Lagrange principle claims that the real displacement field $\bar{u}(t,\xi)$ minimizes functional $\Phi(t)$ on the subset $\Upsilon_\circ(t) \subset \Upsilon$ of displacement fields $\bar{u}^*(t,\xi)$ which satisfy condition (3.2.86). In this case, strain energy densities W_m depend on the first two principal invariants only, $W_m = W_m(I_1, I_2)$. Repeating the calculations carried out in Section 3.2 for an incompressible hyperelastic medium, we obtain

$$
\hat{\sigma}(t) = -p(t)\hat{I} + 2\sum_{m=1}^{M}[\mathcal{X}_m(t,0)\hat{\Theta}_m(t,0) + \int_0^t \frac{\partial \mathcal{X}_m}{\partial \tau}(t,\tau)\hat{\Theta}_m(t,\tau)d\tau]. \quad (3.83)
$$

Here $p(t)$ is pressure, i.e. the Lagrange coefficient for restriction (3.2.86), and

$$
\hat{\Theta}_m(t,\tau) = [\frac{\partial W_m}{\partial I_1}(I_1(t,\tau), I_2(t,\tau)) + I_1(t,\tau)\frac{\partial W_m}{\partial I_2}(I_1(t,\tau), I_2(t,\tau))]\hat{F}_*(t,\tau)
$$
$$
- \frac{\partial W_m}{\partial I_2}(I_1(t,\tau), I_2(t,\tau))\hat{F}_*^2(t,\tau). \quad (3.84)
$$

286

Exercise 3.11. By employing Eqs. (1.2.33), (3.83), and (3.84), show that the constitutive equation of an aging incompressible viscoelastic medium can be presented in the form (3.83) with

$$
\hat{\Theta}_m(t,\tau) = \frac{\partial W_m}{\partial I_1}(I_1(t,\tau), I_2(t,\tau))\hat{F}_*(t,\tau)
$$
$$
- \frac{\partial W_m}{\partial I_2}(I_1(t,\tau), I_2(t,\tau))\hat{F}_*^{-1}(t,\tau). \quad \square \qquad (3.85)
$$

According to Eqs. (3.83) – (3.85), an incompressible aging viscoelastic medium is characterized by an integer M, functions $\mathcal{X}_m(t,s)$, and strain energies $W_m(I_1, I_2)$. Let us consider the simplest case when $M = 1$ and $W_1 = W$ coincides with strain energy density (3.2.103) of the neo-Hookean medium

$$
W = \frac{1}{2}(I_1 - 3). \qquad (3.86)
$$

Substitution of expressions (3.84) and (3.86) into Eq. (3.83) yields the constitutive equation of *the neo-Hookean aging viscoelastic medium*

$$
\hat{\sigma}(t) = -p(t)\hat{I} + \bar{\nabla}\bar{r}^T(t) \cdot [\mathcal{X}(t,0)\hat{I} + \int_0^t \frac{\partial \mathcal{X}}{\partial \tau}(t,\tau)\hat{g}^{-1}(\tau)d\tau] \cdot \bar{\nabla}\bar{r}(t). \qquad (3.87)
$$

Eq. (3.87) predicts correctly experimental data for butyl rubber for deformations of up to 200 per cent, see Drozdov (1994).

3.4. Constitutive Equations for Viscoelastic Media

In this subsection we discuss some approaches to constructing constitutive models in finite viscoelasticity. Since a number of viscoelastic materials with large deformations satisfy the incompressibility condition (e.g. elastomers and rubber-like polymers), we confine ourselves to isotropic incompressible viscoelastic media.

3.4.1. Linear constitutive equations

For incompressible viscoelastic media with small strains, constitutive models (3.57), (3.60), and (3.65) can be presented in the form

$$
\hat{\sigma}(t) = -p(t)\hat{I} + 2\mu(t)[\hat{e}(t) - \int_0^t R(t,\tau)\hat{e}(\tau)d\tau]
$$
$$
= -p(t)\hat{I} + 2[\mathcal{X}(t,t)\hat{e}(t) - \int_0^t \frac{\partial \mathcal{X}}{\partial \tau}(t,\tau)\hat{e}(\tau)d\tau], \qquad (3.88)
$$

287

where $\mu(t)$ is the shear modulus, $R(t, \tau)$ is the shear relaxation kernel, and

$$\mathcal{X}(t, \tau) = \mu(t)[1 - \int_\tau^t R(t, s)ds]. \tag{3.89}$$

By introducing the function

$$H(t, \tau) = \mathcal{X}(t, t)\delta(\tau) - \frac{\partial \mathcal{X}}{\partial \tau}(t, \tau), \tag{3.90}$$

where $\delta(t)$ is *the Dirac delta–function*, we write Eq. (3.88) as follows:

$$\hat{\sigma}(t) = -p(t)\hat{I} + 2 \int_0^t H(t, \tau)\hat{e}(\tau)d\tau. \tag{3.91}$$

To construct a linear constitutive equation with finite strains, the infinitesimal strain tensor $\hat{e}$ in Eq. (3.91) should be replaced by an appropriate strain tensor at finite strains. The only restriction imposed on this procedure is that the constitutive equation derived should preserve the objective form (2.3.15).

One of the first models of this type was suggested by Lodge (1964). According to *the Lodge model*, tensor $\hat{e}$ is replaced by the relative Finger strain tensor $\hat{E}_{F*}$, see Eq. (2.1.36). By employing the incompressibility condition, we find

$$\hat{\sigma}(t) = -p(t)\hat{I} + \int_0^t H(t, \tau)[\hat{F}_*(t, \tau) - \hat{I}]d\tau. \tag{3.92}$$

Since pressure p is an undetermined parameter, Eq. (3.92) can be written as

$$\hat{\sigma}(t) = -p(t)\hat{I} + \int_0^t H(t, \tau)\hat{F}_*(t, \tau)d\tau. \tag{3.93}$$

Eq. (3.93) is a generalization of the constitutive equation of neo-Hookean elastic solid (3.2.103) to viscoelastic media, cf. Exercise 3.2.26. To transform the right-hand side of Eq. (3.93), we substitute Eq. (2.1.46) for the relative Finger tensor and replace the unit tensor by the expression

$$\hat{I} = \bar{\nabla}_0 \bar{r}^T(t) \cdot \hat{g}^{-1}(t) \cdot \bar{\nabla}_0 \bar{r}(t).$$

As a result, we find

$$\hat{\sigma}(t) = \bar{\nabla}_0 \bar{r}^T(t) \cdot [-p(t)\hat{g}^{-1}(t) + \int_0^t H(t, \tau)\hat{g}^{-1}(\tau)d\tau] \cdot \bar{\nabla}_0 \bar{r}(t).$$

Exercise 3.12. Bearing in mind that pressure p is an undetermined parameter, show that this formula coincides with Eq. (3.87). $\square$

Exercise 3.13. Check that the Lodge model preserves the objective form (2.3.15). $\square$

288

Integrating the second term in Eq. (3.93) by parts, we obtain the *Tanaka–Yamamoto–Takano* form of the Lodge model, see Yamamoto (1971), which is also called *the Oldroyd–Walter–Fredrickson model*, see Spriggs et al. (1966),

$$\hat{\sigma}(t) = -p(t)\hat{I} + \int_0^t H_0(t,\tau)\frac{\partial \hat{F}_*}{\partial \tau}(t,\tau)d\tau, \tag{3.94}$$

where $H_0(t,\tau)$ is a kernel of the integral operator.

Coleman & Noll (1961) suggested a linear model, where the infinitesimal strain tensor $\hat{\epsilon}$ is replaced by the relative Almansi strain tensor $\hat{A}_*(t,\tau)$. By using Eqs. (2.1.34) and (2.1.49), we obtain *the first order model of viscoelasticity*

$$\hat{\sigma}(t) = -p(t)\hat{I} - \int_0^t H(t,\tau)[F_*^{-1}(t,\tau) - \hat{I}]d\tau. \tag{3.95}$$

Since pressure p is an undetermined parameter, Eq. (3.95) can be presented as

$$\hat{\sigma}(t) = -p(t)\hat{I} - \int_0^t H(t,\tau)F_*^{-1}(t,\tau)d\tau. \tag{3.96}$$

Exercise 3.14. Check that model (3.96) preserves the objective form (2.3.15).
□

The Ward–Jenkis model provides an extension of Eqs. (3.93) and (3.96), see Spriggs et al. (1966),

$$\hat{\sigma}(t) = -p(t)\hat{I} + \int_0^t [H_1(t,\tau)F_*(t,\tau) - H_2(t,\tau)F_*^{-1}(t,\tau)]d\tau. \tag{3.97}$$

Model (3.97) is rather complicated for applications, since two relaxation kernels $H_k(t,\tau)$ should be found by fitting experimental data. A model with one relaxation kernel is suggested by Tanner (1968),

$$\hat{\sigma}(t) = -p(t)\hat{I} + \int_0^t H(t,\tau)[(1-a)\hat{F}_*(t,\tau) - a\hat{F}_*^{-1}(t,\tau)]d\tau, \tag{3.98}$$

where $a \in [0,1]$ is a material parameter.

According to *the Christensen model*, see Christensen (1980), tensor $\hat{\epsilon}(t)$ in Eq. (3.88) is replaced by the Cauchy strain tensor $\hat{C}(t)$, while tensor $\hat{\epsilon}(\tau)$ in the integrand is replaced by the Cauchy strain tensor $\hat{C}(\tau)$, and additional deformation gradients are introduced in the right-hand side of Eq. (3.89) to preserve objective form (2.3.15) of the constitutive equation

$$\hat{\sigma}(t) = -p(t)\hat{I} + 2\mu(t)\hat{C}(t) - 2\mu\bar{\nabla}_0\bar{r}^T(t) \cdot \int_0^t R(t,\tau)\hat{C}(\tau)d\tau \cdot \bar{\nabla}_0\bar{r}(t).$$

We replace $\hat{C}(t)$ by using Eqs. (2.1.17) and (2.1.32). Since pressure p is an undetermined parameter, we arrive at the formula

$$\hat{\sigma}(t) = -p(t)\hat{I} + \mu(t)\bar{\nabla}_0\bar{r}^T(t) \cdot [\hat{I} - 2\int_0^t R(t,\tau)\hat{C}(\tau)d\tau] \cdot \bar{\nabla}_0\bar{r}(t). \qquad (3.99)$$

Eq. (3.99) describes adequately experimental data for polyisobutylene for deformations of up to 30 per cent, see Christensen (1980).

3.4.2. Constitutive equations in the form of truncated Taylor series

According to *the Weierstrass theorem*, see Green & Rivlin (1957) and Coleman & Noll (1960), any sufficiently smooth tensor valued functional in Eq. (2.3.16) may be approximated by polynomials

$$\hat{\mathcal{G}} = \sum_{m=1}^{\infty} \hat{\mathcal{G}}_m, \qquad (3.100)$$

where

$$\mathcal{G}_m = \int_0^t \cdots \int_0^t R_m(t,\tau_1,\ldots,\tau_m)d\hat{A}_*(t,\tau_1) \cdot \ldots \cdot d\hat{A}_*(t,\tau_m),$$

$\hat{A}_*(t,\tau)$ is the relative Almansi strain tensor, and the integrals are understood in the Stiltjes sense. Substitution of expression (3.100) into Eq. (2.3.16) implies a constitutive model with an infinite number of multiple integrals. For its application, only several terms are taken into account. The theory accounting for linear terms is called *the linear finite viscoelasticity*, see Coleman & Noll (1961) and Pipkin & Rivlin (1961).

More sophisticated models are derived by taking into account nonlinear terms in Eq. (3.100). By including the second order terms, we arrive at the second order theory of viscoelasticity, see Coleman & Noll (1961),

$$\hat{\sigma}(t) = -p(t)\hat{I} - 2\int_0^t H(t,\tau)\hat{A}_*(t,\tau)d\tau$$
$$+ \int_0^t \int_0^t [H_1(t,\tau_1,\tau_2)\hat{A}_*(t,\tau_1) : \hat{A}_*(t,\tau_2)$$
$$+ H_2(t,\tau_1,\tau_2)I_1(\hat{A}_*(t,\tau_1))\hat{A}_*(t,\tau_2)]d\tau_1 d\tau_2,$$

where $H(t,\tau)$, $H_k(t,\tau_1,\tau_2)$ are kernels of the integral operators.

Onogi et al. (1970) suggested the third order theory of viscoelasticity

$$\hat{\sigma}(t) = -p(t)\hat{I} + \int_0^t H_1(t,\tau)\hat{A}_*(t,\tau)d\tau$$
$$+ \int_0^t \int_0^t H_2(t,\tau_1,\tau_2)[\hat{A}_*(t,\tau_1) \cdot \hat{A}_*(t,\tau_2) + \hat{A}_*(t,\tau_2) \cdot \hat{A}_*(t,\tau_1)]d\tau_1 d\tau_2$$

$$+ \int_0^t \int_0^t \int_0^t H_3(t, \tau_1, \tau_2, \tau_3)[\hat{A}_*(t, \tau_1) \cdot \hat{A}_*(t, \tau_2) \cdot \hat{A}_*(t, \tau_3)$$
$$+\hat{A}_*(t, \tau_1) \cdot \hat{A}_*(t, \tau_3) \cdot \hat{A}_*(t, \tau_2) + \hat{A}_*(t, \tau_2) \cdot \hat{A}_*(t, \tau_1) \cdot \hat{A}_*(t, \tau_3)$$
$$+\hat{A}_*(t, \tau_2) \cdot \hat{A}_*(t, \tau_3) \cdot \hat{A}_*(t, \tau_1) + \hat{A}_*(t, \tau_3) \cdot \hat{A}_*(t, \tau_1) \cdot \hat{A}_*(t, \tau_1)$$
$$+\hat{A}_*(t, \tau_3) \cdot \hat{A}_*(t, \tau_2) \cdot \hat{A}_*(t, \tau_1)]d\tau_1 d\tau_2 d\tau_3.$$

A similar expression was proposed by Bernstein (1966) by using polynomials in the relative Cauchy strain tensor $\hat{C}_*(t, \tau)$.

These models are, mainly, of theoretical interest, since experimental determination of several relaxation kernels depending on several variables seems to be a too complicated problem.

3.4.3. BKZ-type constitutive equations

Discussing constitutive models with infinitesimal strains, we noted that constitutive equations in linear viscoelasticity may be derived by replacing material parameters in the constitutive equations for elastic media by integral operators. A similar approach applied to hyperelastic materials leads us to *the BKZ-type constitutive models* in finite viscoelasticity, see Bernstein et al. (1963).

We assume that for an isotropic incompressible viscoelastic medium there is a function U of times t and τ, and the principal invariants $I_k(t, \tau)$ of the relative Finger tensor $\hat{F}_*(t, \tau)$ (an analogue of strain energy density in hyperelasticity). By using this function, we replace the constitutive model of an incompressible hyperelastic solid (3.2.100) by the following equation:

$$\hat{\sigma}(t) = -p(t)\hat{I} + 2 \int_0^t [\frac{\partial U}{\partial I_1}(t, \tau, I_1(t, \tau), I_2(t, \tau))\hat{F}_*(t, \tau)$$
$$-\frac{\partial U}{\partial I_2}(t, \tau, I_1(t, \tau), I_2(t, \tau))\hat{F}_*^{-1}(t, \tau)]d\tau. \qquad (3.101)$$

Experimental validation of Eq. (3.101) was carried out on polyvinil chloride, butyl rubber, and polyisobutylene, see Bernstein et al. (1963) and Zapas & Craft (1965).

We distinguish two versions of the BKZ-type models. The first is based on the following

Principle of separability. Function $U(t, \tau, I_1, I_2)$ characterizing the viscoelastic response equals the product of a function $H(t, \tau)$ which describes viscous properties and a function $W(I_1, I_2)$ which determines the instantaneous (elastic) response

$$U(t, \tau, I_1, I_2) = H(t, \tau)W(I_1, I_2). \qquad \square \qquad (3.102)$$

According to the other version, instead of the separability principle we accept an analogue of Eq. (3.102) either in the form

$$U(t, \tau, I_1, I_2) = H(t, \tau, I_1, I_2) W(I_1, I_2), \qquad (3.103)$$

or in the form

$$U(t, \tau, I_1, I_2) = H(t, \tau, I_2(\hat{D}), I_3(\hat{D})) W(I_1, I_2). \qquad (3.104)$$

Here $I_k(\hat{D})$ is the kth principal invariant of the rate-of-strain tensor $\hat{D}$, while I_k denotes the kth principal invariant of the Finger deformation tensor $\hat{F}$. We recall that for an incompressible medium, $I_1(\hat{D}) = 0$ and $I_3(\hat{F}) = 1$.

First, we assume the separability principle to be fulfilled (for experimental verification of this assertion, see Section 7.4) and suppose that function $H(t, \tau)$ may be presented as

$$H(t, \tau) = \mathcal{X}(t, 0)\delta(\tau) + \frac{\partial \mathcal{X}}{\partial \tau}(t, \tau). \qquad (3.105)$$

Substituting expressions (3.102) and (3.105) into Eq. (3.101) we find

$$
\begin{aligned}
\hat{\sigma}(t) = -p(t)\hat{I} + 2\{ \mathcal{X}(t, 0)[& \frac{\partial W}{\partial I_1}(I_1(t, 0), I_2(t, 0))\hat{F}(t) \\
& - \frac{\partial W}{\partial I_2}(I_1(t, 0), I_2(t, 0))\hat{F}^{-1}(t)] \\
+ \int_0^t \frac{\partial \mathcal{X}}{\partial \tau}(t, \tau)[& \frac{\partial W}{\partial I_1}(I_1(t, \tau), I_2(t, \tau))\hat{F}_*(t, \tau) \\
& - \frac{\partial W}{\partial I_2}(I_1(t, \tau), I_2(t, \tau))\hat{F}_*^{-1}(t, \tau)]d\tau \}.
\end{aligned}
\qquad (3.106)
$$

Comparison of Eqs. (3.83), (3.85) with Eq. (3.106) implies that the BKZ model (3.106) is a particular case of the model of adaptive links. The BKZ constitutive equation corresponds to the case when only one kind of links exists between polymeric molecules.

By substituting strain energy density (3.86) into Eqs. (3.101) and (3.102), we obtain the constitutive equation (3.93) of the Lodge model.

We now concentrate on the constitutive assumption (3.103). Choosing strain energy density of the neo-Hookean medium (3.86), we arrive at *the generalized Lodge model*

$$\hat{\sigma}(t) = -p(t)\hat{I} + \int_0^t H(t, \tau, I_1, I_2)\hat{F}_*(t, \tau)d\tau. \qquad (3.107)$$

By accepting the additional assumption

$$H(t, \tau, I_1, I_2) = H_0(t, \tau)\Xi(I_1, I_2), \qquad (3.108)$$

we transform Eq. (3.107) into *the Wagner model*

$$\hat{\sigma}(t) = -p(t)\hat{I} + \int_0^t H_0(t,\tau)\Xi(I_1,I_2)\hat{F}_*(t,\tau)d\tau, \qquad (3.109)$$

where $H_0(t,\tau)$ is called *memory function*, and $\Xi(I_1,I_2)$ is called *damping function*. Model (3.109) predicts correctly experimental data for polyethylene melts, see Wagner (1976) and Laun (1978).

As common practice, an exponential form is accepted for the function $H_0(t,\tau)$. There is no rational procedure to construct the function $\Xi(I_1,I_2)$. Several expressions for this function are used in applications:
– *the Kaye–Kennett model*, see Kaye & Kennett (1974),

$$\Xi = \exp(-a\sqrt{I_1 - 3}),$$

– *the Wagner model*, see Wagner (1976),

$$\Xi = \exp(-a\sqrt{I_2 - 3}),$$

– *the Winter model*, see Winter (1978),

$$\Xi = \exp(-a\sqrt{\frac{I_2}{3}}),$$

– *the Wagner–Raible–Meissner model*, see Wagner et al. (1979)

$$\Xi = c_1 \exp(-a_1\sqrt{I - 3}) + c_2 \exp(-a_2\sqrt{I - 3}).$$

Here $I = \alpha I_1 + (1 - \alpha)I_2$, and α, a, a_1, a_2, c_1, c_2 are material parameters.

Substituting expression (3.103) into Eq. (3.101) and accepting strain energy density of the Mooney–Rivlin solid (3.2.104) we obtain *the additive functional constitutive law of the first kind*

$$\hat{\sigma}(t) = -p(t)\hat{I} + \int_0^t H(t, I_1(t,\tau), I_2(t,\tau))[c_{10}\hat{F}_*(t,\tau) - c_{01}\hat{F}_*^{-1}(t,\tau)]d\tau. \quad (3.110)$$

Eq. (3.110) shows fair agreement with experimental data for a silicon polymer, see Emery & White (1969).

We do not intend to discuss constitutive models (3.104) in detail, referring to Astarita & Marrucci (1974), and present only three well-known models:
– *the Bogue model* (a generalization of the Lodge model), see Middleman (1969),

$$\hat{\sigma}(t) = -p(t)\hat{I} + \int_0^t H(t, \tau, I_2(\hat{D}))\hat{F}_*(t,\tau)d\tau; \qquad (3.111)$$

293

– the *Carreau model* (a generalization of the Tanner model), see Goldstein (1974) and Macdonald (1976),

$$\hat{\sigma}(t) = -p(t)\hat{I} + \int_0^t H(t, \tau, I_2(\hat{D}(\tau)))[(1-a)\hat{F}_*(t,\tau) - a\hat{F}_*^{-1}(t,\tau)]d\tau; \quad (3.112)$$

– the *OWFS model* (a generalization of the Tanner model and the Oldroyd–Walter–Frederickson model), see Huppler et al. (1967a,b),

$$\hat{\sigma}(t) = -p(t)\hat{I} + \int_0^t H(t, \tau, I_2(\hat{D}(\tau)))[(1-a)\frac{\partial \hat{F}_*}{\partial \tau}(t,\tau) - a\frac{\partial \hat{F}_*^{-1}}{\partial \tau}(t,\tau)]d\tau. \quad (3.113)$$

The mechanical behavior of the BKZ-type models is determined by the derivatives of one function $U(t,\tau,I_1,I_2)$. It is quite natural to generalize Eq. (3.101) and to consider the constitutive equation of an incompressible viscoelastic medium in the form

$$\hat{\sigma}(t) = -p(t)\hat{I} + 2\int_0^t [U_1(t,\tau,I_1(t,\tau),I_2(t,\tau))\hat{F}_*(t,\tau)$$
$$- U_2(t,\tau,I_1(t,\tau),I_2(t,\tau))\hat{F}_*^{-1}(t,\tau)]d\tau, \quad (3.114)$$

where U_1 and U_2 are arbitrary material functions, see Wagner (1977). Eq. (3.114) is called *the single integral constitutive equation*. By setting $U_1 = \frac{1}{2}H$ and $U_2 = 0$, we reduce Eq. (3.114) to the generalized Lodge model (3.107).

It is worth noting an approach to constructing constitutive equations which combines elements of differential models, see Section 2, and BKZ-type models. According to this method, the viscoelastic response is determined by using the Kelvin–Voigt model containing a viscoelastic spring and a viscous dashpot. Stresses in the spring are calculated by Eq. (3.82), while stresses in the dashpot obey the Newton law (2.54). Summing up stresses in the spring and in the dashpot, see Eq. (2.52), we obtain

$$\hat{\sigma}(t) = \frac{2}{\sqrt{I_3(t,0)}}\{\mathcal{X}(t,0)[(\frac{\partial W}{\partial I_1}(I_k(t,0)) + I_1(t,0)\frac{\partial W}{\partial I_2}(I_k(t,0)))\hat{F}(t)$$
$$- \frac{\partial W}{\partial I_2}(I_k(t,0))\hat{F}^2(t) + I_3(t,0)\frac{\partial W}{\partial I_3}(I_k(t,0))\hat{I}]$$
$$+ \int_0^t \frac{\partial \mathcal{X}}{\partial \tau}[(\frac{\partial W}{\partial I_1}(I_k(t,\tau)) + I_1(t,\tau)\frac{\partial W}{\partial I_2}(I_k(t,\tau)))\hat{F}_*(t,\tau)$$
$$- \frac{\partial W}{\partial I_2}(I_k(t,\tau))\hat{F}_*^2(t,\tau) + I_3(t,\tau)\frac{\partial W}{\partial I_3}(I_k(t,\tau))\hat{I}]d\tau\} + 2\eta\hat{D}(t).$$

This equation may be treated as a generalization of the constitutive model (3.62) to finite strains. For a non-aging incompressible viscoelastic medium, it

294

is simplified

$$\hat{\sigma}(t) = -p(t)\hat{I} + 2\eta\hat{D}(t) + 2\{[1 + Q_0(t)][\psi_1(t,0)\hat{F}(t) + \psi_2(t,0)\hat{F}^2(t)]$$
$$- \int_0^t \dot{Q}_0(t-\tau)[\psi_1(t,\tau)\hat{F}_*(t,\tau) + \psi_2(t,\tau)\hat{F}_*^2(t,\tau)]d\tau\}, \quad (3.115)$$

where functions ψ_1 and ψ_2 are determined by Eq. (3.2.98). Eq. (3.115) accounts for two types of the material viscosities: (i) a "weak" viscosity typical of solids, which is characterized by the dependence of the Cauchy stress tensor $\hat{\sigma}(t)$ on the history of deformations; and (ii) a "strong" viscosity typical of liquids, which is described by the dependence of stresses on the rate-of-strain tensor $\hat{D}(t)$. The presence of both types of viscosities is a characteristic feature of non-crosslinked polymers, which demonstrate "solid" response at the initial period of loading and steady "viscous" flow for large times. Constitutive equation (3.115) was employed by Engler (1991) in the study of viscoelastic stability.

3.4.4. Semi-linear constitutive models

For three-dimensional deformations with infinitesimal strains, an analogue of the linear constitutive equation (3.23) may be written as

$$\hat{\sigma}(t) = -p(t)\hat{I} + 2\int_0^t X(t,\tau)\frac{d\hat{e}}{d\tau}(\tau)d\tau, \quad (3.116)$$

where function $X(t,\tau)$ has the form (3.71).

To extend constitutive equation (3.116) to finite strains, Chang et al. (1976) suggested (i) to replace the infinitesimal strain tensor $\hat{e}$ by an appropriate strain tensor at finite strains, and (ii) to transform the obtained equation confining ourselves to affine deformations (4.1.2)

$$\bar{r}(t) = f(t)\bar{r}_0 \cdot \hat{\Lambda},$$

where $f(t)$ is a smooth scalar function, and $\hat{\Lambda}$ is a constant tensor. Differentiation of this equality with respect to Lagrangian coordinates ξ^i implies that $\bar{g}_i(t) = f(t)\bar{g}_{0\,i} \cdot \hat{\Lambda}$, where $\bar{g}_{0\,i}$ and $\bar{g}_i(t)$ are tangent vectors in the initial and actual configurations. Substituting the latter expression into Eq. (2.1.5), we find

$$\bar{\nabla}_0\bar{r}(t) = f(t)\hat{\Lambda}. \quad (3.117)$$

It follows from Eqs. (3.117) and (2.1.44) that

$$\bar{\nabla}_\tau\bar{r}(t) = \frac{f(t)}{f(\tau)}\hat{I}. \quad (3.118)$$

295

Eqs. (3.117), (3.118) together with Eqs. (2.1.25) and (2.1.46) yield

$$\hat{F}(t) = f^2(t)\hat{\Lambda}^T \cdot \hat{\Lambda}, \qquad \hat{g}_*(t,\tau) = \hat{F}_*(t,\tau) = \left(\frac{f(t)}{f(\tau)}\right)^2 \hat{I}. \qquad (3.119)$$

Let us replace tensor $\hat{\epsilon}$ in Eq. (3.116) by the Finger strain tensor $\hat{E}_F$. By using the incompressibility condition and Eq. (2.1.36), we obtain, cf. Eq. (3.94),

$$\hat{\sigma}(t) = -p(t)\hat{I} + \int_0^t \mathcal{X}(t,\tau)\frac{d\hat{F}}{d\tau}(\tau)d\tau. \qquad (3.120)$$

Substitution of expressions (3.119) into Eq. (3.120) implies that

$$
\begin{aligned}
\hat{\sigma}(t) &= -p(t)\hat{I} + \int_0^t \mathcal{X}(t,\tau)\frac{df^2(\tau)}{d\tau}\hat{\Lambda}^T \cdot \hat{\Lambda}\,d\tau \\
&= -p(t)\hat{I} + \frac{1}{2}\int_0^t \mathcal{X}(t,\tau)\{f^2(t)\frac{d}{d\tau}[(\frac{f(\tau)}{f(t)})^2] + [\frac{d}{d\tau}(\frac{f(\tau)}{f(t)})^2]f^2(t)\}\hat{\Lambda}^T \cdot \hat{\Lambda}\,d\tau \\
&= -p(t)\hat{I} + \frac{1}{2}\int_0^t \mathcal{X}(t,\tau)[\hat{F}(t) \cdot \frac{d\hat{g}_*}{d\tau}(t,\tau) + \frac{d\hat{g}_*}{d\tau}(t,\tau) \cdot \hat{F}(t)]d\tau \\
&= -p(t)\hat{I} + \int_0^t \mathcal{X}(t,\tau)[\hat{F}(t) \cdot \frac{d\hat{C}_*}{d\tau}(t,\tau) + \frac{d\hat{C}_*}{d\tau}(t,\tau) \cdot \hat{F}(t)]d\tau, \qquad (3.121)
\end{aligned}
$$

where $\hat{C}_*(t,\tau)$ is the relative Cauchy strain tensor. Eq. (3.121) is called *semilinear constitutive model* with finite strains.

We distinguish two approaches to generalization of Eq. (3.121). According to the first, see Chang et al. (1976) and Bloch et al. (1978), the relative Cauchy strain tensor is replaced by the Eulerian m-tensor, see Eq. (2.1.66). As a result, we arrive at the formula

$$\hat{\sigma}(t) = -p(t)\hat{I} + \int_0^t \mathcal{X}(t,\tau)[\hat{F}^{\frac{m}{2}}(t) \cdot \frac{d\hat{\mathcal{E}}_{E*}^{(m)}}{d\tau}(t,\tau) + \frac{d\hat{\mathcal{E}}_{E*}^{(m)}}{d\tau}(t,\tau) \cdot \hat{F}^{\frac{m}{2}}(t)]d\tau, \qquad (3.122)$$

which is called *the linear viscoelastic model with moderately large deformations*. Eq. (3.122) describes adequately the mechanical response of styrene-butadiene rubber in a wide range of temperatures, see Bloch et al. (1978).

According to the other approach, the second order terms (with respect to the Finger tensor) are introduced into the constitutive relations. We do not intend to discuss this approach in detail, and mention only two models as examples.

The McGuirt–Lianis model implies that, see McGuirt & Lianis (1970),

$$\hat{\sigma}(t) = -p(t)\hat{I} + 2(\frac{\partial W}{\partial I_1} + I_1\frac{\partial W}{\partial I_2})\hat{F}(t) - \frac{\partial W}{\partial I_2}\hat{F}^2(t)$$

$$+\sum_{k=0}^{2}\int_{0}^{t}H_k(t,\tau)[\hat{F}^k(\tau)\cdot\frac{\partial\hat{g}_*}{\partial\tau}(t,\tau)+\frac{\partial\hat{g}_*}{\partial\tau}(t,\tau)\cdot\hat{F}^k(\tau)]d\tau$$

$$+\sum_{k,l=0}^{2}\int_{0}^{t}H_{kl}(t,\tau)\hat{F}^k(\tau)I_1(\hat{F}^l(\tau)\cdot\frac{\partial\hat{g}_*}{\partial\tau}(t,\tau))d\tau, \tag{3.123}$$

where $H_k(t,\tau)$, $H_{kl}(t,\tau)$ are kernels of the integral operators, and $W(I_1,I_2)$ is a strain energy density. Eq. (3.123) demonstrates fair prediction of experimental data for styrene-butadiene rubber.

According to *the DeHoff–Lianis–Goldberg model*, see DeHoff et al. (1966),

$$\hat{\sigma}(t) = -p(t)\hat{I}+2(\frac{\partial W}{\partial I_1}+I_1\frac{\partial W}{\partial I_2})\hat{F}(t)-\frac{\partial W}{\partial I_2}\hat{F}^2(t)$$

$$+\int_{0}^{t}\{[\sum_{k=0}^{2}H_k(t,\tau)\hat{F}^k(\tau)]\cdot\frac{\partial\hat{g}_*}{\partial\tau}(t,\tau)+\frac{\partial\hat{g}_*}{\partial\tau}(t,\tau)\cdot[\sum_{k=0}^{2}H_k(t,\tau)\hat{F}^k(\tau)]\}d\tau$$

$$+\sum_{k=0}^{2}\hat{F}^k(t)\int_{0}^{t}I_1(\frac{\partial\hat{g}_*}{\partial\tau}(t,\tau)\cdot[\sum_{l=0}^{2}H_{kl}(t,\tau)\hat{F}^k(\tau)])d\tau. \tag{3.124}$$

Eq. (3.124) predicts adequately experimental data for polyurethane and ethylene propylene rubber for deformations of up to 200 per cent.

4. Creep and Relaxation Kernels

This Section is concerned with creep and relaxation measures and kernels. First, we provide several examples of creep and relaxation kernels and compare theoretical results with experimental data. Afterward, we formulate and discuss general features of creep and relaxation measures. For simplicity we confine ourselves to uniaxial deformations of specimens.

4.1. Examples of Creep and Relaxation Kernels

In this subsection we suggest several explicit expressions for creep and relaxation measures and discuss their agreement with experimental data.

We distinguish two types of relaxation measures: regular and singular.
Definition. A measure $Q(t,s)$ is called *regular* if it is twice continuously differentiable. If a measure $Q(t,s)$ is only differentiable, and its derivative, the relaxation kernel $R(t,s)$, has an integrable singularity at $t=s$, then $Q(t,s)$ is called *weakly singular*.

4.1.1. Regular kernels

We begin with regular relaxation measures for non-aging viscoelastic media. The simplest model of regular relaxation measures corresponds to the standard viscoelastic solid, see Eq. (2.18),

$$Q_0(t) = -\chi[1 - \exp(-\frac{t}{T})], \qquad (4.1)$$

where χ is the material viscosity, and T is the characteristic time of relaxation. It is assumed that $0 \leq \chi < 1$ and $T > 0$. Differentiation of Eq. (4.1) with the use of Eq. (3.19) yields

$$R(t) = \frac{\chi}{T} \exp(-\frac{t}{T}), \qquad (4.2)$$

where $R(t)$ is the relaxation kernel.

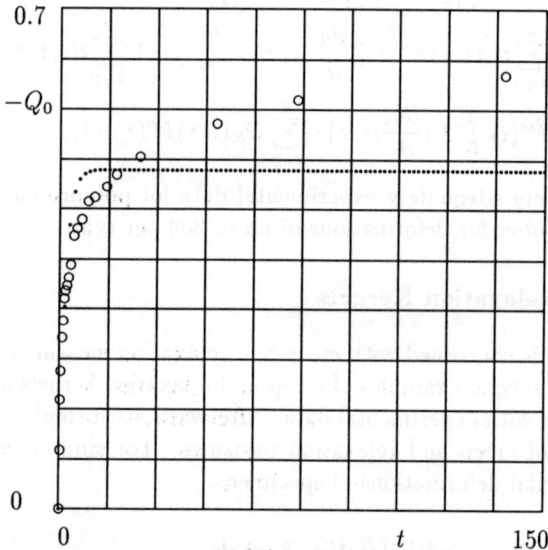

Fig. 4.1. Relaxation measure Q_0 versus time t (min) for nylon-6. Large circles correspond to experimental data obtained by La Mantia et al. (1980). Small circles correspond to function (4.1) with $\chi = 0.475$ and $T = 1.653$.

Eq. (4.1) has the following features attractive for applications:
(i) Eq. (4.1) has a simple mechanical interpretation: it may be modeled by a system consisting of two elastic springs and a dashpot, see Section 2;
(ii) Eq. (4.1) allows the creep kernel to be found in explicit form;
(iii) Eq. (4.1) describes qualitatively the material response observed in experiments for both creep and relaxation. In particular, it implies finite positive limits for the creep and relaxation kernels.

An important deficiency of model (4.1) is poor agreement with experimental data for a number of polymeric materials, see Fig. 4.1.

298

Exercise 4.1. Derive a formula for the creep kernel corresponding to relaxation kernel (4.2). □

A more sophisticated expression for the relaxation kernel is provided by the so-called *Prony series*, i.e. a finite sum of exponential functions, see Soussou et al. (1970),

$$R(t) = \sum_{m=1}^{M} \frac{\chi_m}{T_m} \exp\left(-\frac{t}{T_m}\right). \tag{4.3}$$

Eq. (4.3) with $M \approx 10$ predicts correctly experimental data for a number of media. As common practice, with the growth of m the characteristic times of relaxation T_m increase exponentially, while the characteristic viscosities χ_m decrease, see Table 4.1.

Table 4.1. The characteristic viscosities and the characteristic times of relaxation for polyurethane. The data are adopted from Christensen (1982).

m	χ_m	T_m (hour)
1	0.283	$1.5 \cdot 10^{-5}$
2	0.153	$1.5 \cdot 10^{-4}$
3	0.140	$1.5 \cdot 10^{-3}$
4	0.111	$1.5 \cdot 10^{-2}$
5	0.087	$1.5 \cdot 10^{-1}$
6	0.044	1.5
7	0.034	$1.5 \cdot 10$
8	0.006	$1.5 \cdot 10^{2}$

Creep kernel $K(t)$ corresponding to the relaxation kernel (4.3) is presented in the form similar to Eq. (4.3)

$$K(t) = \sum_{m=1}^{M} \frac{\beta_m}{\tau_m} \exp\left(-\frac{t}{\tau_m}\right), \tag{4.4}$$

where β_m and τ_m are material parameters. Constants τ_m are called the characteristic times of *retardation*.

Exercise 4.2. For $M = 2$, derive connections between material parameters χ_m, T_m on the one hand, and β_m, τ_m on the other hand. □

Unlike χ_m, constants β_m increase with the growth of m, reach their maximum and, afterward, decrease, see Table 4.2.

The essential deficiency of Eqs. (4.3) and (4.4) is a large number of material parameters to be found by fitting experimental data, which leads to a significant reduction in accuracy.

A natural generalization of Eq. (4.3) is an infinite sum of exponential functions which may be presented in the integral form

$$R(t) = \int_0^\infty \frac{\chi(T)}{T} \exp(-\frac{t}{T})dT. \tag{4.5}$$

Function $\chi(T)$ in Eq. (4.5) is called distribution function of relaxation times or *relaxation spectrum*.

Table 4.2. The characteristic viscosities and the characteristic times of retardation for epoxy binder EDT-10. The data are adopted from Kochetkov & Maksimov (1990).

m	β_m	τ_m (hour)
1	$0.823 \cdot 10^{-1}$	$2.75 \cdot 10^{-3}$
2	$0.779 \cdot 10^{-1}$	$2.03 \cdot 10^{-2}$
3	$0.850 \cdot 10^{-1}$	$1.50 \cdot 10^{-1}$
4	$0.951 \cdot 10^{-1}$	1.11
5	0.110	8.20
6	0.129	$6.06 \cdot 10^1$
7	0.154	$4.48 \cdot 10^2$
8	0.263	$3.31 \cdot 10^3$
9	$0.177 \cdot 10^1$	$2.44 \cdot 10^4$
10	$0.212 \cdot 10^1$	$1.81 \cdot 10^5$
11	$0.522 \cdot 10^1$	$1.33 \cdot 10^6$
12	$0.164 \cdot 10^2$	$9.86 \cdot 10^6$
13	$0.207 \cdot 10^2$	$7.28 \cdot 10^7$
14	$0.124 \cdot 10^2$	$5.38 \cdot 10^8$
15	$0.878 \cdot 10^1$	$3.98 \cdot 10^9$
16	$0.663 \cdot 10^1$	$2.94 \cdot 10^{10}$

A similar formula is employed for creep kernel $K(t)$

$$K(t) = \int_0^\infty \frac{\bar{\chi}(T)}{T} \exp(-\frac{t}{T})dT, \tag{4.6}$$

where function $\bar{\chi}(T)$ is called *retardation spectrum*.

For given relaxation and retardation spectra, Eqs. (4.5) and (4.6) demonstrate fair prediction of experimental data. On the other hand, determination of these spectra is an ill-posed problem, and it is extremely difficult to solve it correctly by using standard experimental data. We do not intend to discuss this problem, referring to Tschoegl (1989) and Kaschta & Schwarzl (1994).

To ensure fair agreement with experimental data, Gromov & Miroshnikov (1978) suggested the following generalization of Eq. (4.5):

$$R(t) = \int_0^\infty \frac{\chi(T)}{T} \exp[-\frac{\phi(T)t}{T}]dT, \tag{4.7}$$

where $\chi(T)$ and $\phi(T)$ are material functions to be found. No experimental results have been presented confirming the efficiency of formula (4.7).

Achenbach & Chao (1962) suggested the following expression for the relaxation measure:

$$Q_0(t) = -\chi\{1 - [1 - \frac{1 - \sqrt{1-\chi}}{1 + \sqrt{1-\chi}}\frac{t}{T}]\exp(-\frac{t}{T})\},$$ (4.8)

which is characterized by two material parameters χ and T.

It is of interest to demonstrate the reasons to introduce relaxation measure (4.8). It follows from Eqs. (3.20) and (4.1) that relaxation function $Q_*(t)$ of the standard viscoelastic solid equals

$$Q_*(t) = \delta^2 + (1 - \delta^2)\exp(-\frac{t}{T}),$$

where $\delta^2 = 1 - \chi$.

Exercise 4.3. Derive the following formula for the Laplace transform $\tilde{Q}_*(p)$ of the function $Q_*(t)$:

$$\tilde{Q}_*(p) = \frac{1}{p}\frac{pT + \delta^2}{pT + 1}. \qquad \square$$ (4.9)

As a generalization of expression (4.9), Achenbach & Chao (1962) suggested to consider a relaxation function with the Laplace transform

$$\tilde{Q}_*(p) = \frac{1}{p}(\frac{pT + \delta}{pT + 1})^2.$$ (4.10)

Exercise 4.4. Check that the function

$$Q_*(t) = \delta^2 + (1 - \delta^2)[1 - (\frac{1 - \delta}{1 + \delta})\frac{t}{T}]\exp(-\frac{t}{T})$$ (4.11)

has the Laplace transform (4.10). $\square$

Exercise 4.5. Check that Eq. (4.8) provides the relaxation measure which corresponds to relaxation function (4.11). $\square$

For most viscoelastic materials, their relaxation functions are positive, decreasing, and *concave*. Function (4.8) is neither strictly decreasing nor concave. Moreover, for δ small enough, it can be negative. These features make applications of relaxation measure (4.8) questionable.

The Kohlrausch–William–Watts (stretched exponential) function is a natural generalization of Eq. (4.1)

$$Q_0(t) = -\chi\{1 - \exp[-(\frac{t}{T})^\alpha]\}.$$ (4.12)

Relaxation measure (4.12) is characterized by three material parameters α, χ, and T, see Garbarski (1992) and Scanlan & Janzen (1992) for experimental validation of Eq. (4.12).

As common practice, Eq. (4.12) describes experimental data more adequately than the standard viscoelastic solid (4.1). On the other hand, Eq. (4.12) does not allow the creep kernel $K(t)$ to be expressed in terms of elementary functions. The latter is an essential obstacle for applications of Eq. (4.12) in engineering.

4.1.2. Weakly singular kernels

Let us now consider weakly singular creep and relaxation kernels. We begin with the power-law relaxation measure, see Findley et al. (1989) and Rabotnov (1969)

$$Q_0(t) = -(\frac{t}{T})^\alpha, \tag{4.13}$$

where $\alpha \in (0,1)$ and $T > 0$ are material parameters. Differentiation of Eq. (4.13) implies that

$$R(t) = \frac{\eta t^{\alpha-1}}{\Gamma(\alpha)} = \eta \mathcal{J}_{\alpha-1}(t), \tag{4.14}$$

where $\eta = \alpha\Gamma(\alpha)T^{-\alpha}$, $\mathcal{J}_\alpha(t)$ is the Abel kernel (2.28), and $\Gamma(z)$ is the gamma-function (2.23).

Let us calculate the creep kernel $K(t)$ corresponding to the relaxation kernel (4.14). For this purpose, *the fractional-exponential function* $Z_\alpha(t,\lambda)$ is introduced as the resolvent kernel for the kernel $\mathcal{J}_\alpha(t)$. According to Eqs. (3.39) and (3.41), the unique solution of the Volterra equation

$$x(t) - \lambda \int_0^t \frac{(t-s)^\alpha}{\Gamma(1+\alpha)} x(s)ds = f(t) \tag{4.15}$$

is given by the formula

$$x(t) = f(t) + \lambda \int_0^t Z_\alpha(t-s,\lambda)f(s)ds. \tag{4.16}$$

For $\alpha \in (0,1)$, the function $Z_\alpha(t,\lambda)$ cannot be expressed in terms of elementary functions.

Let us calculate the Laplace transforms of the functions $\mathcal{J}_\alpha(t)$ and $Z_\alpha(t,\lambda)$. Setting $t = p\xi$ in Eq. (2.23), we find

$$\Gamma(1+z) = \int_0^\infty (p\xi)^z \exp(-p\xi)pd\xi = p^{1+z} \int_0^\infty \xi^z \exp(-p\xi)d\xi. \tag{4.17}$$

302

This equality together with Eqs. (3.51) and (2.28) implies that

$$\tilde{\mathcal{J}}_\alpha(p) = \frac{1}{p^{1+\alpha}}. \qquad (4.18)$$

It follows from Eqs. (3.54) and (4.18) that

$$\tilde{Z}_\alpha(p, \lambda) = \frac{1}{p^{1+\alpha} - \lambda}. \qquad (4.19)$$

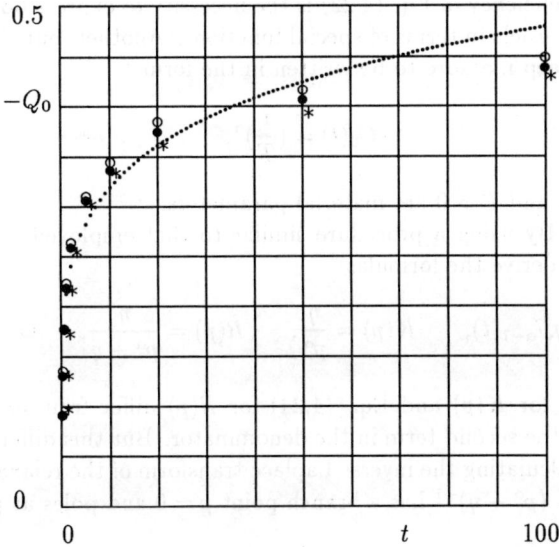

Fig. 4.2. Relaxation measure Q_0 versus time t (min) for polypropylene. Large circles and asterisks correspond to experimental data obtained by Smart & Williams (1972). Unfilled circles: $\epsilon = 0.02$, filled circles: $\epsilon = 0.03$, asterisks: $\epsilon = 0.04$. Small circles correspond to Eq. (4.13) with $\alpha = 0.177$ and $T = 6312.95$ (min).

The creep kernel $K(t)$ corresponding to the relaxation kernel (4.14) is written as

$$K(t) = \eta Z_{\alpha-1}(t, \eta). \qquad (4.20)$$

Eqs. (4.14), (4.18), (4.19), and (4.20) yield

$$\tilde{R}(p) = \frac{\eta}{p^\alpha}, \qquad \tilde{K}(p) = \frac{\eta}{p^\alpha - \eta}. \qquad (4.21)$$

Exercise 4.6. By using the inverse Laplace transform, show that the creep kernel $K(t)$ increases in time with exponential rate. □

303

The relaxation measure (4.13) demonstrates good agreement with experimental data for a number of viscoelastic materials, see Fig. 4.2. Nevertheless, function (4.13) is not widely used in applications, since the exponential rate of growth for the creep measure contradicts experimental data.

Another application of the fractional-exponential function was suggested by Rabotnov (1969). According to the Rabotnov model, the creep measure $K(t)$ is presented as

$$K(t) = \eta Z_{\alpha-1}(t, -\eta), \tag{4.22}$$

where $\alpha \in (0,1)$ and $\eta > 0$ are material parameters, cf. Eq. (4.20).

A serious deficiency of Eq. (4.22) is the necessity to express both the creep and relaxation kernels in terms of special functions. Another, but similar model presumes the creep measure to be written in the form

$$C_0(t) = \left(\frac{t}{T}\right)^{\alpha}, \tag{4.23}$$

where $\alpha \in (0,1)$ and $T > 0$ are material parameters.

Exercise 4.7. By using a procedure similar to that employed for relaxation measure (4.13), derive the formulas

$$K(t) = \eta \mathcal{J}_{\alpha-1}(t), \qquad \tilde{K}(p) = \frac{\eta}{p^{\alpha}}, \qquad \tilde{R}(p) = \frac{\eta}{p^{\alpha} + \eta}. \qquad \square \tag{4.24}$$

Eq. (4.21) for $\tilde{K}(p)$ and Eq. (4.24) for $\tilde{R}(p)$ differ from each other by only the sign of the second term in the denominator. But this difference is very important for calculating the inverse Laplace transform of the relaxation kernel. Function $\psi(p) = (p^{\alpha} + \eta)^{-1}$ has a branch point $p = 0$ and poles at points

$$p_k = \eta^{\frac{1}{\alpha}} \exp \frac{\pi i}{\alpha}(2k - 1) \qquad (k = 0, 1, \ldots),$$

where $i = \sqrt{-1}$. For $\alpha \in (0,1)$ we have $\arg p_0 < \pi$ and $\arg p_k > \pi$ for $k = 1, 2, \ldots$. Therefore, the main branch of the function $\psi(p)$ with $-\pi \le p < \pi$ has no poles. Applying the inverse Laplace transform we obtain

$$R(t) = \frac{\eta}{2\pi i} \int_{c-i\infty}^{c+i\infty} \frac{\exp(pt)dp}{p^{\alpha} + \eta}, \tag{4.25}$$

where c is an arbitrary positive constant.

Exercise 4.8. Show that formula (4.25) is valid for any positive c. $\square$

Since function $\psi(p)$ has the only singular point $p = 0$, the integral of this function along the curve plotted in Fig 4.3 vanishes. By using the standard procedure with reference to *Jordan's lemma*, we find that the integrals along

304

large and small arcs of circles vanish as their radii tend to infinity and zero, respectively. As a result, we obtain

$$R(t) = \frac{\eta}{2\pi i} \int_0^\infty [\frac{1}{x^\alpha \exp(-\pi\alpha i) + \eta} - \frac{1}{x^\alpha \exp(\pi\alpha i) + \eta}] \exp(-xt)dx. \quad (4.26)$$

Exercise 4.9. Derive Eq. (4.26). □

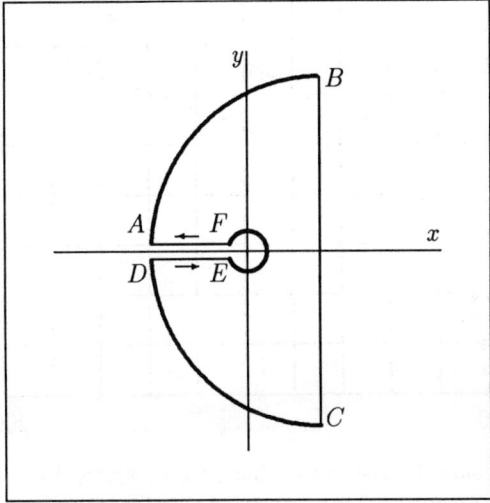

Fig. 4.3. A closed path for calculating integral (4.25).

By using the Euler formula $\exp(\pi\alpha i) = \cos(\pi\alpha) + i\sin(\pi\alpha)$, Eq. (4.26) can be transformed as

$$R(t) = \eta \frac{\sin(\pi\alpha)}{\pi} \int_0^\infty \frac{x^\alpha \exp(-xt)dx}{x^{2\alpha} + 2\eta x^\alpha \cos(\pi\alpha) + \eta^2}.$$

By introducing the new variable $x = \eta^{\frac{1}{\alpha}}u$, we find

$$R(t) = \eta^{\frac{1}{\alpha}} \frac{\sin(\pi\alpha)}{\pi} \int_0^\infty \frac{u^\alpha \exp(-\eta^{\frac{1}{\alpha}}ut)du}{u^{2\alpha} + 2u^\alpha \cos(\pi\alpha) + 1}. \quad (4.27)$$

Eqs. (4.23) and (4.27) are convenient for the explicit analysis and demonstrate fair agreement with experimental data, in particular, for small times, see Fig. 4.4. For large times, Eq. (4.23) forces some discrepancies between experimental and theoretical results, since the rate of increase of the creep measure

(4.23) exceeds the rate of growth observed in experiments. Moreover, creep measure $C_0(t)$ should tend to some finite limit as $t \to \infty$, while formula (4.23) contradicts this assumption.

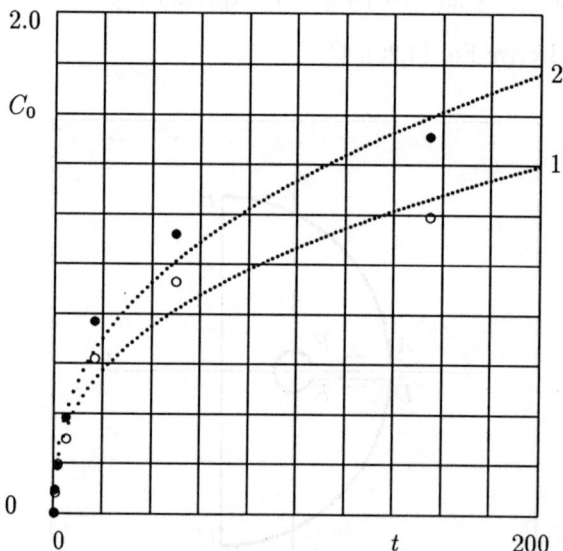

Fig. 4.4. Creep measure C_0 versus time t (min) for a polypropylene monofilament. Large circles correspond to experimental data obtained by Harley & Ward (1965). Unfilled circles: $\sigma = \sigma_1 = 25$ (MPa), filled circles: $\sigma = \sigma_2 = 35$ (MPa). Small circles correspond to power function (4.23). The least square method implies that $\alpha = 0.39$ and $T = 84.26$ (min) for curve 1 and $\alpha = 0.40$ and $T = 48.44$ (min) for curve 2.

A refined version of the creep kernel (4.24) is suggested by Rzhanitsyn (1968)

$$K(t) = \eta \mathcal{J}_{\alpha-1}(t) \exp(-\beta t), \tag{4.28}$$

where $\alpha \in (0,1)$, $\beta > 0$, and $\eta > 0$ are material parameters. Expression (4.28) is called *the generalized fractional-exponential function*.

The presence of the exponential term in Eq. (4.28) does not make construction of the relaxation kernel more sophisticated.

Exercise 4.10. Calculate the Laplace transform of the creep kernel (4.28) and the corresponding relaxation kernel

$$\tilde{K}(p) = \frac{\eta}{(p+\beta)^\alpha}, \qquad \tilde{R}(p) = \frac{\eta}{(p+\beta)^\alpha + \eta}. \qquad \square \tag{4.29}$$

Exercise 4.11. Derive the following formula for the relaxation kernel corresponding to Rzhanitsyn's creep kernel:

$$R(t) = \eta^{\frac{1}{\alpha}} \exp(-\beta t) \frac{\sin(\pi\alpha)}{\pi} \int_0^\infty \frac{u^\alpha \exp(-\eta^{\frac{1}{\alpha}} ut) du}{u^{2\alpha} + 2u^\alpha \cos(\pi\alpha) + 1}. \quad \Box \qquad (4.30)$$

Formally, Rzhanitsyn's kernel (4.28) should describe experimental data more adequately compared with kernel (4.24). However, some experiments show that it does not provide any improvement in comparison with Eq. (4.24), see Garbarski (1992) for a discussion of this issue.

To construct new creep and relaxation kernels, Garbarski (1992) suggested to prescribe explicit expressions for the Laplace transforms of creep and relaxation kernels which allow integral presentations of these kernels to be derived.

As an example, the so-called *root function* is proposed

$$\tilde{K}(p) = \frac{1 + a\sqrt{p}}{1 + a\sqrt{p} + bp}, \qquad (4.31)$$

where a and b are material parameters.

Exercise 4.12. Show that the Laplace transform of the corresponding relaxation kernel can be written as

$$\tilde{R}(p) = \frac{1}{2} \frac{1 + a\sqrt{p}}{1 + a\sqrt{p} + cp}, \qquad (4.32)$$

where $c = \frac{1}{2}b$. $\Box$

Exercise 4.13. By using the inverse Laplace transform, derive the following integral presentations for the creep and relaxation kernels:

$$K(t) = \frac{ab}{\pi} \int_0^\infty \frac{u^{\frac{3}{2}} \exp(-ut) du}{(bu - 1)^2 + a^2 u}, \quad R(t) = \frac{ac}{2\pi} \int_0^\infty \frac{u^{\frac{3}{2}} \exp(-ut) du}{(cu - 1)^2 + a^2 u}. \quad \Box \quad (4.33)$$

Eqs. (4.33) demonstrate good agreement with creep curves for several polymeric materials: polycarbonate "Lexan", "Lustran", and poly(methyl methacrylate) "Plexiglass". Nevertheless, employment of these expressions in engineering practice is questionable due to the complicated form of both creep and relaxation kernels.

It is also worth noting relaxation kernels expressed as infinite series depending on several material parameters. The kernels are singular for some values of these parameters, while they are regular for other values. We do not intend to analyze properties of these kernels in detail, referring to Renardy et al. (1987) and Scanlan & Janzen (1992), and mention these kernels only for completeness.

The Gennes and Rouse relaxation functions are determined as infinite sums of exponential functions. According to *the Gennes model*,

$$Q_*(t) = \frac{8}{\pi^2} \sum_{m=1}^{\infty} \frac{1}{(2m-1)^2} \exp[-(2m-1)^2 \frac{t}{T}], \qquad (4.34)$$

and according to *the Rouse model*,

$$Q_*(t) = \chi \sum_{m=1}^{\infty} \exp(-m^2 \frac{t}{T}), \qquad (4.35)$$

where χ and T are material parameters. Eq. (4.35) contradicts condition (3.21), since the sum diverges at $t = 0$. Relaxation functions (4.34) and (4.35) are employed to describe flows of polymeric melts, and they do not correspond to experimental data for viscoelastic solids.

The following kernels were introduced by Renardy (1982), see also Renardy et al. (1987), for the description of wave propagation in viscoelastic liquids. They are of interest mainly from the mathematical standpoint as examples of relaxation kernels with various types of singularity at point $t = 0$:
– a kernel with a power-type singularity

$$R(t) = \chi \sum_{m=1}^{\infty} \exp(-m^\alpha \frac{t}{T}), \qquad (4.36)$$

– a kernel with a logarithmic singularity

$$R(t) = \chi \sum_{m=1}^{\infty} \exp(-\frac{t}{T} \exp(m)), \qquad (4.37)$$

– a kernel with a log–log singularity

$$R(t) = \chi \sum_{m=1}^{\infty} \exp(-\frac{t}{T} \exp(\exp(n))). \qquad (4.38)$$

Here α, χ, and T are material parameters. To the best of our knowledge, relaxation kernels (4.36) – (4.38) have not been verified experimentally.

4.1.3. Creep and relaxation kernels of aging media

Unlike non-aging materials, only a few expressions have been suggested for creep and relaxation measures of aging media. On the other hand, experimental observations allow the main characteristic features of functions $E(t)$, $C(t, s)$, and $Q(t, s)$ to be formulated for polymeric materials and concrete, see Brinson & Gates (1995), Espinoza & Aklonis (1993), Lee & McKenna (1990a), McKenna & Kovacs (1984), Ross (1958), and Struik (1978, 1987).

Experimental data show that Young's modulus $E(t)$ is positive, increases monotonically in time, and tends to some limiting value $E(\infty)$ which is called *the limiting elastic modulus*. The temporal derivative $\dot{E}(t)$ is non-negative and vanishes as time approaches infinity.

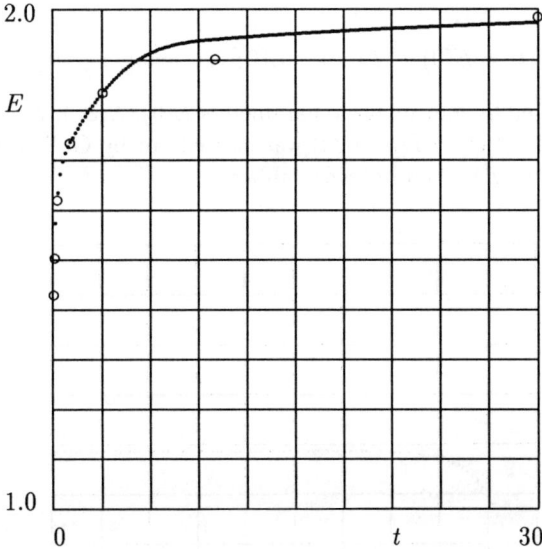

Fig. 4.5. Young's modulus E (MPa) versus time t (days) for polypropylene quenched from 120°C to 20°C. Large circles correspond to experimental data obtained by Struik (1978). Small circles provide their approximation by Eq. (4.39) with $E(0) = 1.23$ (MPa), $E(\infty) = 1.93$ (MPa), and $T = 0.94$ (days).

These assumptions have a simple interpretation in the model of adaptive links. According to this model, function $X(t, \tau)$ is proportional to the number of links being born before instant τ and existing at moment t. For an aging elastic material, $Q(t, t) = 0$ and Eq. (3.14) implies that $X(t, t) = E(t)$, which means that $E(t)$ is the number of elastic links in the system at instant t. The above restrictions on the Young modulus are equivalent to the assumptions that
(i) for any instant $t \geq 0$, the number of links is positive;
(ii) this number tends to some limiting value as time tends to infinity;
(iii) the rate of increase in the number of links is positive and vanishes with the growth of time.
The dependence of Young's modulus on time can be approximated by the exponential function

$$E(t) = E(0) + [E(\infty) - E(0)][1 - \exp(-\frac{t}{T})], \qquad (4.39)$$

309

where $E(0) > 0$ is the initial elastic modulus, $E(\infty) > 0$ is the limiting elastic modulus, and $T > 0$ is the characteristic time of aging. Formula (4.39) provides fair prediction of experimental data for a number of viscoelastic materials, see Fig. 4.5.

A generalization of Eq. (4.39) is the so-called *Kohlrausch–William–Watts formula*

$$E(t) = E(0) + [E(\infty) - E(0)]\{1 - \exp[-(\tfrac{t}{T})^\gamma]\}, \qquad (4.40)$$

which is determined by four material parameters $E(0)$, $E(\infty)$, T, and $\gamma \in (0, 1)$. Experimental validation of Eq. (4.30) was carried out by Gul' et al. (1992) for polyethylene, isobutylene, and nitride rubber.

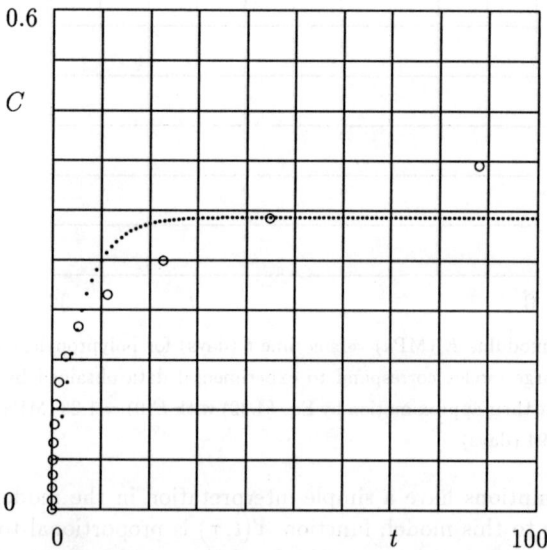

Fig. 4.6. Creep measure $C(t+\rho, \rho)$ (GPa^{-1}) *versus* time t (hours) for polypropylene quenched from 120°C to 20°C. Aging time $\rho = 30$ (days). Large circles correspond to experimental data obtained by Struik (1978). Small circles provide the approximation of experimental results by Eq. (4.41), where the aging function $\phi(s)$ and the rate of creep γ are found by fitting experimental data.

Arutyunyan (1952) suggested the following expression for the creep measure $C(t, s)$:

$$C(t, \tau) = \phi(\tau)\{1 - \exp[-\gamma(t - \tau)]\}, \qquad (4.41)$$

where $\phi(\tau)$ is called *aging function*, and γ is the characteristic rate of creep. Experimental data for plain and reinforced concrete show that the function

310

$\phi(\tau)$ is positive, decreases monotonically in time and tends to some positive limiting value as $\tau \to \infty$. The following expression is proposed for the aging function $\phi(\tau)$, see Arutyunyan (1952):

$$\phi(\tau) = a_0 + \sum_{n=1}^{N} \frac{a_n}{\tau + \tau_n}, \qquad (4.42)$$

where a_n and τ_n are material constants. To the best of our knowledge, only the case $N = 1$ was verified experimentally.

Prokhopovich (1963) suggested another expression for the aging function

$$\phi(\tau) = \sum_{n=0}^{N} A_n \exp(-\beta_n \tau), \qquad (4.43)$$

where A_n and β_n are material constants.

For the creep measure (4.41), the corresponding relaxation measure $Q(t, \tau)$ is calculated explicitly. To find $Q(t, \tau)$, we differentiate Eq. (3.5) twice, use Eqs. (3.6) and (4.41), and arrive at the differential equation

$$\frac{d^2\epsilon}{dt^2} + \gamma \frac{d\epsilon}{dt} = \frac{d}{dt}\left[\frac{1}{E(t)} \frac{d\sigma}{dt}\right] + \frac{\gamma}{E(t)}[1 + E(t)\phi(t)]\frac{d\sigma}{dt} \qquad (4.44)$$

with the initial conditions

$$\sigma(0) = 0, \qquad \frac{d\sigma}{dt}(0) = E(0)\frac{d\epsilon}{dt}(0). \qquad (4.45)$$

Exercise 4.14. Derive Eqs. (4.44) and (4.45). $\square$

Integration of Eq. (4.44) yields

$$\frac{\dot{\sigma}(t)}{E(t)} = \dot{\epsilon}(t) - \gamma \int_0^t E(\tau)\phi(\tau) \exp[-\gamma \int_\tau^t (1 + E(\xi)\phi(\xi))d\xi]\dot{\epsilon}(\tau)d\tau.$$

By integrating this equality once more we derive Eqs. (3.13) and (3.14) with the relaxation measure

$$Q(t, \tau) = -\gamma E(\tau)\phi(\tau) \int_\tau^t E(s) \exp[-\gamma \int_\tau^s (1 + E(\xi)\phi(\xi))d\xi]ds. \qquad (4.46)$$

Despite the presence of explicit expressions for both the creep and relaxation kernels, applications of Eq. (4.41) are limited due to poor agreement with observations, see Fig. 4.6, which demonstrates significant discrepancies between experimental data and their prediction.

As a version of the constitutive equation for an aging viscoelastic medium, Slaughter & Fleck (1993) suggested the model

$$\dot{\sigma} + \frac{\alpha}{t + T}\sigma = E\dot{\epsilon}, \qquad (4.47)$$

where E is a constant elastic modulus, α and T are material parameters.

Exercise 4.15. Show that Eq. (4.47) may be presented in the form (3.5) with the creep measure

$$C(t, \tau) = \frac{\alpha}{E} \ln \frac{t + T}{\tau + T}. \qquad \square \qquad (4.48)$$

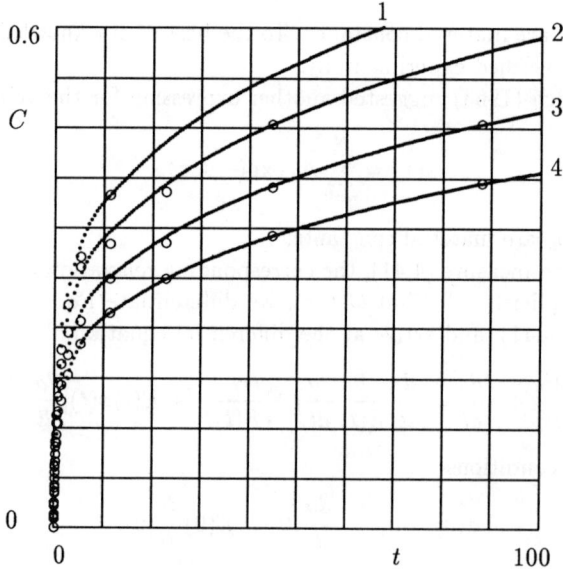

Fig. 4.7. Creep measure $C(t+\rho, \rho)$ (GPa^{-1}) *versus* time t (hours) for polypropylene quenched from 120°C to 20°C. Large circles correspond to experimental data obtained by Struik (1978). Curve 1: $\rho = 1$ (day), curve 2: $\rho = 3$ (days), curve 3: $\rho = 10$ (days), curve 4: $\rho = 30$ (days). Small circles of curve 4 provide the approximation of experimental data by cubic splines. Other curves plotted by small circles are derived from curve 4 by multiplication by an appropriate constant η. Curve 1: $\eta = 1.550$, curve 2: $\eta = 1.380$, curve 3: $\eta = 1.171$.

To derive a constitutive model which ensures good agreement with experimental data, we, first, consider creep curves $C = C(t + \rho, \rho)$ for various ages ρ, see Fig. 4.7. The experimental data demonstrate affinity of the creep curves $C(t + \rho, \rho)$ corresponding to various ages ρ of the material. From the mathematical standpoint, this implies that

$$C(t, \tau) = \phi(\tau) f(t - \tau), \qquad (4.49)$$

where $\phi(\tau)$ is an aging function, and $f(t)$ is *a creep function*. From the physical standpoint, Eq. (4.49) means that the processes of creep and aging are independent of each other.

It follows from Eqs. (4.41) and (4.49) that Arutyunyan's presentation imposes a specific restriction on the creep function, requiring it to be exponential. Experimental data show that this form of the creep function, which is extremely convenient for applications, fails to be true.

To introduce an additional assumption regarding material functions we plot the functions

$$r_1(t) = 1 - \frac{E(0)}{E(t)}, \qquad r_2(t) = 1 - \frac{\phi(t)}{\phi(0)}. \tag{4.50}$$

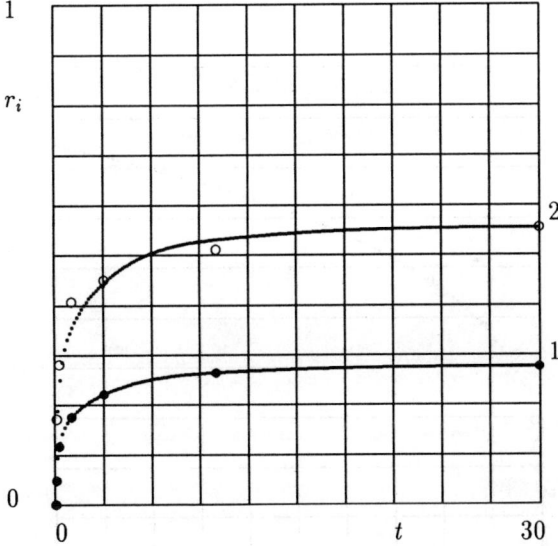

Fig. 4.8. Ratios $r_1 = 1 - E(0)/E(t)$ (filled circles) and $r_2 = 1 - \phi(t)/\phi(0)$ (unfilled circles) *versus* aging time t (days) for polypropylene quenched from 120°C to 20°C. Large circles correspond to experimental data obtained by Struik (1978). Small circles provide the approximations of functions $r_i(t)$ by cubic splines. Curve 1 corresponds to the function $r_1(t)$, curve 2 is obtained from curve 1 by multiplication by $a = 2.01$.

Fig. 4.8 demonstrates affinity of the curves $r_i(t)$

$$r_2(t) = a r_1(t), \tag{4.51}$$

where a is a material constant. Substitution of expressions (4.50) into Eq. (4.51) yields

$$\phi(t) = \phi(0) \frac{(1-a)E(t) + aE(0)}{E(t)}. \tag{4.52}$$

According to Eq. (4.49), only the product $\phi(\tau)f(t-\tau)$ has some physical meaning. Therefore, the limiting value of the aging function $\phi(\tau)$ may be chosen arbitrarily. For convenience, we set

$$\phi(\infty) = 1. \qquad (4.53)$$

It follows from this equality and Eq. (4.52) that

$$\phi(\tau) = \frac{E(\infty)}{E(\tau)} \frac{(1-a)E(\tau)+aE(0)}{(1-a)E(\infty)+aE(0)}. \qquad (4.54)$$

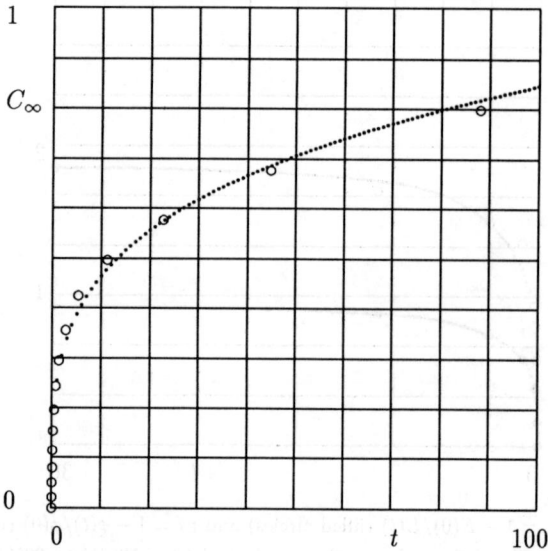

Fig. 4.9. Limiting creep measure C_∞ *versus* time t (hours) for polypropylene quenched from $120°C$ to $20°C$. Large circles correspond to experimental data obtained by Struik (1978). Small circles provide the approximation of experimental data by Eq. (4.23). The least square method implies that $\alpha = 0.261$ and $T = 190.69$ (hours).

Eqs. (3.6), (4.49), and (4.54) imply the formula for the creep kernel of an aging viscoelastic material

$$\mathcal{Y}(t,\tau) = -\frac{\partial}{\partial \tau} \{\frac{1}{E(\tau)}[1 + E(\infty)\frac{(1-a)E(\tau)+aE(0)}{(1-a)E(\infty)+aE(0)} f(t-\tau)]\}. \qquad (4.55)$$

In order to transform Eq. (4.55), we consider the response of the aged viscoelastic medium when ρ tends to infinity. It is quite natural to presume that

314

this response is described by the constitutive equation (3.11) of a non-aging viscoelastic medium with the creep kernel in convolution

$$\mathcal{Y}_\infty(t, \tau) = -\frac{1}{E(\infty)} \frac{\partial C_\infty(t - \tau)}{\partial \tau},$$ (4.56)

where $C_\infty(t)$ is the limiting creep measure of the aged medium. Comparison of Eqs. (4.55) and (4.56) implies that

$$f(t) = \frac{1}{E(\infty)} C_\infty(t).$$

Finally, substitution of this expression into Eq. (4.55) yields

$$\mathcal{Y}(t, \tau) = -\frac{\partial}{\partial \tau} \left\{ \frac{1}{E(\tau)} [1 + \frac{(1 - a)E(\tau) + aE(0)}{(1 - a)E(\infty) + aE(0)} C_\infty(t - \tau)] \right\}.$$ (4.57)

Presentation (4.57) for the creep kernel of an aging viscoelastic material is characterized by two functions of time $E(t)$ and $C_\infty(t)$ and a constant a. The function $E(t)$ characterizes the aging process and is independent of creep. The function $C_\infty(t)$ characterizes the creep only and coincides with the creep measure of the aged material. Only the constant a establishes a correspondence between creep and aging.

From the engineering standpoint, Eq. (4.57) allows the material functions to be determined extremely simply: function $E(t)$ is found from instant (elastic) experiments during the period of aging, while function $C_\infty(t)$ is obtained from creep experiments for the aged material, when the aging process is completed.

The function $C_\infty(t)$ may be approximated by the power function (4.23). Fig. 4.9 demonstrates excellent agreement between experimental data and their theoretical prediction with the use of Eq. (4.23).

4.2. Properties of Creep and Relaxation Measures

We begin our consideration with conditions imposed on relaxation measures of non-aging viscoelastic media. Afterwards, we extend these conditions to relaxation measures and creep measures of aging viscoelastic materials.

4.2.1. Relaxation measures of non-aging viscoelastic media

Let us consider a regular relaxation measure $Q_0(t)$ and suppose that the following conditions are fulfilled:

$$-1 < Q_0(\infty) < Q_0(t) \leq 0, \qquad Q_0(0) = 0,$$ (4.58)
$$\dot{Q}_0(t) < 0, \quad \dot{Q}_0(\infty) = 0,$$ (4.59)
$$\ddot{Q}_0(t) > 0, \quad \ddot{Q}_0(\infty) = 0.$$ (4.60)

Traditionally, conditions (4.58) – (4.60) are treated as main restrictions imposed on the function $Q_0(t)$. To explain their mechanical meaning we consider a specimen deformed according to the program

$$\epsilon(t) = \begin{cases} 0 & 0 \leq t < \tau, \\ \epsilon_0 & \tau \leq t < \infty, \end{cases} \tag{4.61}$$

where ϵ_0 and τ are positive constants. Substitution of Eq. (4.61) into Eq. (3.18) implies the following formulas for the stress $\sigma(t)$ and its derivative $\dot{\sigma}(t)$:

$$\sigma(t) = \begin{cases} 0 & 0 \leq t < \tau, \\ E[1 + Q_0(t) - Q_0(\tau)]\epsilon_0 & \tau \leq t < \infty, \end{cases}$$

$$\dot{\sigma}(t) = \begin{cases} 0 & 0 \leq t < \tau, \\ E\dot{Q}_0(t)\epsilon_0 & \tau \leq t < \infty. \end{cases} \tag{4.62}$$

It follows from Eqs. (4.58), (4.59), and the first equality (4.62) that the stress $\sigma(t)$ in a viscoelastic material is less than the stress in the corresponding elastic material (the stress relaxation phenomenon). The stress tends to some positive limiting value as $t \to \infty$ (limited relaxation). Eq. (4.60) and the other equality (4.62) imply that the rate of relaxation $|\dot{\sigma}(t)|$ decreases monotonically and tends to zero as $t \to \infty$.

Consider now the deformation program

$$\epsilon(t) = \epsilon_0 \delta(t - \tau), \tag{4.63}$$

where $\delta(t)$ is the Dirac delta-function. Eqs. (3.18) and (4.63) imply that

$$\sigma(t) = \begin{cases} 0 & 0 \leq t < \tau, \\ E\dot{Q}_0(t - \tau)\epsilon_0 & \tau < t < \infty, \end{cases}$$

$$\dot{\sigma}(t) = \begin{cases} 0 & 0 \leq t < \tau, \\ E\ddot{Q}_0(t - \tau)\epsilon_0 & \tau < t < \infty. \end{cases} \tag{4.64}$$

It follows from Eqs. (4.59), (4.60), and the first equality (4.64) that the stress is negative and tends to zero as $t \to \infty$. If we present the Dirac function as a limit of a sequence of rectangular impulses with growing intensities and decreasing durations, the negativity of the stress may be treated as a consequence of the principle of fading memory: the effect of the falling branch of any impulse is stronger than the effect of the lifting branch. According to the first equality (4.64), the absolute value $|\sigma(t)|$ of the stress decreases monotonically. It follows from the other equality (4.64) and Eq. (4.60) that the rate of relaxation $|\dot{\sigma}(t)|$ decreases monotonically and tends to zero as $t \to \infty$.

Instead of the restrictions on relaxation measure $Q_0(t)$, the following inequalities are used for the function $H_0(t) = Q_0(t) - Q_0(\infty)$, see e.g. Hrusa & Nohel (1985), MacCamy (1977):

$$H_0(t) \geq 0, \qquad \dot{H}_0(t) \leq 0, \qquad \ddot{H}_0(t) \geq 0. \tag{4.65}$$

The first inequality (4.65) implies the non-negative definiteness of the function $H_0(t)$, the second means its monotonicity, and the third yields the convexity of $H_0(t)$.

Following Pipkin (1972), Beris & Edwards (1993) considered smooth relaxation measures and imposed an infinite sequence of constitutive inequalities

$$(-1)^n \frac{d^n H_0}{dt^n}(t) \geq 0 \qquad (n = 0, 1, \ldots). \tag{4.66}$$

To explain the mechanical meaning of Eq. (4.66) Beris & Edwards (1993) introduced loading programs in the form of the so-called nth *derivatives of the Dirac function*. These functions are treated as an idealization of a sequence of $n + 1$ strain steps with alternating signs when the length of any step tends to zero.

Exercise 4.16. Provide an explanations of the physical meaning of Eq. (4.66) for $n > 2$. □

By employing *the Bernstein theorem*, we obtain from Eq. (4.66) that there is a non-negative definite function $\chi(T)$ such that

$$H_0(t) = \int_0^\infty \chi(T) \exp(-\frac{t}{T}) dT. \tag{4.67}$$

Differentiation of expression (4.67) yields Eq. (4.5) with an additional condition of non-negativity for the relaxation spectrum $\chi(T)$. Since the inverse assertion regarding the connection between Eqs. (4.66) and (4.67) is also valid, inequalities (4.66) may eliminate relaxation measures which are incorrect physically.

Exercise 4.17. Provide an interpretation of the non-negativity of function $\chi(T)$ in Eq. (4.67) in the framework of the model of adaptive links. □

Definition. An integrable function $a(t)$ is *of positive type* if for any square integrable complex function $\eta(t)$

$$\Re \int_0^t \eta(s) ds \int_0^s a(s - \tau)\bar{\eta}(\tau) d\tau \geq 0, \tag{4.68}$$

where $\Re$ stands for the real part of a complex number $p = \alpha + i\omega$, $i = \sqrt{-1}$, and the superscript bar denotes complex conjugate.

Proposition 4.1. Let $a(t)$ be an integrable function with $\int_0^\infty |a(t)| dt < \infty$. The function $a(t)$ is of positive type if and only if for any real ω

$$\Re \tilde{a}(i\omega) \geq 0, \tag{4.69}$$

where the tilde stands for the Laplace transform.

Proof. First, we show that the positivity of $a(t)$ implies Eq. (4.69). For this purpose, we choose a function η in the form

$$\eta(t) = \begin{cases} 0 & t \in (-\infty, 0), \\ \exp(pt) & t \in [0, T], \\ 0 & t \in (T, \infty), \end{cases} \tag{4.70}$$

where p is a complex parameter with a positive real part α, and T is a real parameter. Substitution of expression (4.70) into Eq. (4.68) implies that for any $t \geq T$

$$\begin{aligned}
\Re \int_0^t \eta(s)ds \int_0^s a(s-\tau)\bar{\eta}(\tau)d\tau &= \Re \int_0^T \eta(s)ds \int_0^s a(s-\tau)\bar{\eta}(\tau)d\tau \\
&= \Re \int_0^T \eta(s)ds \int_0^s a(s)\bar{\eta}(s-\tau)d\tau \\
&= \Re \int_0^T \exp(ps)ds \int_0^s a(\tau)\exp[\bar{p}(s-\tau)]d\tau \\
&= \Re \int_0^T \exp[(p+\bar{p})s]ds \int_0^s a(\tau)\exp(-\bar{p}\tau)d\tau \\
&= \Re \int_0^T \exp(2\alpha s)ds \int_0^s a(\tau)\exp(-p\tau)d\tau. \tag{4.71}
\end{aligned}$$

Introducing the notation

$$\psi_1(t) = 2\alpha\exp(-2\alpha t), \qquad \psi_2(t) = \int_0^t a(s)\exp(-ps)ds, \tag{4.72}$$

we rewrite Eq. (4.71) as follows:

$$\Re \int_0^t \eta(s)ds \int_0^s a(s-\tau)\bar{\eta}(\tau)d\tau = \frac{1}{2\alpha}\exp(2\alpha T)\Re \int_0^T \psi_1(T-s)\psi_2(s)ds. \tag{4.73}$$

Eq. (4.73) together with inequality (4.68) yields

$$\Re \int_0^T \psi_1(T-s)\psi_2(s)ds \geq 0 \tag{4.74}$$

for any p and T. The integral in Eq. (4.74) may be presented as

$$\Re \int_0^T \psi_1(T-s)\psi_2(s)ds = I_1 + I_2 + I_3, \tag{4.75}$$

where

$$I_1 = \Re \int_0^{T_0} \psi_1(T-s)\psi_2(s)ds, \qquad I_2 = \Re \int_{T_0}^T \psi_1(T-s)\tilde{a}(p)ds,$$

$$I_3 = \Re \int_{T_0}^T \psi_1(T-s)[\psi_2(s) - \tilde{a}(p)]ds, \tag{4.76}$$

318

and T_0 is an arbitrary real number, $T_0 < T$. It follows from Eq. (4.72) that $\psi_2(\infty) = \tilde{a}(p)$. Therefore, for any positive $\varepsilon > 0$ there is a T_0 such that for any $s \geq T_0$

$$|\psi_2(s) - \tilde{a}(p)| < \varepsilon.$$

This inequality together with Eq. (4.76) implies that

$$|I_3| \leq \varepsilon |\int_{T_0}^{T} \psi_1(T - s)ds| = \varepsilon |\int_{0}^{T-T_0} \psi_1(s)ds| \leq 2\alpha\varepsilon \int_{0}^{\infty} \exp(-2\alpha s)ds = \varepsilon. \quad (4.77)$$

Calculation of I_2 yields

$$I_2 = \Re\tilde{a}(p) \int_{T_0}^{T} \psi_1(T - s)ds = \Re\tilde{a}(p) \int_{0}^{T-T_0} \psi_1(s)ds$$

$$= 2\alpha\Re\tilde{a}(p) \int_{0}^{T-T_0} \exp(-2\alpha s)ds = \Re\tilde{a}(p)\{1 - \exp[-2\alpha(T - T_0)]\}. \quad (4.78)$$

Finally, substitution of expression (4.72) into the first Eq. (4.76) implies that

$$|I_1| \leq |\int_{0}^{T_0} \psi_1(T - s)\psi_2(s)ds| = 2\alpha \exp(-2\alpha T) \int_{0}^{T_0} |\psi_2(s)| \exp(2\alpha s)ds. \quad (4.79)$$

It follows from Eqs. (4.77) – (4.79) that for any $\varepsilon > 0$, any $T_0 > 0$, and any complex $p = \alpha + i\omega$ with $\alpha > 0$ there is a positive $T_1 > T_0$ such that for any $T > T_1$

$$|I_1| \leq \varepsilon, \qquad |I_2 - \Re\tilde{a}(p)| \leq \varepsilon, \qquad |I_3| \leq \varepsilon. \quad (4.80)$$

Eqs. (4.74), (4.75), and (4.80) imply that for any $\alpha > 0$ and any ω

$$\Re\tilde{a}(\alpha + i\omega) \geq 0. \quad (4.81)$$

Since $\tilde{a}(p)$ is a continuous function of complex variable p, inequality (4.69) follows from Eq. (4.81).

We now prove the inverse assertion that Eq. (4.69) implies that function $a(t)$ is of positive type. Let $T > 0$ be a positive constant, and $\eta(t)$ a square integrable complex function. Introduce a new function

$$\eta_T(t) = \begin{cases} \eta(t) & t \in [0, T], \\ 0 & \text{otherwise.} \end{cases}$$

Evidently,

$$\Re \int_{0}^{t} \eta(s)ds \int_{0}^{s} a(s - \tau)\bar{\eta}(\tau)d\tau = \Re \int_{0}^{\infty} \eta_t(s)ds \int_{0}^{s} a(s - \tau)\bar{\eta}_t(\tau)d\tau. \quad (4.82)$$

It follows from Proposition 3.1 and *the Parseval theorem* that

$$\Re \int_0^\infty \eta_t(s)ds \int_0^s a(s-\tau)\bar{\eta}_t(\tau)d\tau = \frac{1}{2\pi}\Re \int_{-\infty}^\infty \tilde{\eta}_t(iw)\tilde{a}(iw)\overline{\tilde{\eta}_t(iw)}dw$$

$$= \frac{1}{2\pi}\int_{-\infty}^\infty \Re\tilde{a}(iw)|\tilde{\eta}_t(iw)|^2 dw. \qquad (4.83)$$

According to Eqs. (4.82) and (4.83), Eq. (4.69) implies inequality (4.68). □

Definition. An integrable function $a(t)$ is *of strong positive type* if there is a positive constant δ such that function $a_0(t) = a(t) - \delta\exp(-t)$ is of positive type.

Proposition 4.2. An integrable function $a(t)$ is of strong positive type if and only if there is a positive δ such for any w

$$\Re\tilde{a}(iw) \geq \frac{\delta}{1+w^2}. \qquad (4.84)$$

Proof. Let us calculate the Laplace transform of function $a_0(t)$

$$\tilde{a}_0(p) = \tilde{a}(p) - \delta\int_0^\infty \exp[-(1+p)t]dt = \tilde{a}(p) - \frac{\delta}{1+p}.$$

Exercise 4.18. Derive the following formula:

$$\Re\tilde{a}_0(iw) = \Re\tilde{a}(iw) - \frac{\delta}{1+w^2}. \qquad \square \qquad (4.85)$$

According to Proposition 4.1, function $a(t)$ is of strong positive type if and only if $\Re\tilde{a}_0(iw) \geq 0$. Substitution of expression (4.85) into this inequality implies the desired assertion. □

Proposition 4.3. Let a function $a(t)$ be triple continuously differentiable and for any $t \geq 0$ satisfy the inequalities

$$a(t) \geq 0, \qquad \dot{a}(t) \leq 0, \qquad \ddot{a}(t) \geq 0. \qquad (4.86)$$

Let the function $a(t)$ and its derivatives vanish at infinity, and $\ddot{a}(t) > 0$ on a set of positive measure. Then $a(t)$ is of strong positive type.

Proof. Let us estimate the integral

$$\tilde{a}(iw) = \int_0^\infty a(t)\exp(-iwt)dt. \qquad (4.87)$$

For $w = 0$, integration by parts yields

$$\tilde{a}(0) = \int_0^\infty a(t)dt = ta(t)|_0^\infty - \int_0^\infty \dot{a}(t)tdt$$

$$= [ta(t) - \frac{1}{2}\dot{a}(t)t^2]_0^\infty + \frac{1}{2}\int_0^\infty t^2\ddot{a}(t)dt = \frac{1}{2}\int_0^\infty t^2\ddot{a}(t)dt > \delta,$$

where δ is a positive constant. Since function $\tilde{a}(i\omega)$ is continuous at $\omega = 0$, there is a positive Ω_1 such that for any $|\omega| \leq \Omega_1$

$$\Re\tilde{a}(i\omega) \geq \frac{\delta_1}{1+\omega^2}, \tag{4.88}$$

where $\delta_1 = \frac{1}{2}\delta$. By using similar reasoning, we obtain for $\omega \neq 0$

$$\tilde{a}(i\omega) = \frac{1}{i\omega}a(0) - \frac{1}{\omega^2}\dot{a}(0) - \frac{1}{\omega^2}\int_0^\infty \ddot{a}(t)\exp(-i\omega t)dt. \tag{4.89}$$

Since

$$\int_0^\infty \ddot{a}(t)dt = \dot{a}(\infty) - \dot{a}(0) = -a(0),$$

Eq. (4.89) can be written as

$$\tilde{a}(i\omega) = \frac{1}{i\omega}a(0) + \frac{1}{\omega^2}\int_0^\infty \ddot{a}(t)[1 - \exp(-i\omega t)]dt.$$

Therefore,

$$\Re\tilde{a}(i\omega) = \frac{1}{\omega^2}\int_0^\infty \ddot{a}(t)(1 - \cos\omega t)dt. \tag{4.90}$$

Integration by parts implies that

$$\int_0^\infty \ddot{a}(t)\cos\omega t\,dt = -\frac{1}{\omega}\int_0^\infty \frac{d^3 a}{dt^3}(t)\sin\omega t\,dt. \tag{4.91}$$

Hence,

$$\lim_{\omega\to\infty}\int_0^\infty \ddot{a}(t)\cos\omega t\,dt = 0.$$

It follows from this equality and Eq. (4.90) that there is a positive Ω_2 such that for any $|\omega| > \Omega_2$

$$\Re\tilde{a}(i\omega) > \frac{\delta_1}{\omega^2} \geq \frac{\delta_1}{1+\omega^2}. \tag{4.92}$$

Finally, for $\Omega_1 < |\omega| < \Omega_2$ inequality

$$\Re\tilde{a}(i\omega) \geq \frac{\delta_1}{1+\omega^2} \tag{4.93}$$

follows from the positivity of $\ddot{a}(t)$ on a set of positive measure and Eq. (4.90). The desired assertion follows from Eqs. (4.88), (4.92), (4.93) and Proposition 4.2. $\square$

In the proofs of Propositions 4.1 – 4.3 we impose strong restrictions on the smoothness of function $a(t)$. These limitations may be significantly weakened, see Gripenberg et al. (1990).

Applying the above results to function $H_0(t)$ we obtain that inequalities (4.65) together with the additional assumption regarding the positivity of $\ddot{H}_0(t)$ on a set of positive measure yield

$$\Re\tilde{H}_0(i\omega) \geq \frac{\delta}{1+\omega^2}. \tag{4.94}$$

Since

$$\Re\tilde{H}_0(i\omega) = \int_0^\infty H_0(t)\cos\omega t\, dt = -\frac{1}{\omega}\int_0^\infty \dot{Q}_0(t)\sin\omega t\, dt, \tag{4.95}$$

Eq. (4.94) implies the strong negativity of *the Fourier sine transform* of the relaxation measure $Q_0(t)$. The latter condition plays the key role in the study of existence and uniqueness for solutions to boundary value problems in linear viscoelasticity, see Fabrizio & Morro (1992).

The textbook is concentrated on regular relaxation measures. Nevertheless, we should make some remarks regarding restrictions imposed on *singular relaxation measures*. Renardy (1988) assumed that there is a positive constant κ such that for any ω

$$|\Re\tilde{Q}_*(i\omega)| \geq \kappa|\Im\tilde{Q}_*(i\omega)|, \tag{4.96}$$

where $\Im$ stands for the imaginary part of a complex number, and $Q_*(t)$ is a relaxation function. Gripenberg (1993, 1994) assumed function $Q_*(t)$ to be locally integrable, of positive type, and to satisfy the inequality

$$\Re\tilde{Q}_*(p) \geq \kappa|\Im\tilde{Q}_*(p)| \tag{4.97}$$

for any complex $p = \alpha + i\omega$ with $\alpha > 0$.

Let us now return to regular relaxation measures and suppose additionally that there are positive constants T_1 and T_2, $T_1 \leq T_2$, such that for any $t \geq 0$

$$\frac{1}{T_2} \leq \frac{\ddot{Q}_0(t)}{|\dot{Q}_0(t)|} \leq \frac{1}{T_1}. \tag{4.98}$$

Proposition 4.4. Let conditions (4.58) – (4.60) and (4.98) be fulfilled. Then for any $t \geq 0$

$$|Q_0(\infty)|\exp(-\frac{t}{T_1}) \leq Q_0(t) - Q_0(\infty) \leq |Q_0(\infty)|\exp(-\frac{t}{T_2}). \qquad \square \tag{4.99}$$

Proof. It follows from Eqs. (4.59) and (4.98) that

$$-\frac{1}{T_2}\dot{Q}_0(t) \leq \ddot{Q}_0(t) \leq -\frac{1}{T_1}\dot{Q}_0(t). \tag{4.100}$$

Integration of Eq. (4.100) from t to infinity with the use of Eq. (4.58) and (4.59) yields

$$\frac{1}{T_2}[Q_0(t) - Q_0(\infty)] \leq |\dot{Q}_0(t)| \leq \frac{1}{T_1}[Q_0(t) - Q_0(\infty)]. \tag{4.101}$$

Eqs. (4.100) and (4.101) imply that

$$\frac{1}{T_2^2}[Q_0(t) - Q_0(\infty)] \leq \ddot{Q}_0(t) \leq \frac{1}{T_1^2}[Q_0(t) - Q_0(\infty)]. \tag{4.102}$$

Rewrite Eq. (4.101) in the form

$$-\frac{dt}{T_1} \leq \frac{dQ_0}{Q_0 - Q_0(\infty)} \leq -\frac{dt}{T_2}.$$

Integrating this inequality from 0 to t and using Eq. (4.58) we derive Eq. (4.99). □

Function (4.1) satisfies the conditions of Proposition 4.4 with $T_1 = T_2 = T$ and provides a typical example of a relaxation measure $Q_0(t)$ which tends to its limiting value $Q_0(\infty)$ with the exponential rate.

4.2.2. Relaxation measures of aging viscoelastic media

Let us now formulate some properties of relaxation kernels and measures for aging viscoelastic materials. We begin with regular relaxation measures and assume that for any $0 \leq \tau \leq t < \infty$ function (3.14) satisfies the following conditions:

$$\mathcal{X}(t,t) = E(t), \qquad 0 < \mathcal{X}(t,\tau) \leq E(\tau), \tag{4.103}$$

$$\frac{\partial \mathcal{X}}{\partial t}(t,\tau) \leq 0, \qquad \frac{\partial \mathcal{X}}{\partial \tau}(t,\tau) \geq 0, \qquad \frac{\partial^2 \mathcal{X}}{\partial t \partial \tau}(t,\tau) \leq 0. \tag{4.104}$$

To explain the mechanical meaning of restrictions (4.103) and (4.104), we recall that in the model of adaptive links function $\mathcal{X}(t,\tau)$ is treated as the number of links being born before instant τ and existing at moment t, while the derivative $\partial \mathcal{X}/\partial \tau(t,\tau)$ equals the rate of creation for new links.

Eq. (4.103) means that for any moments τ and t, $0 < \tau < t$, there are links arisen before the instant τ and existing at the instant t. The number of these links in an aging viscoelastic medium is less than or equal to the number of links in an appropriate aging elastic material. This assumption may be treated as a postulate of *limited creep*.

The first inequality (4.104) means that the number of links arisen before the instant τ and existing at the current instant t decreases in t due to their breakage. The third inequality (4.104) implies that the rate of "manufacturing" new links which exist up to the moment t diminishes in time due to the failure of internal connections. The second inequality (4.104) expresses the fact that the number of links at the instant t arisen before the instant τ increases in τ. From the physical standpoint, this means that the number of links arising in a fixed interval of time grows with the length of this interval.

To formulate an additional condition, similar to Eq. (4.98), we assume that there are positive constants T_1 and T_2, $T_1 < T_2$, such that for any $0 \leq \tau \leq t$

$$\frac{1}{T_2^2} \frac{\partial Q}{\partial \tau}(t,\tau) \leq \frac{\partial^3 Q}{\partial t^2 \partial \tau}(t,\tau) \leq \frac{1}{T_1^2} \frac{\partial Q}{\partial \tau}(t,\tau). \tag{4.105}$$

Inequality (4.105) was suggested by Dafermos (1970) in the study of asymptotic stability in linear viscoelasticity. For a non-aging medium, Eq. (4.105) may be written as follows:

$$\frac{1}{T_1^2} \frac{dQ_0}{dt}(t) \leq \frac{d^3 Q_0}{dt^3}(t) \leq \frac{1}{T_2^2} \frac{dQ_0}{dt}(t). \tag{4.106}$$

Exercise 4.19. By integrating Eq. (4.106) from t to infinity with the use of Eqs. (4.59) and (4.60), deduce Eq. (4.98) from Eq. (4.106). $\square$

4.2.3. Creep measures of aging viscoelastic media

In this subsection we formulate some properties of regular creep measures for aging viscoelastic media:

$$C(t,\tau) > 0, \qquad (0 \leq \tau < t), \tag{4.107}$$

$$C(t,t) = 0, \qquad (t \geq 0), \tag{4.108}$$

$$\frac{\partial C}{\partial t}(t,\tau) \geq 0, \qquad (0 \leq \tau \leq t), \tag{4.109}$$

$$\lim_{t \to \infty} C(t,\tau) = \phi(\tau), \qquad (\tau \geq 0), \tag{4.110}$$

$$\lim_{t \to \infty} \frac{\partial C}{\partial t}(t,\tau) = 0, \qquad (\tau \geq 0), \tag{4.111}$$

$$\frac{\partial C}{\partial \tau}(t,\tau) < 0, \qquad (0 \leq \tau < t), \tag{4.112}$$

$$\frac{\partial C}{\partial \tau}(t,\tau_1) < \frac{\partial C}{\partial \tau}(t,\tau_2), \qquad (0 \leq \tau_1 < \tau_2 \leq t). \tag{4.113}$$

To discuss the mechanical meaning of these inequalities, we consider the piecewise constant loading program

$$\sigma(t) = \begin{cases} 0 & 0 \leq t < \tau, \\ \sigma_0 & t \geq \tau, \end{cases} \tag{4.114}$$

where σ_0 and τ are positive constants. Substitution of expression (4.114) into Eq. (3.5) yields

$$\epsilon(t) = \begin{cases} 0 & 0 \leq t < \tau, \\ \sigma_0[E^{-1}(\tau) + C(t,\tau)] & t \geq \tau. \end{cases} \qquad (4.115)$$

Inequality (4.107) determines the creep phenomenon: for a fixed stress $\sigma = \sigma_0$, the strain ϵ in a viscoelastic specimen exceeds the strain in an appropriate elastic specimen. Eq. (4.108) means that the material creep vanishes at the instant when external load is applied to a specimen. It follows from Eq. (4.109) that the creep deformation increases in time. According to Eq. (4.110), the strain tends to some finite limiting value which depends on the instant τ when the specimen was loaded. Eq. (4.111) means that the rate of creep decreases in time and tends to zero. Eq. (4.112) together with the corresponding property of Young's modulus implies that the derivative of the strain with respect to the instant of loading τ is negative, i.e. the strain increases with the growth of time between the moment of loading τ and the current instant t. Finally, Eq. (4.113) means that the rate of creep decreases with the growth of the interval between the instant of loading τ and the current instant t.

Chapter 7

Boundary problems in finite viscoelasticity

This Chapter is concerned with some boundary problems in finite and infinitesimal viscoelasticity. In Section 1, we consider the Rayleigh problem for a linear non-aging viscoelastic medium and demonstrate that the smoothness of the relaxation kernel affects significantly the propagation of singularities. Section 2 deals with acceleration waves in an aging viscoelastic medium, i.e. solutions of the governing equations which are continuous and have continuous first derivatives, while the second derivatives may have finite jumps of discontinuity on a wave front. It is shown that for shear motions acceleration waves of any amplitude decay in time, whereas for elongation motions acceleration waves with small amplitudes decay exponentially in time, while waves with relatively large amplitudes grow and tend to infinity on finite intervals of time. Section 3 is concerned with radial oscillations of an incompressible viscoelastic thick-walled spherical shell. We demonstrate the effect of initial conditions on the period of natural oscillations for an elastic shell with finite strains, and establish a correspondence between small forced oscillations of viscoelastic shells made of materials with finite and infinite memories. In Section 4, quasi-static torsion of a circular cylinder is studied. We derive relationships which connect torque and compressive force on the one hand, and the twist angle on the other, and validate the obtained formulas by using experimental data for butyl rubber. The obtained results confirm the constitutive model of adaptive links based on the separation principle. Section 5 deals with homogeneous deformations of an incompressible viscoelastic medium loaded on the interval $(-\infty, 0]$. At instant $t = 0$, external loads are reduced, and the problem consists in the qualitative description of an elastic recoil. The set of homogeneous deformations is divided into classes (elongation motions, shear-free motions, etc.), and we analyze whether the recoil belongs to the same class as the motion under loading. It is shown that for some deformation histories, the body recovers to a state in which it has never been under loading. Section 6 is concerned with continuous growth of a viscoelastic medium with finite strains. The growth means a monotonic increase of the body's mass due to the material influx from the environment. This process is modeled as successive accretion of thin layers on a surface of the growing body. It is shown that final stresses in a growing solid depend on the loading history, the rate of the mass influx, and rheological properties of the medium.

1. The Rayleigh Problem

In this Section we consider *the Rayleigh problem* for a linear non-aging viscoelastic medium and demonstrate that the smoothness of relaxation kernels affects significantly the propagation of *singularities*. Namely, it is shown for non-crosslinked polymers with positive Newtonian viscosity that any singularity is smoothed and the speed of its propagation is infinite. If Newtonian viscosity vanishes, then two cases should be distinguished. For a smooth relaxation kernel $R(t)$, a singularity propagates with a finite speed and its amplitude decays exponentially. For a weakly singular relaxation kernel $R(t)$ with a logarithmic singularity at $t = 0$, any initial singularity is smoothed, while the corresponding solution may only be continuous, but not infinitely differentiable.

1.1. Formulation of the Problem

Let us consider one-dimensional motion of a linear non-aging viscoelastic medium. The medium is in its natural state and occupies half-space $x \geq 0$. Here $\{x^i\}$ are Cartesian coordinates with unit vectors $\bar{e}_i$, and $x = x^1$. At instant $t = 0$, the boundary surface $x = 0$ begins to move in direction x. The displacement of the boundary is assumed to be time-independent, and equal to u_0. Body forces and surface tractions vanish.

The displacement vector $\bar{u} = u^i \bar{e}_i$ has only one non-zero component $u^1 = u(t, x)$. Calculation of the tensor of infinitesimal strains $\hat{e} = \epsilon^{ij} \bar{e}_i \bar{e}_j$ implies that it has only one non-zero component

$$\epsilon^{11} = \frac{\partial u}{\partial x}(t, x).$$

It follows from this equality that

$$\epsilon = \frac{\partial u}{\partial x}(t, x), \qquad e_{11} = \frac{2}{3}\frac{\partial u}{\partial x}(t, x), \tag{1.1}$$

where ϵ is the first invariant of the tensor $\hat{e}$ and e_{11} is the only non-zero component of the deviator $\hat{e}$ of the strain tensor $\hat{e}$.

327

The medium obeys the constitutive equation of a non-aging viscoelastic material (6.3.62)

$$\sigma(t) = 3K\epsilon(t), \qquad \hat{s}(t) = 2G[\hat{e}(t) - \int_0^t R(t-s)\hat{e}(s)ds] + 2\eta\frac{d\hat{e}}{dt}(t), \qquad (1.2)$$

where K and G are volume and shear moduli of elasticity, η is Newtonian viscosity, $R(t) = -\dot{Q}_0(t)$ is a relaxation kernel, $Q_0(t)$ is a relaxation measure, and the superscript dot denotes differentiation with respect to time. We assume that $R(t)$ is absolutely integrable, and $Q_0(t)$ satisfies inequalities (6.4.58) – (6.4.60).

Exercise 1.1. By substituting expressions (1.1) into Eq. (1.2), find the following non-zero components of the stress tensor $\hat{\sigma} = \sigma^{ij}\bar{e}_i\bar{e}_j$:

$$\sigma^{11} = (K + \frac{4}{3}G)\frac{\partial u}{\partial x}(t,x) - \frac{4}{3}G\int_0^t R(t-s)\frac{\partial u}{\partial x}(s,x)ds + \frac{4}{3}\eta\frac{\partial^2 u}{\partial t\partial x}(t,x),$$

$$\sigma^{22} = \sigma^{33} = K\frac{\partial u}{\partial x}(t,x). \qquad \Box \quad (1.3)$$

In the absence of body forces, the motion equation is written as

$$\rho\frac{\partial^2\bar{u}}{\partial t^2} = \bar{\nabla}\cdot\hat{\sigma},$$

where ρ is material density, $\bar{\nabla}$ denotes the Hamilton operator in the initial configuration, and the dot stands for the inner product. Substituting expressions (1.3) into this equality we obtain the following *integro-differential equation* for the function $u(t,x)$:

$$\rho\frac{\partial^2 u}{\partial t^2}(t,x) = (K + \frac{4}{3}G)\frac{\partial^2 u}{\partial x^2}(t,x) + \frac{4\eta}{3}\frac{\partial^3 u}{\partial t\partial x^2}(t,x) - \frac{4G}{3}\int_0^t R(t-s)\frac{\partial^2 u}{\partial x^2}(s,x)ds.$$
$$(1.4)$$

Initial and boundary conditions for Eq. (1.4) are written as

$$u(0,x) = 0, \qquad \frac{\partial u}{\partial t}(0,x) = 0 \qquad (x > 0) \qquad (1.5)$$

$$u(t,0) = u_0 \qquad (t > 0). \qquad (1.6)$$

Our objective is to analyze the behavior of solutions of Eq. (1.4) satisfying conditions (1.5) and (1.6) under various assumptions regarding the relaxation kernel $R(t)$.

1.2. The Elastic Problem

First, we recall the solution for an elastic medium with $\eta = 0$ and $R(t) = 0$.

328

Exercise 1.2. Check that the solution of Eq. (1.4)

$$\rho \frac{\partial^2 u}{\partial t^2}(t, x) = (K + \frac{4}{3}G)\frac{\partial^2 u}{\partial x^2}(t, x)$$

with initial and boundary conditions (1.5) and (1.6) has the form

$$u = u_0 \mathcal{H}(\sqrt{\frac{3K + 4G}{3\rho}}t - x),\qquad(1.7)$$

where $\mathcal{H}(t)$ is the Heaviside function (6.1.5). □

According to Eq. (1.7), the solution has a discontinuity, which propagates with a fixed speed

$$c = \sqrt{\frac{3K + 4G}{3\rho}}.\qquad(1.8)$$

The amplitude of the discontinuity $A = u_0$ remains uniform in time and equal to the initial jump of discontinuity.

1.3. The Viscous Problem

We now consider the solution for a viscous liquid with $K = 0$, $G = 0$, and $R(t) = 0$. Eq. (1.4) implies that

$$\rho \frac{\partial^2 u}{\partial t^2}(t, x) = \frac{4\eta}{3}\frac{\partial^3 u}{\partial t \partial x^2}(t, x).$$

Integration of this equation with initial condition (1.5) yields

$$\rho \frac{\partial u}{\partial t}(t, x) = \frac{4\eta}{3}\frac{\partial^2 u}{\partial x^2}(t, x).\qquad(1.9)$$

We seek a solution of Eq. (1.9) as a function $u = U(\xi)$ of one variable $\xi = x/\sqrt{t}$. Substitution of this expression into Eq. (1.9) implies that

$$\frac{d^2 U}{d\xi^2} + \frac{3\rho}{8\eta}\xi \frac{dU}{d\xi} = 0.\qquad(1.10)$$

Introducing the new unknown function $v(\xi) = dU(\xi)/d\xi$, we write Eq. (1.10) as follows:

$$\frac{dv}{v} = -\frac{3\rho}{8\eta}\xi d\xi.$$

Integration of this equation implies that

$$\frac{dU}{d\xi} = C \exp(-\frac{3\rho}{16\eta}\xi^2),\qquad(1.11)$$

329

where C is a constant. Integrating Eq. (1.11) once more, we obtain

$$U(\xi) = C \int_\xi^\infty \exp(-\frac{3\rho}{16\eta}z^2)dz + C_1, \qquad (1.12)$$

where C_1 is a constant.

To determine constants C and C_1 we employ conditions (1.5) and (1.6). For a function of the argument ξ, these equalities can be presented as

$$U(0) = u_0, \qquad U(\infty) = 0. \qquad (1.13)$$

Exercise 1.3. By substituting expression (1.12) into Eq. (1.13) and using the well-known formula

$$\int_0^\infty \exp(-\frac{z^2}{2})dx = \sqrt{\frac{\pi}{2}},$$

show that

$$C = u_0\sqrt{\frac{3\rho}{4\pi\eta}}, \qquad C_1 = 0. \quad \square \qquad (1.14)$$

Finally, substitution of expressions (1.14) into Eq. (1.12) yields

$$u(t,x) = u_0\sqrt{\frac{3\rho}{4\pi\eta}} \int_{x/\sqrt{t}}^\infty \exp(-\frac{3\rho}{16\eta}z^2)dz. \qquad (1.15)$$

For any $t > 0$, solution (1.15) is non-zero and analytic in the domain $x \geq 0$. This means that the speed of propagation for initial perturbations is infinite, and any initial discontinuity is smoothed out for any $t > 0$.

1.4. The Viscoelastic Problem with a Positive Newtonian Viscosity

Eqs. (1.7) and (1.15) present two different types of solutions to evolutionary problems, typical of *hyperbolic* and *parabolic* equations, respectively. Our objective now is to analyze the behavior of solutions in the general case, when both elastic and viscous effects are taken into account. For this purpose, we use the Laplace method.

In this subsection we confine ourselves to the case $\eta > 0$ which is typical of non-crosslinked polymers. The case of crosslinked viscoelastic materials with $\eta = 0$ will be the subject of the next subsections.

Our study is based on the following well-known assertion: if the inverse Laplace transform of a function is absolutely integrable along a vertical line in the right half-plane, then this function is continuous.

Calculation of the Laplace transform of Eq. (1.4) implies that

$$3\rho p^2 \tilde{u}(p,x) = [3K + 4G + 4p\eta - 4G\tilde{R}(p)]\frac{\partial^2 \tilde{u}}{\partial x^2}(p,x), \qquad (1.16)$$

where $\tilde{R}(p)$ is the Laplace transform of the relaxation kernel. The function $\tilde{R}(p)$ is determined in the right half-plane, since $R(t)$ is absolutely integrable.

Exercise 1.4. By using Proposition 6.3.1, derive formula (1.16). □

Calculating the Laplace transform of the boundary condition (1.6), we find

$$\tilde{u}(p,0) = \frac{u_0}{p}. \tag{1.17}$$

To derive another boundary condition for Eq. (1.16) we assume additionally that for any $p = \alpha + i\omega$ with $\alpha > 0$

$$\tilde{u}(p,\infty) = 0. \tag{1.18}$$

An arbitrary solution of Eq. (1.16) may be written as follows:

$$\tilde{u}(p,x) = A_1(p)\exp(\sqrt{a(p)}x) + A_2(p)\exp(-\sqrt{a(p)}x), \tag{1.19}$$

where

$$a(p) = \frac{3\rho p^2}{3K + 4G + 4p\eta - 4G\tilde{R}(p)}. \tag{1.20}$$

Substitution of expression (1.19) into Eqs. (1.17) and (1.18) implies that

$$\tilde{u}(p,x) = \frac{u_0}{p}\exp[-\sqrt{a(p)}x]. \tag{1.21}$$

Exercise 1.5. Check that expression (1.21) satisfies Eq. (1.16) with boundary conditions (1.17) and (1.18). □

We rewrite Eq. (1.20) as follows:

$$a(p) = \frac{3\rho p}{4\eta[1 + \phi(p)]}, \tag{1.22}$$

where

$$\phi(p) = \frac{1}{4\eta p}[3K + 4G - 4G\tilde{R}(p)]. \tag{1.23}$$

Let us now show that for any $\alpha > 0$

$$\lim_{\omega \to \pm\infty} \phi(p) = 0. \tag{1.24}$$

For this purpose, we calculate $\tilde{R}(0)$. It follows from Eqs. (6.3.19) and (6.4.58) that

$$\tilde{R}(0) = \int_0^\infty R(t)dt = -\int_0^\infty \dot{Q}_0(t)dt = -Q_0(\infty) + Q_0(0) = -Q_0(\infty). \tag{1.25}$$

331

Substitution of expression (1.25) into Eq. (1.23) implies that

$$\phi(p) = \frac{1}{4\eta p}\{3K + 4G[1 + Q_0(\infty)] + 4G[\tilde{R}(0) - \tilde{R}(p)]\}. \tag{1.26}$$

Function $H_0(t) = Q_0(t) - Q_0(\infty)$ is continuous, bounded, and $H_0(\infty) = 0$. Therefore, its Laplace transform $\tilde{H}_0(p)$ is an analytical function in the right half-plane. Integrating by parts, we obtain

$$\begin{aligned}
\tilde{H}_0(p) &= \int_0^\infty H_0(t)\exp(-pt)dt \\
&= \frac{1}{p}[-H_0(t)\exp(-pt)|_0^\infty + \int_0^\infty \dot{H}_0(t)\exp(-pt)dt] \\
&= \frac{1}{p}[-Q_0(\infty) + \int_0^\infty \dot{Q}_0(t)\exp(-pt)dt] \\
&= \frac{1}{p}[\tilde{R}(0) - \int_0^\infty R(t)\exp(-pt)dt] = \frac{1}{p}[\tilde{R}(0) - \tilde{R}(p)]. \tag{1.27}
\end{aligned}$$

Since function $H_0(t)$ is continuous and bounded, function $H_0(t)\exp(-\alpha t)$ is absolutely integrable for any $\alpha > 0$. Thus, according to *the Riemann–Lebesgue lemma,*

$$\lim_{\omega \to \pm\infty} \int_0^\infty H_0(t)\exp(-\alpha t)\sin(\omega t)dt = 0,$$

$$\lim_{\omega \to \pm\infty} \int_0^\infty H_0(t)\exp(-\alpha t)\cos(\omega t)dt = 0.$$

Therefore, for any $p = \alpha + i\omega$ with $\alpha > 0$

$$\lim_{\omega \to \pm\infty} \tilde{H}_0(p) = 0. \tag{1.28}$$

The desired equality (1.24) follows from Eqs. (1.26), (1.27), and (1.28). □

It follows from Eqs. (1.21), (1.22), and (1.24) that function $\tilde{u}(p, x)$ is analytical in the right half-plane. By applying the inverse Laplace transform, we obtain

$$u(t, x) = \frac{u_0}{2\pi i}\int_{\beta-i\infty}^{\beta+i\infty} \frac{1}{p}\exp[pt - \sqrt{a(p)}x]dp, \tag{1.29}$$

where β is an arbitrary positive quantity. Substitution of expression (1.22) into Eq. (1.29) implies that

$$u(t, x) = \frac{u_0}{2\pi i}\int_{\beta-i\infty}^{\beta+i\infty} \frac{1}{p}\exp[pt - \sqrt{\frac{3\rho p}{4\eta(1 + \phi(p))}}x]dp. \tag{1.30}$$

Let us differentiate formally the function $u(t, x)$ with respect to t and x

$$\frac{\partial^{k+l} u}{\partial t^k \partial x^l}(t, x) = \frac{(-1)^l u_0}{2\pi i} \int_{\beta-i\infty}^{\beta+i\infty} p^{k-1} [\frac{3\rho p}{4\eta(1 + \phi(p))}]^{\frac{1}{2}}$$

$$\times \exp[pt - \sqrt{\frac{3\rho p}{4\eta(1 + \phi(p))}} x] dp. \tag{1.31}$$

It follows from Eq. (1.24) that for any $x > 0$ function

$$(-1)^l p^{k-1} [\frac{3\rho p}{4\eta(1 + \phi(p))}]^{\frac{1}{2}} \exp[-x\sqrt{\frac{3\rho p}{4\eta(1 + \phi(p))}}]$$

tends to zero faster than any negative power of p. Thus, its inverse Laplace transform is a continuous function of t and x. The latter together with Eq. (1.31) means that the function $u(t, x)$ is smooth in the domain $t > 0$ and $x > 0$. Therefore, for $\eta > 0$ any solution to the Rayleigh problem is infinitely differentiable, and *shock waves* in a linear viscoelastic medium are impossible.

1.5. The Viscoelastic Problem with $\eta = 0$ and a Smooth Relaxation Kernel

Let us now consider the Rayleigh problem for a linear viscoelastic medium with $\eta = 0$ and a smooth relaxation kernel $R(t)$. The smoothness means here that the function $R(t)$ is infinitely differentiable, and the series

$$\sum_{n=1}^{\infty} R^{(n)}(0) z^n$$

converges for any $|z| \le \delta$ with a positive δ. Here $R^{(n)}(t)$ stands for the nth derivative of function $R(t)$. One-dimensional motion of a *nonlinear* viscoelastic medium with a smooth relaxation kernel was studied by Dafermos (1986).

By integrating by parts, we obtain

$$\tilde{R}(p) = \int_0^{\infty} R(t) \exp(-pt) dt$$

$$= -\frac{1}{p} R(t) \exp(-pt)|_0^{\infty} + \frac{1}{p} \int_0^{\infty} \dot{R}(t) \exp(-pt) dt$$

$$= \frac{1}{p} R(0) - \frac{1}{p^2} \dot{R}(t) \exp(-pt)|_0^{\infty} + \frac{1}{p^2} \int_0^{\infty} \ddot{R}(t) \exp(-pt) dt$$

$$= \ldots = \sum_{n=1}^{\infty} \frac{R^{(n)}(0)}{p^n}. \tag{1.32}$$

Substitution of expression (1.32) into Eq. (1.20) yields

$$\sqrt{a(p)} = p\sqrt{\frac{3\rho}{3K + 4G}}[1 - \frac{4G}{3K + 4G} \sum_{n=1}^{\infty} \frac{R^{(n)}(0)}{p^n}]^{-\frac{1}{2}}$$

333

$$= p\sqrt{\frac{3\rho}{3K+4G}}[1 + \frac{2G}{3K+4G}\frac{R(0)}{p} + \psi(\frac{1}{p^2})]$$

$$= p\sqrt{\frac{3\rho}{3K+4G}} + \frac{2G}{3K+4G}\sqrt{\frac{3\rho}{3K+4G}}R(0) + \psi_1(\frac{1}{p}), \quad (1.33)$$

where $\psi(z)$ and $\psi_1(z)$ are analytical functions with $\psi(0) = 0$ and $\psi_1(0) = 0$.
It follows from Eqs. (1.8), (1.29), and (1.33) that

$$u(t,x) = \frac{u_0}{2\pi i}\exp[-\frac{2GR(0)}{3K+4G}\frac{x}{c}]\int_{\beta-i\infty}^{\beta+i\infty}\frac{1}{p}\exp[p(t-\frac{x}{c})]\exp[-\psi_1(\frac{1}{p})x]dp$$

$$= \frac{u_0}{2\pi i}\exp[-\frac{2GR(0)}{3K+4G}\frac{x}{c}]\int_{\beta-i\infty}^{\beta+i\infty}\frac{1}{p}\exp[p(t-\frac{x}{c})]dp + V(t,x), \quad (1.34)$$

where

$$V(t,x) = \frac{u_0}{2\pi i}\exp[-\frac{2GR(0)}{3K+4G}\frac{x}{c}]$$

$$\times \int_{\beta-i\infty}^{\beta+i\infty}\frac{1}{p}\exp(pt)\{\exp[-(\frac{p}{c}+\psi_1(\frac{1}{p}))x] - \exp(-\frac{p}{c}x)\}dp. \quad (1.35)$$

For any positive β, the integral in Eq. (1.35) converges absolutely and uniformly with respect to x in any bounded domain. Therefore, the function $V(t,x)$ is continuous in t and x.

Exercise 1.6. Check that for any t

$$\frac{1}{2\pi i}\int_{\beta-i\infty}^{\beta+i\infty}\frac{1}{p}\exp(pt)dp = \mathcal{H}(t). \quad \square$$

By using this formula, we can calculate the integral in Eq. (1.34) explicitly

$$u(t,x) = u_0\mathcal{H}(t-\frac{x}{c})\exp[-\frac{2GR(0)}{3K+4G}\frac{x}{c}] + V(t,x). \quad (1.36)$$

Eq. (1.36) implies that for $\eta = 0$ and a smooth relaxation kernel, any solution to the Rayleigh problem has a jump of discontinuity across *the wave front*. The amplitude of discontinuity

$$\mathcal{A} = u_0\exp[-\frac{2GR(0)}{3K+4G}\frac{x}{c}]$$

can be transformed as follows:

$$\mathcal{A} = u_0\exp[-\frac{2GR(0)}{3K+4G}t]. \quad (1.37)$$

334

It follows from Eq. (1.37) that unlike an elastic medium, where the amplitude of discontinuity remains unchanged, in a viscoelastic medium the amplitude of discontinuity decreases exponentially in time.

Instead of material parameters K and G it is convenient to employ Young's modulus E and Poisson's ratio ν which are connected with elastic moduli by formulas (6.3.59). Substituting expressions (6.3.59) into Eq. (1.37) we obtain

$$\mathcal{A} = u_0 \exp[-\frac{1 - 2\nu}{3(1 - \nu)} R(0)t].$$

According to this formula, the rate of decrease in the amplitude of discontinuity is determined by the initial value of the relaxation kernel $R(0)$ and Poisson's ratio ν only. For example, for the standard viscoelastic solid (6.4.2) with $R(0) = \chi/T$, the rate of decrease in the amplitude of discontinuity is proportional to the material viscosity χ and is inversely proportional to the characteristic time of relaxation T.

1.6. The Viscoelastic Problem with $\eta = 0$ and a Weakly Singular Relaxation Kernel

In order to demonstrate that a singularity of the relaxation kernel $R(t)$ at $t = 0$ increases the smoothness of the solution $u(t, x)$, we consider kernel (6.4.37) with a logarithmic singularity

$$R(t) = \chi \sum_{n=1}^{\infty} \exp[-\frac{t}{T} \exp(n)],$$

where χ and T are material parameters. The general case is studied by using similar arguments, see Pruss (1993) for details.

Calculation of the Laplace transform of $R(t)$ yields

$$\tilde{R}(p) = \chi T \sum_{n=1}^{\infty} \frac{1}{pT + \exp(n)}. \tag{1.38}$$

To study the behavior of $\tilde{R}(p)$ as $p \to \infty$ we employ the following
Proposition 1.1. Let $f(t)$ be a continuously differentiable function. Then

$$|\int_0^{\infty} f(t)dt - \sum_{n=1}^{\infty} f(n)| \leq \int_0^{\infty} |\dot{f}(t)|dt. \tag{1.39}$$

Proof. Let us fix a positive integer n and calculate $\int_n^{n+1} f(t)dt$. By introducing the new variable $\tau = t - n$ and integrating by parts, we obtain

$$\int_n^{n+1} f(t)dt = \int_0^1 f(\tau + n)d\tau = f(\tau + n)\tau|_0^1 - \int_0^1 \dot{f}(\tau + n)\tau d\tau$$

$$= f(n + 1) - \int_0^1 \dot{f}(\tau + n)\tau d\tau.$$

Hence,

$$\left| \int_n^{n+1} f(t)dt - f(n) \right| \le \left| \int_0^1 \dot{f}(\tau + n)\tau d\tau \right|.$$

The integral in the right-hand side is estimated with the use of the Cauchy inequality

$$\left| \int_n^{n+1} f(t)dt - f(n) \right| \le \int_0^1 |\dot{f}(\tau + n)|\tau d\tau$$

$$\le \int_0^1 |\dot{f}(\tau + n)|d\tau = \int_n^{n+1} |\dot{f}(t)|dt. \tag{1.40}$$

Eq. (1.39) follows from Eq. (1.40) and the Cauchy inequality. $\square$

By applying Proposition 1.1 to the function $\tilde{R}(p)$, we obtain

$$\left| \tilde{R}(p) - \chi T \int_0^\infty \frac{dt}{pT + \exp(t)} \right| \le \chi T \int_0^\infty \frac{\exp(t)dt}{[pT + \exp(t)]^2}. \tag{1.41}$$

To calculate the integrals in Eq. (1.41) we introduce the new variable $z = \exp(t)$. For example, we find

$$\int_0^\infty \frac{dt}{\exp(t) + pT} = \int_1^\infty \frac{dz}{z(z + pT)} = \frac{1}{pT} \int_1^\infty (\frac{1}{z} - \frac{1}{z + pT})dz$$

$$= -\frac{1}{pT} \ln(1 + \frac{pT}{z})|_1^\infty = \frac{1}{pT} \ln(1 + pT). \tag{1.42}$$

Exercise 1.7. By using a similar procedure, show that

$$\int_0^\infty \frac{\exp(t)dt}{[pT + \exp(t)]^2} = \frac{1}{pT + 1}. \qquad \square \tag{1.43}$$

Substitution of expressions (1.42) and (1.43) into Eq. (1.41) implies that

$$\tilde{R}(p) = \frac{\chi}{p} \ln(1 + pT) + \psi_2(p) \tag{1.44}$$

where

$$|\psi_2(p)| \le \frac{\chi T}{pT + 1}.$$

It follows from Eqs. (1.20) and (1.44) that

$$\sqrt{a(p)} = p\sqrt{\frac{3\rho}{3K + 4G}}[1 - \frac{4G\chi}{(3K + 4G)p} \ln(1 + pT) - \frac{4G}{3K + 4G}\psi_2(p)]^{-\frac{1}{2}}$$

$$= p\sqrt{\frac{3\rho}{3K + 4G}}[1 + \frac{2G\chi}{(3K + 4G)p} \ln(1 + pT) + \frac{2G}{3K + 4G}\psi_3(\frac{1}{p})], \tag{1.45}$$

336

where $\psi_3(z)$ is an analytical function with $\psi_3(0) = 0$. Substitution of expression (1.45) into Eq. (1.29) yields

$$u(t, x) = \frac{u_0}{2\pi i} \int_{\beta - i\infty}^{\beta + i\infty} p^{-1 - \frac{2G\chi}{3K+4G}\frac{x}{c}} \exp[p(t - \frac{x}{c})] \psi_4(\frac{1}{p}) dp, \qquad (1.46)$$

where $\psi_4(z)$ is an analytical function.

For any $x > 0$, the integral of the function $p^{-1 - \frac{2G\chi}{3K+4G}\frac{x}{c}}$ converges absolutely. Thus, Eq. (1.46) implies that any solution $u(t, x)$ is continuous in the domain $t > 0$ and $x > 0$. Therefore, the presence of the logarithmic singularity in the relaxation kernel leads to smoothing any discontinuity in the initial data. Moreover, the smoothness of any solution increases with the growth of x (i.e. with the growth of time t). This feature distinguishes the behavior of a viscoelastic medium with a singular kernel from the behavior of purely elastic and purely viscous media.

2. Acceleration Waves

This Section deals with acceleration waves in an aging viscoelastic medium. *An acceleration wave* is a solution of the governing equations which is continuous, has continuous derivatives with respect to spatial coordinates and time, while the second derivatives are assumed to have finite jumps of discontinuity on a smooth moving surface (wave front).

Acceleration waves in non-aging viscoelastic media with finite strains were studied by Coleman et al. (1965) and Coleman & Gurtin (1965) under specific constitutive assumptions. It has been shown that the behavior of discontinuities differs essentially depending on the amplitude of perturbations. Acceleration waves with small amplitudes decay exponentially in time, while waves with relatively large amplitudes grow and tend to infinity in finite intervals of time. Such a behavior may be treated as a development of acceleration waves into more strong singularities, for example, into shock waves, see Renardy et al. (1987) for a discussion of this issue.

We concentrate on acceleration waves in aging media. Two basic solutions are considered: elongation motion (subsection 1) and shear motion (subsection 2). These solutions are chosen owing to their simplicity and importance for applications. A governing equation for a jump of discontinuity on a wave front is derived in subsection 3. It is shown that for shear motions, acceleration waves of any amplitude decay in time, whereas for elongation motions, both regimes of growth and decay of acceleration waves occur. For an aging medium, the

337

front speed depends essentially on the material aging and increases in time.

2.1. Elongation Motion

Let us consider a viscoelastic medium which is in its natural state and occupies the whole space. At instant $t = 0$, an elongation motion begins, which is described by equations (4.5.1)

$$x^1 = X^1 + u(t, X^1), \qquad x^2 = X^2, \qquad x^3 = X^3. \tag{2.1}$$

Here $\{X^i\}$ and $\{x^i\}$ are Cartesian coordinates in the initial and actual configurations with unit vectors $\bar{e}_i$, $u(t, X^1)$ is an unknown function.

The radius-vector of a point with Cartesian coordinates $X = \{X^i\}$ in the actual configuration equals $\bar{r}(t, X) = [X^1 + u(t, X^1)]\bar{e}_1 + X^2\bar{e}_2 + X^3\bar{e}_3$. Tangent vectors of the main and dual bases are calculated by formulas (4.5.3) and (4.5.5). Substituting these expressions into Eq. (2.1.44), we find the relative deformation gradient

$$\bar{\nabla}_s \bar{r}(t) = \bar{g}^i(s)\bar{g}_i(t) = \kappa(t, s)\bar{e}_1\bar{e}_1 + \bar{e}_2\bar{e}_2 + \bar{e}_3\bar{e}_3, \tag{2.2}$$

where

$$\kappa(t, s) = [1 + \frac{\partial u}{\partial X^1}(t)][1 + \frac{\partial u}{\partial X^1}(s)]^{-1}. \tag{2.3}$$

Here and below argument X^1 is omitted for simplicity. Substitution of expression (2.2) into Eq. (2.1.46) implies the following formulas for the relative Finger tensor:

$$\hat{F}_*(t, s) = \kappa^2(t, s)\bar{e}_1\bar{e}_1 + \bar{e}_2\bar{e}_2 + \bar{e}_3\bar{e}_3,$$
$$\hat{F}_*^2(t, s) = \kappa^4(t, s)\bar{e}_1\bar{e}_1 + \bar{e}_2\bar{e}_2 + \bar{e}_3\bar{e}_3. \tag{2.4}$$

Exercise 2.1. Show that the characteristic polynomial for tensor $\hat{F}_*$ is

$$\det(\hat{F}_* - \lambda\hat{I}) = \det \begin{bmatrix} \kappa^2 - \lambda & 0 & 0 \\ 0 & 1 - \lambda & 0 \\ 0 & 0 & 1 - \lambda \end{bmatrix}$$
$$= -\lambda^3 + (2 + \kappa^2)\lambda^2 - (1 + 2\kappa^2)\lambda + \kappa^2,$$

where $\hat{I}$ is the unit tensor. □

Since the principal invariants of tensor $\hat{F}_*$ coincide (up to their signs) with coefficients of the characteristic polynomial, we find

$$I_1(\hat{F}_*) = 2 + \kappa^2, \qquad I_2(\hat{F}_*) = 1 + 2\kappa^2, \qquad I_3(\hat{F}_*) = \kappa^2. \tag{2.5}$$

The material response is assumed to obey the constitutive equation of an aging viscoelastic solid (6.3.82). Substitution of expressions (2.4) and (2.5) into Eq. (6.3.82) yields the following formula for the Cauchy stress tensor $\hat{\sigma}$:

$$\hat{\sigma} = \sigma_{11}\bar{e}_1\bar{e}_1 + \sigma_{22}\bar{e}_2\bar{e}_2 + \sigma_{33}\bar{e}_3\bar{e}_3. \tag{2.6}$$

The non-zero components of the tensor $\hat{\sigma}$ are calculated as follows:

$$\sigma_{11}(t) = \frac{2}{\kappa(t,0)}[\mathcal{X}(t,0)\Psi_1(t,0)\kappa^2(t,0) + \int_0^t \frac{\partial \mathcal{X}}{\partial s}(t,s)\Psi_1(t,s)\kappa^2(t,s)ds],$$

$$\sigma_{22}(t) = \sigma_{33}(t) = \frac{2}{\kappa(t,0)}\{\mathcal{X}(t,0)[\Psi_1(t,0) + (\kappa^2(t,0) - 1)\Psi_2(t,0)]$$

$$+ \int_0^t \frac{\partial \mathcal{X}}{\partial s}(t,s)[\Psi_1(t,s) + (\kappa^2(t,s) - 1)\Psi_2(t,s)]ds\}, \quad (2.7)$$

where

$$\Psi_1 = W_1 + 2W_2 + W_3, \qquad \Psi_2 = W_2 + W_3,$$

$$W_k(t,s) = \frac{\partial W}{\partial I_k}(I_1(\hat{F}(t,s)), I_2(\hat{F}(t,s)), I_3(\hat{F}(t,s))). \quad (2.8)$$

We assume that
(i) strain energy density $W(I_1, I_2, I_3)$ satisfies the restrictions formulated in Section 3.2. In particular, Eq. (3.2.23) is valid;
(ii) function $\mathcal{X}(t,s)$ satisfies conditions (6.4.103) and (6.4.104) and may be presented in the dimensionless form similar to Eq. (6.3.71)

$$\mathcal{X}(t,s) = \beta(s) + Q(t,s), \quad (2.9)$$

where $\beta(s) = 1$ and $Q(t,s) = Q_0(t-s)$ for a non-aging medium.

The displacement vector equals $\bar{u} = u(t)\bar{e}_1$. Differentiation of this expression yields

$$\bar{a} = \frac{\partial^2 u}{\partial t^2}(t)\bar{e}_1, \quad (2.10)$$

where $\bar{a}$ is the acceleration vector. It follows from mass conservation law (2.2.10) and Eq. (2.5) that

$$\rho(t) = \frac{\rho_0}{\kappa(t,0)}, \quad (2.11)$$

where ρ_0 and ρ are mass densities in the initial and actual configurations.
Exercise 2.2. By using Eqs. (1.3.4), (4.5.5), and (2.6), show that

$$\bar{\nabla}_t \cdot \hat{\sigma}(t) = [\frac{\bar{e}_1}{\kappa(t,0)}\frac{\partial}{\partial X^1} + \bar{e}_2\frac{\partial}{\partial X^2} + \bar{e}_3\frac{\partial}{\partial X^3}]\cdot\hat{\sigma}(t) = \frac{1}{\kappa(t,0)}\frac{\partial \sigma_{11}}{\partial X^1}(t)\bar{e}_1. \quad \square \ (2.12)$$

External body and surface forces are assumed to vanish. Substitution of expressions (2.10) – (2.12) into the motion equation $\rho\bar{a}(t) = \bar{\nabla}_t \cdot \hat{\sigma}(t)$ yields

$$\rho_0\frac{\partial^2 u}{\partial t^2}(t) = \frac{\partial}{\partial X^1}\{\frac{2}{\kappa(t,0)}[\mathcal{X}(t,0)\Psi_1(t,0)\kappa^2(t,0)$$

$$+ \int_0^t \frac{\partial \mathcal{X}}{\partial s}(t,s)\Psi_1(t,s)\kappa^2(t,s)ds]\}. \quad (2.13)$$

339

Eq. (2.5) implies that function W depends on κ only:

$$W(I_1, I_2, I_3) = W(2 + \kappa^2, 1 + 2\kappa^2, \kappa^2) = \mathcal{W}(\kappa). \tag{2.14}$$

It follows from Eqs. (2.8) and (2.14) that $\mathcal{W}' = 2\kappa\Psi_1$, where the prime denotes the differentiation. Substitution of this expression into Eq. (2.13) yields

$$\rho_0 \frac{\partial^2 u}{\partial t^2}(t) = \frac{\partial}{\partial X^1}[\mathcal{X}(t,0)\mathcal{W}'(\kappa(t,0)) + \int_0^t \frac{\partial \mathcal{X}}{\partial s}(t,s)\mathcal{W}'(\kappa(t,s))\frac{\kappa(t,s)}{\kappa(t,0)}ds]. \tag{2.15}$$

Eq. (2.15) will be studied below in subsection 3, while we now analyze the other type of one-dimensional motions, namely, simple shear.

2.2. Shear Motion

Shear motion of a hyperelastic medium was analyzed in Section 4.5. This motion is described by Eqs. (4.5.22)

$$x^1 = X^1 + u(t, X^2), \qquad x^2 = X^2, \qquad x^3 = X^3, \tag{2.16}$$

where $u(t, X^2)$ is an unknown function. Comparing Eqs. (2.1) and (2.16) we see that elongation and shear motions differ from each other by the spatial argument of the unknown function. For elongation motions, the displacement vector depends on the Cartesian coordinate along which the displacement occurs, whereas for shear motions, this vector depends on a coordinate orthogonal to the direction of displacement.

In the actual configuration, the radius-vector of a point with Cartesian coordinates $\{X^i\}$ equals $\bar{r} = [X^1 + u(t, X^2)]\bar{e}_1 + X^2\bar{e}_2 + X^3\bar{e}_3$. Tangent vectors are determined by Eqs. (4.5.24) and (4.5.26). Substitution of these expressions into Eq. (2.1.44) implies that

$$\bar{\nabla}_s \bar{r}(t) = \bar{e}_1\bar{e}_1 + \bar{e}_2\bar{e}_2 + \bar{e}_3\bar{e}_3 + \kappa(t,s)\bar{e}_2\bar{e}_1, \tag{2.17}$$

where

$$\kappa(t,s) = \frac{\partial u}{\partial X^2}(t) - \frac{\partial u}{\partial X^2}(s). \tag{2.18}$$

Here and below argument X^2 is omitted for simplicity. Substitution of expression (2.17) into Eq. (2.1.46) yields

$$\begin{aligned}
\hat{F}_*(t,s) &= [1 + \kappa^2(t,s)]\bar{e}_1\bar{e}_1 + \bar{e}_2\bar{e}_2 + \bar{e}_3\bar{e}_3 + \kappa(t,s)(\bar{e}_1\bar{e}_2 + \bar{e}_2\bar{e}_1), \\
\hat{F}_*^2(t,s) &= [(1 + \kappa^2(t,s))^2 + \kappa^2(t,s)]\bar{e}_1\bar{e}_1 + [1 + \kappa^2(t,s)]\bar{e}_2\bar{e}_2 \\
&\quad + \bar{e}_3\bar{e}_3 + \kappa(t,s)[2 + \kappa^2(t,s)](\bar{e}_1\bar{e}_2 + \bar{e}_2\bar{e}_1).
\end{aligned} \tag{2.19}$$

340

Exercise 2.3. Derive the characteristic polynomial for the relative Finger tensor

$$\det(\hat{F}_* - \lambda\hat{I}) = \det\begin{bmatrix} 1+\kappa^2-\lambda & \kappa & 0 \\ \kappa & 1-\lambda & 0 \\ 0 & 0 & 1-\lambda \end{bmatrix}$$

$$= -\lambda^3 + (3+\kappa^2)\lambda^2 - (3+\kappa^2)\lambda + 1. \quad \square$$

Since the principal invariants coincide (up to their signs) with coefficients of the characteristic polynomial, we find

$$I_1(\hat{F}_*) = 3 + \kappa^2, \qquad I_2(\hat{F}_*) = 3 + \kappa^2, \qquad I_3(\hat{F}_*) = 1. \tag{2.20}$$

Eqs. (2.20) imply that shear motion is isochoric, i.e. it preserves the volume element.

Substitution of expressions (2.19) and (2.20) into the constitutive equation (6.3.82) yields

$$\hat{\sigma}(t) = \sigma_{11}\bar{e}_1\bar{e}_1 + \sigma_{22}\bar{e}_2\bar{e}_2 + \sigma_{33}\bar{e}_3\bar{e}_3 + \sigma_{12}(\bar{e}_1\bar{e}_2 + \bar{e}_2\bar{e}_1) \tag{2.21}$$

with

$$\sigma_{11}(t) = 2\{\mathcal{X}(t,0)[\Psi_1(t,0) + \kappa^2(t,0)\Psi_0(t,0)] \\ + \int_0^t \frac{\partial\mathcal{X}}{\partial s}(t,s)[\Psi_1(t,s) + \Psi_0(t,s)\kappa^2(t,s)]ds\},$$

$$\sigma_{22}(t) = 2[\mathcal{X}(t,0)\Psi_1(t,0) + \int_0^t \frac{\partial\mathcal{X}}{\partial s}(t,s)\Psi_1(t,s)ds],$$

$$\sigma_{33}(t) = 2\{\mathcal{X}(t,0)[\Psi_1(t,0) + W_2(t,0)\kappa^2(t,0)] \\ + \int_0^t \frac{\partial\mathcal{X}}{\partial s}(t,s)[\Psi_1(t,s) + W_2(t,s)\kappa^2(t,s)]ds\},$$

$$\sigma_{12}(t) = 2[\mathcal{X}(t,0)\Psi_0(t,0)\kappa(t,0) + \int_0^t \frac{\partial\mathcal{X}}{\partial s}(t,s)\Psi_0(t,s)\kappa(t,s)ds]. \tag{2.22}$$

Here $\Psi_1(t,s)$ has the form (2.8) and

$$\Psi_0(t,s) = W_1(t,s) + W_2(t,s). \tag{2.23}$$

Differentiation of the displacement vector $\bar{u} = u(t)\bar{e}_1$ with respect to time implies Eq. (2.10). Since shear motion is isochoric, $\rho(t) = \rho_0$.

Exercise 2.4. By using Eqs. (1.3.4), (4.5.26), and (2.21), show that

$$\bar{\nabla}_t \cdot \hat{\sigma}(t) = [(\bar{e}_1 - \frac{\partial u}{\partial X^2}(t)\bar{e}_2)\frac{\partial}{\partial X^1} + \bar{e}_2\frac{\partial}{\partial X^2} + \bar{e}_3\frac{\partial}{\partial X^3}] \cdot \hat{\sigma}(t)$$

$$= \frac{\partial\sigma_{12}}{\partial X^2}(t)\bar{e}_1 + \frac{\partial\sigma_{22}}{\partial X^2}(t)\bar{e}_2. \quad \square$$

Substitution of these expressions into the motion equation

$$\rho(t)\bar{a}(t) = \bar{\nabla}_t \cdot \hat{\sigma}(t) + \rho(t)\bar{B}(t)$$

yields

$$\rho_0 \frac{\partial^2 u}{\partial t^2}(t) = \frac{\partial \sigma_{12}}{\partial X^2} + \rho_0 B_1(t), \qquad 0 = \frac{\partial \sigma_{22}}{\partial X^2} + \rho_0 B_2(t), \qquad 0 = \rho_0 B_3(t), \quad (2.24)$$

where $\bar{B} = \bar{B}(t, X^i)$ is a body force: $\bar{B} = B_1\bar{e}_1 + B_2\bar{e}_2 + B_3\bar{e}_3$.

It follows from the third equation (2.24) that $B_3 = 0$ for shear motions, whereas B_2 should be non-zero in order to satisfy the second equation (2.24). Let $B_2 = B_2(t, X^2)$ be chosen in such a way as to turn the second equation (2.24) into an identity, and to make component B_1 vanish. Then the only governing equation is

$$\rho_0 \frac{\partial^2 u}{\partial t^2}(t) = 2\frac{\partial}{\partial X^2}[\mathcal{X}(t,0)\Psi_0(t,0)\kappa(t,0) + \int_0^t \frac{\partial \mathcal{X}}{\partial s}(t,s)\Psi_0(t,s)\kappa(t,s)ds]. \quad (2.25)$$

It follows from Eq. (2.20) that strain energy density W depends on κ only:

$$W(I_1, I_2, I_3) = W(3 + \kappa^2, 3 + \kappa^2, 1) = \mathcal{W}(\kappa). \quad (2.26)$$

Eqs. (2.23) and (2.26) imply that $\mathcal{W}'(\kappa) = 2\kappa\Psi_0$. Substitution of this expression into Eq. (2.25) yields

$$\rho_0 \frac{\partial^2 u}{\partial t^2}(t) = \frac{\partial}{\partial X^2}[\mathcal{X}(t,0)\mathcal{W}'(\kappa(t,0)) + \int_0^t \frac{\partial \mathcal{X}}{\partial s}(t,s)\mathcal{W}'(\kappa(t,s))ds]. \quad (2.27)$$

Eqs. (2.15) and (2.27) may be written as follows:

$$\rho_0 \frac{\partial^2 u}{\partial t^2}(t,\xi) = \frac{\partial}{\partial \xi}[X(t,0)H_0(\frac{\partial u}{\partial \xi}(t)) + \int_0^t \frac{\partial X}{\partial s}(t,s)H_1(\frac{\partial u}{\partial \xi}(t), \frac{\partial u}{\partial \xi}(s))ds]. \quad (2.28)$$

Here

$$\xi = X^1, \qquad H_0(v_1) = \mathcal{W}'(1 + v_1), \qquad H_1(v_1, v_2) = \frac{1}{1 + v_2}\mathcal{W}'(\frac{1 + v_1}{1 + v_2}) \quad (2.29)$$

for elongation, and

$$\xi = X^2, \qquad H_0(v_1) = \mathcal{W}'(v_1), \qquad H_1(v_1, v_2) = \mathcal{W}'(v_1 - v_2) \quad (2.30)$$

for shear. In formulas (2.29) and (2.30) we utilize the notation:

$$v_1 = \frac{\partial u}{\partial \xi}(t), \qquad v_2 = \frac{\partial u}{\partial \xi}(s).$$

Our objective now is to study particular solutions of Eq. (2.28) in the form of acceleration waves. At first sight these solutions, which have smooth first derivatives and demonstrate jumps of discontinuity in the second derivatives (see their precise definition below), seem rather exotic. Nevertheless, acceleration waves play an important role in nonlinear viscoelasticity, first, because they provide explicit particular solutions of governing equations, and second, since they present examples of the classical solutions with the minimal smoothness: if we introduce weaker assumptions, and consider solutions with jumps of discontinuity in the first derivatives as well, we should seek the so-called *generalized solutions* which satisfy governing equations in the integral sense. The latter solutions are of special interest in the theory of shock waves, but they are beyond the scope of the textbook.

2.3. One-dimensional Acceleration Waves

Definition. A solution $u(t, \xi)$ of Eq. (2.28) which is continuous, has continuous derivatives for any $t \geq 0$ and $x > 0$, and which has finite jumps of discontinuity for the second derivatives on an unknown curve $t = \Gamma(\xi)$, is called *acceleration wave*. The curve $t = \Gamma(\xi)$ is called its *wave front*.

We do not intend to discuss the existence and uniqueness of acceleration waves, referring to Coleman et al. (1965) and Renardy et al. (1987). In this subsection we assume that an acceleration wave exists for any $t \geq 0$. Our objective is to derive explicit expressions for jumps of discontinuity of the second derivatives of function $u(t, \xi)$ on the wave front.

We confine ourselves to the wave propagation into a medium at rest, and assume that $u(t, \xi) = 0$ for $t > \Gamma(x)$. Denote by $\bar{\tau} = \{1, \Gamma'\}$ a tangent vector and by $\bar{n} = \{\Gamma', -1\}$ a normal vector to the wave front, see Fig. 2.1. For any function $f(t, x)$ we use the notation

$$[f] = f^+ - f^-,$$

where f^+ and f^- denote the limiting values of the function f on the wave front as we approach curve Γ being in the domain of motion $(+)$ and in the domain of rest $(-)$, respectively.

Since solution $u(t, \xi)$ and its first derivatives are continuous on the wave front, we find

$$[u] = 0, \qquad [\frac{\partial u}{\partial t}] = 0, \qquad [\frac{\partial u}{\partial \xi}] = 0. \tag{2.31}$$

The derivatives of functions $[\partial u/\partial t]$ and $[\partial u/\partial \xi]$ with respect to the tangent vector $\bar{\tau}$ vanish. Since

$$\frac{\partial}{\partial \tau} = \frac{\partial}{\partial \xi} + \Gamma'(\xi) \frac{\partial}{\partial t},$$

343

this means that

$$[\frac{\partial^2 u}{\partial t \partial \xi}] + \Gamma'(\xi)[\frac{\partial^2 u}{\partial t^2}] = 0, \qquad [\frac{\partial^2 u}{\partial \xi^2}] + \Gamma'(\xi)[\frac{\partial^2 u}{\partial t \partial \xi}] = 0. \qquad (2.32)$$

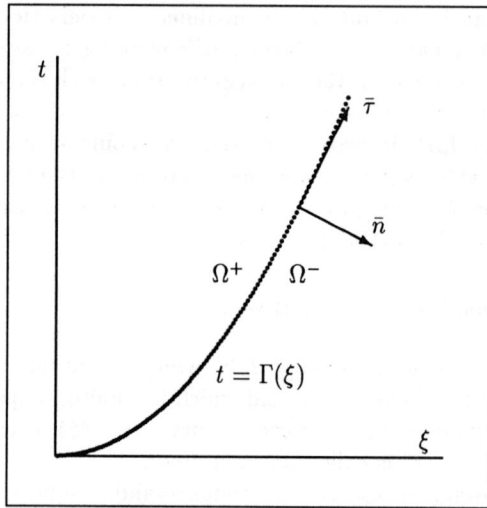

Fig. 2.1. A wave front for a one-dimensional motion.

Let us replace the derivative Γ' in Eq. (2.32) by the wave speed. For this purpose we fix a point ξ_0 on the wave front at instant $t = 0$ and denote by $\xi(t, \xi_0)$ its position at an arbitrary moment $t > 0$. Since the point belongs to the wave front, we have $t = \Gamma(\xi(t, \xi_0))$. Differentiation of this equality with respect to time yields

$$1 = \Gamma'(\xi)\frac{\partial \xi}{\partial t}. \qquad (2.33)$$

The wave speed c is defined as the speed of a material point which moves together with the wave front

$$c = \frac{\partial \xi}{\partial t}(t, \xi_0). \qquad (2.34)$$

Substitution of expression (2.34) into Eq. (2.33) implies that

$$c = \frac{1}{\Gamma'(\xi)}. \qquad (2.35)$$

Eq. (2.32) together with Eq. (2.35) yields

$$[\frac{\partial^2 u}{\partial \xi^2}] = \frac{1}{c^2}[\frac{\partial^2 u}{\partial t^2}], \qquad [\frac{\partial^2 u}{\partial t \partial \xi}] = -\frac{1}{c}[\frac{\partial^2 u}{\partial t^2}]. \qquad (2.36)$$

344

Eq. (2.28) can be written as

$$\rho_0 \frac{\partial^2 u}{\partial t^2}(t) = \mathcal{X}(t,0)\frac{dH_0}{dv_1}(\frac{\partial u}{\partial \xi}(t))\frac{\partial^2 u}{\partial \xi^2}(t)$$

$$+ \int_0^t \frac{\partial \mathcal{X}}{\partial s}(t,s)\{\frac{\partial H_1}{\partial v_1}(\frac{\partial u}{\partial \xi}(t),\frac{\partial u}{\partial \xi}(s))\frac{\partial^2 u}{\partial \xi^2}(t)$$

$$+ \frac{\partial H_1}{\partial v_2}(\frac{\partial u}{\partial \xi}(t),\frac{\partial u}{\partial \xi}(s))\frac{\partial^2 u}{\partial \xi^2}(s)\}ds, \tag{2.37}$$

where functions H_i are determined by Eqs. (2.29) and (2.30). According to Eq. (2.31), we should set

$$\frac{\partial u}{\partial t} = 0, \qquad \frac{\partial u}{\partial \xi} = 0 \tag{2.38}$$

on the wave front $t = \Gamma(\xi)$, as well as set

$$\frac{\partial u}{\partial \xi}(s,\xi) = 0, \qquad \frac{\partial^2 u}{\partial \xi^2}(s,\xi) = 0 \tag{2.39}$$

for $s < t = \Gamma(\xi)$. It follows from Eqs. (2.38) and (2.39) that on the wave front Eq. (2.37) takes the form

$$\rho_0(\frac{\partial^2 u}{\partial t^2})^+ = \{\mathcal{X}(t,0)\frac{dH_0}{dv_1}(0) + \int_0^t \frac{\partial \mathcal{X}}{\partial s}(t,s)\frac{\partial H_1}{\partial v_1}(0,0)ds\}(\frac{\partial^2 u}{\partial \xi^2})^+.$$

Calculating the integral, we obtain

$$\rho_0(\frac{\partial^2 u}{\partial t^2})^+ = \{\mathcal{X}(t,0)\frac{dH_0}{dv_1}(0) + (\mathcal{X}(t,t) - \mathcal{X}(t,0))\frac{\partial H_1}{\partial v_1}(0,0)\}(\frac{\partial^2 u}{\partial \xi^2})^+.$$

Bearing in mind that for any non-negative integers m and n

$$(\frac{\partial^{m+n} u}{\partial t^m \partial \xi^n})^- = 0, \tag{2.40}$$

we find

$$[\frac{\partial^2 u}{\partial t^2}] = \frac{1}{\rho_0}\{\mathcal{X}(t,0)\frac{dH_0}{dv_1}(0) + (\mathcal{X}(t,t) - \mathcal{X}(t,0))\frac{\partial H_1}{\partial v_1}(0,0)\}[\frac{\partial^2 u}{\partial \xi^2}]. \tag{2.41}$$

Comparison of Eqs. (2.36) and (2.41) implies that

$$c^2(t) = \frac{1}{\rho_0}\{\mathcal{X}(t,0)\frac{dH_0}{dv}(0) + (\mathcal{X}(t,t) - \mathcal{X}(t,0))\frac{\partial H_1}{\partial v_1}(0,0)\}. \tag{2.42}$$

According to formula (2.42), the wave speed c is a function of time. For non-aging media, c turns into a constant.

345

Let us introduce the derivative of a smooth function $f(t, \xi)$ along the wave front. For this purpose, we consider a point $\xi = \xi(t, \xi_0)$ which moves together with the wave front, and calculate the total derivative with respect to time of the function $f(t, \xi(t, \xi_0))$

$$\frac{df}{dt}(t, \xi(t, \xi_0)) = \frac{\partial f}{\partial t}(t, \xi(t, \xi_0)) + \frac{\partial f}{\partial \xi}(t, \xi(t, \xi_0))\frac{\partial \xi}{\partial t}(t, \xi_0).$$

Substituting expression (2.34) into this formula we obtain

$$\frac{df}{dt}(t, \xi) = \frac{\partial f}{\partial t}(t, \xi) + c(t)\frac{\partial f}{\partial \xi}(t, \xi). \tag{2.43}$$

Let us apply Eq. (2.43) to functions $\partial^2 u/\partial t^2$ and $\partial^2 u/\partial t\partial\xi$

$$\frac{d}{dt}\left[\frac{\partial^2 u}{\partial t^2}\right] = \left[\frac{\partial^3 u}{\partial t^3}\right] + c\left[\frac{\partial^3 u}{\partial t^2\partial\xi}\right],$$
$$\frac{d}{dt}\left[\frac{\partial^2 u}{\partial t\partial\xi}\right] = \left[\frac{\partial^3 u}{\partial t^2\partial\xi}\right] + c\left[\frac{\partial^3 u}{\partial t\partial\xi^2}\right]. \tag{2.44}$$

Excluding the term $[\partial^3 u/\partial t^2\partial\xi]$ from Eqs. (2.44) we find

$$\frac{d}{dt}\left[\frac{\partial^2 u}{\partial t^2}\right] - c\frac{d}{dt}\left[\frac{\partial^2 u}{\partial t\partial\xi}\right] = \left[\frac{\partial^3 u}{\partial t^3}\right] - c^2\left[\frac{\partial^3 u}{\partial t\partial\xi^2}\right].$$

It follows from this equality and the second equation (2.36) that

$$2\frac{d}{dt}\left[\frac{\partial^2 u}{\partial t^2}\right] = \frac{\dot{c}}{c}\left[\frac{\partial^2 u}{\partial t^2}\right] + \left[\frac{\partial^3 u}{\partial t^3}\right] - c^2\left[\frac{\partial^3 u}{\partial t\partial\xi^2}\right]. \tag{2.45}$$

In order to calculate the right-hand side of Eq. (2.45), we differentiate Eq. (2.37) with respect to time. As a result, we arrive at the formula

$$\rho_0\frac{\partial^3 u}{\partial t^3} = \frac{\partial \mathcal{X}}{\partial t}(t, 0)L_0(t) + \mathcal{X}(t, 0)L_1(t) + \frac{\partial \mathcal{X}}{\partial s}(t, t)L_2(t)$$
$$+ \int_0^t \left\{\frac{\partial^2 \mathcal{X}}{\partial t\partial s}(t, s)M_1(t, s) + \frac{\partial \mathcal{X}}{\partial s}(t, s)M_2(t, s)\right\}ds. \tag{2.46}$$

where

$$L_0(t) = \frac{dH_0}{dv_1}\left(\frac{\partial u}{\partial \xi}(t)\right)\frac{\partial^2 u}{\partial \xi^2}(t),$$

$$L_1(t) = \frac{d^2 H_0}{dv_1^2}\left(\frac{\partial u}{\partial \xi}(t)\right)\frac{\partial^2 u}{\partial t\partial\xi}(t)\frac{\partial^2 u}{\partial \xi^2}(t) + \frac{dH_0}{dv_1}\left(\frac{\partial u}{\partial \xi}(t)\right)\frac{\partial^3 u}{\partial t\partial\xi^2}(t),$$

346

$$L_2(t) = \{\frac{\partial H_1}{\partial v_1}(\frac{\partial u}{\partial \xi}(t), \frac{\partial u}{\partial \xi}(t)) + \frac{\partial H_1}{\partial v_2}(\frac{\partial u}{\partial \xi}(t), \frac{\partial u}{\partial \xi}(t))\}\frac{\partial^2 u}{\partial \xi^2}(t),$$

$$M_1(t,s) = \frac{\partial H_1}{\partial v_1}(\frac{\partial u}{\partial \xi}(t), \frac{\partial u}{\partial \xi}(s))\frac{\partial^2 u}{\partial \xi^2}(t) + \frac{\partial H_1}{\partial v_2}(\frac{\partial u}{\partial \xi}(t), \frac{\partial u}{\partial \xi}(s))\frac{\partial^2 u}{\partial \xi^2}(s),$$

$$M_2(t,s) = \frac{\partial^2 H_1}{\partial v_1^2}(\frac{\partial u}{\partial \xi}(t), \frac{\partial u}{\partial \xi}(s))\frac{\partial^2 u}{\partial t \partial \xi}(t)\frac{\partial^2 u}{\partial \xi^2}(t)$$

$$+\frac{\partial H_1}{\partial v_1}(\frac{\partial u}{\partial \xi}(t), \frac{\partial u}{\partial \xi}(s))\frac{\partial^3 u}{\partial t \partial \xi^2}(t) + \frac{\partial^2 H_1}{\partial v_1 \partial v_2}(\frac{\partial u}{\partial \xi}(t), \frac{\partial u}{\partial \xi}(s))\frac{\partial^2 u}{\partial t \partial \xi}(t)\frac{\partial^2 u}{\partial \xi^2}(s).$$

In these formulas, the derivative of function $\mathcal{X}(t,s)$ with respect of the first argument is denoted as $\partial \mathcal{X}/\partial t$, and the derivative of this function with respect to the second argument is denoted as $\partial \mathcal{X}/\partial s$.

Eq. (2.46) together with Eqs. (2.38) and (2.39) implies that

$$\rho_0(\frac{\partial^3 u}{\partial t^3})^+ = \{\mathcal{X}(t,0)\frac{dH_0}{dv_1}(0) + (\mathcal{X}(t,t) - \mathcal{X}(t,0))\frac{\partial H_1}{\partial v_1}(0,0)\}(\frac{\partial^3 u}{\partial t \partial \xi^2})^+$$

$$+\{\frac{\partial \mathcal{X}}{\partial t}(t,0)\frac{dH_0}{dv_1}(0) + \frac{\partial \mathcal{X}}{\partial s}(t,t)(\frac{\partial H_1}{\partial v_1}(0,0) + \frac{\partial H_1}{\partial v_2}(0,0))$$

$$+(\frac{\partial \mathcal{X}}{\partial t}(t,t) - \frac{\partial \mathcal{X}}{\partial t}(t,0))\frac{\partial H_1}{\partial v_1}(0,0)\}(\frac{\partial^2 u}{\partial \xi^2})^+$$

$$+\{\mathcal{X}(t,0)\frac{d^2 H_0}{dv_1^2}(0) + (\mathcal{X}(t,t) - \mathcal{X}(t,0))\frac{\partial^2 H_1}{\partial v_1}(0,0)\}(\frac{\partial^2 u}{\partial t \partial \xi})^+(\frac{\partial^2 u}{\partial \xi^2})^+.$$

Finally, taking into account Eqs. (2.36) and (2.42), we find

$$[\frac{\partial^3 u}{\partial t^3}] - c^2[\frac{\partial^3 u}{\partial t \partial \xi^2}]$$

$$= \frac{1}{\rho_0 c^2}\langle\{\frac{\partial \mathcal{X}}{\partial t}(t,0)\frac{dH_0}{dv_1}(0) + \frac{\partial \mathcal{X}}{\partial s}(t,t)(\frac{\partial H_1}{\partial v_1}(0,0) + \frac{\partial H_1}{\partial v_2}(0,0))$$

$$+(\frac{\partial \mathcal{X}}{\partial t}(t,t) - \frac{\partial \mathcal{X}}{\partial t}(t,0))\frac{\partial H_1}{\partial v_1}(0,0)\}[\frac{\partial^2 u}{\partial t^2}]$$

$$-\frac{1}{c}\{\mathcal{X}(t,0)\frac{d^2 H_0}{dv_1^2}(0) + (\mathcal{X}(t,t) - \mathcal{X}(t,0))\frac{\partial^2 H_1}{\partial v_1}(0,0)\}[\frac{\partial^2 u}{\partial t^2}]^2\rangle. \qquad (2.47)$$

Substitution of expression (2.47) into Eq. (2.45) implies that

$$\frac{d}{dt}[\frac{\partial^2 u}{\partial t^2}] = \Theta_1(t)[\frac{\partial^2 u}{\partial t^2}] - \Theta_2(t)[\frac{\partial^2 u}{\partial t^2}]^2, \qquad (2.48)$$

where

$$\Theta_1(t) = \frac{\dot{c}(t)}{c(t)} + \frac{1}{\rho_0 c^2(t)}\{\frac{\partial \mathcal{X}}{\partial t}(t,0)\frac{dH_0}{dv_1}(0)$$

347

$$+\frac{\partial \mathcal{X}}{\partial s}(t,t)(\frac{\partial H_1}{\partial v_1}(0,0) + \frac{\partial H_1}{\partial v_2}(0,0)) + (\frac{\partial \mathcal{X}}{\partial t}(t,t) - \frac{\partial \mathcal{X}}{\partial t}(t,0))\frac{\partial H_1}{\partial v_1}(0,0)\}, \quad (2.49)$$

$$\Theta_2(t) = \frac{1}{\rho_0 c^3(t)}\{\mathcal{X}(t,0)\frac{d^2 H_0}{dv_1^2}(0) + (\mathcal{X}(t,t) - \mathcal{X}(t,0))\frac{\partial^2 H_1}{\partial v_1}(0,0)\}. \quad (2.50)$$

Exercise 2.5. Deduce from Eqs. (2.42) and (2.49) that

$$\Theta_1(t) = 3\frac{\dot{c}(t)}{c(t)} + \frac{1}{\rho_0 c^2(t)}\frac{\partial \mathcal{X}}{\partial s}(t,t)\frac{\partial H_1}{\partial v_2}(0,0). \quad \square$$

Eq. (2.48) is a nonlinear differential equation for the function $g = [\partial u^2/\partial t^2]$

$$\frac{dg}{dt} = \Theta_1 g - \Theta_2 g^2, \quad (2.51)$$

with coefficients $\Theta_1(t)$ and $\Theta_2(t)$ which are assumed to be smooth functions of time t.

Exercise 2.6. Prove that for constant coefficients Θ_1 and Θ_2, any solution $g(t)$ of Eq. (2.51) satisfies the equality

$$|\Theta_2 - \frac{\Theta_1}{g(t)}| = |\Theta_2 - \frac{\Theta_1}{g(0)}|\exp(\Theta_1 t)$$

for $\Theta_1 \neq 0$, and the equality

$$\frac{1}{g(t)} = \frac{1}{g(0)} + \Theta_2 t$$

for $\Theta_1 = 0$. $\square$

Exercise 2.7. Check that for $g(t) \neq 0$, function $h(t) = g^{-1}(t)$ satisfies the linear non-homogeneous equation

$$\frac{dh}{dt} = -\Theta_1 h + \Theta_2, \quad h(0) = g^{-1}(0). \quad \square \quad (2.52)$$

Exercise 2.8. Show that the solution of Eq. (2.52) may be presented in the form

$$h(t) = \{h(0) + \int_0^t \Theta_2(s)\exp(\int_0^s \Theta_1(\tau)d\tau)ds\}\exp(-\int_0^t \Theta_1(s)ds),$$

which implies that

$$g(t) = g(0)\{1 + g(0)\int_0^t \Theta_2(s)\exp(\int_0^s \Theta_1(\tau)d\tau)ds\}^{-1}\exp(\int_0^t \Theta_1(s)ds). \quad (2.53)$$

Proposition 2.1. Suppose that

$$\sup_{t\geq 0}\left|\int_0^t \Theta_2(s)\exp\left(\int_0^s \Theta_1(\tau)d\tau\right)ds\right| < \infty. \tag{2.54}$$

Then any solution with a sufficiently small initial datum

$$|g(0)| < \frac{1}{2\sup_{t\geq 0}\left|\int_0^t \Theta_2(s)\exp\left(\int_0^s \Theta_1(\tau)d\tau\right)ds\right|} \tag{2.55}$$

exists for any instant $t \geq 0$, while solutions with sufficiently large initial data, for example, with

$$g(0) = -\frac{2}{\left|\int_0^\infty \Theta_2(s)\exp\left(\int_0^s \Theta_1(\tau)d\tau\right)ds\right|},$$

reach infinity at finite moments of time.

Proof. We prove the first assertion of Proposition 2.1. The other assertion is left to the reader as an exercise. It follows from Eq. (2.55) that for any $t \geq 0$

$$\left|1 + g(0)\int_0^t \Theta_2(s)\exp\left(\int_0^s \Theta_1(\tau)d\tau\right)ds\right|$$

$$\geq 1 - |g(0)|\sup_{t\geq 0}\left|\int_0^t \Theta_2(s)\exp\left(\int_0^s \Theta_1(\tau)d\tau\right)ds\right|$$

$$> 1 - \frac{\sup_{t\geq 0}\left|\int_0^t \Theta_2(s)\exp\left(\int_0^s \Theta_1(\tau)d\tau\right)ds\right|}{2\sup_{t\geq 0}\left|\int_0^t \Theta_2(s)\exp\left(\int_0^s \Theta_1(\tau)d\tau\right)ds\right|} = \frac{1}{2}.$$

This estimate together with Eq. (2.53) implies that for any $t \geq 0$

$$|g(t)| \leq 2|g(0)|\exp\left(\int_0^t \Theta_1(s)ds\right)$$

$$< \frac{\exp\left(\int_0^t \Theta_1(s)ds\right)}{\sup_{t\geq 0}\left|\int_0^t \Theta_2(s)\exp\left(\int_0^s \Theta_1(\tau)d\tau\right)ds\right|}. \qquad \square$$

Proposition 2.2. Suppose that functions $\Theta_1(t)$ and $\Theta_2(t)$ are continuous in $[0,\infty)$ and have finite limits

$$\lim_{t\to\infty}\Theta_1(t) = \Theta_1^0 < 0, \qquad \lim_{t\to\infty}\Theta_2(t) = \Theta_2^0 \neq 0. \tag{2.56}$$

Then any solution of Eq. (2.51) with a sufficiently small initial datum tends to zero as $t \to \infty$, whereas solutions with relatively large initial data reach infinity at finite moments of time.

Proof. To prove this assertion we should demonstrate that Eqs. (2.56) imply inequality (2.54) and refer to Proposition 2.1. It follows from Eq. (2.56) that there is a positive T such that for any $t > T$

$$\Theta_1(t) \leq \frac{\Theta_1^0}{2}, \qquad |\Theta_2(t)| \leq 2|\Theta_2^0|.$$

The finiteness of the left-hand side of Eq. (2.54) follows from the estimates

$$\sup_{t\geq 0}|\int_0^t\Theta_2(s)\exp(\int_0^s\Theta_1(\tau)d\tau)ds|$$

$$\leq|\int_0^T\Theta_2(s)\exp(\int_0^s\Theta_1(\tau)d\tau)ds|+|\int_T^t\Theta_2(s)\exp(\int_0^s\Theta_1(\tau)d\tau)ds|$$

$$\leq|\int_0^T\Theta_2(s)\exp(\int_0^s\Theta_1(\tau)d\tau)ds|+2|\Theta_2^0|\int_T^t\exp(\frac{1}{2}\Theta_1^0s)ds$$

$$\leq|\int_0^T\Theta_2(s)\exp(\int_0^s\Theta_1(\tau)d\tau)ds|+4|\frac{\Theta_2^0}{\Theta_1^0}|\exp(\frac{1}{2}\Theta_1^0T)<\infty.\quad\square$$

For $\Theta_2^0=0$, we should distinguish two cases: when the function $\Theta_2(t)$ is non-zero on a set of positive measure, and when this function equals zero identically.

Exercise 2.9. Prove that if the function $\Theta_2(t)$ is non-zero on a set of a positive measure, then Proposition 2.2 remains valid. $\square$

Exercise 2.10. Show that if $\Theta_1^0<0$ and $\Theta_2(t)=0$ identically, then any solution of Eq. (2.51) tends to zero as $t\to\infty$. $\square$

2.4. Acceleration Waves for Shear and Elongation Motions

Let us now return to the study of elongation and shear motions and discuss which conclusions may be deduced from the above analysis of Eq. (2.51).

First, we study shear motion. It follows from Eq. (2.26) that

$$W'=2\kappa\Delta W,\quad W''=2\Delta W+4\kappa^2\Delta^2W,\quad W'''=12\kappa\Delta^2W+8\kappa^3\Delta^3W,\quad(2.57)$$

where

$$\Delta=\frac{\partial}{\partial I_1}+\frac{\partial}{\partial I_2}.$$

Substitution of expressions (2.57) into Eq. (2.30) implies that

$$\frac{dH_0}{dv_1}(0)=\mu,\qquad\frac{d^2H_0}{dv_1^2}(0)=0,$$

$$\frac{\partial H_1}{\partial v_1}(0,0)=-\frac{\partial H_1}{\partial v_2}(0,0)=\mu,\qquad\frac{\partial^2H_1}{\partial v_1^2}(0,0)=0,\qquad(2.58)$$

where $\mu=\frac{1}{2}\Delta W(3,3,1)$, cf. Eq. (3.2.29). Eqs. (2.42) and (2.58) yield

$$c^2(t)=\frac{\mu}{\rho_0}\mathcal{X}(t,t).$$

Substitution of expression (2.9) into this equality implies that

$$c(t)=\sqrt{\frac{\beta(t)\mu}{\rho_0}}.\qquad(2.59)$$

For non-aging media with $\beta(t) = 1$, Eq. (2.59) coincides with the classical formula for the speed of $S-waves$ in linear elasticity, see Sneddon & Berry (1958).

By substituting expressions (2.9), (2.58), and (2.59) into Eqs. (2.49) and (2.50), we obtain

$$\Theta_1(t) = \frac{1}{\beta(t)}\{\frac{\dot{\beta}(t)}{2} + \frac{\partial Q}{\partial t}(t,t)\}, \qquad \Theta_2(t) = 0. \tag{2.60}$$

By using Eqs. (2.59) and (2.60) and the constitutive restrictions on elastic moduli and relaxation kernels, we arrive at the following conclusions:
(i) speed c of shear acceleration waves in an aging medium increases in time and tends to some limiting value as $t \to \infty$. The speed depends on the first derivatives of strain energy density W in the natural configuration (in terms of the Lame coefficient μ) and on the initial density ρ_0, and it is independent of viscous properties of the medium, as well as of the amplitudes of waves;
(ii) for

$$\lim_{t\to\infty} \frac{\partial Q}{\partial t}(t,t) < 0, \tag{2.61}$$

any shear acceleration wave in an aging viscoelastic medium decays in time exponentially. Indeed, according to the properties of elastic moduli for aging media, $\lim_{t\to\infty} \beta(t) = 0$, and our assertion follows from Exercise 2.10;
(iii) from the engineering standpoint, the exponential decay in amplitudes of acceleration waves means an opportunity to neglect discontinuous solutions after some transition period. Eq. (2.61) provides a condition for such a simplification. For a non-aging medium, Eq. (2.61) turns into the inequality $\dot{Q}_0(0) < 0$, which is valid for any viscoelastic material with a regular relaxation kernel, see Eq. (6.4.59).

Let us now consider elongation motion. Calculating the derivatives of function W we obtain formulas (2.57) with

$$\Delta = \frac{\partial}{\partial I_1} + 2\frac{\partial}{\partial I_2} + \frac{\partial}{\partial I_3}.$$

Substitution of expressions (2.57) into Eq. (2.29) implies that

$$\frac{dH_0}{dv_1}(0) = \lambda + 2\mu, \qquad \frac{d^2 H_0}{dv_1^2}(0) = 3(\lambda + 2\mu) + \Lambda,$$

$$\frac{\partial H_1}{\partial v_1}(0,0) = -\frac{\partial H_1}{\partial v_2}(0,0) = \lambda + 2\mu, \qquad \frac{\partial^2 H_1}{\partial v_1^2}(0,0) = 3(\lambda + 2\mu) + \Lambda. \tag{2.62}$$

Here we use Eqs. (3.2.23), (3.2.30) and set

$$8\Delta^3 W(3,3,1) = \Lambda. \tag{2.63}$$

Exercise 2.11. Show that $\Lambda = 0$ in the linear theory of viscoelasticity. $\square$

Substitution of expressions (2.62) into Eq. (2.42) yields

$$c(t) = \sqrt{\frac{(\lambda + 2\mu)\beta(t)}{\rho_0}}. \tag{2.64}$$

For non-aging media with $\beta(t) = 1$, Eq. (2.64) coincides with the classical formula for the speed of *P–waves* in linear elasticity, see Sneddon & Berry (1958).

Exercise 2.12. By substituting expressions (2.62) – (2.64) into Eqs. (2.49) and (2.50), derive Eq. (2.60) for $\Theta_1(t)$ and the following formula for $\Theta_2(t)$:

$$\Theta_2(t) = \left(3 + \frac{\Lambda}{\lambda + 2\mu}\right)\sqrt{\frac{\rho_0}{(\lambda + 2\mu)\beta(t)}}. \quad \square \tag{2.65}$$

Eqs. (2.60), (2.64), and (2.65) together with Proposition 2.2 lead to the following conclusions:

(i) speed c of elongation acceleration waves in an aging medium increases in time and tends to some limiting value as $t \to \infty$. Unlike shear waves, the speed depends on the second derivatives of strain energy density W in the natural configuration (in terms of the Lame parameters λ and μ);

(ii) if inequality (2.61) holds and

$$\frac{\Lambda}{\lambda + 2\mu} > -3, \tag{2.66}$$

then we may distinguish elongation acceleration waves with small and large amplitudes. In an aging viscoelastic medium, acceleration waves with small amplitudes decay in time exponentially, while acceleration waves with large amplitudes grow. The increase in the amplitude of acceleration waves may be a cause for transition of relatively smooth solutions, such as acceleration waves, into stronger discontinuities as shock waves;

(iii) from the engineering standpoint, elongation acceleration waves may be treated as unstable. The latter means that these waves either decay exponentially in time and become undetectable in experiments, or they turn into shock waves within a finite interval of time.

The above analysis demonstrates typical features of any solution with a "weak" discontinuity. By using the same approach, we may also study solutions with discontinuities in third, fourth, etc. derivatives, propagating into a viscoelastic medium at rest. The material aging affects the dependence of the wave speed on time. Formally, the material aging may change the whole picture of wave propagating, if inequality (2.61) is not fulfilled, but, to the best of our knowledge, viscoelastic materials with the property

$$\lim_{t \to \infty} \frac{\partial Q}{\partial t}(t, t) = 0$$

have not been observed in experiments.

3. Radial Oscillations of a Sphere

This Section deals with radial *oscillations* of an incompressible viscoelastic thick-walled spherical *shell*. We derive the governing integro-differential equation for radial oscillations. This equation is simplified by assuming that the thickness of the shell is sufficiently small. We analyze natural oscillations of an elastic shell with finite strains and small forced oscillations of a viscoelastic shell. Finally, we establish a correspondence between periodic solutions (oscillations) of linear integro-differential equations with finite and infinite memories.

3.1. Formulation of the Problem

Let us consider a viscoelastic hollow sphere (thick-walled shell) which occupies the domain $\{R_1 \leq R \leq R_2, \ 0 \leq \Theta < 2\pi, \ 0 \leq \Phi \leq \pi\}$ in the initial stress-free configuration. Here $\{R, \Theta, \Phi\}$ are spherical coordinates in the initial configuration with unit vectors $\bar{e}_R$, $\bar{e}_\Theta$, and $\bar{e}_\Phi$. At instant $t = 0$, an internal pressure $p_1(t)$ and an external pressure $p_2(t)$ are applied to the sphere, see Fig. 3.1. Owing to the symmetry of loading, spherical coordinates $\{r, \theta, \phi\}$ in the actual configuration are expressed in terms of the spherical coordinates in the initial configuration as follows:

$$r = f(t, R), \qquad \theta = \Theta, \qquad \phi = \Phi, \tag{3.1}$$

where f is an unknown function. The radius-vector $\bar{r}$ in the actual configuration of a point with Lagrangian coordinates $\{R, \Theta, \Phi\}$ equals

$$\bar{r} = f(t, R)\bar{e}_R. \tag{3.2}$$

We differentiate expression (3.2) with respect to spatial coordinates and use formulas (1.1.2) and (1.1.6). As a result, we obtain tangent vectors

$$\bar{g}_1 = h(t)\bar{e}_R, \qquad \bar{g}_2 = f(t)\bar{e}_\Theta, \qquad \bar{g}_3 = f(t)\sin\Theta\,\bar{e}_\Phi, \tag{3.3}$$

where $h(t) = \partial f(t)/\partial R$. For simplicity, argument R is omitted.
Exercise 3.1. Derive expressions for dual vectors

$$\bar{g}^1 = \frac{1}{h(t)}\bar{e}_R, \qquad \bar{g}^2 = \frac{1}{f(t)}\bar{e}_\Theta, \qquad \bar{g}^3 = \frac{1}{f(t)\sin\Theta}\bar{e}_\Phi. \qquad \Box \tag{3.4}$$

It follows from Eqs. (3.3), (3.4) and (2.1.44) that

$$\bar{\nabla}_s \bar{r}(t) = \frac{h(t)}{h(s)}\bar{e}_R\bar{e}_R + \frac{f(t)}{f(s)}(\bar{e}_\Theta\bar{e}_\Theta + \bar{e}_\Phi\bar{e}_\Phi). \tag{3.5}$$

353

Substitution of expression (3.5) for the relative deformation gradient into Eqs. (2.1.46) implies formulas for the relative Finger tensor

$$\hat{F}_*(t,s) = (\frac{h(t)}{h(s)})^2 \bar{e}_R \bar{e}_R + (\frac{f(t)}{f(s)})^2 (\bar{e}_\Theta \bar{e}_\Theta + \bar{e}_\Phi \bar{e}_\Phi),$$

$$\hat{F}_*^2(t,s) = (\frac{h(t)}{h(s)})^4 \bar{e}_R \bar{e}_R + (\frac{f(t)}{f(s)})^4 (\bar{e}_\Theta \bar{e}_\Theta + \bar{e}_\Phi \bar{e}_\Phi). \tag{3.6}$$

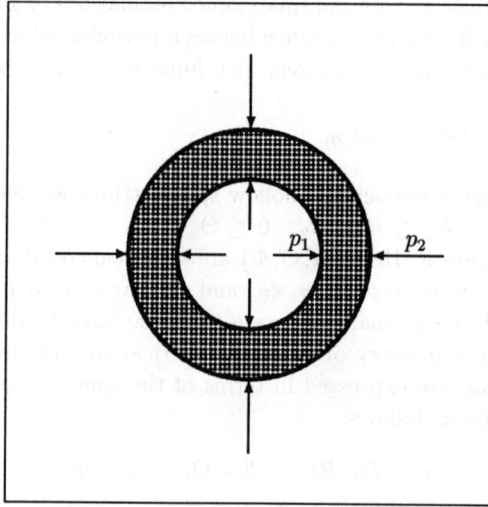

Fig. 3.1. A viscoelastic hollow sphere under internal and external pressure.

Exercise 3.2. Check that the principal invariants of the relative Finger tensor $\hat{F}_*(t,s)$ equal

$$I_1(t,s) = (\frac{h(t)}{h(s)})^2 + 2(\frac{f(t)}{f(s)})^2, \qquad I_2(t,s) = [2(\frac{h(t)}{h(s)})^2 + (\frac{f(t)}{f(s)})^2](\frac{f(t)}{f(s)})^2,$$

$$I_3(t,s) = (\frac{h(t)}{h(s)})^2(\frac{f(t)}{f(s)})^4. \qquad \square \ (3.7)$$

Substituting the third equation (3.7) into Eq. (3.2.86), we find

$$\frac{\partial f}{\partial R}(t,R) = (\frac{R}{f(t,R)})^2. \tag{3.8}$$

Integration of this equality implies that

$$f^3(t,R) = R^3 + C(t), \tag{3.9}$$

354

where $C = C(t)$ is a function to be found. Therefore,

$$f(t, R) = [R^3 + C(t)]^{\frac{1}{3}}, \qquad h(t, R) = (\frac{R}{f(t, R)})^2. \qquad (3.10)$$

It follows from Eqs. (3.6), (3.7), and (3.10) that

$$\hat{F}_*(t, s) = (\frac{f(s)}{f(t)})^4 \bar{e}_R \bar{e}_R + (\frac{f(t)}{f(s)})^2 (\bar{e}_\Theta \bar{e}_\Theta + \bar{e}_\Phi \bar{e}_\Phi),$$

$$\hat{F}_*^2(t, s) = (\frac{f(s)}{f(t)})^8 \bar{e}_R \bar{e}_R + (\frac{f(t)}{f(s)})^4 (\bar{e}_\Theta \bar{e}_\Theta + \bar{e}_\Phi \bar{e}_\Phi),$$

$$I_1(t, s) = (\frac{f(s)}{f(t)})^4 + 2(\frac{f(t)}{f(s)})^2, \qquad I_2(t, s) = (\frac{f(t)}{f(s)})^4 + 2(\frac{f(s)}{f(t)})^2. \qquad (3.11)$$

Substitution of expressions (3.11) into the constitutive equations of an aging, incompressible, viscoelastic medium (6.3.83) and (6.3.84) with $M = 1$ implies that

$$\hat{\sigma} = \sigma_{RR} \bar{e}_R \bar{e}_R + \sigma_{\Theta\Theta} \bar{e}_\Theta \bar{e}_\Theta + \sigma_{\Phi\Phi} \bar{e}_\Phi \bar{e}_\Phi. \qquad (3.12)$$

Here $\hat{\sigma}$ is the Cauchy stress tensor with the non-zero components

$$\sigma_{RR} = -p(t) + 2\{\mathcal{X}(t, 0)[(\frac{R}{f(t)})^4 W_1(t, 0) + 2(\frac{R}{f(t)})^2 W_2(t, 0)]$$

$$+ \int_0^t \frac{\partial \mathcal{X}}{\partial s}(t, s)[(\frac{f(s)}{f(t)})^4 W_1(t, s) + 2(\frac{f(s)}{f(t)})^2 W_2(t, s)]ds\},$$

$$\sigma_{\Theta\Theta} = \sigma_{\Phi\Phi} = -p(t) + 2\{\mathcal{X}(t, 0)[(\frac{f(t)}{R})^2 W_1(t, 0)$$

$$+ ((\frac{R}{f(t)})^2 + (\frac{f(t)}{R})^4) W_2(t, 0)] + \int_0^t \frac{\partial \mathcal{X}}{\partial s}(t, s)[(\frac{f(t)}{f(s)})^2 W_1(t, s)$$

$$+ ((\frac{f(s)}{f(t)})^2 + (\frac{f(t)}{f(s)})^4) W_2(t, s)]ds\}, \qquad (3.13)$$

$p = p(t, R)$ is a pressure, $W(I_1, I_2)$ is a strain energy density satisfying constitutive restrictions imposed in Section 3.2,

$$\psi_1(t, s) = W_1(t, s) + I_1(t, s) W_2(t, s), \qquad \psi_2(t, s) = -W_2(t, s),$$

$$W_k(t, s) = \frac{\partial W}{\partial I_k}(I_1(\hat{F}(t, s)), I_2(\hat{F}(t, s))).$$

Function $\mathcal{X}(t, s)$ satisfies conditions (6.4.103) and (6.4.104) and is presented in the form (2.9), where $\beta(t)$ is a dimensionless shear modulus, and $Q(t, s)$ is a regular relaxation measure. For a non-aging medium, we set $\beta(t) = 1$ and $Q(t, s) = Q_0(t - s)$.

Exercise 3.3. By using Eq. (3.13), show that

$$\sigma_{RR}(t) - \sigma_{\Theta\Theta}(t) = 2\{\mathcal{X}(t,0)[(\frac{R}{f(t)})^4 - (\frac{f(t)}{R})^2][W_1(t,0) + (\frac{f(t)}{R})^2 W_2(t,0)]$$

$$+ \int_0^t \frac{\partial \mathcal{X}}{\partial s}(t,s)[(\frac{f(s)}{f(t)})^4 - (\frac{f(t)}{f(s)})^2][W_1(t,s) + (\frac{f(t)}{f(s)})^2 W_2(t,0)]ds\}. \qquad \square \ (3.14)$$

The displacement vector equals $\bar{u} = [f(t,R) - R]\bar{e}_R$.

Exercise 3.4. By differentiating this equality and using Eq. (3.10), derive the following expression for the acceleration vector:

$$\bar{a} = \frac{1}{3f^2(t)}[\ddot{C}(t) - \frac{2(\dot{C}(t))^2}{3f^3(t)}]\bar{e}_R. \qquad \square \qquad (3.15)$$

Exercise 3.5. By using Eqs. (1.3.4), (3.4), (3.10), and (3.12), show that

$$\bar{\nabla}_t \cdot \hat{\sigma}(t) = [\frac{1}{h(t)} \frac{\partial \sigma_{RR}}{\partial R}(t) + \frac{2}{f(t)}(\sigma_{RR} - \sigma_{\Theta\Theta}(t))]\bar{e}_R$$

$$+ (\sigma_{\Theta\Theta} - \sigma_{\Phi\Phi})\frac{\cos\Theta}{f(t)\sin\Theta}\bar{e}_\Theta. \qquad \square \qquad (3.16)$$

In the absence of body forces, Eqs. (3.15) and (3.16) together with the motion equation $\rho(t)\bar{a}(t) = \bar{\nabla}_t \cdot \hat{\sigma}(t)$ yield

$$\frac{\rho_0}{3r^2}[\ddot{C}(t) - \frac{2(\dot{C}(t))^2}{3r^3}] = \frac{\partial \sigma_{RR}}{\partial r}(t) + \frac{2}{r}[\sigma_{RR}(t) - \sigma_{\Theta\Theta}(t)], \qquad (3.17)$$

where ρ_0 and ρ are mass densities in the initial and actual configurations. In order to derive Eq. (3.17) we employ formula (3.1) and the incompressibility condition (3.2.87).

Let us integrate Eq. (3.17) from $r_1 = f(t,R_1)$ to $r_2 = f(t,R_2)$. By using the boundary conditions

$$\sigma_{RR}(t,r_1) = -p_1(t), \qquad \sigma_{RR}(t,r_2) = -p_2(t), \qquad (3.18)$$

we find

$$\frac{\rho_0}{3}[\ddot{C}(t)(\frac{1}{r_1(t)} - \frac{1}{r_2(t)}) - \frac{(\dot{C}(t))^2}{6}(\frac{1}{r_1^4(t)} - \frac{1}{r_2^4(t)})]$$

$$= p_0(t) + 2\int_{r_1(t)}^{r_2(t)} \frac{\sigma_{RR}(t) - \sigma_{\Theta\Theta}(t)}{r}dr,$$

where $p_0(t) = p_1(t) - p_2(t)$. We replace the "new" variable r by the "old" variable R under the sign of integral. By employing Eq. (3.10), we obtain

$$\frac{\rho_0}{3}[\ddot{C}(t)(\frac{1}{r_1(t)} - \frac{1}{r_2(t)}) - \frac{(\dot{C}(t))^2}{6}(\frac{1}{r_1^4(t)} - \frac{1}{r_2^4(t)})]$$

$$= p_0(t) + 2\int_{R_1}^{R_2}[\sigma_{RR}(t) - \sigma_{\Theta\Theta}(t)]\frac{R^2 dR}{f^3(t)}. \qquad (3.19)$$

356

Finally, substitution of expression (3.14) into Eq. (3.19) implies that

$$\rho_0[\ddot{C}(t)(\frac{1}{r_1(t)} - \frac{1}{r_2(t)}) - \frac{(\dot{C}(t))^2}{6}(\frac{1}{r_1^4(t)} - \frac{1}{r_2^4(t)})] = 3p_0(t)$$

$$+12\int_{R_1}^{R_2}\{\mathcal{X}(t,0)[(\frac{R}{f(t)})^4 - (\frac{f(t)}{R})^2][W_1(t,0) + (\frac{f(t)}{R})^2 W_2(t,0)]$$

$$+\int_0^t \frac{\partial\mathcal{X}}{\partial s}(t,s)[(\frac{f(s)}{f(t)})^4 - (\frac{f(t)}{f(s)})^2][W_1(t,s) + (\frac{f(t)}{f(s)})^2 W_2(t,0)]ds\}\frac{R^2 dR}{f^3(t)}. \quad (3.20)$$

Eq. (3.20) is a nonlinear integro-differential equation for the unknown function $C(t)$. Since Eq. (3.20) contains the second derivative of the unknown function with respect to time, initial conditions $C(0)$ and $\dot{C}(0)$ should be prescribed to solve this equation uniquely.

It is convenient to transform Eq. (3.20) by using its first integral. For this purpose, we employ the following assertions, which are provided here as exercises.

Exercise 3.6. By direct calculation with the use of Eq. (3.10), show that

$$\frac{d}{dt}\{\rho_0(\frac{dC}{dt}(t))^2[\frac{1}{r_1}(t) - \frac{1}{r_2}(t)]\} = 2\rho_0\frac{dC}{dt}(t)\{\frac{d^2C}{dt^2}(t)[\frac{1}{r_1}(t) - \frac{1}{r_2}(t)]$$

$$-\frac{1}{6}(\frac{dC}{dt}(t))^2[\frac{1}{r_1^4}(t) - \frac{1}{r_2^4}(t)]\}. \quad \square \quad (3.21)$$

It follows from Eq. (3.11) that for any strain energy density $W(I_1, I_2)$

$$\frac{\partial W}{\partial t}(I_1(t,s), I_2(t,s)) = W_1(t,s)\frac{\partial}{\partial t}[(\frac{f(s)}{f(t)})^4 + 2(\frac{f(t)}{f(s)})^2]$$

$$+W_2(t,s)\frac{\partial}{\partial t}[(\frac{f(t)}{f(s)})^4 + 2(\frac{f(s)}{f(t)})^2]. \quad (3.22)$$

Exercise 3.7. By using Eqs. (3.10) and (3.22), check that

$$\frac{\partial W}{\partial t}(I_1(t,s), I_2(t,s)) = \frac{4}{3f^3(t)}\frac{dC}{dt}(t)[(\frac{f(t)}{f(s)})^2 - (\frac{f(s)}{f(t)})^4]$$

$$\times[W_1(t,s) + (\frac{f(t)}{f(s)})^2 W_2(t,s)]. \quad \square \quad (3.23)$$

We multiply Eq. (3.20) by $\dot{C}(t)$ and employ Eqs. (3.21) and (3.23). As a result, we obtain

$$\frac{d}{dt}[\rho_0(\frac{dC}{dt}(t))^2(\frac{1}{r_1(t)} - \frac{1}{r_2(t)})] - 6p_0(t)\frac{dC}{dt}(t)$$

$$+18\int_{R_1}^{R_2}[\mathcal{X}(t,0)\frac{\partial W}{\partial t}(t,0) + \int_0^t\frac{\partial\mathcal{X}}{\partial s}(t,s)\frac{\partial W}{\partial t}(t,s)ds]R^2 dR = 0. \quad (3.24)$$

357

For an arbitrary strain energy density W, our chances to find an explicit solution of Eq. (3.24) appear to be extremely low, and some additional assumptions should be introduced.

3.2. Radial Oscillations of a Thin-walled Shell

Following Zhong-Heng & Solecki (1963), we confine ourselves to radial oscillations of a *thin-walled* spherical shell, when the initial thickness $R_2 - R_1$ is essentially less than the internal radius R_1. To formulate this restriction we introduce the small parameter

$$\eta = \frac{R_2 - R_1}{R_1},$$

and suppose additionally that external forces are also proportional to the small parameter: $p_0(t) = \eta P(t)$.

Direct calculations with the use of Eq. (3.10) imply that

$$\frac{1}{r_1(t)} - \frac{1}{r_2(t)} = \frac{1}{R_1}\{[1 + z(t)]^{-\frac{1}{3}} - [(1 + \eta)^3 + z(t)]^{-\frac{1}{3}}\}, \tag{3.25}$$

where $z(t) = C(t)/R_1^3$. Expanding the right-hand side of Eq. (3.25) into a series in η and neglecting terms of the second order compared with η, we find

$$\frac{1}{r_1(t)} - \frac{1}{r_2(t)} = \frac{\eta}{R_1[1 + z(t)]^{\frac{4}{3}}}. \tag{3.26}$$

With the same level of accuracy, we may write

$$\int_{R_1}^{R_2} [\mathcal{X}(t,0)\frac{\partial W}{\partial t}(t,0) + \int_0^t \frac{\partial \mathcal{X}}{\partial s}(t,s)\frac{\partial W}{\partial t}(t,s)ds]R^2 dR$$

$$= \eta R_1^3[\mathcal{X}(t,0)\frac{\partial W^0}{\partial t}(t,0) + \int_0^t \frac{\partial \mathcal{X}}{\partial s}(t,s)\frac{\partial W^0}{\partial t}(t,s)ds], \tag{3.27}$$

where

$$W^0(t,s) = W(I_1^0(t,s), I_2^0(t,s)), \tag{3.28}$$

and the superscript zero means that the principal invariants are calculated at point $R = R_1$. It follows from Eq. (3.11) that

$$I_1^0(t,s) = (\frac{1 + z(t)}{1 + z(s)})^{-\frac{4}{3}} + 2(\frac{1 + z(t)}{1 + z(s)})^{\frac{2}{3}},$$

$$I_2^0(t,s) = (\frac{1 + z(t)}{1 + z(s)})^{\frac{4}{3}} + 2(\frac{1 + z(t)}{1 + z(s)})^{-\frac{2}{3}}. \tag{3.29}$$

Substitution of expressions (3.26) and (3.27) into Eq. (3.24) implies the following integro-differential equation:

$$\rho_0 R_1^2 \frac{d}{dt}[(1 + z(t))^{-\frac{4}{3}}(\frac{dz}{dt}(t))^2] - 6P(t)\frac{dz}{dt}(t)$$

$$+18[\mathcal{X}(t,0)\frac{\partial W^0}{\partial t}(t,0) + \int_0^t \frac{\partial \mathcal{X}}{\partial s}(t,s)\frac{\partial W^0}{\partial t}(t,s)ds] = 0. \tag{3.30}$$

To solve Eq. (3.30) explicitly, an additional simplification is necessary. We consider two problems of particular interest for applications. First, we neglect the material viscosity and analyze natural oscillations of an elastic shell with finite strains, when Eq. (3.30) is reduced to a nonlinear ordinary differential equation. Afterward, we study linear forced oscillations of a viscoelastic shell with infinitesimal strains, when Eq. (3.30) may be linearized in the vicinity of the initial configuration.

3.3. Natural Oscillations of an Elastic Shell

Let us consider *natural* oscillations of an elastic hollow sphere with finite strains. According to Eq. (2.9), $\mathcal{X}(t,s) = 1$ for an elastic medium, and the integral in the left-hand side of Eq. (3.30) vanishes. To study natural oscillations we should neglect external forces and set $P(t) = 0$. Owing to this equality, the second term in the left-hand side of Eq. (3.30) vanishes. As a result, Eq. (3.30) is reduced to the following equation:

$$\frac{d}{dt}\{\rho_0 R_1^2[(1 + z(t))^{-\frac{4}{3}}(\frac{dz}{dt}(t))^2] + 18W^0(t,0)\} = 0.$$

Integration of this equality implies that

$$\rho_0 R_1^2[(1 + z(t))^{-\frac{4}{3}}(\frac{dz}{dt}(t))^2] + 18W^0(t,0) = E_0, \tag{3.31}$$

where E_0 is a constant determined by initial conditions.

It is convenient to present $W^0(t,0)$ as a function of $z(t)$ only. It follows from Eq. (3.29) that

$$W^0(t,0) = \frac{1}{18}w(z(t)), \tag{3.32}$$

where

$$w(z) = 18W((1 + z)^{-\frac{4}{3}} + 2(1 + z)^{\frac{2}{3}}, (1 + z)^{\frac{4}{3}} + 2(1 + z)^{-\frac{2}{3}}).$$

Substitution of expression (3.32) into Eq. (3.31) yields

$$\sqrt{\rho_0}R_1\frac{dz}{dt} = (1 + z)^{\frac{2}{3}}[E_0 - w(z)]^{\frac{1}{2}}.$$

359

The latter equation can be presented in the form

$$\sqrt{\rho_0} R_1 \frac{dz}{(1+z)^{\frac{2}{3}}[E_0 - w(z)]^{\frac{1}{2}}} = dt. \qquad (3.33)$$

Let us consider periodic oscillations of an elastic shell with *a period* T, when function $z(t)$ changes from z_1 to z_2, where z_1 and z_2 are solutions of the equation

$$w(z) = E_0. \qquad (3.34)$$

Exercise 3.8. Show that points z_1 and z_2 correspond to the maximal *deflations* and *inflations* of the shell, when velocities of its points vanish. Prove that $z_1 < 0$ and $z_2 > 0$. □

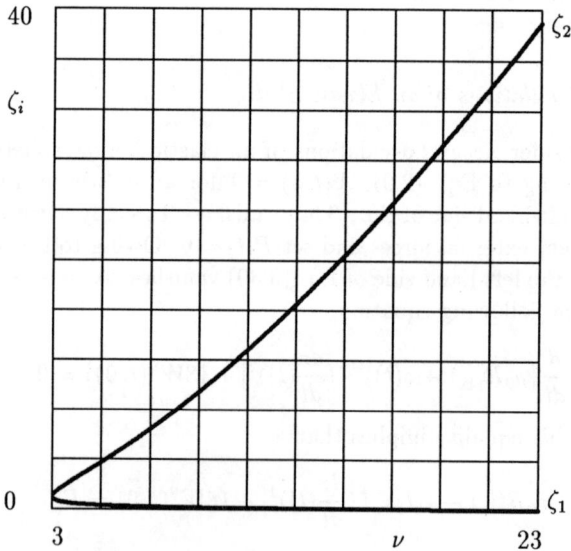

Fig. 3.2. The dimensionless parameters ζ_1 and ζ_2 *versus* parameter ν.

During one period of oscillations, function $z(t)$ increases from z_1 to z_2, and, afterward, decreases from z_2 to z_1. Integration of Eq. (3.33) implies that

$$T = 2\sqrt{\rho_0} R_1 \int_{z_1}^{z_2} \frac{dz}{(1+z)^{\frac{2}{3}}[E_0 - w(z)]^{\frac{1}{2}}}. \qquad (3.35)$$

As an example, let us consider a shell made of the neo-Hookean elastic medium with strain energy density (3.2.103).

360

Exercise 3.9. By substituting expression (3.2.103) into Eq. (3.34), show that $z_i = \zeta_i - 1$, where ζ_1 and ζ_2 are solutions of the nonlinear algebraic equation

$$2\zeta^2 - \nu\zeta^{\frac{4}{3}} + 1 = 0$$

with $\nu = 3 + E_0/(9\mu)$. $\square$

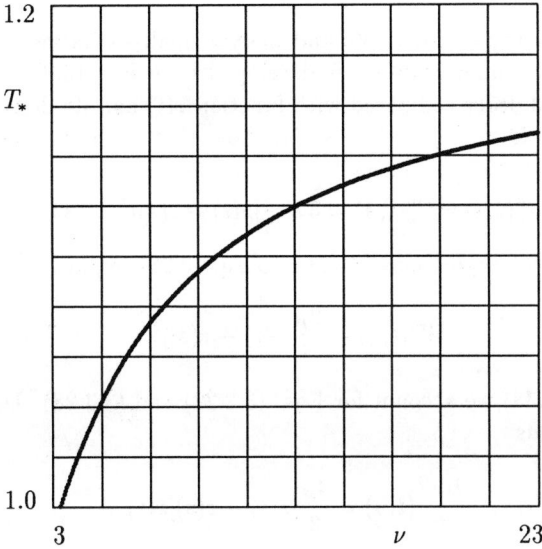

Fig. 3.3. The dimensionless period of natural oscillations $T_* = T/\nu_1$ *versus* parameter ν.

The dimensionless parameters ζ_i are plotted *versus* ν in Fig. 3.2. Parameter ζ_1 decreases monotonically and tends to zero, while parameter ζ_2 increases with the growth of ν, i.e. with the growth of the initial energy. It follows from Eq. (3.35) that

$$T = \frac{2}{3}\sqrt{\frac{\rho_0}{\mu}}R_1 \int_{\zeta_1}^{\zeta_2}[\nu\zeta^{\frac{4}{3}} - (1 + 2\zeta^2)]^{-\frac{1}{2}}d\zeta. \qquad (3.36)$$

According to Eq. (3.36), period T of natural oscillations is determined by two constants: $\nu_1 = \pi R_1\sqrt{\rho_0/(3\mu)}$ and ν. The constant ν_1 is independent of initial conditions, and it is determined by material and geometrical parameters. The physical meaning of ν_1 will be revealed below, when we consider small oscillations of a spherical shell. Since the constant ν is determined by initial conditions, period T of natural oscillations depends on initial conditions as well. This dependence is plotted in Fig. 3.3. The period of natural oscillations increases with the growth of energy at the initial instant. However, this dependence is rather weak, at least for the neo-Hookean medium, and should be taken into account only for oscillations with large amplitudes (when initial speeds and/or displacements are sufficiently large).

3.4. Small Oscillations of a Non-aging Viscoelastic Shell

Let us now return to Eq. (3.30) and analyze small oscillations of a viscoelastic shell. The governing equation is developed from Eq. (3.30) by neglecting terms of the third order compared with function $z(t)$ and its derivatives.

Exercise 3.10. By using Eqs. (3.29), show that with the desired level of accuracy

$$I_1^0(t, s) = I_2^0(t, s) = 3 + \frac{4}{3}[z(t) - z(s)]^2. \quad \square \tag{3.37}$$

Substitution of expressions (3.37) into Eq. (3.28) implies that

$$W^0(t, s) = \frac{2\mu}{3}[z(t) - z(s)]^2. \tag{3.38}$$

To derive Eq. (3.38) we account for Eqs. (3.2.22) and (3.3.28). Differentiation of Eq. (3.38) yields

$$\frac{\partial W^0}{\partial t}(t, s) = \frac{4\mu}{3}[z(t) - z(s)]\dot{z}(t).$$

Exercise 3.11. By substituting this expression into Eq. (3.30) and neglecting terms of the third order compared with $z(t)$ and its derivatives, derive the following equation:

$$\rho_0 R_1^2 \ddot{z}(t) + 12\mu\{\mathcal{X}(t, 0)z(t) + \int_0^t \frac{\partial \mathcal{X}}{\partial s}(t, s)[z(t) - z(s)]ds\} = 3P(t). \quad \square \tag{3.39}$$

We now confine ourselves to non-aging viscoelastic media.

Exercise 3.12. Show that for non-aging materials, Eq. (3.39) implies that

$$\rho_0 R_1^2 \ddot{z}(t) + 12\mu[z(t) + \int_0^t \dot{Q}_0(t - s)z(s)ds] = 3P(t). \quad \square \tag{3.40}$$

In the absence of external forces, natural oscillations of an elastic shell with $Q_0(t) = 0$ obey the differential equation

$$\rho_0 R_1^2 \ddot{z}(t) + 12\mu z(t) = 0.$$

Exercise 3.13. Show that *frequency $\omega_\star$* and period $T_\star$ of small natural oscillations of an elastic spherical shell equal

$$\omega_\star = \sqrt{\frac{12\mu}{\rho_0 R_1^2}}, \qquad T_\star = \frac{2\pi}{\omega_\star} = \pi R_1 \sqrt{\frac{\rho_0}{3\mu}}. \qquad \square \qquad (3.41)$$

It follows from Eq. (3.41) that the parameter ν_1 which determines period T of natural oscillations for a neo-Hookean shell at finite strains coincides with the period of small oscillations $T_\star$ for the corresponding linear elastic shell with infinitesimal strains.

Let us now consider *forced* oscillations of a viscoelastic shell with small strains under time-periodic pressure. We assume that

$$P(t) = P_0 \exp(i\omega t), \qquad (3.42)$$

where P_0 is an amplitude and ω is a frequency of external loads. It is natural to seek a solution of Eq. (3.40) in the form similar to Eq. (3.42)

$$z(t) = z_0 \exp(i\omega t), \qquad (3.43)$$

where z_0 is an unknown amplitude.

It is easy to check that substitution of expressions (3.42) and (3.43) into Eq. (3.40) does not lead to an identity. To obtain an identity, zero as the lower bound of integration in formula (3.40) should be replaced by $-\infty$. As a result, we find

$$\rho_0 R_1^2 \ddot{z}(t) + 12\mu[z(t) + \int_{-\infty}^t \dot{Q}_0(t-s)z(s)ds] = 3P(t). \qquad (3.44)$$

The question why Eq. (3.40) may be replaced by Eq. (3.44) will be discussed in detail later. We presume here that this replacement may be carried out, and derive some conclusions from this procedure.

Substitution of expressions (3.42) and (3.43) into Eq. (3.44) yields

$$-\rho_0 R_1^2 \omega^2 z_0 \exp(i\omega t) + 12\mu z_0[\exp(i\omega t) + \int_{-\infty}^t \dot{Q}_0(t-s)\exp(i\omega s)ds]$$
$$= 3P_0 \exp(i\omega t). \qquad (3.45)$$

Introducing the new variable $\tau = t - s$, we find

$$\int_{-\infty}^t \dot{Q}_0(t-s)\exp(i\omega s)ds = \int_0^\infty \dot{Q}_0(\tau)\exp[i\omega(t-\tau)]d\tau$$
$$= \exp(i\omega t)\int_0^\infty \dot{Q}_0(\tau)\exp(-i\omega\tau)d\tau.$$

Substitution of this expression into Eq. (3.45) implies that

$$\Delta(\omega)z_0 = \frac{P_0}{4\mu}. \qquad (3.46)$$

Here

$$\Delta(\omega) = 1 + \int_0^\infty \dot{Q}_0(s) \exp(-i\omega s) ds - \frac{\omega^2}{\omega_*^2}, \tag{3.47}$$

where ω_* is determined by formula (3.41).

It follows from Eq. (3.47) that for an elastic shell with $Q_0(t) = 0$, function $\Delta(\omega)$ vanishes when $\omega = \omega_*$, i.e. when the frequency of external load coincides with the natural frequency of the shell. Eq. (3.46) implies that in this case amplitude z_0 of forced oscillations tends to infinity. This phenomenon is called *resonance*.

Let us now show that for non-aging viscoelastic media, resonance is impossible. For this purpose, we present complex function $\Delta(\omega)$ as follows:

$$\Delta(\omega) = \Delta_R(\omega) + i\Delta_I(\omega), \tag{3.48}$$

where

$$\Delta_R(\omega) = 1 + \int_0^\infty \dot{Q}_0(s) \cos(\omega s) ds - \frac{\omega^2}{\omega_*^2}, \quad \Delta_I(\omega) = -\int_0^\infty \dot{Q}_0(s) \sin(\omega s) ds. \tag{3.49}$$

Proposition 3.1. Let a function $Q_0(t)$ be sufficiently smooth and satisfy conditions (6.4.58) – (6.4.60). Then there is a positive Δ_0 such that for any ω

$$|\Delta(\omega)|^2 \geq \Delta_0^2. \tag{3.50}$$

Proof. Since

$$|\Delta(\omega)|^2 = \Delta_R^2(\omega) + \Delta_I^2(\omega),$$

it suffices to prove that there exists a $\Delta_0 > 0$ such that for any ω either $\Delta_R(\omega) > \Delta_0$ or $\Delta_I(\omega) > \Delta_0$. It follows from Eq. (3.49) together with Eqs. (6.4.58) and (6.4.59) that

$$\Delta_R(0) = 1 + \int_0^\infty \dot{Q}_0(s) ds = 1 + Q_0(\infty) > 0.$$

Therefore, there is an $\omega_1 > 0$, such that $\Delta_R(\omega) > 0$ for any $|\omega| \leq \omega_1$.

According to *the Riemann–Lebesgue lemma*, for any smooth function $Q_0(t)$

$$\lim_{\omega \to \pm\infty} \int_0^\infty \dot{Q}_0(s) \cos(\omega s) ds = 0. \tag{3.51}$$

Eqs. (3.49) and (3.51) imply that

$$\lim_{\omega \to \pm\infty} \Delta_R(\omega) = -\infty.$$

Therefore, there is an $\omega_2 > \omega_1$, such that $|\Delta_R(\omega)| > 0$ for any $|\omega| \geq \omega_2$.

Finally, we consider the interval $[\omega_1, \omega_2]$. It follows from Eqs. (3.49), (6.4.94), and (6.4.95) that for any ω from this interval

$$\Delta_I(\omega) = \omega \Re \tilde{H}_0(i\omega) \geq \frac{\delta \omega}{1 + \omega^2} > \frac{\delta \omega_1}{1 + \omega_2^2} > 0,$$

where δ is determined by Eq. (6.4.94). $\square$

Returning to forced oscillations of a viscoelastic shell, we find that for any frequency ω

$$|z_0| \leq \frac{P_0}{4\mu\Delta_0}. \tag{3.52}$$

Eq. (3.52) means that for any time-periodic external load, the corresponding solution in the form (3.43) is bounded, and resonance in a linear viscoelastic shell is impossible.

3.5. Oscillations in Systems with Finite and Infinite Memory

To complete our analysis, we should prove an assertion regarding the opportunity to replace the lower bound of integration in Eq. (3.40). We derive this assertion as a corollary of a more general theorem in the theory of Volterra integro-differential equations in convolution, see Burton (1985). To develop it, we present Eq. (3.40) in the Cauchy form

$$\dot{Z}(t) = AZ(t) + \int_0^t B(t-s)Z(s)ds + \Pi(t), \tag{3.53}$$

where

$$Z(t) = \begin{bmatrix} z(t) \\ \dot{z}(t) \end{bmatrix}, \quad A = \begin{bmatrix} 0 & 1 \\ -\frac{12\mu}{\rho_0 R_1^2} & 0 \end{bmatrix},$$

$$B(t) = \begin{bmatrix} 0 & 0 \\ -\frac{12\mu}{\rho_0 R_1^2} \dot{Q}(t) & 0 \end{bmatrix}, \quad \Pi(t) = \begin{bmatrix} 0 \\ \frac{3P(t)}{\rho_0 R_1^2} \end{bmatrix},$$

Below we analyze linear integro-differential equation (3.53) for an n-vector function $Z(t)$ under the following assumptions:
(i) $n \times n$-matrix A is constant;
(ii) $n \times n$-matrix function $B(t)$ is continuous and $|B| = \int_0^\infty |B(t)|dt < \infty$;
(iii) n-vector function $\Pi(t)$ is continuous and periodic in time with a period T_0.

Our objective is to prove the following

Proposition 3.2. Let $Z(t)$ be a solution of Eq. (3.53) bounded on $[0, \infty)$. Then there is an increasing *sequence* of integers $\{n_i\}$ such that the sequence

$\{Z(t + n_iT_0)\}$ converges uniformly on any interval $[0, \tau_*]$ to a function $Y(t)$ which satisfies the equation

$$\dot{Y}(t) = AY(t) + \int_{-\infty}^{t} B(t - s)Y(s)ds + \Pi(t). \quad \square \qquad (3.54)$$

Proof. We divide the proof into four steps.

Step 1. Any continuous and periodic function $\Pi(t)$ is bounded on $[0, \infty)$. Since the functions $Z(t)$ and $\Pi(t)$ are bounded, there is a positive δ_0 such that for any $t \geq 0$

$$\sup_{t \geq 0} |Z(t)| \leq \delta_0, \qquad \sup_{t \geq 0} |\Pi(t)| \leq \delta_0. \qquad (3.55)$$

It follows from Eqs. (3.53), (3.55), and the Cauchy inequality that for any $t \geq 0$ and any positive integer n_i

$$|\dot{Z}(t + n_iT_0)| \leq |A|\, |Z(t + n_iT_0)| + \int_0^{t+n_iT_0} |B(t + n_iT_0 - s)|\, |Z(s)|ds$$

$$+|\Pi(t + n_iT_0)| \leq [1 + |A| + \int_0^{t+n_iT_0} |B(t + n_iT_0 - s)|ds]\delta_0. \quad (3.56)$$

We introduce the new variable $\tau = t + n_iT_0 - s$ and estimate the integral as follows:

$$\int_0^{t+n_iT_0} |B(t + n_iT_0 - s)|ds = \int_0^{t+n_iT_0} |B(\tau)|d\tau \leq \int_0^{\infty} |B(\tau)|d\tau = |B|.$$

Substitution of this expression into Eq. (3.56) implies that

$$|\dot{Z}(t + n_iT_0)| \leq (1 + |A| + |B|)\delta_0. \qquad (3.57)$$

It follows from Eqs. (3.55) and (3.57) that the sequence $\{Z(t + n_iT_0)\}$ is *uniformly bounded* and *equicontinuous* on any interval $[0, \tau_*]$. Therefore, it has a subsequence which converges uniformly on $[0, \tau_*]$ to a continuous function $Y(t)$. Without loss of generality, we assume that the sequence $\{Z(t+n_iT_0)\}$ converges to $Y(t)$ uniformly on $[0, \tau_*]$.

Step 2. Eq. (3.57) implies that the sequence $\{\dot{Z}(t+n_iT_0)\}$ is uniformly bounded on $[0, \tau_*]$. Let us show that it is equicontinuous as well. It follows from Eq. (3.53) that for any $t \geq 0$ and any positive integer n_i

$$\dot{Z}(t + n_iT_0) = AZ(t + n_iT_0) + \int_0^{t+n_iT_0} B(t + n_iT_0 - s)Z(s)ds + \Pi(t + n_iT_0)$$

$$= AZ(t + n_iT_0) + \int_{-n_iT_0}^{t} B(t - \tau)Z(\tau + n_iT_0)ds + \Pi(t). \qquad (3.58)$$

366

Thus, for any $t \in [0, \tau_*]$ and any positive integers n_i and $n_k > n_i$ we obtain

$$|\dot{Z}(t + n_k T_0) - \dot{Z}(t + n_i T_0)| \leq |A|\,|Z(t + n_k T_0) - Z(t + n_i T_0)|$$
$$+ \int_{-n_i T_0}^{t} |B(t - \tau)|\,|Z(\tau + n_k T_0) - Z(\tau + n_i T_0)|d\tau$$
$$+ \int_{-n_k T_0}^{-n_i T_0} |B(t - \tau)|\,|Z(\tau + n_k T_0)|d\tau.$$

Using Eq. (3.55), we find from this inequality

$$|\dot{Z}(t + n_k T_0) - \dot{Z}(t + n_i T_0)| \leq \delta_0 \int_{-n_k T_0}^{-n_i T_0} |B(t - \tau)|d\tau$$
$$+ [|A| + \int_{-n_i T_0}^{t} |B(t - \tau)|d\tau]\ \sup_{s \in [0,\tau_*]} |Z(s + n_k T_0) - Z(s + n_i T_0)|. \qquad (3.59)$$

The first integral is estimated as follows:

$$\int_{-n_k T_0}^{-n_i T_0} |B(t - \tau)|d\tau = \int_{t + n_i T_0}^{t + n_k T_0} |B(s)|ds \leq \int_{t + n_i T_0}^{\infty} |B(s)|ds.$$

It follows from condition (ii) that this integral is less than an arbitrary positive quantity, provided n_i is sufficiently large. The second integral is estimated as

$$\int_{-n_i T_0}^{t} |B(t - \tau)|d\tau = \int_{0}^{t + n_i T_0} |B(s)|ds \leq \int_{0}^{\infty} |B(s)|ds = |B|.$$

Since the sequence $\{Z(t + n_i T_0)\}$ is equicontinuous on $[0, \tau_*]$, we obtain from Eq. (3.59) that the sequence $\{\dot{Z}(t + n_i T_0)\}$ is equicontinuous as well. Therefore, it has a subsequence which converges to a continuous function $Y_0(t)$ uniformly on $[0, \tau_*]$. Without loss of generality we assume that the sequence $\{\dot{Z}(t + n_i T_0)\}$ converges to $Y_0(t)$.

Step 3. Let us prove that

$$Y_0(t) = \dot{Y}(t). \qquad (3.60)$$

Indeed, for any $t \geq 0$ and any positive integer n_i we have

$$Z(t + n_i T_0) = Z(0) + \int_{0}^{t + n_i T_0} \dot{Z}(s)ds = Z(0) + \int_{-n_i T_0}^{t} \dot{Z}(\tau + n_i T_0)d\tau.$$

The left-hand side of this equality tends to $Y(t)$, while its right-hand side tends to $Z(0) + \int_{-\infty}^{t} Y_0(\tau)d\tau$ as $n_i \to \infty$. Therefore, we can write

$$Y(t) = Z(0) + \int_{-\infty}^{t} Y_0(\tau)d\tau.$$

Differentiation of this equality yields Eq. (3.60). Hence, the sequence $\{\dot{Z}(t + n_i T_0)\}$ converges to $\dot{Y}(t)$ uniformly on $[0, \tau_*]$.

Step 4. The uniform limit of the left-hand side of Eq. (3.58) is $\dot{Y}(t)$. The uniform limit of the right-hand side of Eq. (3.58) is

$$AY(t) + \int_{-\infty}^{t} B(t - \tau)Y(\tau)d\tau + \Pi(t).$$

This ensures that the limiting function $Y(t)$ satisfies Eq. (3.54). $\square$

By applying Proposition 3.2 to Eq. (3.40) we obtain that under assumptions (6.4.58) – (6.4.60) for any bounded (with the first derivative) solution $z(t)$ of Eq. (3.40) there is a sequence of positive integers $\{n_i\}$ such that $z(t + 2\pi n_i/\omega)$ and $\dot{z}(t + 2\pi n_i/\omega)$ tend to the corresponding solutions of Eq. (3.44) uniformly on any finite interval of time.

3.6. Concluding Remarks

In this Section nonlinear radial oscillations of a thick-walled viscoelastic shell have been analyzed. It is shown that
(i) for an elastic shell with finite strains (when the material viscosity is neglected), the period of natural oscillations depends on the amplitude (in terms of the initial data) and increases with the growth of the initial energy. This phenomenon is typical of large deformations, whereas the period of natural oscillations at infinitesimal strains is independent of their amplitude;
(ii) for an elastic shell with infinitesimal strains, the amplitude of forced oscillations grows and tends to infinity, when the frequency of external load tends to the natural frequency of the shell (resonance). This phenomenon is typical of elastic media, whereas for viscoelastic media the resonance is impossible provided the relaxation measure satisfies constitutive restrictions (6.4.58) – (5.4.60).

4. Torsion of a Circular Cylinder

This Section is concerned with quasi-static *torsion* of a circular cylinder. We derive governing equations which combine torque M and compressive force P on the one hand, and twist α per unit length on the other. These equations are validated by comparison with experimental data for a cylinder made of butyl rubber. The obtained results show fair agreement between experimental data and their prediction with the use of some constitutive model based on the separability principle.

4.1. Formulation of the Problem

Let a circular cylinder with length l and radius R_0 be in its natural (stress-free) state. At instant $t = 0$, a torque $M(t)$ and a compressive load $P(t)$ are applied to the ends of the cylinder and force the following deformation:

$$r = R, \qquad \phi = \Phi + \alpha Z, \qquad z = Z. \tag{4.1}$$

Here $\{R, \Phi, Z\}$ and $\{r, \phi, z\}$ are cylindrical coordinates in the initial and actual configurations with unit vectors $\bar{e}_R$, $\bar{e}_\Phi$, $\bar{e}_Z$ and $\bar{e}_r$, $\bar{e}_\phi$, $\bar{e}_z$, respectively, $\alpha = \alpha(t)$ is twist per unit length.

In the actual configuration, the radius-vector $\bar{r}$ of a point with Lagrangian coordinates $\{R, \Phi, Z\}$ equals

$$\bar{r} = R\bar{e}_r + Z\bar{e}_z.$$

Differentiation of this equality with the use of Eqs. (1.1.1) and (1.1.6) yields

$$\bar{g}_1 = \bar{e}_r, \qquad \bar{g}_2 = r\bar{e}_\phi, \qquad \bar{g}_3 = \alpha r\bar{e}_\phi + \bar{e}_z. \tag{4.2}$$

Exercise 4.1. Show that dual vectors are calculated as

$$\bar{g}^1 = \bar{e}_r, \qquad \bar{g}^2 = \frac{1}{r}\bar{e}_\phi - \alpha\bar{e}_z, \qquad \bar{g}^3 = \bar{e}_z. \qquad \square \tag{4.3}$$

Eqs. (4.2) and (4.3) together with Eq. (2.1.44) imply that

$$\bar{\nabla}_s\bar{r}(t) = \bar{e}_r\bar{e}_r + \bar{e}_\phi\bar{e}_\phi + \bar{e}_z\bar{e}_z + r[\alpha(t) - \alpha(s)]\bar{e}_z\bar{e}_\phi. \tag{4.4}$$

It follows from Eqs. (4.4) and (2.1.46) that

$$\hat{F}_*(t, s) = \bar{e}_r\bar{e}_r + [1 + (\alpha(t) - \alpha(s))^2 r^2]\bar{e}_\phi\bar{e}_\phi + \bar{e}_z\bar{e}_z$$
$$+[\alpha(t) - \alpha(s)]r(\bar{e}_\phi\bar{e}_z + \bar{e}_z\bar{e}_\phi),$$
$$\hat{F}_*^2(t, s) = \bar{e}_r\bar{e}_r + [(1 + (\alpha(t) - \alpha(s))^2 r^2)^2 + (\alpha(t) - \alpha(s))^2 r^2]\bar{e}_\phi\bar{e}_\phi$$
$$+[1 + (\alpha(t) - \alpha(s))^2 r^2]\bar{e}_z\bar{e}_z$$
$$+(\alpha(t) - \alpha(s))[2 + (\alpha(t) - \alpha(s))^2 r^2]r(\bar{e}_\phi\bar{e}_z + \bar{e}_z\bar{e}_\phi). \tag{4.5}$$

In the cylindrical coordinates, tensor $\hat{F}_*(t, s)$ can be presented in the matrix form as

$$\hat{F}_*(t, s) = \begin{bmatrix} 1 & 0 & 0 \\ 0 & 1 + [\alpha(t) - \alpha(s)]^2 r^2 & [\alpha(t) - \alpha(s)]r \\ 0 & [\alpha(t) - \alpha(s)]r & 1 \end{bmatrix}.$$

Exercise 4.2. Derive the following characteristic equation for the relative Finger tensor:

$$\det[\hat{F}_*(t, s) - \lambda\hat{I}] = -\lambda^3 + \{3 + [\alpha(t) - \alpha(s)]^2 r^2\}\lambda^2$$
$$-\{3 + [\alpha(t) - \alpha(s)]^2 r^2\}\lambda + 1. \qquad \square$$

Since the principal invariants of tensor $\hat{F}_*(t,s)$ coincide (up to their signs) with coefficients of the characteristic equation, we find

$$I_1(t,s) = I_2(t,s) = 3 + [\alpha(t) - \alpha(s)]^2 r^2, \qquad I_3(t,s) = 1. \qquad (4.6)$$

The latter equality (4.6) shows that torsion is an isochoric deformation which preserves the volume element.

The material behavior obeys the constitutive equations of an aging incompressible viscoelastic medium (6.3.83) and (6.3.84) with $M = 1$. Substitution of expressions (4.5) and (4.6) into Eqs. (6.3.83) and (6.3.84) implies that

$$\hat{\sigma} = \sigma_{rr}\bar{e}_r\bar{e}_r + \sigma_{\phi\phi}\bar{e}_\phi\bar{e}_\phi + \sigma_{zz}\bar{e}_z\bar{e}_z + \sigma_{\phi z}(\bar{e}_\phi\bar{e}_z + \bar{e}_z\bar{e}_\phi). \qquad (4.7)$$

Here $\hat{\sigma}$ is the Cauchy stress tensor with the non-zero components

$$\sigma_{rr} = -p(t,r) + 2\{\mathcal{X}(t,0)[W_1(t,0) + (2 + \alpha^2(t)r^2)W_2(t,0)]$$
$$+ \int_0^t \frac{\partial \mathcal{X}}{\partial s}(t,s)[W_1(t,s) + (2 + (\alpha(t) - \alpha(s))^2 r^2)W_2(t,s)]ds\},$$
$$\sigma_{\phi\phi} = -p(t,r) + 2\{\mathcal{X}(t,0)[(1 + \alpha^2(t)r^2)W_1(t,0) + (2 + \alpha^2(t)r^2)W_2(t,0)]$$
$$+ \int_0^t \frac{\partial \mathcal{X}}{\partial s}(t,s)[(1 + (\alpha(t) - \alpha(s))^2 r^2)W_1(t,s)$$
$$+ (2 + (\alpha(t) - \alpha(s))^2 r^2)W_2(t,s)]ds\},$$
$$\sigma_{zz} = -p(t,r) + 2\{\mathcal{X}(t,0)[W_1(t,0) + 2W_2(t,0)]$$
$$+ \int_0^t \frac{\partial \mathcal{X}}{\partial s}(t,s)[W_1(t,s) + 2W_2(t,s)]ds\},$$
$$\sigma_{r\phi} = 2r\{\mathcal{X}(t,0)\alpha(t)[W_1(t,0) + W_2(t,0)]$$
$$+ \int_0^t \frac{\partial \mathcal{X}}{\partial s}(t,s)[\alpha(t) - \alpha(s)][W_1(t,s) + W_2(t,s)]ds\}, \qquad (4.8)$$

where $p = p(t,R)$ is a pressure, $W(I_1, I_2)$ is a strain energy density,

$$\psi_1(t,s) = W_1(t,s) + I_1(t,s)W_2(t,s), \qquad \psi_2(t,s) = -W_2(t,s),$$
$$W_k(t,s) = \frac{\partial W}{\partial I_k}(I_1(\hat{F}(t,s)), I_2(\hat{F}(t,s))).$$

Function $\mathcal{X}(t,s)$ satisfies conditions (6.4.103) and (6.4.104) and is presented in the form (2.9), where $\beta(t)$ is a dimensionless shear modulus, and $Q(t,s)$ is a regular relaxation measure. For a non-aging material, we set $\beta(t) = 1$ and $Q(t,s) = Q_0(t-s)$.

It follows from Eqs. (4.8) that

$$\sigma_{\phi\phi}(t,r) = \sigma_{rr}(t,r) + 2r^2\{\mathcal{X}(t,0)\alpha^2(t)W_1(t,0)$$

370

$$+ \int_0^t \frac{\partial \mathcal{X}}{\partial s}(t,s)(\alpha(t) - \alpha(s))^2 W_1(t,s)ds\},$$

$$\sigma_{zz}(t,r) = \sigma_{rr}(t,r) - 2r^2\{\mathcal{X}(t,0)\alpha^2(t)W_2(t,0)$$

$$+ \int_0^t \frac{\partial \mathcal{X}}{\partial s}(t,s)(\alpha(t) - \alpha(s))^2 W_2(t,s)ds\}. \tag{4.9}$$

Exercise 4.3. Show that in the absence of body forces, the only equilibrium equation (2.2.29) is written as

$$\frac{\partial \sigma_{rr}}{\partial r} + \frac{\sigma_{rr} - \sigma_{\phi\phi}}{r} = 0. \qquad \square \tag{4.10}$$

By integrating Eq. (4.10) from r to R_0 and using Eq. (4.9) and the boundary condition

$$\sigma_{rr}|_{r=R_0} = 0,$$

we obtain

$$\sigma_{rr}(t,r) = -2 \int_r^{R_0} [\mathcal{X}(t,0)\alpha^2(t)W_1(t,0)$$

$$+ \int_0^t \frac{\partial \mathcal{X}}{\partial s}(t,s)(\alpha(t) - \alpha(s))^2 W_1(t,s)ds]\rho d\rho. \tag{4.11}$$

We now calculate torque M and compressive force P by the formulas

$$M = 2\pi \int_0^{R_0} \sigma_{\phi z} r^2 dr, \qquad P = -2\pi \int_0^{R_0} \sigma_{zz} r dr. \tag{4.12}$$

Substitution of Eqs. (4.8), (4.9), and (4.11) into Eq. (4.12) implies that

$$M(t) = 4\pi \int_0^{R_0} \{\mathcal{X}(t,0)\alpha(t)[W_1(t,0) + W_2(t,0)]$$

$$+ \int_0^t \frac{\partial \mathcal{X}}{\partial s}(t,s)[\alpha(t) - \alpha(s)][W_1(t,s) + W_2(t,s)]ds\}r^3 dr, \tag{4.13}$$

$$P(t) = 4\pi \int_0^{R_0} \{[\mathcal{X}(t,0)\alpha^2(t)W_2(t,0) + \int_0^t \frac{\partial \mathcal{X}}{\partial s}(t,s)(\alpha(t) - \alpha(s))^2 W_2(t,s)ds]r^2$$

$$+ \int_r^{R_0} [\mathcal{X}(t,0)\alpha^2(t)W_1(t,0) + \int_0^t \frac{\partial \mathcal{X}}{\partial s}(t,s)(\alpha(t) - \alpha(s))^2 W_1(t,s)ds]\rho d\rho\}r dr. \tag{4.14}$$

Changing the order of integration in Eq. (4.14), we find

$$P(t) = 2\pi \int_0^{R_0} \{\mathcal{X}(t,0)\alpha^2(t)[W_1(t,0) + 2W_2(t,0)]$$

$$+ \int_0^t \frac{\partial \mathcal{X}}{\partial s}(t,s)(\alpha(t) - \alpha(s))^2 [W_1(t,s) + 2W_2(t,s)]ds\}r^3 dr. \tag{4.15}$$

For a given function $\alpha(t)$, Eqs. (4.13) and (4.15) determine torque $M(t)$ and compressive force $P(t)$. For creep experiments, when twist angle α per unit length remains uniform in time

$$\alpha(t) = \begin{cases} 0, & t \leq 0, \\ \alpha, & t > 0, \end{cases}$$

these formulas may be simplified

$$M(t) = 4\pi\alpha\mathcal{X}(t,0) \int_0^{R_0} [\frac{\partial W}{\partial I_1}(3 + \alpha^2 r^2, 3 + \alpha^2 r^2)$$
$$+ \frac{\partial W}{\partial I_2}(3 + \alpha^2 r^2, 3 + \alpha^2 r^2)]r^3 dr,$$

$$P(t) = 2\pi\alpha^2\mathcal{X}(t,0) \int_0^{R_0} [\frac{\partial W}{\partial I_1}(3 + \alpha^2 r^2, 3 + \alpha^2 r^2)$$
$$+ 2\frac{\partial W}{\partial I_2}(3 + \alpha^2 r^2, 3 + \alpha^2 r^2)]r^3 dr. \tag{4.16}$$

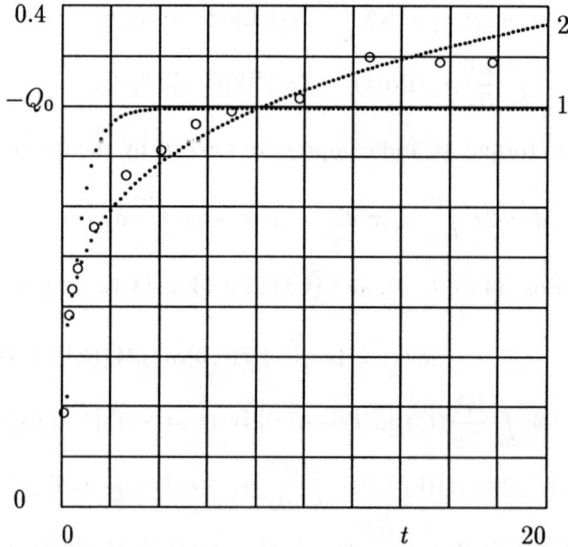

Fig. 4.1. Relaxation measure Q_0 *versus* time t (sec). Large circles correspond to experimental data obtained by Hausler and Sayir (1995) for torsion of a butyl rubber specimen at infinitesimal strains with twist angle $\vartheta = 0.0081$. Small circles correspond to exponential function (4.21) with $\chi_1 = 0.318$ and $\gamma_1 = 1.610$ (curve 1), and to power function (4.22) with $\chi_2 = 0.205$ and $\gamma_2 = 0.212$ (curve 2).

For given functions $M(t)$ and $P(t)$, Eqs. (4.16) may be treated as integral equations for strain energy density $W(I_1, I_2)$. The problem of determining the function $W(I_1, I_2)$ from Eqs. (4.16) is ill-posed, and it is difficult to presume that its regular solution may be obtained.

4.2. Comparison of Theoretical and Experimental Results

In this subsection we compare the above theoretical results with experimental data obtained by Hausler & Sayir (1995) for torsion of a butyl rubber specimen with moderately large strains, when twist $\theta = \alpha l$ is sufficiently small. Expanding function W in series in $I_1 - 3$ and $I_2 - 3$ and neglecting terms of higher order of magnitude compared with ϑ, we find

$$M(t) = 2J\frac{\vartheta}{l}\mathcal{X}(t,0)(C_1 + C_2), \qquad P(t) = J(\frac{\vartheta}{l})^2\mathcal{X}(t,0)(C_1 + 2C_2), \quad (4.17)$$

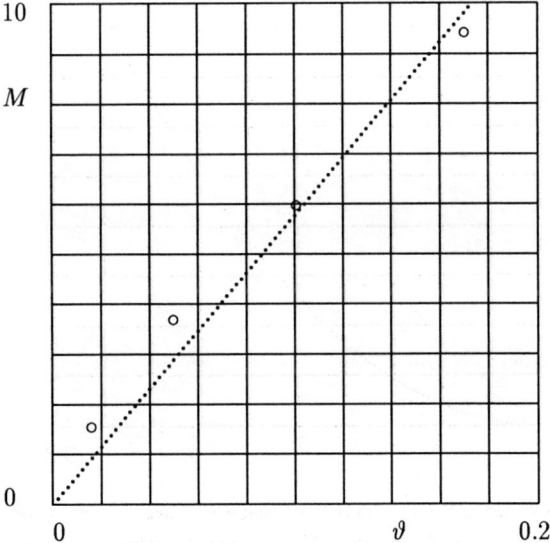

Fig. 4.2. Torque M (N·m) *versus* twist angle ϑ. Large circles correspond to experimental data obtained by Hausler & Sayir (1995). Small circles provide the approximation by linear function (4.23) with $a_M = 57.68$ (N·m).

where

$$J = \frac{\pi}{2}R_0^4, \qquad C_k = \frac{\partial W}{\partial I_k}(3,3), \qquad k = 1,2.$$

Exercise 4.4. Check that Eqs. (4.17) coincide with the governing equations for torsion of a circular cylinder made of a viscoelastic medium with the Mooney–Rivlin strain energy density (3.2.104). □

For a non-aging viscoelastic medium, Eqs. (4.17) are written as

$$M(t) = 2J\frac{\vartheta}{l}[1 + Q_0(t)](C_1 + C_2), \quad P(t) = J(\frac{\vartheta}{l})^2[1 + Q_0(t)](C_1 + 2C_2). \quad (4.18)$$

According to *the separability principle*, see Section 6.3, viscous and elastic effects are independent of each other. This means that the relaxation measure $Q_0(t)$ is independent of strains and stresses. This function may be found from experimental data for torsion at infinitesimal strains, since experiments with small strains are essentially simpler than experiments with large deformations.

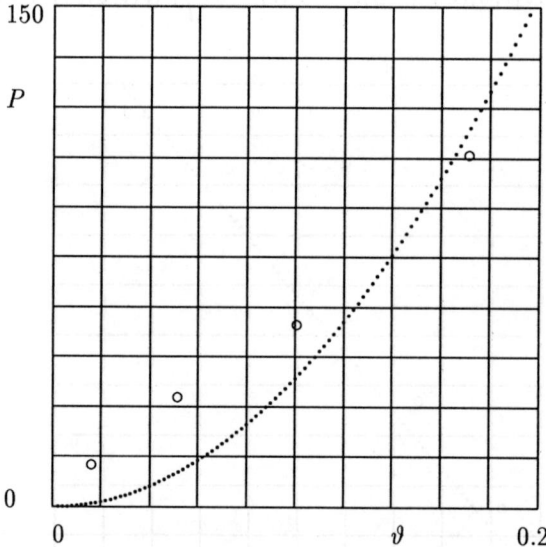

Fig. 4.3. Compressive force P (N) *versus* twist angle ϑ. Large circles correspond to experimental data obtained by Hausler & Sayir (1995). Small circles provide approximation by quadratic function (4.24) with $a_P = 3857.55$ (N).

For infinitesimal strains, torque $M(t)$ is connected with twist ϑ by the equality

$$M(t) = \mu J\frac{\vartheta}{l}[1 + Q_0(t)], \qquad (4.19)$$

where $\mu = 2(C_1 + C_2)$ is the shear modulus, see Eq. (3.2.28). It follows from Eq. (4.19) that

$$Q_0(t) = \frac{M(t)}{M(0)} - 1. \qquad (4.20)$$

374

The relaxation measure Q_0 *versus* time t is plotted in Fig. 4.1 for a butyl rubber specimen at $\vartheta = 0.00181$. We approximate function $Q_0(t)$ by the exponential function

$$Q_0(t) = -\chi_1[1 - \exp(-\gamma_1 t)], \tag{4.21}$$

and by the power function

$$Q_0(t) = -\chi_2 t^{\gamma_2}. \tag{4.22}$$

Parameters χ_k and γ_k are found by fitting experimental data by the least square method.

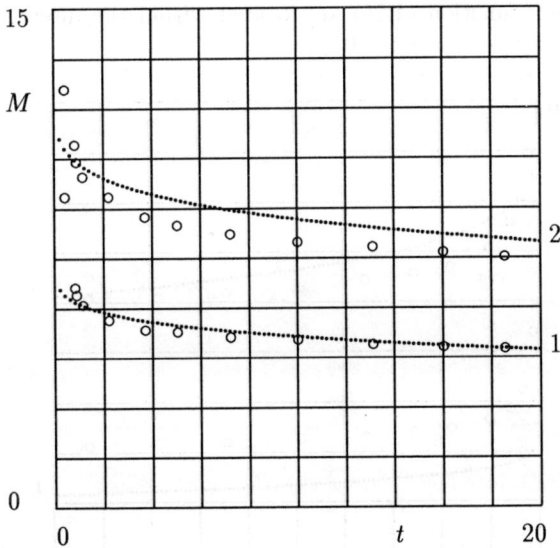

Fig. 4.4. Torque M (N·m) *versus* time t (sec). Curve 1: twist angle $\gamma = 0.102$, curve 2: twist angle $\gamma = 0.172$. Large circles correspond to experimental data obtained by Hausler & Sayir (1995). Small circles are derived by using Eqs. (4.18).

Exponential function (4.21) demonstrates large discrepancies between experimental points and their prediction, whereas power function (4.22) ensures good agreement with experimental data. Below we employ the power function with parameters $\chi_2 = 0.21$ and $\gamma_2 = 0.21$ as relaxation measure $Q_0(t)$ for butyl rubber.

To find parameters C_1 and C_2 we consider M and P values obtained at $\tau = 2$ (sec) for various ϑ values, and approximate them by the linear function

$$M(\tau) = a_M \vartheta, \tag{4.23}$$

and by the quadratic function

$$P(\tau) = a_P \vartheta^2. \tag{4.24}$$

Experimental data and their predictions are plotted in Figs. 4.2 and 4.3. The results show good agreement between experimental data for torque M and their prediction with the use of linear dependence (4.23) with $a_M = 57.68$ (N·m). Fitting of experimental data for compressive force P is worse. The least square method provides the value $a_P = 3857.55$ (N). Data obtained by Hausler & Sayir (1995) demonstrate large differences between observations for various specimens. The discrepancies between experimental data and their approximation by quadratic function (4.24) are located within the interval of errors in measurements (at least for $\vartheta > 0.05$).

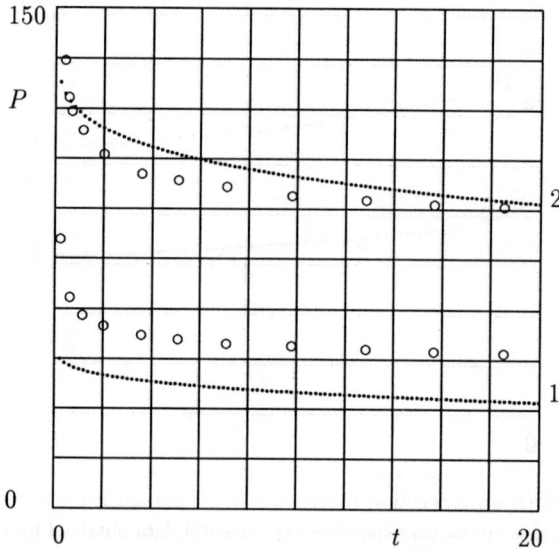

Fig. 4.5. Compressive force P (N) *versus* time t (sec). Curve 1: twist angle $\gamma = 0.102$, curve 2: twist angle $\gamma = 0.172$. Large circles correspond to experimental data obtained by Hausler & Sayir (1995). Small circles are derived by using Eqs. (4.18).

Exercise 4.5. By using Eqs. (4.18), (4.23), and (4.24), check that

$$C_1 = \frac{l(a_M - a_P l)}{J[1 + Q_0(\tau)]}, \qquad C_2 = \frac{l(2a_P l - a_M)}{2J[1 + Q_0(\tau)]}. \qquad \square \tag{4.25}$$

As common practice, parameters C_1 and C_2 are positive. This assertion together with Eqs. (4.25) implies that

$$a_P l < a_M < 2a_P l.$$

376

The experimental data by Hausler & Sayir (1995) show that this inequality is not fulfilled. Calculations according to formulas (4.25) yield $C_1 = -2.03$ (MPa) and $C_2 = 5.04$ (MPa). Despite the negative value of C_1, these results are acceptable, since the shear modulus μ is positive.

It is of special interest to compare experimental results for moderately large deformations with their prediction based on Eqs. (4.18). Data for torque M and compressive load P are plotted in Figs. 4.4 and 4.5. These figures demonstrate good agreement between experimental results and their prediction by using model (6.3.83) and (6.3.84). In particular, an excellent fitting is obtained for torque M at relatively small twist ($\gamma = 0.102$). Prediction of compressive load P leads to small discrepancies between theory and experiments, which may be explained by errors in measurements of P at instant τ, see Fig. 4.3.

Discrepancies between experimental data and their prediction by the integral model (6.3.83) and (6.3.84) as well as by the differential model (6.2.65) derived by Hausler & Sayir (1995) are similar. The difference consists in the number of parameters used in models. Model (6.3.83) and (6.3.84) contains only 4 parameters: χ_2, γ_2 for the relaxation measure $Q_0(t)$, and C_1 and C_2 for the strain energy density $W(I_1, I_2)$, while model (6.2.65) contains 10 parameters which are determined by fitting experimental data.

To the best of our knowledge, four is the minimal number of the material parameters in constitutive models for incompressible viscoelastic media with finite strains: two parameters of relaxation measure characterize the material viscosity and the characteristic time of relaxation, while two parameters of strain energy density characterize its dependence on the first and second principal invariants, respectively. The above results show that the integral constitutive model based on the separability principle describes adequately torsion with moderate strains.

4.3. Concluding Remarks

This Section deals with quasi-static torsion of a circular cylinder made of an incompressible viscoelastic material with finite strains. The following results are derived:
(i) in the framework of the constitutive model (6.3.83) and (6.3.84), this problem is universal, i.e. its explicit solution (4.13) and (4.15) is derived for an arbitrary strain energy density $W(I_1, I_2)$;
(ii) experimental data for butyl rubber confirm the separability principle which states that the material viscosity is independent of elastic response. By employing this principle, we may reduce significantly the number of experiments necessary to validate constitutive models, and simplify these experiments. In particular, only the standard relaxation tests with infinitesimal strains are neces-

sary to determine relaxation measure $Q_0(t)$, and the standard time-independent test should be carried out to find strain energy density $W(I_1, I_2)$. Data of the latter test may be obtained at any instant τ and recalculated according to algorithm (4.25).

5. Homogeneous Deformations

This Section is concerned with homogeneous deformations of an incompressible viscoelastic medium. The medium is loaded on the interval $(-\infty, 0]$ in such a way that homogeneous time-dependent deformations occur in it. From the mathematical standpoint, this means that the relative gradient of deformations $\bar{\nabla}_s \bar{r}(t)$ is independent of spatial coordinates for $-\infty < s \le t \le 0$. At instant $t = 0$, external loads are reduced, and the problem consists in finding *an elastic recoil*.

The question of interest is to provide a qualitative description of the quasistatic *recovery*. We divide the whole spectrum of homogeneous deformations into classes (elongation motions, shear-free motions, etc.) and analyze whether the recoil belongs to the same class as the motion under loading. The most surprising result is that for some deformation histories, the body recovers to a state in which it has never been under loading. The exposition is close to Engler (1991).

5.1. Formulation of the Problem

Let us consider a viscoelastic medium which occupies a region Ω in its natural state. The medium is loaded on the interval $(-\infty, 0]$ in such a way that only *homogeneous* deformations occur in it. This means that for any moment t

$$\bar{r}(t) = \hat{\Lambda}(t) \cdot \bar{r}_{-\infty}, \tag{5.1}$$

where $\bar{r}(t)$ and $\bar{r}_{-\infty}$ are radius-vectors in the actual and initial configurations, respectively, and $\hat{\Lambda}(t)$ is a time-dependent tensor. The only difference between our standard notation and the notation of this Section is that we replace the initial moment from 0 to $-\infty$.

Differentiation of Eq. (5.1) implies that

$$\bar{g}_i(t) = \hat{\Lambda}(t) \cdot \bar{g}_{-\infty \, i} = \bar{g}_{-\infty \, i} \cdot \hat{\Lambda}^T(t), \tag{5.2}$$

where $\bar{g}_{-\infty \, i}$ and $\bar{g}_i(t)$ are tangent vectors in the initial and actual configurations, and the superscript T denotes transpose.

Exercise 5.1. Check that dual vectors are calculated as follows:

$$\bar{g}^i(t) = (\hat{\Lambda}^{-1}(t))^T \cdot \bar{g}^i_{-\infty} = \bar{g}^i_{-\infty} \cdot \hat{\Lambda}^{-1}(t). \quad \square \tag{5.3}$$

378

Substitution of expressions (5.2) and (5.3) into Eq. (2.1.44) yields

$$\bar{\nabla}_s \bar{r}(t) = \bar{g}^i(s)\hat{\Lambda}(t) \cdot \bar{g}_{-\infty\,i} = \bar{g}^i(s)\bar{g}_{-\infty\,i} \cdot \hat{\Lambda}^T(t)$$
$$= (\hat{\Lambda}^{-1}(s))^T \cdot \hat{I} \cdot \hat{\Lambda}^T(t) = [\hat{\Lambda}(t) \cdot \hat{\Lambda}^{-1}(s)]^T, \tag{5.4}$$

where $\hat{I}$ is the unit tensor. It follows from Eqs. (5.4) and (2.1.46) that the relative Finger tensor equals

$$\hat{F}_*(t, s) = \hat{\Lambda}(t) \cdot \hat{\Lambda}^{-1}(s) \cdot (\hat{\Lambda}^{-1}(s))^T \cdot \hat{\Lambda}^T(t). \tag{5.5}$$

Multiplying Eq. (5.5) by itself, we obtain

$$\hat{F}_*^2(t, s) = \hat{\Lambda}(t) \cdot \hat{\Lambda}^{-1}(s) \cdot (\hat{\Lambda}^{-1}(s))^T \cdot \hat{\Lambda}^T(t) \cdot \hat{\Lambda}(t) \cdot \hat{\Lambda}^{-1}(s) \cdot (\hat{\Lambda}^{-1}(s))^T \cdot \hat{\Lambda}^T(t). \tag{5.6}$$

Differentiation of Eq. (5.1) with respect to time implies that

$$\bar{v}(t) = \frac{d\hat{\Lambda}}{dt}(t) \cdot \bar{r}_{-\infty}.$$

Exercise 5.2. By applying the gradient operator to this equality and using Eq. (5.3), show that

$$\bar{\nabla}_t \bar{v}(t) = (\hat{\Lambda}^T(t))^{-1} \cdot (\frac{d\hat{\Lambda}}{dt}(t))^T. \quad \square \tag{5.7}$$

Substitution of expression (5.7) into Eq. (1.4.25) yields

$$\hat{D}(t) = \frac{1}{2}[\frac{d\hat{\Lambda}}{dt}(t) \cdot \hat{\Lambda}^{-1}(t) + (\hat{\Lambda}^T(t))^{-1} \cdot (\frac{d\hat{\Lambda}}{dt}(t))^T]. \tag{5.8}$$

We assume that the Cauchy stress tensor $\hat{\sigma}(t)$ satisfies constitutive equation (6.3.115). Replacing the initial moment $t = 0$ by $t = -\infty$, we write

$$\hat{\sigma}(t) = -p(t)\hat{I} + 2\eta\hat{D}(t)$$
$$+2\{[1 + Q_0(\infty)][\psi_1(t, -\infty)\hat{F}(t, -\infty) + \psi_2(t, -\infty)\hat{F}^2(t, -\infty)]$$
$$- \int_{-\infty}^{t} \dot{Q}_0(t - s)[\psi_1(t, s)\hat{F}(t, s) + \psi_2(t, s)\hat{F}^2(t, s)]ds\}. \tag{5.9}$$

Here $p = p(t)$ is a pressure, $W(I_1, I_2)$ is a strain energy density,

$$\psi_1(t, s) = W_1(t, s) + I_1(t, s)W_2(t, s), \qquad \psi_2(t, s) = -W_2(t, s),$$
$$W_k(t, s) = \frac{\partial W}{\partial I_k}(I_1(\hat{F}(t, s)), I_2(\hat{F}(t, s))),$$

$Q_0(t)$ is a relaxation measure satisfying conditions (6.4.58) – (6.4.60), and the superscript dot denotes differentiation with respect to time.

For any homogeneous deformation, tensor $\hat{\sigma}(t)$ is independent of spatial coordinates. Therefore, it satisfies the equilibrium equation (2.2.21), which, in the absence of body forces, is written as $\bar{\nabla}_t \cdot \hat{\sigma}(t) = 0$. For any $t \geq 0$, surface traction is absent, and in order to satisfy the boundary conditions in stresses, we set

$$\hat{\sigma}(t) = 0.$$

Substitution of expression (5.9) into this equality yields

$$
\begin{aligned}
p(t)\hat{I} &- 2\eta \hat{D}(t) \\
= 2\{[1 + Q_0(\infty)][\psi_1(t, -\infty)\hat{F}(t, -\infty) &+ \psi_2(t, -\infty)\hat{F}^2(t, -\infty)] \\
- \int_{-\infty}^{t} \dot{Q}_0(t - s)[\psi_1(t, s)\hat{F}(t, s) &+ \psi_2(t, s)\hat{F}^2(t, s)]ds\}.
\end{aligned}
\tag{5.10}
$$

We multiply Eq. (5.10) by $\hat{\Lambda}^{-1}(t)$ from the left, and by $(\hat{\Lambda}^{-1}(t))^T$ from the right. Afterward, we substitute expressions (5.5), (5.6), and (5.8) into the obtained equality. As a result, we find

$$
\begin{aligned}
& p(t)\hat{\Lambda}^{-1}(t) \cdot (\hat{\Lambda}^{-1}(t))^T \\
& -\eta \hat{\Lambda}^{-1}(t) \cdot [(\hat{\Lambda}^{-1}(t))^T \cdot (\frac{d\hat{\Lambda}}{dt}(t))^T + \frac{d\hat{\Lambda}}{dt}(t) \cdot \hat{\Lambda}^{-1}(t)] \cdot (\hat{\Lambda}^{-1}(t))^T \\
& = 2\{[1 + Q_0(\infty)][\psi_1(t, -\infty)\hat{I} + \psi_2(t, -\infty)\hat{\Lambda}^T(t) \cdot \hat{\Lambda}(t)] \\
& - \int_{-\infty}^{t} \dot{Q}_0(t - s)[\psi_1(t, s)\hat{\Lambda}^{-1}(s) \cdot (\hat{\Lambda}^{-1}(s))^T \\
& +\psi_2(t, s)\hat{\Lambda}^{-1}(s) \cdot (\hat{\Lambda}^{-1}(s))^T \cdot \hat{\Lambda}^T(t) \cdot \hat{\Lambda}(t) \cdot \hat{\Lambda}^{-1}(s) \cdot (\hat{\Lambda}^{-1}(s))^T]ds\}.
\end{aligned}
\tag{5.11}
$$

To derive Eq. (5.11) we use the equality $\hat{\Lambda}(-\infty) = \hat{I}$, which means that the initial configuration is stress-free.

Introduce the notation

$$\hat{\Gamma}(t) = \hat{\Lambda}^{-1}(t) \cdot (\hat{\Lambda}^{-1}(t))^T.
\tag{5.12}$$

Exercise 5.3. Check that tensor $\hat{\Gamma}(t)$ is inverse to the Cauchy deformation tensor from the initial to the actual configuration

$$\hat{g}(t) = \bar{\nabla}_{-\infty}\bar{r}(t) \cdot (\bar{\nabla}_{-\infty}\bar{r}(t))^T. \quad \square$$

Let us calculate the derivative of $\Gamma(t)$ with respect to time

$$\frac{d\hat{\Gamma}}{dt}(t) = \frac{d\hat{\Lambda}^{-1}}{dt}(t) \cdot (\hat{\Lambda}^{-1}(t))^T + \hat{\Lambda}^{-1}(t) \cdot (\frac{d\hat{\Lambda}^{-1}}{dt}(t))^T.
\tag{5.13}$$

It follows from the identity $\hat{\Lambda}(t) \cdot \hat{\Lambda}^{-1}(t) = \hat{I}$ that

$$\frac{d\hat{\Lambda}}{dt}(t) \cdot \hat{\Lambda}^{-1}(t) + \hat{\Lambda}(t) \cdot \frac{d\hat{\Lambda}^{-1}}{dt}(t) = 0,$$

which implies that

$$\frac{d\hat{\Lambda}^{-1}}{dt}(t) = -\hat{\Lambda}^{-1}(t) \cdot \frac{d\hat{\Lambda}}{dt}(t) \cdot \hat{\Lambda}^{-1}(t). \tag{5.14}$$

Substitution of expression (5.14) into Eq. (5.13) yields

$$\frac{d\hat{\Gamma}}{dt}(t) = -\hat{\Lambda}^{-1}(t) \cdot [\frac{d\hat{\Lambda}}{dt}(t) \cdot \hat{\Lambda}^{-1}(t) + (\hat{\Lambda}^{-1}(t))^T \cdot (\frac{d\hat{\Lambda}}{dt}(t))^T] \cdot (\hat{\Lambda}^{-1}(t))^T. \tag{5.15}$$

By using Eqs. (5.12) and (5.15), we write Eq. (5.11) as follows:

$$\eta\frac{d\hat{\Gamma}}{dt}(t) + p(t)\hat{\Gamma}(t) = 2\{[1 + Q_0(\infty)][\psi_1(t, -\infty)\hat{I} + \psi_2(t, -\infty)\hat{\Gamma}^{-1}(t)]$$
$$- \int_{-\infty}^{t} \dot{Q}_0(t - s)[\psi_1(t, s)\hat{\Gamma}(s) + \psi_2(t, s)\hat{\Gamma}(s) \cdot \hat{\Gamma}^{-1}(t) \cdot \hat{\Gamma}(s)]ds\}. \tag{5.16}$$

Eq. (5.16) is a nonlinear integro-differential equation which describes homogeneous deformations of a viscoelastic medium. Our objective is to study elastic recovery after the load is reduced. For this purpose, we assume that function $\hat{\Gamma}(t)$ for $t < 0$, as well as $\hat{\Gamma}(0)$ are given:

$$\hat{\Gamma}(t) = \hat{\Theta}(t) \quad (-\infty < t < 0), \quad \hat{\Gamma}(0) = \hat{\Theta}_0. \tag{5.17}$$

We suppose that the medium is in its natural state before the deformation

$$\hat{\Theta}(-\infty) = \hat{I}, \tag{5.18}$$

and that $\hat{\Theta}(0) \neq \hat{\Theta}_0$. A solution of Eq. (5.16) is a continuously differentiable tensor function which satisfies Eq. (5.16) for any $t \geq 0$, and which belongs to the set $\mathcal{N}$ of symmetrical, positive definite tensors of the second rank with the unit determinant. Since any tensor of the second rank may be presented by a (3×3)-matrix in a fixed basis, we treat $\mathcal{N}$ as the set of symmetrical, positive definite (3×3)-matrices with the unit determinant, referring to some Cartesian coordinate frame $\{x^i\}$.

Exercise 5.4. Show that the set $\mathcal{N}$ is not a vector space. □

We do not intend to analyze the existence and uniqueness of solutions to the initial problem (5.16), (5.17) and assume that for any $t \geq 0$, the unique solution $\hat{\Gamma}(t)$ of Eqs. (5.16) and (5.17) belongs to $\mathcal{N}$ provided the initial conditions $\hat{\Theta}(t)$ and $\hat{\Theta}_0$ belong to $\mathcal{N}$ as well. Our aim is to describe subsets N of the vector space $\mathcal{M}$ of real (3×3)-matrices which (i) are of interest for applications, and (ii) have the same property as the set $\mathcal{N}$: for any $t \geq 0$, any solution $\hat{\Gamma}(t) \in N \cap \mathcal{N}$ provided $\hat{\Theta}(t), \hat{\Theta}_0 \in N \cap \mathcal{N}$.

In subsection 2 we list several subsets of $\mathcal{M}$ which correspond to particular homogeneous deformations, and prove an appropriate assertion (Proposition 5.1). Regretfully, we cannot derive the most important results for an arbitrary strain energy density $W(I_1, I_2)$, and confine ourselves to the Knowles viscoelastic materials (3.2.114) with strain energy density W depending on I_1 only, $W = W(I_1)$. For the Knowles media, Eq. (5.16) turns into the linear integro-differential equation

$$\eta \frac{d\hat{\Gamma}}{dt}(t) + p(t)\hat{\Gamma}(t) = 2[1 + Q_0(\infty)]W_1(t, -\infty)\hat{I}$$
$$-2\int_{-\infty}^{t} \dot{Q}_0(t-s)W_1(t,s)\hat{\Gamma}(s)ds, \qquad (5.19)$$

with $W_1 = dW/dI_1$.

5.2. Examples of Homogeneous Deformations

Let us list several subsets of $\mathcal{M}$ which are of special interest for applications:
(i) set M_1 of *shear-free deformations* is determined as a set of matrices

$$\hat{g} = \begin{bmatrix} \lambda_1 & 0 & 0 \\ 0 & \lambda_2 & 0 \\ 0 & 0 & \lambda_3 \end{bmatrix} \qquad (5.20)$$

with positive λ_i. Particular cases of shear-free deformations are *filament stretching*

$$\lambda_1 = \lambda \geq 1, \qquad \lambda_2 = \lambda_3 = \lambda^{-\frac{1}{2}} \leq 1, \qquad (5.21)$$

and *sheet stretching*

$$\lambda_1 = \lambda \leq 1, \qquad \lambda_2 = \lambda_3 = \lambda^{-\frac{1}{2}} \geq 1. \qquad (5.22)$$

According to Exercise 5.3, for any $\hat{g} \in M_1$ the corresponding matrix $\hat{\Theta}$ belongs to the set N_1 of matrices inverse to $\hat{g}$. It is easy to check that the set N_1 consists of matrices which have the form (5.20) as well.
Exercise 5.5. Prove that N_1 is a subspace of $\mathcal{M}$. $\square$
Exercise 5.6. Show that the set $N_1 \cap \mathcal{N}$ consists of matrices (5.20) which satisfy the conditions $\lambda_i > 0$ and $\lambda_1 \lambda_2 \lambda_3 = 1$. $\square$
(ii) set M_2 of *shear deformations* is determined as a set of matrices

$$\hat{g} = \begin{bmatrix} 1 & a & b \\ a & 1+a^2 & ab+c \\ b & ab+c & 1+b^2+c^2 \end{bmatrix}, \qquad (5.23)$$

where a, b, and c are non-negative quantities.

Exercise 5.7. Show that $\det \hat{g} = 1$ for arbitrary a, b, and c in Eq. (5.23). □

Exercise 5.8. Check that for *simple shear*, see Eq. (4.2.1), $a = \kappa$, $b = c = 0$, where κ is the angle of shear. Show that in this case, the set M_2 consists of matrices

$$\hat{g} = \begin{bmatrix} 1 & \kappa & 0 \\ \kappa & 1+\kappa^2 & 0 \\ 0 & 0 & 1 \end{bmatrix}. \qquad \square \qquad (5.24)$$

Exercise 5.9. By using Exercise 5.3 and Eq. (5.24), derive the corresponding matrix $\hat{\Theta}$

$$\hat{\Theta} = \begin{bmatrix} 1+\kappa^2 & -\kappa & 0 \\ -\kappa & 1 & 0 \\ 0 & 0 & 1 \end{bmatrix}. \qquad (5.25)$$

Show that the matrix $\hat{\Theta}$ belongs to the subspace $N_2 \subset \mathcal{M}$ of matrices

$$\begin{bmatrix} a & -b & 0 \\ -b & c & 0 \\ 0 & 0 & c \end{bmatrix} \qquad (5.26)$$

which are characterized by parameters a, b, and c. □

Exercise 5.10. Check that the set $N_2 \cap \mathcal{N}$ consists of matrices (5.26) which satisfy the condition $c(ac - b^2) = 1$. □

(iii) set M_3 of *deformations with orthogonal embedded plane and vector* is determined as a set of matrices

$$\hat{g} = \begin{bmatrix} g_{11} & g_{12} & 0 \\ g_{12} & g_{22} & 0 \\ 0 & 0 & g_{33} \end{bmatrix}, \qquad (5.27)$$

where

$$g_{11} > 0, \qquad g_{11}g_{22} - g_{12}^2 > 0, \qquad g_{33} > 0. \qquad (5.28)$$

Exercise 5.11. Check that there is a vector and a plane embedded into a medium, which remain orthogonal throughout the deformation history provided $\hat{g}(t)$ belongs to M_3. □

(iv) set M_4 of *deformations with two orthogonal embedded planes* is determined as a set of matrices

$$\hat{g} = \begin{bmatrix} g_{11} & g_{12} & 0 \\ g_{12} & g_{22} & g_{23} \\ 0 & g_{23} & g_{33} \end{bmatrix}, \qquad (5.29)$$

which satisfy conditions (5.28).

Exercise 5.12. Check that there are two planes embedded into a medium, which remain orthogonal throughout the deformation history provided $\hat{g}(t)$ belongs to M_4. □

Proposition 5.1. Let N be a subspace of $\mathcal{M}$, and the initial data $\hat{\Theta}(t)$ and $\hat{\Theta}_0$ belong to $N \bigcap \mathcal{N}$ for any $t < 0$. Then the solution $\hat{\Gamma}(t)$ of Eq. (5.19) belongs to $N \bigcap \mathcal{N}$ for any $t \geq 0$.

Proof. We will prove this assertion for $\eta > 0$. The proof for $\eta = 0$ is similar, and it is left to the reader as an exercise.

For any matrix $\hat{A}$, let $\|\hat{A}\|$ be the Euclidean norm of $\hat{A}$, and $d(\hat{A}, N)$ *the distance* from $\hat{A}$ to a set N

$$d(\hat{A}, N) = \inf_{\hat{B} \in N} \|\hat{A} - \hat{B}\|.$$

Evidently, $d(\hat{A}, N) = 0$ if and only if $\hat{A} \in N$. For any linear subspace N, we have $d(\hat{I}, N) = 0$.

We integrate Eq. (5.19) from 0 to t and obtain

$$\eta \hat{\Gamma}(t) + \int_0^t p(s)\hat{\Gamma}(s)ds = \eta \hat{\Theta}_0 + 2[1 + Q_0(\infty)] \int_0^t W_1(s, -\infty)ds \hat{I}$$

$$- \int_0^t [\int_{-\infty}^0 \dot{Q}_0(\tau - s)W_1(\tau, s)\hat{\Theta}(s)ds + \int_0^\tau \dot{Q}_0(\tau - s)W_1(\tau, s)\hat{\Gamma}(s)ds]d\tau. \quad (5.30)$$

Introduce the notation $u(t) = d(\hat{\Gamma}(t), N)$. Calculation of the distance between the left- and right-hand sides of Eq. (5.30) and the space N implies that

$$\eta u(t) \leq \int_0^t |p(s)|u(s)ds + \int_0^t d\tau \int_0^\tau |\dot{Q}_0(\tau - s)|\, |W_1(\tau, s)|u(s)ds$$

$$= \int_0^t P(t, s)u(s)ds, \quad (5.31)$$

where

$$P(t, s) = |p(t)| + \int_s^t |\dot{Q}_0(\tau - s)|\, |W_1(\tau, s)|d\tau.$$

By applying *Gronwall's inequality* to Eq. (5.31), we find $u(t) = 0$, which means that $\hat{\Gamma}(t) \in N$ for any $t \geq 0$. $\square$

Proposition 5.1 implies that if an initial homogeneous deformation $\hat{g}(t)$ for $t \leq 0$ belongs to a set of matrices M, and the set of inverse matrices N is a subspace of $\mathcal{M}$, then during recovery a viscoelastic medium performs a deformation which belongs to a set of matrices inverse to $N \bigcap \mathcal{N}$.

5.3. Properties of Positive Definite Symmetrical Matrices

For further analysis we should derive some properties of matrices which belong to set $\mathcal{N}$. For convenience, appropriate statements are formulated and proved in this subsection. Our study of homogeneous deformations will proceed in subsection 4.

Proposition 5.2. Let $\hat{a}(t)$ be a continuously differentiable on $[0, \infty)$, symmetrical (3×3)-matrix function with the unit determinant,

$$\det \hat{a}(t) = 1. \tag{5.32}$$

Then for any $t \geq 0$ we have

$$\frac{d\hat{a}}{dt}(t) : \hat{a}^{-1}(t) = 0. \tag{5.33}$$

Proof. Eq. (5.33) characterizes a property of symmetrical matrices of an arbitrary finite dimension. In particular case of (3×3)-matrices, to derive Eq. (5.33) we use direct calculations which are not too cumbersome.

Let a symmetrical matrix $\hat{a}(t)$ have the form

$$\hat{a} = \begin{bmatrix} a_{11} & a_{12} & a_{13} \\ a_{12} & a_{22} & a_{23} \\ a_{13} & a_{23} & a_{33} \end{bmatrix}.$$

Then

$$\frac{d\hat{a}}{dt} = \begin{bmatrix} \dot{a}_{11} & \dot{a}_{12} & \dot{a}_{13} \\ \dot{a}_{12} & \dot{a}_{22} & \dot{a}_{23} \\ \dot{a}_{13} & \dot{a}_{23} & \dot{a}_{33} \end{bmatrix},$$

and

$$\hat{a}^{-1} = \begin{bmatrix} a_{22}a_{33} - a_{23}^2 & -(a_{12}a_{33} - a_{13}a_{23}) & a_{12}a_{23} - a_{13}a_{22} \\ -(a_{12}a_{33} - a_{13}a_{23}) & a_{11}a_{33} - a_{13}^2 & -(a_{11}a_{23} - a_{12}a_{13}) \\ a_{12}a_{23} - a_{13}a_{22} & -(a_{11}a_{23} - a_{12}a_{13}) & a_{11}a_{22} - a_{12}^2 \end{bmatrix}.$$

By direct calculations, we find

$$\frac{d\hat{a}}{dt} : \hat{a}^{-1} = I_1\left(\frac{d\hat{a}}{dt} \cdot \hat{a}^{-1}\right)$$

$$= \dot{a}_{11}(a_{22}a_{33} - a_{23}^2) - \dot{a}_{12}(a_{12}a_{33} - a_{13}a_{23}) + \dot{a}_{13}(a_{12}a_{23} - a_{13}a_{22})$$

$$- \dot{a}_{12}(a_{12}a_{33} - a_{13}a_{23}) + \dot{a}_{22}(a_{11}a_{33} - a_{13}^2) - \dot{a}_{23}(a_{11}a_{23} - a_{12}a_{13})$$

$$+ \dot{a}_{13}(a_{12}a_{23} - a_{13}a_{22}) - \dot{a}_{23}(a_{11}a_{23} - a_{12}a_{13}) + \dot{a}_{33}(a_{11}a_{22} - a_{12}^2). \tag{5.34}$$

Exercise 5.13. Show that the right-hand side of Eq. (5.34) equals

$$\frac{d}{dt} \det \hat{a},$$

and, according to Eq. (5.32), it vanishes. $\square$

Proposition 5.3. Let $\hat{a}$ and $\hat{b}$ be symmetrical, positive definite (3×3)-matrices with the unit determinants. Then

$$I_1(\hat{a} \cdot \hat{b}^{-1}) \geq 3, \tag{5.35}$$

where the equality holds if and only if $\hat{a} = \hat{b}$.

Proof. Since matrices $\hat{a}$ and $\hat{b}$ are symmetrical and positive definite, there is a basis where they may be presented as

$$\hat{a} = \begin{bmatrix} 1 & 0 & 0 \\ 0 & 1 & 0 \\ 0 & 0 & 1 \end{bmatrix}, \qquad \hat{b} = \begin{bmatrix} b_1 & 0 & 0 \\ 0 & b_2 & 0 \\ 0 & 0 & b_3 \end{bmatrix}. \tag{5.36}$$

Here b_i are positive quantities which satisfy the equality

$$b_1 b_2 b_3 = 1. \tag{5.37}$$

Since $I_1(\hat{a} \cdot \hat{b}^{-1})$ is an invariant quantity, it can be calculated in the basis where matrices $\hat{a}$ and $\hat{b}$ have the form (5.36). Therefore,

$$I_1(\hat{a} \cdot \hat{b}^{-1}) = \frac{1}{b_1} + \frac{1}{b_2} + \frac{1}{b_3}. \tag{5.38}$$

Let us minimize the function

$$R(b_1, b_2, b_3) = \frac{1}{b_1} + \frac{1}{b_2} + \frac{1}{b_3}$$

on the set $\{b_i > 0\}$ with constraint (5.37). By using the standard procedure, we introduce the Lagrange multiplier μ and replace this problem of minimization by the problem of unconditional minimization for the function

$$R_\mu(b_1, b_2, b_3) = \frac{1}{b_1} + \frac{1}{b_2} + \frac{1}{b_3} + \mu(b_1 b_2 b_3 - 1).$$

Calculating the derivatives of R_μ with respect to b_i and equating them to zero, we obtain

$$-\frac{1}{b_1^2} + \mu b_2 b_3 = 0, \qquad -\frac{1}{b_2^2} + \mu b_1 b_3 = 0, \qquad -\frac{1}{b_3^2} + \mu b_1 b_2 = 0. \tag{5.39}$$

Exercise 5.14. Show that the only solution of Eqs. (5.37) and (5.39) is

$$b_1 = b_2 = b_3 = \mu = 1. \qquad \square \tag{5.40}$$

Exercise 5.15. Show that the function $R(b_1, b_2, b_3)$ reaches its minimal value at point (5.40). $\square$

Therefore, for any matrices $\hat{a}$ and $\hat{b}$

$$I_1(\hat{a} \cdot \hat{b}^{-1}) = R(b_1, b_2, b_3) \geq R(1,1,1) = 3. \tag{5.41}$$

The equality in Eq. (5.41) is fulfilled for $b_1 = b_2 = b_3 = 1$ only, i.e. when $\hat{a} = \hat{b}$.
$\square$

5.4. Elastic Recovery after Homogeneous Deformations

In this subsection we proceed to study homogeneous deformations of a viscoelastic medium. First, we derive an explicit expression for pressure $p(t)$. For this purpose, we multiply Eq. (5.19) by $\hat{\Gamma}(t)$ and apply Proposition 5.2 to the obtained equality. As a result, we find

$$3p(t) = 2[1 + Q_0(\infty)]W_1(t, -\infty)I_1(\hat{\Gamma}^{-1}(t))$$
$$-2\int_{-\infty}^{t} \dot{Q}_0(t-s)W_1(t,s)I_1(\hat{\Gamma}(s) \cdot \hat{\Gamma}^{-1}(t))ds. \tag{5.42}$$

It follows from Eqs. (5.5) and (5.12) that

$$I_1(\hat{\Gamma}(s) \cdot \hat{\Gamma}^{-1}(t)) = I_1(\hat{\Lambda}^{-1}(s) \cdot (\hat{\Lambda}^{-1}(s))^T \cdot \hat{\Lambda}^T(t) \cdot \hat{\Lambda}(t))$$
$$= I_1(\hat{\Lambda}(t) \cdot \hat{\Lambda}^{-1}(s) \cdot (\hat{\Lambda}^{-1}(s))^T \cdot \hat{\Lambda}^T(t)) = I_1(\hat{F}_*(t,s)) = I_1(t,s). \tag{5.43}$$

Substitution of expression (5.43) into Eq. (5.42) with the use of Eq. (5.18) implies that

$$3p(t) = 2[1 + Q_0(\infty)]W_1(t, -\infty)I_1(t, -\infty)$$
$$-2\int_{-\infty}^{t} \dot{Q}_0(t-s)W_1(t,s)I_1(t,s)ds. \tag{5.44}$$

Proposition 5.4. Let $\hat{B}$ be an arbitrary (3×3)-matrix, such that $|\hat{B} : \hat{\Theta}(t)| + |\hat{B} : \hat{\Theta}_0|$ does not equal zero identically. Suppose that
(i) there is an interval $[t_1, t_2] \subset (-\infty, 0]$ such that for any $t \in [t_1, t_2]$

$$\Theta(t) \neq \Theta_0; \tag{5.45}$$

(ii) there are $\alpha_1 \leq 0$ and $\alpha_2 \geq 0$ such that for any $t < 0$

$$\alpha_1 \leq \hat{B} : \hat{\Theta}(t) \leq \alpha_2, \tag{5.46}$$

and

$$\alpha_1 \leq \hat{B} : \hat{\Theta}_0 \leq \alpha_2; \tag{5.47}$$

(iii) function $Q_0(t)$ satisfies conditions (6.4.58) and (6.4.59);

387

(iv) strain energy density $W(I_1)$ satisfies the inequality

$$\frac{dW}{dI_1}(I_1) > 0 \quad (I_1 \geq 3); \tag{5.48}$$

(v) Newtonian viscosity η is positive.
Then for any $t > 0$ we have

$$\alpha_1 < \hat{B} : \hat{\Gamma}(t) < \alpha_2. \tag{5.49}$$

Proof. Let us multiply Eq. (5.19) by $\hat{B}$ and introduce the notation $w(t) = \hat{B} : \hat{\Gamma}(t)$. With the use of Eqs. (5.17) and (5.18), we find

$$\eta \frac{dw}{dt}(t) + p(t)w(t) = 2[1 + Q_0(\infty)]W_1(t, -\infty)w(-\infty)$$
$$-2 \int_{-\infty}^{t} \dot{Q}_0(t-s)W_1(t,s)w(s)ds. \tag{5.50}$$

Substitution of expression (5.44) into Eq. (5.50) implies that

$$\eta \frac{dw}{dt}(t) = 2[1 + Q_0(\infty)]W_1(t, -\infty)[w(-\infty) - \frac{I_1(t, -\infty)}{3}w(t)]$$
$$-2 \int_{-\infty}^{t} \dot{Q}_0(t-s)W_1(t,s)[w(s) - \frac{I_1(t,s)}{3}w(t)]ds. \tag{5.51}$$

We derive the inequality $\hat{B} : \hat{\Gamma}(t) < \alpha_2$ in the right-hand side of Eq. (5.49). The inequality in the left-hand side of Eq. (5.49) is developed by using a similar procedure and it is left to the reader as an exercise.

Let us assume that there is a $T > 0$ such that

$$w(T) = \alpha_2, \quad w(t) \leq \alpha_2 \quad (t < T), \tag{5.52}$$

and show that this assumption leads us to a contradiction. First, we consider the case $\alpha_2 = 0$. It follows from Eqs. (5.51) and (5.52) that

$$\eta \frac{dw}{dt}(T) = 2[1 + Q_0(\infty)]W_1(T, -\infty)w(-\infty)$$
$$-2 \int_{-\infty}^{T} \dot{Q}_0(T-s)W_1(T,s)w(s)ds. \tag{5.53}$$

According to conditions (iii) and (iv), the terms $1 + Q_0(\infty)$, $-\dot{Q}_0(t)$, and $W_1(T, s)$ are positive. The terms $w(-\infty)$ and $w(s)$ are non-positive, and they do not equal zero identically. Since $\eta > 0$ we obtain

$$\frac{dw}{dt}(T) < 0, \tag{5.54}$$

388

which contradicts Eq. (5.52).

We now consider the case $\alpha_2 > 0$. It follows from Eqs. (5.51) and (5.52) that

$$\eta \frac{dw}{dt}(T) = 2\alpha_2[1 + Q_0(\infty)]W_1(T, -\infty)[\frac{w(-\infty)}{\alpha_2} - \frac{I_1(T, -\infty)}{3}]$$

$$-2\alpha_2 \int_{-\infty}^{T} \dot{Q}_0(T-s)W_1(T,s)[\frac{w(s)}{\alpha_2} - \frac{I_1(t,s)}{3}]ds. \qquad (5.55)$$

Eq. (5.52) implies that $w(s)/\alpha_2 \leq 1$ for any $s \in (-\infty, T]$. According to condition (i) and Proposition 5.3, $I_1(t, s) > 3$ on the interval $[t_1, t_2]$. Since the terms $1 + Q_0(\infty)$, $-\dot{Q}_0(t)$, and $W_1(T, s)$ are positive, Eq. (5.55) implies inequality (5.54), which contradicts Eq. (5.52). Therefore, our hypothesis (5.52) is not true, and $w(t) < \alpha_2$ for any $t > 0$. $\square$

Remark. Proposition 5.4 is proved for a viscoelastic medium with $\eta > 0$. For media with $\eta = 0$, this assertion remains true. In order to prove it we should introduce additional hypotheses regarding the behavior of relaxation measure $Q_0(t)$, as well as to use more sophisticated estimates. We leave the case $\eta = 0$ to the reader as an exercise, referring to Engler (1991), where an appropriate proof is provided for a viscoelastic liquid with $Q_0(\infty) = -1$.

5.5. Examples

For simplicity, we confine ourselves to continuous histories of deformations with $\hat{\Theta}(0) = \hat{\Theta}_0$.

First, we assume that a shear-free deformation (5.20) occurs in a viscoelastic medium on the interval $(-\infty, 0]$. It follows from Exercise 5.5 and Proposition 5.1 that the medium performs also a shear-free deformation after the load is reduced and for any $t \geq 0$

$$\hat{\Gamma}(t) = \begin{bmatrix} a(t) & 0 & 0 \\ 0 & b(t) & 0 \\ 0 & 0 & c(t) \end{bmatrix}, \qquad (5.56)$$

where $a(t)$, $b(t)$, and $c(t)$ are positive functions.

Let us consider filament stretching (5.21) with $\sup_{t \in (-\infty, 0]} \lambda(t) = \lambda^0$. Choosing the matrix $\hat{B}$ in the form

$$\hat{B} = \begin{bmatrix} 1 & 0 & 0 \\ 0 & 0 & 0 \\ 0 & 0 & 0 \end{bmatrix},$$

and applying Proposition 5.4, we obtain that for any $t \geq 0$

$$1 < a(t) < \lambda^0.$$

389

Exercise 5.16. Derive a similar inequality for the recoil after sheet stretching (5.22). □

We now assume that simple shear (5.24) occurs in a viscoelastic medium on the interval $(-\infty, 0]$. It follows from Exercise 5.9 and Proposition 5.1 that for any $t \geq 0$, matrix $\hat{\Gamma}(t)$ has the form (5.26)

$$\hat{\Gamma}(t) = \begin{bmatrix} a(t) & -b(t) & 0 \\ -b(t) & c(t) & 0 \\ 0 & 0 & c(t) \end{bmatrix},$$

where $a(t)$, $b(t)$, and $c(t)$ are functions to be found. Calculation of the inverse matrix implies that

$$\hat{g}(t) = \begin{bmatrix} c^2(t) & b(t)c(t) & 0 \\ b(t)c(t) & b^2(t) + c^{-1}(t) & 0 \\ 0 & 0 & c^{-1}(t) \end{bmatrix}. \tag{5.57}$$

Exercise 5.17. Check that Eq. (5.57) provides the inverse matrix for $\hat{\Gamma}(t)$. □

Proposition 5.4 together with Eq. (5.25) implies that the following inequalities are valid for any $t > 0$:

$$0 < a(t) < 1 + \kappa_0^2, \qquad 0 < b(t) < \kappa_0, \qquad 0 < c(t) < 1, \tag{5.58}$$

where $\kappa_0 = \sup_{t \in (-\infty, 0]} |\kappa(t)|$. To derive Eqs. (5.58) we employ matrices $\hat{B}_i$ in the form

$$\hat{B}_1 = \begin{bmatrix} 1 & 0 & 0 \\ 0 & 0 & 0 \\ 0 & 0 & 0 \end{bmatrix}, \qquad \hat{B}_2 = \begin{bmatrix} 0 & -\frac{1}{2} & 0 \\ -\frac{1}{2} & 0 & 0 \\ 0 & 0 & 0 \end{bmatrix}, \qquad \hat{B}_3 = \begin{bmatrix} 0 & 0 & 0 \\ 0 & 0 & 0 \\ 0 & 0 & 1 \end{bmatrix}.$$

Diagonal elements of the Cauchy deformation tensor $\hat{g}$ in Cartesian coordinates $\{x^i\}$ characterize elongation (contraction) along unit vectors $\bar{e}_i$. It follows from Eqs. (5.57) and (5.58) that during the recovery process, the specimen expands in the $\bar{e}_2$ direction, expands (by a smaller amount) in the $\bar{e}_3$ direction, and contracts in the $\bar{e}_1$ direction. This behavior of the specimen differs from its behavior under loading: for simple shear neither elongation nor contraction can be seen in $\bar{e}_2$ and $\bar{e}_3$ directions. This means that the sample recovers to states where it has never been under loading.

To the best of our knowledge, there are no experiments on viscoelastic solids where the obtained results were tested. However, experiments on viscoelastic liquids demonstrate elongation and contraction along all the directions $\bar{e}_i$ during recovery after simple shear, see Lodge (1964).

5.6. Concluding Remarks

The recovery process is studied after homogeneous deformations of a viscoelastic medium. The governing equations are derived for the description of an elastic recoil. Two basic assertions (Propositions 5.1 and 5.4) are proved regarding the elastic recovery. It is shown that for different classes of homogeneous deformations, the medium demonstrates different behavior after the load is reduced. For example, for shear-free motions, the initial deformation as well as the elastic recoil belong to the same class of deformations, while for simple shear, the sample recovers to states where it has never been under loading.

6. Accretion of Viscoelastic Solids

This Section is concerned with continuous growth of a viscoelastic medium with finite strains. The growth means a monotonic increase in the body's mass due to the material influx from the environment. This process is modeled as successive *accretion* of thin layers on a surface of *the growing body*. Since successive layers (*built-up portions*) are applied to the deformed boundary, final stresses depend on the rate of accretion and on the loading history.

The problem of accretion originated in the sixties. It is in the focus of attention due to a wide range of applications: from building of *dams and embankments*, see Goodman & Brown (1963) and Cristiano & Chantranuluck (1974), to creation of *self-gravitating planets*, see Brown & Goodman (1963) and Arutyunyan & Drozdov (1984), from manufacturing of *thin films*, see Hearn et al. (1986) and Tsai & Dillon (1987), to *consolidation of metallic droplets*, see Mathur et al. (1989), from *solidification of adhesive layers*, see Duong & Knauss (1993a,b), to *winding composite pressure vessels*, see Drozdov (1994) and Drozdov & Kalamkarov (1995).

Our objective is to derive a model for the description of *continuous accretion* and to study stress distribution in a growing cylindrical *pressure vessel* with finite strains. We demonstrate that the final stress distribution in a growing body depends on (i) the loading history, (ii) the rate of the material influx, and (iii) rheological properties of material.

6.1. A Mathematical Model of Continuous Accretion

Let us consider a viscoelastic medium which is in its natural state and occupies a domain Ω^0 with a boundary Γ^0. At instant $t = 0$, external forces are applied to the medium, and continuous accretion begins.

For an arbitrary moment $t \geq 0$, the growing body in the actual configuration occupies a domain $\Omega(t)$ with a boundary $\Gamma(t)$. The surface $\Gamma(t)$ is divided into three connected parts. On a part $\Gamma_u(t)$, displacements are prescribed. On a part $\Gamma_\sigma(t)$, a surface traction is given. Finally, on the part $\Gamma_g(t) = \Gamma(t) \setminus (\Gamma_u(t) \bigcup \Gamma_\sigma(t))$ continuous accretion of material occurs on an interval $[0, T]$, see Fig. 6.1.

For continuous accretion, during the interval $[s, s + ds]$ a built-up portion (layer) with volume (thickness) proportional to ds joins the growing body. The natural state of the built-up portion may differ from the current state of the accreting surface $\Gamma_g(s)$. Therefore, this portion should be previously deformed to join the growing body. After the built-up portion joins the accreting surface, it immediately merges with the growing medium. For any $t \in [0, T]$, the main body together with built-up portions which join it on the interval $[0, t)$ is modeled as a monolithic solid.

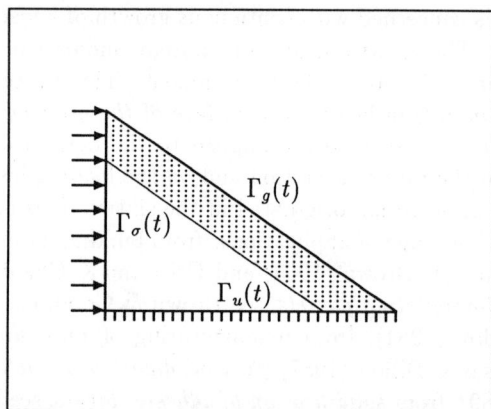

Fig. 6.1. A growing body under loading.

To describe the accretion process, we introduce three basic configurations of a growing body.

The first is *the reference configuration*, where we fix Lagrangian coordinates $\xi = \{\xi^i\}$ and postulate a plan (schedule) of the growth. For definiteness, we assume that for any $t \in [0, T]$, a growing body occupies a domain $\Omega^0(t)$ with a boundary $\Gamma^0(t) = \Gamma_u^0(t) \bigcup \Gamma_\sigma^0(t) \bigcup \Gamma_g^0(t)$ in the reference configuration. To formulate a plan of growth we determine which points on boundaries of built-up portions and on the accreting surface $\Gamma_g^0(t)$ merge with each other at any moment $t \in [0, T]$.

The second is the natural configuration where stresses in an element of the growing medium vanish, and where this element remains unless external forces are applied.

For the main body at $t = 0$, the reference configuration coincides with the natural configuration, while for built-up portions these configurations can differ from each other. For any accreted layer, its natural configuration should be determined either by prescribing an appropriate strain tensor for transition

392

from the reference configuration to the natural configuration, or by introducing the corresponding stress tensor (*preloading*). For a given constitutive model, these two approaches are equivalent. For definiteness, we employ the former (geometrical) approach and determine Finger's tensor $\hat{F}_*(\xi)$ for transition from the reference configuration to the natural configuration for any point $\xi \in \Omega^0(T)$.

The third is the actual configuration occupied by the growing body at the current moment t under the action of external forces. As a rule, this configuration should be found by solving the governing equations. It is characterized either by the deformation gradient for transition from the reference configuration to the actual configuration $\bar{\nabla}_0 \bar{r}(t, \xi)$, or by the corresponding Finger tensor $\hat{F}(t, \xi)$.

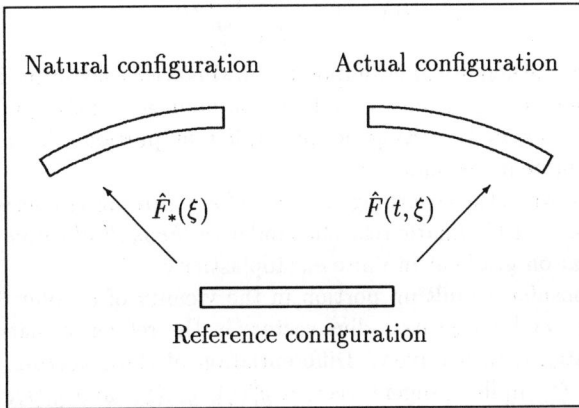

Fig. 6.2. Reference, natural, and actual configurations of a built-up portion.

As common practice, tensor $\hat{F}_*(\xi)$ is not prescribed explicitly, and it is given as a function of $\hat{F}(\kappa(\xi), \xi)$, where $\kappa(\xi)$ is the moment when the built-up portion in the vicinity of a point ξ merges with the growing body. For example, for the growth without pretension we set $\hat{F}_*(\xi) = \hat{F}(\kappa(\xi), \xi)$ for any $\xi \in \Omega^0(T)$.

The process of continuous accretion is modeled as a limit of the following process of *discrete* growth. We divide the interval $[0, T]$ by moments $t_k = k\Delta$ ($k = 0, \ldots, N$), where $\Delta = T/N$. At moment t_k, the growing medium occupies a domain $\Omega^0(t_k)$ in the reference configuration. During the interval $[t_k, t_{k+1}]$, it merges with a built-up portion which occupies a domain $\Delta\Omega^0(t_k)$ in the reference configuration, and they together create a new monolithic solid which occupies a domain $\Omega^0(t_{k+1})$ in the reference configuration.

For the built-up portion $\Delta\Omega^0(t_k)$, transition from the reference configuration to the natural configuration is assumed to be a real deformation and to

393

satisfy some *compatibility condition* which means that the tensor $\hat{F}_*(\xi)$ may be expressed in terms of a displacement vector $\bar{u}_*(\xi)$ for $\xi \in \Delta\Omega^0(t_k)$. Continuous accretion is treated as a limit of the process of discrete accretion as $N \to \infty$, $\Delta \to 0$, and the volumes of domains $\Delta\Omega^0(t_k)$ tend to zero. As a result, for continuous accretion Finger's tensor $\hat{F}_*(\xi)$ needs not satisfy any compatibility condition. This will be demonstrated below for winding of a circular cylinder.

For aging viscoelastic media, we may distinguish two instants: a moment $\varpi(\xi)$ when a built-up portion in the vicinity of a point ξ is manufactured, and a moment $\kappa(\xi)$ when this portion joins the growing body, cf Eq. (6.3.26). For simplicity, we assume that these moments coincide with each other. This means that any built-up portion merges with the growing body immediately after its creation

$$\kappa(\xi) = \begin{cases} 0 & \xi \in \Omega^0(0), \\ t & \xi \in \Gamma_g^0(t). \end{cases} \tag{6.1}$$

From the mathematical standpoint, growing viscoelastic media subjected to aging provide an important example of non-homogeneously aging solids with a specific inhomogeneity arising because different portions are manufactured (created) at different instants.

In order to write the governing equations for a growing viscoelastic body, we derive an important kinematic relation similar to *the multiplicative presentation* of the deformation gradient in finite elastoplasticity.

Let us consider a built-up portion in the vicinity of a point ξ and denote by $\bar{r}^0(\xi)$, $\bar{r}_*(\xi)$, and $\bar{r}(t,\xi)$ its radius-vectors in the reference, natural, and actual configurations, respectively. Differentiation of these vectors with respect to coordinate ξ^i implies tangent vectors $\bar{g}_i^0(\xi)$, $\bar{g}_{*\,i}(\xi)$, and $\bar{g}_i(t,\xi)$. By using Eqs. (1.1.14), dual vectors $\bar{g}^{0\,i}(\xi)$, $\bar{g}_*^i(\xi)$, and $\bar{g}^i(t,\xi)$ can be calculated. The corresponding deformation gradients equal

$$\bar{\nabla}^0\bar{r}_* = \bar{g}^{0\,i}\bar{g}_{*\,i}, \qquad \bar{\nabla}^0\bar{r} = \bar{g}^{0\,i}\bar{g}_i, \qquad \bar{\nabla}_*\bar{r} = \bar{g}_*^i\bar{g}_i. \tag{6.2}$$

Proposition 6.1. The following multiplicative presentation is valid:

$$\bar{\nabla}_*\bar{r} = (\bar{\nabla}^0\bar{r}_*)^{-1} \cdot \bar{\nabla}^0\bar{r}. \tag{6.3}$$

Proof. Let us calculate the expression in the right-hand side of Eq. (6.3) with the use of Eqs. (6.2) and (2.1.5)

$$(\bar{\nabla}^0\bar{r}_*)^{-1} \cdot \bar{\nabla}^0\bar{r} = \bar{g}_*^i\bar{g}_i^0 \cdot \bar{g}^{0\,j}\bar{g}_j = \bar{g}_*^i\bar{g}_i = \bar{\nabla}_*\bar{r}. \qquad \square$$

Substitution of expression (6.3) into Eq. (2.1.25) yields

$$\begin{aligned}
\hat{F}^*(t,\xi) &= \bar{\nabla}_*\bar{r}^T(t,\xi) \cdot \bar{\nabla}_*\bar{r}(t,\xi) \\
&= \bar{\nabla}^0\bar{r}^T(t,\xi) \cdot (\bar{\nabla}^0\bar{r}_*^{-1}(\xi))^T \cdot \bar{\nabla}^0\bar{r}_*^{-1}(\xi) \cdot \bar{\nabla}^0\bar{r}(t,\xi) \\
&= \bar{\nabla}^0\bar{r}^T(t,\xi) \cdot (\bar{\nabla}^0\bar{r}_*(\xi) \cdot \bar{\nabla}^0\bar{r}_*^T(\xi))^{-1} \cdot \bar{\nabla}^0\bar{r}(t,\xi) \\
&= \bar{\nabla}^0\bar{r}^T(t,\xi) \cdot \hat{g}_*^{-1}(\xi) \cdot \bar{\nabla}^0\bar{r}(t,\xi),
\end{aligned} \tag{6.4}$$

where $\hat{F}^*(t,\xi)$ is the Finger tensor for transition from the natural configuration to the actual configuration (not to be confused with $\hat{F}_*(\xi)$), and $\hat{g}_*(\xi)$ is the Cauchy tensor for transition from the reference configuration to the natural configuration.

Kinematic relationships (6.3) and (6.4) for growing media introduce the only new element into the system of governing equations. Besides Eq. (6.4), this system combines Eq. (2.1.46) for the relative Finger tensor $\hat{F}(t,s)$ for transition from the actual configuration at instant s to the actual configuration at instant t (unlike previous Sections, this tensor is denoted here by $\hat{F}(t,s)$ instead of $\hat{F}_*(t,s)$), the motion equation (2.2.19), and the constitutive equation (2.3.25), where the Finger tensor $\hat{F}(t,\xi)$ for transition from the initial to the actual configuration should be replaced by tensor $\hat{F}^*(t,\xi)$. In particular, for an incompressible viscoelastic medium with constitutive equations (6.3.83), (6.3.84) and $M = 1$ we write

$$\hat{\sigma}(t,\xi) = -p(t,\xi)\hat{I} + 2\{\mathcal{X}(t-\kappa(\xi),0)[\psi_1^*(t,\xi)\hat{F}^*(t,\xi) + \psi_2^*(t,\xi)\hat{F}^{*2}(t,\xi)]$$
$$+ \int_{\kappa(\xi)}^t \frac{\partial \mathcal{X}}{\partial s}(t-\kappa(\xi),s-\kappa(\xi))[\psi_1(t,s,\xi)\hat{F}(t,s,\xi) + \psi_2(t,s,\xi)\hat{F}^2(t,s,\xi)]ds\}. \quad (6.5)$$

Here $\hat{\sigma}(t,\xi)$ is the Cauchy stress tensor, $p = p(t,\xi)$ is a pressure, $W(I_1,I_2)$ is a strain energy density,

$$\psi_1^*(t,\xi) = W_1^*(t,\xi) + I_1^*(t,\xi)W_2^*(t,\xi), \qquad \psi_2^*(t,\xi) = -W_2^*(t,\xi),$$
$$\psi_1(t,s,\xi) = W_1(t,s,\xi) + I_1(t,s,\xi)W_2(t,s,\xi), \qquad \psi_2(t,s,\xi) = -W_2(t,s,\xi),$$
$$W_k^*(t,\xi) = \frac{\partial W}{\partial I_k}(I_1^*(t,\xi),I_2^*(t,\xi)), \qquad W_k(t,s) = \frac{\partial W}{\partial I_k}(I_1(t,s,\xi),I_2(t,s,\xi)),$$
$$I_k^*(t,\xi) = I_k(\hat{F}^*(t,\xi)), \qquad I_k(t,s,\xi) = I_k(\hat{F}(t,s,\xi)), \quad (6.6)$$

and function $\mathcal{X}(t,s)$ has the form (2.9) and satisfies conditions (6.4.103) and (6.4.104).

The boundary conditions on $\Gamma_u(t)$ and $\Gamma_\sigma(t)$ remain unchanged. On the accreting surface $\Gamma_g(t)$, we set

$$\bar{n} \cdot \hat{\sigma}|_{\Gamma_g(t)} = 0, \quad (6.7)$$

which means the absence of surface traction on $\Gamma_g(t)$. Here $\bar{n}$ is the unit normal to Γ.

To demonstrate the advantages of the above model of growing viscoelastic media, we analyze winding of a circular cylinder. This problem is of special interest in the mechanics of composite pressure vessels and pipes.

6.2. The Lame Problem for a Growing Cylinder

Let us consider a hollow circular cylinder with length l, internal radius a_0

and external radius a_1. The cylinder is located between two rigid plates so that it cannot deform in the axial direction. Friction between the cylinder and the plates is neglected.

Points of the cylinder are marked by cylindrical coordinates $\{R, \Theta, Z\}$ in the reference configuration, by cylindrical coordinates $\{R_*, \Theta_*, Z_*\}$ in the natural configuration, and by cylindrical coordinates $\{r, \theta, z\}$ in the actual configuration. Unit vectors of these coordinate frames are denoted as $\bar{e}_R$, $\bar{e}_\Theta$, $\bar{e}_Z$, $\bar{e}_{*R}$, $\bar{e}_{*\Theta}$, $\bar{e}_{*Z}$, and $\bar{e}_r$, $\bar{e}_\theta$, $\bar{e}_z$, respectively.

At instant $t = 0$, a pressure $p_0(t)$ is applied to the internal surface $R = a_0$ of the cylinder, while the material accretion begins on the external surface $R = a_1$. Due to the mass influx on the interval $[0, T]$, the external radius $a(t)$ of the growing cylinder in the reference configuration changes according to the law

$$a = a(t), \qquad a(0) = a_1, \qquad a(T) = a_2.$$

At instant $t \in [0, T]$, the growing cylinder occupies the domain

$$\{a_0 \le R \le a(t), \qquad 0 \le \Theta < 2\pi, \qquad 0 \le Z \le l\}$$

in the reference configuration. During the interval $[t, t + dt]$, *the cylindrical shell*

$$\{a(t) \le R \le a(t) + da(t), \qquad 0 \le \Theta < 2\pi, \qquad 0 \le Z \le l\}$$

joins the cylinder. The built-up shell is previously deformed by an internal pressure so that its internal radius under the action of this pressure coincides with the external radius of the growing body. Afterward, the shell is wrapped around the growing cylinder and immediately merges with it.

In the absence of body forces, *axi-symmetrical* deformation in the plane $Z = $ const. occurs both in the growing body and in built-up portions. Transitions from the reference configuration to the natural and actual configurations are described by the formulas

$$\begin{aligned} r &= f(t, R), & \theta &= \Theta, & z &= Z, \\ R_* &= f_*(R), & \Theta_* &= \Theta, & Z_* &= Z, \end{aligned} \tag{6.8}$$

where $f(t, R)$ and $f_*(R)$ are functions to be found.

The radius-vectors $\bar{r}^0$ and $\bar{r}$ in the reference configuration and in the actual configuration equal

$$\bar{r}^0 = R\bar{e}_R + Z\bar{e}_Z, \quad \bar{r} = f(t, R)\bar{e}_r + Z\bar{e}_z.$$

Differentiation of these expressions with the use of Eqs. (1.1.1) and (6.8) yields

$$\begin{aligned} \bar{g}_1^0 &= \bar{e}_R, & \bar{g}_2^0 &= R\bar{e}_\Theta, & \bar{g}_3^0 &= \bar{e}_Z, \\ \bar{g}_1(t) &= h(t)\bar{e}_r, & \bar{g}_2(t) &= f(t)\bar{e}_\theta, & \bar{g}_3(t) &= \bar{e}_z, \end{aligned} \tag{6.9}$$

where $h(t) = \partial f / \partial R(t)$, and argument R is omitted for simplicity.

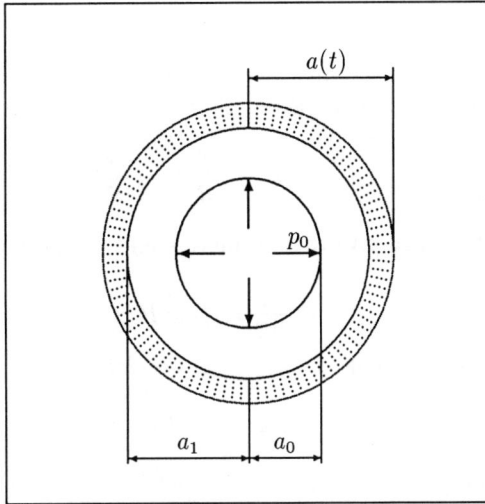

Fig. 6.3. A growing viscoelastic cylinder under internal pressure.

Exercise 6.1. By using Eq. (6.9), show that

$$\bar{g}^{0\,1} = \bar{e}_R, \qquad \bar{g}^{0\,2} = \frac{1}{R}\bar{e}_\Theta, \qquad \bar{g}^{0\,3} = \bar{e}_Z,$$

$$\bar{g}^1(t) = \frac{1}{h(t)}\bar{e}_r, \qquad \bar{g}^2 = \frac{1}{f(t)}\bar{e}_\theta, \qquad \bar{g}^3 = \bar{e}_z. \qquad \square \qquad (6.10)$$

Eqs. (6.9) and (6.10) together with formulas (2.1.5) and (2.1.44) imply that

$$\bar{\nabla}^0 \bar{r}(t) = h(t)\bar{e}_R\bar{e}_r + \frac{f(t)}{R}\bar{e}_\Theta\bar{e}_\theta + \bar{e}_Z\bar{e}_z,$$

$$\bar{\nabla}_s \bar{r}(t) = \frac{h(t)}{h(s)}\bar{e}_r\bar{e}_r + \frac{f(t)}{f(s)}\bar{e}_\theta\bar{e}_\theta + \bar{e}_z\bar{e}_z. \qquad (6.11)$$

Substitution of expression (6.11) into Eq. (2.1.25) yields

$$\hat{F}(t) = h^2(t)\bar{e}_r\bar{e}_r + \left(\frac{f(t)}{R}\right)^2\bar{e}_\theta\bar{e}_\theta + \bar{e}_z\bar{e}_z, \qquad (6.12)$$

where $\hat{F}(t)$ is Finger's tensor for transition from the reference configuration to the actual configuration at moment t.

397

Exercise 6.2. Derive the characteristic polynomial for the Finger tensor

$$\det(\hat{F} - \lambda \hat{I}) = -\lambda^3 + [h^2 + (\frac{f}{R})^2 + 1]\lambda^2 - [h^2 + (\frac{f}{R})^2 + (\frac{fh}{R})^2]\lambda + (\frac{fh}{R})^2. \quad \square$$

Since the principal invariants of the Finger tensor coincide (up to their signs) with coefficients of the characteristic equation, we obtain

$$I_1(\hat{F}) = h^2 + (\frac{f}{R})^2 + 1, \quad I_2(\hat{F}) = h^2 + (\frac{f}{R})^2 + (\frac{fh}{R})^2, \quad I_3(\hat{F}) = (\frac{fh}{R})^2. \quad (6.13)$$

Expression (6.13) together with the incompressibility condition (3.2.86) implies that

$$f(t, R)\frac{\partial f}{\partial R}(t, R) = R. \tag{6.14}$$

Integration of Eq. (6.14) yields

$$f^2(t, R) = R^2 + C(t), \tag{6.15}$$

where $C(t)$ is a function to be found.

Substitution of expression (6.14) into Eqs. (6.11) and (6.12) yields

$$\bar{\nabla}^0 \bar{r}(t, R) = \frac{R}{f(t, R)}\bar{e}_R\bar{e}_r + \frac{f(t, R)}{R}\bar{e}_\Theta\bar{e}_\theta + \bar{e}_Z\bar{e}_z. \tag{6.16}$$

We treat continuous growth as a limit of a discrete accretion process when built-up portions merge with the growing body at moments t_k ($k = 0, \ldots, N$). Applying the above reasoning to transition from the reference configuration to the natural configuration of a built-up portion $\Omega^0(t_k)$, we find

$$f_*^2(R) = R^2 + C_{*k}, \quad a(t_k) \le R < a(t_{k+1}), \tag{6.17}$$

where C_{*k} is an appropriate constant. Similar to Eq. (6.16), we obtain that

$$\bar{\nabla}^0 \bar{r}_*(R) = \frac{R}{f_*(R)}\bar{e}_R\bar{e}_{*R} + \frac{f_*(R)}{R}\bar{e}_\Theta\bar{e}_{*\Theta} + \bar{e}_Z\bar{e}_{*Z}. \tag{6.18}$$

Exercise 6.3. Derive formulas (6.17) and (6.18). $\square$

Introduce the function

$$C_*^{(N)}(R) = \begin{cases} C_{*0} & a(t_0) \le R < a(t_1), \\ C_{*1} & a(t_1) \le R < a(t_2), \\ \ldots & \ldots, \\ C_{*N-1} & a(t_{N-1}) \le R < a(t_N). \end{cases}$$

Denote by $C_*(R)$ the limit of $C_*^{(N)}(R)$ as $N \to \infty$. It follows from Eq. (6.17) that for the continuous process of accretion

$$f_*^2(R) = R^2 + C_*(R),\qquad(6.19)$$

while the deformation gradient preserves the form (6.18).

Exercise 6.4. Check that

$$(\bar\nabla^0 \bar r_*(R))^{-1} = \frac{f_*(R)}{R}\bar e_{*R}\bar e_R + \frac{R}{f_*(R)}\bar e_{*\Theta}\bar e_\Theta + \bar e_{*Z}\bar e_Z. \qquad\square\qquad(6.20)$$

Substitution of expressions (6.16) and (6.20) into Eq. (6.3) implies that

$$\bar\nabla_* \bar r(t) = \frac{f_*}{f(t)}\bar e_{*R}\bar e_r + \frac{f(t)}{f_*}\bar e_{*\Theta}\bar e_\theta + \bar e_{*Z}\bar e_z. \qquad(6.21)$$

It follows from Eqs. (2.1.25), (2.2.46), (6.11), and (6.21) that the relative Finger tensor $\hat F(t, s)$ and the Finger tensor $\hat F^*(t)$ are calculated as follows:

$$\hat F^*(t) = \left(\frac{f_*}{f(t)}\right)^2 \bar e_r \bar e_r + \left(\frac{f(t)}{f_*}\right)^2 \bar e_\theta \bar e_\theta + \bar e_z \bar e_z,$$

$$= \frac{R^2 + C_*(R)}{R^2 + C(t)}\bar e_r \bar e_r + \frac{R^2 + C(t)}{R^2 + C_*(R)}\bar e_\theta \bar e_\theta + \bar e_z \bar e_z,$$

$$\hat F(t, s) = \left(\frac{f(s)}{f(t)}\right)^2 \bar e_r \bar e_r + \left(\frac{f(t)}{f(s)}\right)^2 \bar e_\theta \bar e_\theta + \bar e_z \bar e_z$$

$$= \frac{R^2 + C(s)}{R^2 + C(t)}\bar e_r \bar e_r + \frac{R^2 + C(t)}{R^2 + C(s)}\bar e_\theta \bar e_\theta + \bar e_z \bar e_z. \qquad(6.22)$$

Exercise 6.5. Show that

$$\hat F^{*\,2}(t) = \left(\frac{R^2 + C_*(R)}{R^2 + C(t)}\right)^2 \bar e_r \bar e_r + \left(\frac{R^2 + C(t)}{R^2 + C_*(R)}\right)^2 \bar e_\theta \bar e_\theta + \bar e_z \bar e_z,$$

$$\hat F^2(t, s) = \left(\frac{R^2 + C(s)}{R^2 + C(t)}\right)^2 \bar e_r \bar e_r + \left(\frac{R^2 + C(t)}{R^2 + C(s)}\right)^2 \bar e_\theta \bar e_\theta + \bar e_z \bar e_z. \qquad\square\qquad(6.23)$$

Exercise 6.6. Derive the following formulas for the principal invariants of the Finger tensors:

$$I_1(\hat F^*(t)) = I_2(\hat F^*(t)) = \frac{R^2 + C(t)}{R^2 + C_*(R)} + \frac{R^2 + C_*(R)}{R^2 + C(t)} + 1,$$

$$I_1(\hat F(t, s)) = I_2(\hat F(t, s)) = \frac{R^2 + C(t)}{R^2 + C(s)} + \frac{R^2 + C(s)}{R^2 + C(t)} + 1. \qquad\square\qquad(6.24)$$

Substitution of expressions (6.22) – (6.24) into the constitutive equation (6.5) yields

$$\hat{\sigma} = \sigma_{rr}\bar{e}_r\bar{e}_r + \sigma_{\theta\theta}\bar{e}_\theta\bar{e}_\theta + \sigma_{zz}\bar{e}_z\bar{e}_z. \tag{6.25}$$

Here

$$\sigma_{rr}(t,R) = -p(t,R) + 2\{\mathcal{X}(t - \kappa(R), 0)[\psi_1^*(t,R)$$
$$+ \psi_2^*(t,R)\frac{R^2 + C_*(R)}{R^2 + C(t)}]\frac{R^2 + C_*(R)}{R^2 + C(t)}$$
$$+ \int_{\kappa(R)}^t \frac{\partial \mathcal{X}}{\partial s}(t - \kappa(R), s - \kappa(R))[\psi_1(t,s,R)$$
$$+ \psi_2(t,s,R)\frac{R^2 + C(s)}{R^2 + C(t)}]\frac{R^2 + C(s)}{R^2 + C(t)}ds\},$$

$$\sigma_{\theta\theta}(t,R) = -p(t,R) + 2\{\mathcal{X}(t - \kappa(R), 0)[\psi_1^*(t,R)$$
$$+ \psi_2^*(t,R)\frac{R^2 + C(t)}{R^2 + C_*(R)}]\frac{R^2 + C(t)}{R^2 + C_*(R)}$$
$$+ \int_{\kappa(R)}^t \frac{\partial \mathcal{X}}{\partial s}(t - \kappa(R), s - \kappa(R))[\psi_1(t,s,R)$$
$$+ \psi_2(t,s,R)\frac{R^2 + C(t)}{R^2 + C(s)}]\frac{R^2 + C(t)}{R^2 + C(s)}ds\},$$

$$\sigma_{zz}(t,R) = -p(t,R) + 2\{\mathcal{X}(t - \kappa(R), 0)[\psi_1^*(t,R) + \psi_2^*(t,R)]$$
$$+ \int_{\kappa(R)}^t \frac{\partial \mathcal{X}}{\partial s}(t - \kappa(R), s - \kappa(R))[\psi_1(t,s,R) + \psi_2(t,s,R)]ds\}. \tag{6.26}$$

Exercise 6.7. Show that Eqs. (6.26) imply that

$$\sigma_{\theta\theta} - \sigma_{rr} = 2\{\mathcal{X}(t - \kappa(R), 0)[\psi_1^*(t,R)(\frac{R^2 + C(t)}{R^2 + C_*(R)} - \frac{R^2 + C_*(R)}{R^2 + C(t)})$$
$$+ \psi_2^*(t,R)((\frac{R^2 + C(t)}{R^2 + C_*(R)})^2 - (\frac{R^2 + C_*(R)}{R^2 + C(t)})^2)]$$
$$+ \int_{\kappa(R)}^t \frac{\partial \mathcal{X}}{\partial s}(t - \kappa(R), s - \kappa(R))[\psi_1(t,s,R)(\frac{R^2 + C(t)}{R^2 + C(s)} - \frac{R^2 + C(s)}{R^2 + C(t)})$$
$$+ \psi_2(t,s,R)((\frac{R^2 + C(t)}{R^2 + C(s)})^2 - (\frac{R^2 + C(s)}{R^2 + C(t)})^2)]ds\}. \tag{6.27}$$

In the absence of body forces, the only equilibrium equation (2.2.29) for axi-symmetrical deformation is

$$\frac{\partial \sigma_{rr}}{\partial r} + \frac{1}{r}(\sigma_{rr} - \sigma_{\theta\theta}) = 0. \tag{6.28}$$

Let us now derive boundary conditions on the lateral surfaces of the cylinder. For internal surface $R = a_0$ we have $\bar{n} = -\bar{e}_r$ and $\bar{b} = p_0\bar{e}_r$, where $\bar{b}$ is surface traction. For external surface $R = a(t)$ we employ Eq. (6.7). By using Eqs. (2.2.20) and (6.25), we obtain

$$\sigma_{rr}|_{R=a_0} = -p_0(t), \qquad \sigma_{rr}|_{R=a(t)} = 0. \tag{6.29}$$

Integration of Eq. (6.28) from $r = f(t, a_0)$ to $r = f(t, a(t))$ with the use of Eq. (6.29) implies that

$$p_0(t) + \int_{f(t,a_0)}^{f(t,a(t))} \frac{\sigma_{rr} - \sigma_{\theta\theta}}{r} dr = 0.$$

We replace the "new" variable r by the "old" variable R with the use of Eqs. (6.8) and (6.15). Afterward, we substitute expression (6.27) into the obtained equation and find

$$
2 \int_{a_0}^{a(t)} \{ \mathcal{X}(t - \kappa(R), 0)[\psi_1^*(t, R)(\frac{R^2 + C(t)}{R^2 + C_*(R)} - \frac{R^2 + C_*(R)}{R^2 + C(t)})
$$
$$
+ \psi_2^*(t, R)((\frac{R^2 + C(t)}{R^2 + C_*(R)})^2 - (\frac{R^2 + C_*(R)}{R^2 + C(t)})^2)]
$$
$$
+ \int_{\kappa(R)}^{t} \frac{\partial \mathcal{X}}{\partial s}(t - \kappa(R), s - \kappa(R))[\psi_1(t, s, R)(\frac{R^2 + C(t)}{R^2 + C(s)} - \frac{R^2 + C(s)}{R^2 + C(t)})
$$
$$
+ \psi_2(t, s, R)((\frac{R^2 + C(t)}{R^2 + C(s)})^2 - (\frac{R^2 + C(s)}{R^2 + C(t)})^2)]ds \} \frac{RdR}{R^2 + C(t)} = p_0(t). \tag{6.30}
$$

For given functions $C_*(R)$ and $p_0(t)$, Eq. (6.30) is a nonlinear *integral* equation for function $C(t)$. We recall that $C_*(R)$ characterizes preloading in accreted layers. This function vanishes for the main body (i.e. for $a_0 \le R \le a_1$), and it may be chosen arbitrarily for built-up portions.

6.3. Analysis of Particular Cases

In this subsection, some results of numerical analysis are presented for Eq. (6.30). For definiteness, we confine ourselves to the neo-Hookean viscoelastic material with strain energy density (3.2.103).

6.3.1. A non-growing elastic cylinder

By setting $\mathcal{X}(t, s) = 1$, we simplify Eq. (6.30)

$$
2 \int_{a_0}^{a(t)} [\psi_1^*(t, R)(\frac{R^2 + C(t)}{R^2 + C_*(R)} - \frac{R^2 + C_*(R)}{R^2 + C(t)})
$$
$$
+ \psi_2^*(t, R)((\frac{R^2 + C(t)}{R^2 + C_*(R)})^2 - (\frac{R^2 + C_*(R)}{R^2 + C(t)})^2)] \frac{RdR}{R^2 + C(t)} = p_0(t). \tag{6.31}
$$

Using Eqs. (6.6) and (3.2.103), we obtain

$$\int_{a_0}^{a(t)} [\frac{R^2 + C(t)}{R^2 + C_*(R)} - \frac{R^2 + C_*(R)}{R^2 + C(t)}] \frac{R\,dR}{R^2 + C(t)} = \frac{p_0(t)}{\mu}. \qquad (6.32)$$

In the absence of accretion, we set $a(t) = a_1$. Since $C_*(R) = 0$, Eq. (6.32) is written as

$$\int_{a_0^2}^{a_1^2} (\frac{x + C}{x} - \frac{x}{x + C}) \frac{dx}{x + C} = \frac{2p_0}{\mu}, \qquad (6.33)$$

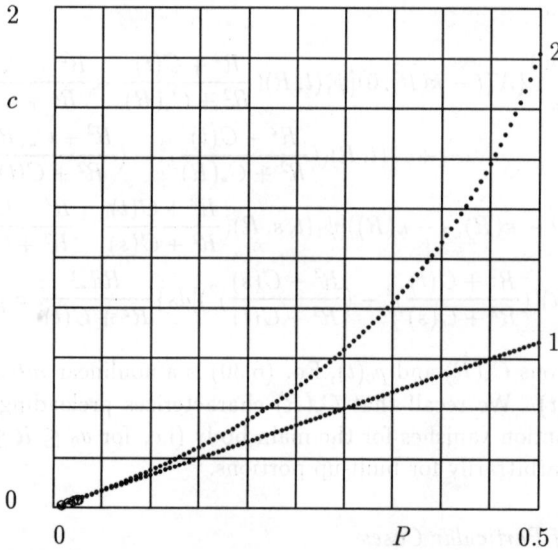

Fig. 6.4. The dimensionless radial displacement $c = C/a_0^2$ *versus* the dimensionless internal pressure $P = p_0/\mu$. Large circles correspond to experimental data obtained by Derman et al. (1978) for synthetic rubber at room temperature. Small circles present results of calculations by using linear theory (6.34) (curve 1), and nonlinear theory (6.33) (curve 2).

where $x = R^2$. For infinitesimal strains, when $C/a_0^2 \ll 1$, Eq. (6.33) implies that

$$C \int_{a_0^2}^{a_1^2} \frac{dx}{x^2} = \frac{p_0}{\mu}.$$

Finally, calculating the integral, we obtain

$$C = \frac{p_0}{\mu(a_0^{-2} - a_1^{-2})}. \qquad (6.34)$$

For finite strains, Eq. (6.33) may be solved only numerically. Nevertheless, it is of interest to compare C values for small and large deformations, see Fig. 6.4, where the results are plotted for $a_1 = 2a_0$.

The dimensionless radial displacement $c = C/a_0^2$ increases monotonically with an increase in the dimensionless internal pressure $P = p_0/\mu$. The rate of growth in C values for finite strains exceeds the corresponding rate for infinitesimal strains. For a given internal pressure p_0, the radial displacements calculated by employing the nonlinear theory (6.33) exceed significantly the displacements calculated in the framework of linear elasticity (6.34). For sufficiently small deformations, results of linear and nonlinear theories coincide and demonstrate excellent prediction of experimental data for synthetic rubber.

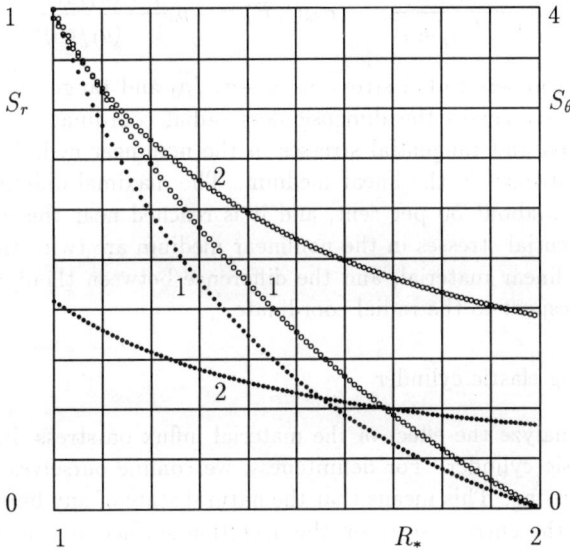

Fig. 6.5. The dimensionless radial stress $S_r = -\sigma_{rr}/p_0$ (curves 1) and the dimensionless tangential stress $S_\theta = \sigma_{\theta\theta}/p_0$ (curves 2) *versus* the dimensionless radial coordinate $R_* = R/a_0$. Calculations are carried out for $a_1 = 2a_0$ and $p_0 = 0.5\mu$. Filled circles correspond to the linear theory, while unfilled circles correspond to the nonlinear theory.

Exercise 6.8. By integrating Eq. (6.28) from $f(t, R)$ to $f(t, a_1)$, show that for a neo-Hookean cylinder

$$\sigma_{rr}(t, R) = -\mu \int_R^{a(t)} \left[\frac{R_1^2 + C(t)}{R_1^2 + C_*(R_1)} - \frac{R_1^2 + C_*(R_1)}{R_1^2 + C(t)}\right] \frac{R_1 dR_1}{R_1^2 + C(t)}. \qquad \Box \qquad (6.35)$$

403

Exercise 6.9. Check that for a non-growing cylinder, Eq. (6.35) is written as

$$\sigma_{rr}(t, R) = -\frac{\mu}{2} \int_{R^2}^{a_1^2} [\frac{x + C(t)}{x} - \frac{x}{x + C(t)}] \frac{dx}{x + C(t)}. \qquad \square \qquad (6.36)$$

Exercise 6.10. Show that for a non-growing cylinder, Eq. (6.27) implies that

$$\sigma_{\theta\theta}(t, R) = \sigma_{rr}(t, R) + \mu[\frac{R^2 + C(t)}{R^2} - \frac{R^2}{R^2 + C(t)}]. \qquad \square \qquad (6.37)$$

Exercise 6.11. By using Eqs. (6.34), (6.36), and (6.37), derive the following formulas for infinitesimal strains:

$$\sigma_{rr}(t, R) = -p_0 \frac{1 - (a_1/R)^2}{1 - (a_1/a_0)^2}, \qquad \sigma_{\theta\theta}(t, R) = -p_0 \frac{1 + (a_1/R)^2}{1 - (a_1/a_0)^2}. \qquad \square \qquad (6.38)$$

The dimensionless radial stress $S_r = -\sigma_{rr}/p_0$ and tangential stress $S_\theta = \sigma_{\theta\theta}/p_0$ are plotted *versus* the dimensionless radial coordinate $R_* = R/a_0$ in Fig. 6.5. Radial and tangential stresses in the nonlinear cylinder exceed the corresponding stresses in the linear medium. The maximal difference between radial stresses is about 30 per cent, and it is reached near the middle of the cylinder. Tangential stresses in the nonlinear medium are twice the tangential stresses in the linear material, and the difference between them is practically uniform with respect to the radial coordinate.

6.3.2. A growing elastic cylinder

We now analyze the effect of the material influx on stress distribution in a growing elastic cylinder. For definiteness, we confine ourselves to accretion without prestressing. This means that the natural state of any built-up portion coincides with the current state on the accretion surface at the instant when the portion merges with the growing body. Formally, this condition may be formulated as

$$C_*(R) = \begin{cases} 0 & a_0 \le R \le a_1, \\ C(\kappa(R)) & a_1 < R \le a_2. \end{cases} \qquad (6.39)$$

Substitution of expression (6.39) into Eq. (6.32) implies that

$$\int_{a_0}^{a_1} [\frac{R^2 + C(t)}{R^2} - \frac{R^2}{R^2 + C(t)}] \frac{RdR}{R^2 + C(t)}$$
$$+ \int_{a_1}^{a(t)} [\frac{R^2 + C(t)}{R^2 + C(\kappa(R))} - \frac{R^2 + C(\kappa(R))}{R^2 + C(t)}] \frac{RdR}{R^2 + C(t)} = \frac{p_0(t)}{\mu}. \qquad (6.40)$$

By employing the new variables $x = R^2$ in the first integral and $s = \kappa(R)$ in the other integral, we rewrite Eq. (6.40) as

$$\int_{a_0^2}^{a_1^2} [\frac{x + C(t)}{x} - \frac{x}{x + C(t)}]\frac{dx}{x + C(t)}$$
$$+2 \int_0^t [\frac{a^2(s) + C(t)}{a^2(s) + C(s)} - \frac{a^2(s) + C(s)}{a^2(s) + C(t)}]\frac{a(s)\dot{a}(s)ds}{a^2(s) + C(t)} = \frac{2p_0(t)}{\mu}, \qquad (6.41)$$

where the superscript dot denotes the differentiation with respect to time.

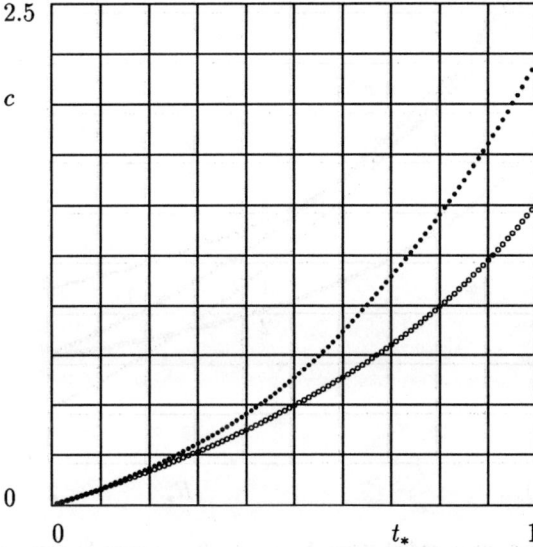

Fig. 6.6. The dimensionless displacement $c = C/a_0^2$ versus the dimensionless time $t_* = t/T$. Unfilled circles correspond to "rapid" accretion (6.43), filled circles correspond to "slow" accretion (6.44).

Exercise 6.12. By differentiating Eq. (6.41), show that

$$\dot{C}(t) = \frac{\dot{p}_0(t)}{\mu}[\int_{a_0^2}^{a_1^2} \frac{x dx}{(x + C(t))^3} + 2 \int_0^t \frac{(a^2(s) + C(s))a(s)\dot{a}(s)ds}{(a^2(s) + C(t))^3}]^{-1}. \qquad \square \ (6.42)$$

To study the effect of the material influx on the stress distribution, we assume that $a_1 = 2a_0$, $a_2 = 3a_0$, and $p_0(t) = 0.6\mu t_*$, where $t_* = t/T$, and consider two regimes of accretion:

$$u_*(t_*) = \begin{cases} 12.5 & 0 \le t_* \le 0.4, \\ 0 & 0.4 < t_* \le 1, \end{cases} \qquad (6.43)$$

405

and

$$u_*(t_*) = \begin{cases} 0 & 0 \leq t_* \leq 0.6, \\ 12.5 & 0.6 < t_* \leq 1. \end{cases} \qquad (6.44)$$

Here $u_* = uT/(\pi l a_0^2)$, where $u(t) = 2\pi a(t)\dot{a}(t)l$ is the rate of accretion (the volume of built-up portions which merge with the growing cylinder per unit time). Regime (6.43) is referred to as "rapid" growth, while regime (6.44) is referred to as "slow" growth.

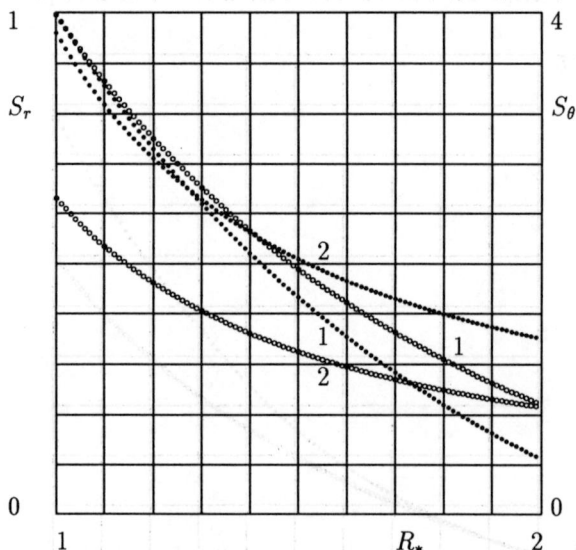

Fig. 6.7. The dimensionless radial stress $S_r = -\sigma_{rr}/p_0$ (curves 1) and the dimensionless tangential stress $S_\theta = \sigma_{\theta\theta}/p_0$ (curves 2) versus the dimensionless radial coordinate $R_* = R/a_0$. Unfilled circles correspond to "rapid" accretion (6.43), filled circles correspond to "slow" accretion (6.44).

The dependence $c = c(t)$ is plotted in Fig. 6.6. The radial displacements corresponding to the "rapid" accretion are less than the radial displacements for the "slow" material influx.

The dimensionless radial stress $S_r = -\sigma_{rr}/p_0$ and the dimensionless tangential stress $S_\theta = \sigma_{\theta\theta}/p_0$ are plotted versus the dimensionless radial coordinate $R_* = R/a_0$ in Figs. 6.7 and 6.8. Radial stresses for the "rapid" accretion are less than radial stresses for the "slow" growth. For both regimes of the material influx, function $S_r(R_*)$ decreases monotonically from 1 to 0. Tangential stresses have jumps of discontinuity on the interface $R = a_1$ between the main body and the region of growth. In the main body, tangential stresses corresponding to the "slow" growth exceed tangential stresses corresponding to the "rapid" growth, whereas in the accreting region, tangential stresses for the "slow" growth are less than tangential stresses for the "rapid" accretion.

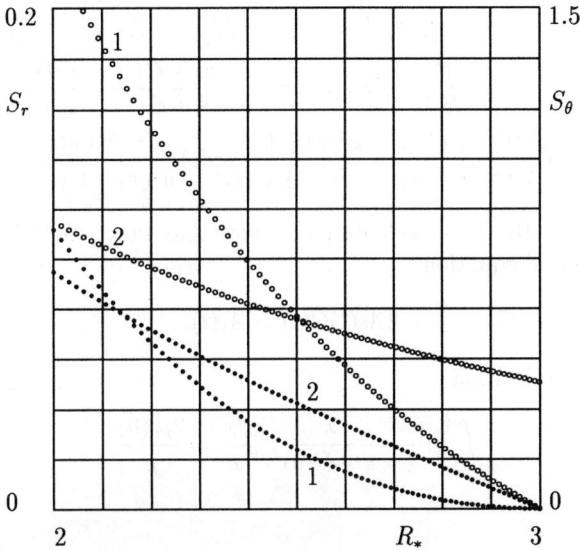

Fig. 6.8. The dimensionless radial stress $S_r = -\sigma_{rr}/p_0$ (curves 1) and the dimensionless tangential stress $S_\theta = \sigma_{\theta\theta}/p_0$ (curves 2) *versus* the dimensionless radial coordinate $R_* = R/a_0$. Unfilled circles correspond to "rapid" accretion (6.43), filled circles correspond to "slow" accretion (6.44).

6.3.3. A growing viscoelastic cylinder

Let us now consider continuous accretion of a non-aging viscoelastic cylinder, when Eq. (6.30) is written as

$$\int_{a_0}^{a(t)} \{[1 + Q_0(t - \kappa(R))][\frac{R^2 + C(t)}{R^2 + C_*(R)} - \frac{R^2 + C_*(R)}{R^2 + C(t)}]$$
$$- \int_{\kappa(R)}^{t} \dot{Q}_0(t - s)[\frac{R^2 + C(t)}{R^2 + C(s)} - \frac{R^2 + C(s)}{R^2 + C(t)}]ds\} \frac{RdR}{R^2 + C(t)} = \frac{p_0(t)}{\mu}.$$

We divide the interval $[a_0, a(t)]$ into two subintervals: $[a_0, a_1]$ and $[a_1, a(t)]$. On the first interval, $\kappa(R) = 0$, and we introduce the new variable $x = R^2$. On the other interval, we set $R = a(\tau)$, where τ is a new variable. As a result, we

407

arrive at the equality

$$\int_{a_0^2}^{a_1^2} \{[1 + Q_0(t)][\frac{x + C(t)}{x} - \frac{x}{x + C(t)}]$$

$$- \int_0^t \dot{Q}_0(t - s)[\frac{x + C(t)}{x + C(s)} - \frac{x + C(s)}{x + C(t)}]ds\}\frac{dx}{x + C(t)}$$

$$+ 2 \int_0^t \{[1 + Q_0(t - \tau)][\frac{a^2(\tau) + C(t)}{a^2(\tau) + C(\tau)} - \frac{a^2(\tau) + C(\tau)}{a^2(\tau) + C(t)}]$$

$$- \int_\tau^t \dot{Q}_0(t - s)[\frac{a^2(\tau) + C(t)}{a^2(\tau) + C(s)} - \frac{a^2(\tau) + C(s)}{a^2(\tau) + C(t)}]ds\}\frac{a(\tau)\dot{a}(\tau)d\tau}{a^2(\tau) + C(t)} = \frac{2p_0(t)}{\mu}. \quad (6.45)$$

Exercise 6.13. By direct calculations, show that function $C(t)$ satisfies the integro-differential equation

$$2A_0(t)\dot{C}(t) = A_1(t) \quad (6.46)$$

with the initial condition

$$\int_{a_0^2}^{a_1^2} [1 - (\frac{x}{x + C(0)})^2]\frac{dx}{x} = \frac{2p_0(0)}{\mu}.$$

Here

$$A_0(t) = \int_{a_0^2}^{a_1^2} \{[1 + Q_0(t)]x - \int_0^t \dot{Q}_0(t - s)[x + C(s)]ds\}\frac{dx}{[x + C(t)]^3}$$

$$+ 2 \int_0^t \{[1 + Q_0(t - \tau)][a^2(\tau) + C(\tau)]$$

$$- \int_\tau^t \dot{Q}_0(t - s)[a^2(\tau) + C(s)]ds\}\frac{a(\tau)\dot{a}(\tau)d\tau}{[a^2(\tau) + C(t)]^3},$$

$$A_1(t) = \frac{2\dot{p}_0(t)}{\mu} - \int_{a_0^2}^{a_1^2} \{\dot{Q}_0(t)[1 - (\frac{x}{x + C(t)})^2]\frac{1}{x}$$

$$- \int_0^t \ddot{Q}_0(t - s)[1 - (\frac{x + C(s)}{x + C(t)})^2]\frac{ds}{x + C(s)}\}dx$$

$$- 2 \int_0^t \{\dot{Q}_0(t - \tau)[1 - (\frac{a^2(\tau) + C(\tau)}{a^2(\tau) + C(t)})^2]\frac{1}{a^2(\tau) + C(\tau)}$$

$$- \int_\tau^t \ddot{Q}_0(t - s)[1 - (\frac{a^2(\tau) + C(s)}{a^2(\tau) + C(t)})^2]\frac{ds}{a^2(\tau) + C(s)}\}a(\tau)\dot{a}(\tau)d\tau. \quad \Box \quad (6.47)$$

Numerical analysis of integro-differential equation (6.46) is more convenient compared with integral equation (6.45).

To study the effect of accretion on stress distribution in a growing cylinder we find function $C(t)$ from Eq. (6.46) and substitute the obtained result into an analog of Eq. (6.35)

$$\sigma_{rr}(t, R) = -p_0(t)$$

$$+\mu \int_{a_0}^{R} \{[1 + Q_0(t - \kappa(R_1))][\frac{R_1^2 + C(t)}{R_1^2 + C_*(R_1)} - \frac{R_1^2 + C_*(R_1)}{R_1^2 + C(t)}]$$

$$-\int_{\kappa(R_1)}^{t} \dot{Q}_0(t - s)[\frac{R_1^2 + C(t)}{R_1^2 + C(s)} - \frac{R_1^2 + C(s)}{R_1^2 + C(t)}]ds\} \frac{R_1 dR_1}{R_1^2 + C(t)}. \qquad (6.48)$$

Afterward, the stresses $\sigma_{\theta\theta}(t, R)$ and $\sigma_{zz}(t, R)$ are found from Eqs. (6.26) and (6.27).

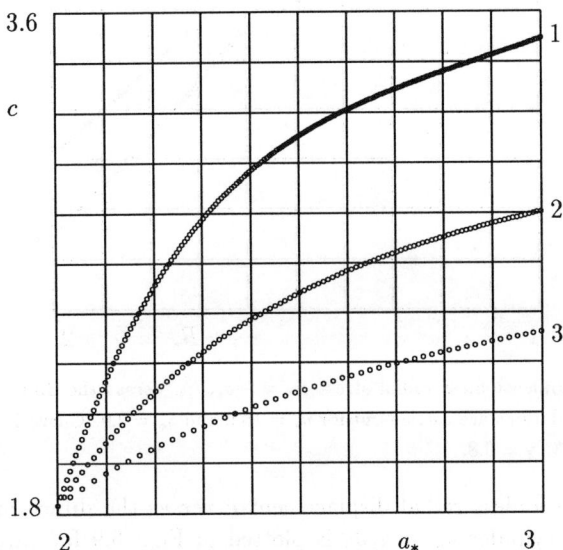

Fig. 6.9. The dimensionless displacement $c = C/a_0^2$ versus the dimensionless external radius $a_* = a/a_0$. Calculations are carried out for the standard viscoelastic solid with $\chi = 0.3$ and $\gamma_* = 2.0$. Curve 1: $u_* = 2.5$, curve 2: $u_* = 5.0$, curve 3: $u_* = 10.0$.

Exercise 6.14. Derive Eq. (6.48). □

The problem under consideration is essentially multiparametrical, and we do not intend to provide an exhausting analysis of the influence of its parameters on stresses in a growing cylinder. We confine ourselves to a cylinder made of the standard viscoelastic solid with relaxation measure (6.4.1) under the action of a time-independent internal pressure p_0.

Our objective is to study the effect of the dimensionless rate of mass influx u_*, as well as of the material viscosity χ and the dimensionless rate of relaxation

$\gamma_* = \gamma T$ on the dimensionless radial displacements $c = C/a_0^2$ and the dimensionless radial stresses $S_r = -\sigma_{rr}/p_0$. We assume that $a_1 = 2a_0$, $a_2 = 3a_0$, and the rate of accretion u_* is time-independent. For definiteness, we set $p_0 = 0.5\mu$.

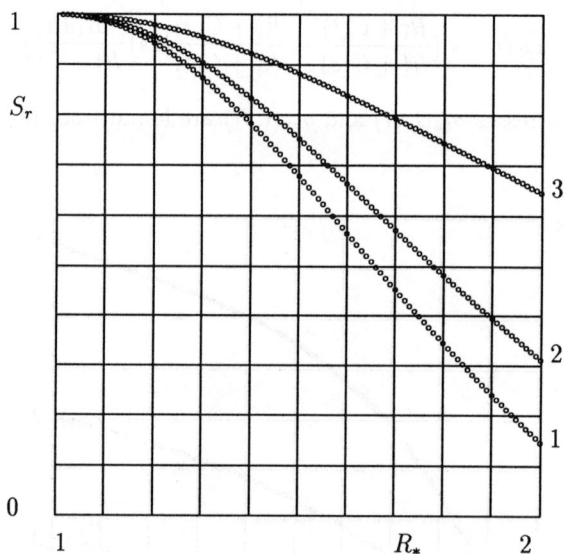

Fig. 6.10. The dimensionless radial stress $S_r = -\sigma_{rr}/p_0$ *versus* the dimensionless radius $R_* = R/a_0$. Calculations are carried out for $u_* = 10.0$ and $\gamma_* = 2.0$. Curve 1: $\chi = 0.3$, curve 2: $\chi = 0.5$, curve 3: $\chi = 0.8$.

The dimensionless radial displacement c *versus* the dimensionless radius of the growing cylinder $a_* = a/a_0$ is plotted in Fig. 6.9 for various rates of accretion. Radial displacements decrease monotonically with an increase in the rate of accretion. This phenomenon is quite natural, since at any instant $t \in [0, T]$, the growth of the rate of accretion leads to an increase in the thickness of the cylinder and, in turn, to a decrease of additional displacements caused by the material creep.

The dimensionless radial stress S_r *versus* the dimensionless radial coordinate $R_* = R/a_0$ is plotted in Figs. 6.10 and 6.11.

Comparison of Figs. 6.5 and 6.7 with Fig. 6.10 shows that radial stresses in a growing viscoelastic cylinder exceed radial stresses in a non-growing elastic cylinder, as well as in a growing elastic cylinder. At first sight, this seems paradoxical, since stresses in a non-growing viscoelastic body should be less than stresses in an appropriate elastic body due to the stress relaxation. In a growing medium, the stress relaxation is not the only physical phenomenon which determines distribution of stresses. Another phenomenon is the external pressure from the accreted part of the body. For a growing viscoelastic cylinder, this pressure increases monotonically in time. This increase may be explained by the material creep which leads to additional radial displacements of the main body. These displacements are restrained by built-up portions which produce an increasing pressure on the main cylinder. Fig. 6.10 shows that the creep flow plays the key role in this process: with the growth of viscosity χ, the dimensionless radial stresses increase.

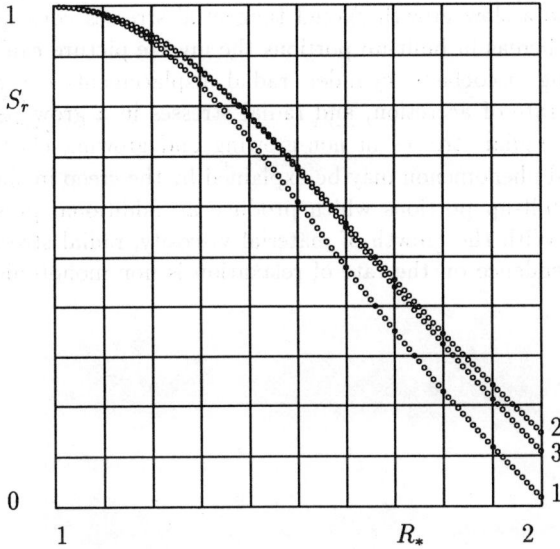

Fig. 6.11. The dimensionless radial stress $S_r = -\sigma_{rr}/p_0$ versus the dimensionless radius $R_* = R/a_0$. Calculations are carried out for $u_* = 10.0$ and $\chi = 0.3$. Curve 1: $\gamma_* = 0.1$, curve 2: $\gamma_* = 2.0$, curve 3: $\gamma_* = 20.0$.

The influence of the characteristic time of relaxation is essentially weaker. The dependence of radial stresses on γ_* is non-monotonic: for small γ_* values, the stresses increase with the growth of γ_*, while for large γ_* values, the stresses decrease, see Fig. 6.11. This dependence is rather feeble, and it may be neglected in applications.

6.4. Concluding Remarks

A mathematical model is suggested for continuous accretion of viscoelastic solids. The key idea of the model consists in the use of three basic configurations: reference, natural, and actual. Governing equations are derived for the

411

description of a growing viscoelastic medium subjected to aging. These equations are applied to analyze stresses and displacements in a growing circular cylinder. It is shown that

(i) radial and tangential stresses in a non-growing elastic cylinder with finite strains exceed the corresponding stresses in an appropriate cylinder with infinitesimal strains;

(ii) in a growing elastic cylinder, radial stresses are continuous, while tangential stresses may have finite jumps of discontinuity provided the rate of accretion is a piece-wise continuous function of time. In the main body, tangential stresses corresponding to a slow growth exceed tangential stresses corresponding to a rapid growth, whereas in built-up portions the inverse picture can be seen;

(iii) in a growing viscoelastic cylinder, radial displacements decrease with an increase in the rate of accretion, and radial stresses in a growing viscoelastic cylinder exceed radial stresses in non-growing and growing elastic cylinders. This paradoxical phenomenon may be explained by the creep in the main body restrained by built-up portions which produce an additional pressure on the main cylinder. With the growth of material viscosity, radial stresses increase, while their dependence on the rate of relaxation is non-monotonic and rather weak.

References

1. Ablowitz, M.J. and Segur, H. (1981) *Solitons and the Inverse Scattering Transform*. SIAM, Philadelphia.

2. Achenbach, J.D. and Chao, C.C. (1962) A three-parameter viscoelastic model particularly suited for dynamic problems. *J. Mech. Phys. Solids.* **10**, 245-252.

3. Adeyeri, J.B., Krizek, R.J., and Achenbach, J.D. (1970) Multiple integral description of the nonlinear viscoelastic behavior of a clay soil. *Trans. Soc. Rheol.* **14**, 375-392.

4. Agarwal, V.K. (1979) On finite anti-plane shear for compressible elastic circular tube. *J. Elasticity.* **9**, 311-319.

5. Aklonis, J.J., MacKnight, W.J., and Shen, M. (1972) *Introduction to Polymer Viscoelasticity*. Wiley, New-York.

6. Alberola, N.D., Lesueur, D., Granier, V., and Joanicot, M. (1995) Viscoelasticity of porous composite materials: experimentals and theory. *Polymer Composites.* **16**, 170-179.

7. Alexander, H. (1968) A constitutive relation for rubber-like materials. *Int. J. Engng. Sci.* **6**, 549-563.

8. Antman, S.S. (1972) The theory of rods. *Handbuch der Physik.* **6a/2**, Springer-Verlag, Berlin.

9. Antman, S.S. (1973) Nonuniqueness of equilibrium states for bars in tension. *J. Math. Anal. Appl.* **44**, 333-349.

10. Antman, S.S. (1979) The eversion of thick spherical shells. *Arch. Rational Mech. Anal.* **70**, 113-123.

11. Antman, S.S. (1983) Regular and singular problems for large elastic deformations of tubes, wedges, and cylinders. *Arch. Rational Mech. Anal.* **83**, 1-52.

12. Arutyunyan, N.K. (1952) *Some Problems in the Creep Theory*. Gostechizdat, Moscow (in Russian).

13. Arutyunyan, N.K. and Drozdov, A.D. (1984) Accreting gravitating viscoelastic sphere with finite deformations. *Mech. Solids.* **19**, 121-134.

14. Arutyunyan, N.K., Drozdov, A.D., and Naumov, V.E. (1987) *Mechanics of Growing Viscoelastoplastic Solids*. Nauka, Moscow (in Russian).

413

15. Arutyunyan, N.K. and Drozdov, A.D. (1992) Phase transitions in nonhomogeneous, aging, viscoelastic bodies. *Int. J. Solids Struct.* **29**, 783-797.

16. Astarita, G. and Marrucci, G. (1974) *Principles of Non-Newtonian Fluid Mechanics.* McGraw-Hill, London.

17. Atkin, R.J. and Fox, N. (1980) *An Introduction to the Theory of Elasticity.* Longman, London.

18. Bagley, R.L. and Torvik, P.J. (1983) A theoretical basis for the application of fractional calculus to viscoelasticity. *J. Rheol.* **27**, 201-210.

19. Bagley, R.L. and Torvik, P.J. (1986) On the fractional calculus model of viscoelastic behavior. *J. Rheol.* **30**, 133-155.

20. Ball, J.M. (1977) Convexity conditions and existence theorems in nonlinear elasticity. *Arch. Rational Mech. Anal.* **63**, 337-403.

21. Barenblatt, G.I., Volodchenkov, V.A., Kershtein, I.M., and Pavlov, D.Ja. (1974) Isothermal necking in polymers. Comparison with creep. *Mechanika Tverdogo Tela.* **5**, 144-156 (in Russian).

22. Batra, R.C. (1972) On non-classical boundary conditions. *Arch. Rational Mech. Anal.* **48**, 163-191.

23. Bazant, Z.P. (ed.) (1988) *Mathematical Modelling of Creep and Shrinkage of Concrete.* Wiley, New-York.

24. Beris, A.N. and Edwards, B.J. (1993) On the admissibility criteria for linear viscoelasticity kernels. *Rheol. Acta.* **32**, 505-510.

25. Bernstein, B., Kearsley, E.A., and Zapas, L.J. (1963) A study of stress relaxation with finite strains. *Trans. Soc. Rheol.* **7**, 391-410.

26. Bernstein, B. (1966) Time-dependent behavior of an incompressible elastic fluid. Some homogeneous deformation histories. *Acta Mech.* **2**, 329-354.

27. Bertilsson, H., Delin, M., Kubat, J., Rychwalski, W.R., and Kubat, M.J. (1993) Strain rates and volume changes during short-term creep of PC and PMMA. *Rheol. Acta.* **32**, 361-369.

28. Bicerano, J. (1993) *Prediction of Polymer Properties.* Marcel Dekker, New-York.

29. Bird, R.B., Armstrong, R.C., and Hassager, O. (1977) *Dynamics of Polymeric Fluids.* **1**. Wiley, New-York.

30. Blatz, P.J. (1960) Application of finite elasticity theory in predicting the performance of solid propeliane rocket motors. *SM Report 60-25.* California Institute of Technology. Pasadena.

31. Blatz, P.J. and Ko, W.L. (1962) Application of finite elastic theory to the deformation of rubbery materials. *Trans. Soc. Rheol.* **6**, 223-251.

32. Blatz, P.J. (1971) On the thermostatic behavior of elastomers. In *Polymer Networks. Structure and Mechanical Properties.* Plenum Press, New-York, 23-45.

33. Bloch, R., Chang, W.V., and Tschoegl, N.W. (1978) The behavior of rub-

berlike materials in moderately large deformations. *J. Rheol.* **22**, 1-32.

34. Brinson, L.C. and Gates, T.S. (1995) Effects of physical aging on long term creep of polymers and polymer matrix composites. *Int. J. Solids Struct.* **32**, 827-846.

35. Brown, C.B., and Goodman, L.E. (1963) Gravitational stresses in accreted bodies. *Proc. Roy. Soc. London.* **A276**, 571-576.

36. Bugakov, I.I. (1989) Principle of superposition as the basis for the nonlinear constitutive equations for media with memory. *Mech. Solids.* **24**, 85-92.

37. Burgess, I.W. and Levinson, M. (1972) The instability of slightly compressible rectangular rubberlike solids under biaxial loadings. *Int. J. Solids Struct.* **8**, 133-148.

38. Burton, T.A. (1985) *Stability and Periodic Solutions of Ordinary and Functional Differential Equations.* Academic Press, Orlando.

39. Busemann, H., Ewalds, G., and Shephard, G.C. (1963) Convex bodies and convexity on Grassmann cones. *Math. Ann.* **151**, 1-41.

40. Carroll, M.M. (1988) Finite strain solutions in compressible isotropic elasticity. *J. Elasticity.* **20**, 65-92.

41. Chang, W.V., Bloch, R., and Tschoegl, N.W. (1976) On the theory of the viscoelastic behavior of soft polymers in moderately large deformations. *Rheol. Acta.* **15**, 367-378.

42. Christensen, R.M. (1980) A nonlinear theory of viscoelasticity for application to elastomers. *Trans. ASME. J. Appl. Mech.* **47**, 762-768.

43. Christensen, R.M. (1982) *Theory of Viscoelasticity. An Introduction.* Academic Press, New-York.

44. Ciarlet, P.G. and Geymonat, G. (1982) Sur les lois de comportement en elasticite non-lineaire compressible. *C.R. Acad. Sci. Paris.* **295**, 423-426.

45. Ciarlet, P.G. (1988) *Mathematical Elasticity.* North-Holland, Amsterdam.

46. Coleman, B.D. and Noll, W. (1960) An approximation theorem for functionals with applications to continuum mechanics. *Arch. Rational Mech. Anal.* **6**, 355-370.

47. Coleman, B.D. and Noll, W. (1961) Foundations of linear viscoelasticity. *Rev. Modern Phys.* **33**, 239-249.

48. Coleman, B.D. (1964) Thermodynamics of materials with memory. *Arch. Rational Mech. Anal.* **17**, 1-46.

49. Coleman, B.D., Gurtin, M.E., and Herrera, I.R. (1965) Waves in materials with memory, 1. The velocity of one-dimensional shock and acceleration waves. *Arch. Rational Mech. Anal.* **19**, 1-19.

50. Coleman, B.D., and Gurtin, M.E. (1965) Waves in materials with memory, 2. On the growth and decay of one-dimensional acceleration waves. *Arch. Rational Mech. Anal.* **19**, 239-265.

51. Coleman, B.D., Markowitz, H., and Noll, W. (1966) *Viscometric Flows of*

Non-Newtonian Fluids. Springer-Verlag, New-York.

52. Coleman, B.D. and Mizel, V.J. (1968) On the general theory of fading memory. *Arch. Rational Mech. Anal.* **29**, 18-31.

53. Christiano, P.P., and Chantranuluck, S. (1974) Retaining wall under action of accreted backfill. *J. Geotechn. Eng. Div. Amer. Soc. Civil Eng. Proc.* **100**, 471-476.

54. Dafermos, C.M. (1970) An abstract Volterra equation with applications to linear viscoelasticity. *J. Diff. Eqns.* **7**, 554-569.

55. Dafermos, C.M. (1986) Development of singilarities in the motion of materials with fading memory. *Arch. Rational Mech. Anal.* **91**, 193-205.

56. DeHoff, P.H., Lianis, G., and Goldberg, W. (1966) An experimental program for finite linear viscoelasticity. *Trans. Soc. Rheol.* 10(1966) 385-397.

57. Deligianni, D.D., Maris, A., and Missirlis, Y.F. (1994) Stress relaxation behavior of trabecular bone specimens. *J. Biomech.* **27**, 1469-1476.

58. Derman, D., Zaphir, Z., and Bodner, S.R. (1978) Nonlinear anelastic behavior of a synthetic rubber at finite strains. *J. Rheol.* **22**, 239-258.

59. Drozdov, A. (1993) On constitutive laws for ageing viscoelastic materials at finite strains. *Europ. J. Mech. A/Solids.* **12**, 305-324.

60. Drozdov, A. (1994) Accretion of viscoelastic bodies at finite strains. *Mech. Research Comm.* **21**, 329-334.

61. Drozdov, A.D. and Kolmanovskii, V.B. (1994) *Stability in Viscoelasticity.* North-Holland, Amsterdam.

62. Drozdov, A.D. and Kalamkarov, A.L. (1995) Optimization of the winding for composite pressure vessels. *Int. J. Pressure Vessels Pipes.* **62**, 69-81.

63. Duong, C.N. and Knauss, W.G. (1993) The effect of thermoviscoelastic residual stresses in adhesive bonds: 1. Stress analysis. *SM Report 93-37B.* California Institute of Technology. Pasadena. 2. Fracture analysis. *SM Report 93-38.* California Institute of Technology. Pasadena.

64. Duvaut, G. and Lions, J.L. (1976) *Inequalities in Mechanics and Physics.* Springer-Verlag, Berlin.

65. Ek, C.-G., Kubat, J., and Rigdahl, M. (1987) Stress relaxation, creep, and internal stresses in high density polyethylene filled with calcium carbonate. *Rheol. Acta.* **26**, 55-63.

66. Emery, A.H. and White, M.L. (1969) A single-integral constitutive equation. *Trans. Soc. Rheol.* **13**, 103-109.

67. Engler, H. (1989) Global regular solution to the dynamic antiplane shear problem in nonlinear viscoelasticity. *Math. Z.* **202**, 251-259.

68. Engler, H. (1991) A matrix Volterra integrodifferential equation occurring in polymer rheology. *Pacific J. Math.* **149**, 25-60.

69. Ericksen, J.L. (1955) Inversion of a perfectly elastic spherical shell. *Z. Angew. Math. Mech.* **35**, 382-385.

416

70. Ericksen, J.L. (1955) Deformations possible in every compressible, isotropic, perfectly elastic body. *J. Math. Phys.* **34**, 126-128.

71. Ericksen, J.L. (1991) *Introduction to the Thermodynamics of Solids.* Chapman and Hall, New-York.

72. Eringen, A.C. (1967) *Mechanics of Continua.* Wiley, New-York.

73. Eshelby, J.D. (1975) The elastic energy-momentum tensor. *J. Elasticity.* **5**, 321-335.

74. Espinoza, A. and Aklonis, J.J. (1993) Modeling viscoelastic behavior of glasses during physical aging. *Polym. Eng. Sci.* **33**, 486-496.

75. Fabrizio, M. and Morro, A. (1992) *Mathematical Problems in Linear Viscoelasticity.* SIAM Studies in Appl. Math., Philadelphia.

76. Feigl, K., Ottinger, H.C., and Meissner, J. (1993) A failure of a class of K–BKZ equations based on principal stretches. *Rheol. Acta.* **32**, 438-446.

77. Ferry, J.D. (1980) *Viscoelastic Properties of Polymers.* Wiley, New-York.

78. Fichera, G. (1972) Existence theorems in elasticity. *Handbuch der Physik.* **6a/2**, Springer-Verlag, Berlin, 347-389.

79. Findley, W.N. (1987) 26-year creep and recovery of poly(vinil chloride) and polyethylene. *Polym. Eng. Sci.* **27**, 582-585.

80. Findley, W.N., Lai, J.S., and Onaran, K. (1989) *Creep and Relaxation of Nonlinear Viscoelastic Materials.* Dover Publications, New-York.

81. Fitzgerald, J.E. (1980) Tensorial Hencky measure of strain and strain rate for finite deformations. *J. Appl. Phys.* **51**, 5111-5115.

82. Flory, P.J. and Tatara, Y.I. (1975) The elastic free energy and the elastic equation of state: elongation and swelling of polydimethylsiloxane networks. *J. Polym. Sci. Polym. Phys. Ed.* **13**, 683-702.

83. Fonseca, I. (1989) Interfacial energy and the Maxwell rule. *Arch. Rational Mech. Anal.* **106**, 63-95.

84. Fosdick, R.L. and Serrin, J. (1979) On the impossibility of linear Cauchy and Piola–Kirchhoff constitutive theories for stress in solids. *J. Elasticity.* **9**, 83-89.

85. Friedrich, C. (1991) Relaxation and retardation functions of the Maxwell model with fractional derivatives. *Rheol. Acta.* **30**, 151-158.

86. Garbarski, J. (1992) The application of an exponential-type function for the modeling of viscoelasticity of solid polymers. *Polym. Eng. Sci.* **32**, 107-114.

87. Gent, A.N. & Thomas, A.G. (1958) Forms for the stored (strain) energy function for vulcanized rubber. *J. Polym. Sci.* **28**, 625-638.

88. Giesekus, H. (1994) *Phanomenologische Rheologie – eine Einfuhrung.* Springer -Verlag, Berlin.

89. Giesekus, H. (1995) An alternative approach to the linear theory of viscoelasticity and some characteristic effects being distinctive of the type of material. *Rheol. Acta.* **34**, 2-11.

90. Glockle, W.G. and Nonnenmacher, T.F. (1994) Fractional relaxation and the time-temperature superposition principle. *Rheol. Acta.* **33**, 337-343.

91. Glucklich, J. and Landel, R.F. (1977) Strain energy function of styrene butadiene rubber and the effect of temperature. *J. Polym. Sci. Polym. Phys. Ed.* **15**, 2185-2199.

92. Goldstein, C. (1974) Transient and steady shear behavior of SBR Polymers. *Trans. Soc. Rheol.* **18**, 357-369.

93. Goodman, L.E. and Brown, C.B. (1963) Dead load stresses and the instability of slopes. *J. Soil Mech. Foundat. Div. Amer. Soc. Civil Eng. Proc.* **89**, 103-134.

94. Gramespacher, H. and Meissner, J. (1995) Reversal of recovery direction during creep recovery of polymer blends. *J. Rheol.* **39**, 151-160.

95. Green, A.E. and Rivlin, R.S. (1957) The mechanics of nonlinear materials with memory. 1. *Arch. Rational Mech. Anal.* **1**, 1-21.

96. Green, A.E. and Adkins, J.E. (1970) *Large Elastic Deformations and Non-Linear Continuum Mechanics.* Oxford Univ. Press, Oxford.

97. Green, A.E. and Naghdi, P.M. (1995) A unified procedure for construction of theories of deformable media. 1. Classical continuum physics. *Proc. Roy. Soc. London.* **448**, 335-356. 2. Generalized continua. *Proc. Roy. Soc. London.* **448**, 357-377. 3. Mixtures of interacting continua. *Proc. Roy. Soc. London.* **448**, 379-388.

98. Green, M.S. and Tobolsky, A.V. (1946) A new approach to the theory of relaxing polymeric media. *J. Chem. Phys.* **14**, 80-92.

99. Gripenberg, G., Londen, S.-O., and Staffans, O. (1990) *Volterra Integral and Functional Equations.* Cambridge Univ. Press, Cambridge.

100. Gripenberg, G. (1993) Global existence of solutions of Volterra integrodifferential equations of parabolic type. *J. Diff. Eqns.* **102**, 382-390.

101. Gripenberg, G. (1994) On the uniqueness and nonuniqueness of weak solutions of hyperbolic-parabolic Volterra equations. *Diff. Integral Eqns.* **7**, 509-522.

102. Gromov, V.G. and Miroshnikov, V.P. (1978) Effect of thermomechanical coupling in the theory of viscoelasticity. *Soviet Phys. Doklady.* **23**, 434-436.

103. Gul', V.E., Yanovskii, Yu.G., Shamraevskaya, T.V., and Litvinenko, O.A. (1992) Correlation of the relaxation and strain-strength properties of elastomers. *Mech. Composite Mater.* **28**, 484-488.

104. Gurtin, M.E. and Martins, L.C. (1976) Cauchy's theorem in classical physics. *Arch. Rational Mech. Anal.* **60**, 305-328.

105. Gurtin, M.E. (1981) *An Introduction to Continuum Mechanics.* Academic Press, New-York.

106. Gurtin, M.E. (1983) Two-phase deformations of elastic solids. *Arch. Rational Mech. Anal.* **84**, 1-29.

107. Gurtin, M.E. and Spear, K. (1983) On the relationship between the logarithmic strain rate and the stretching tensor. *Int. J. Solids Struct.* **19**, 437-444.

108. Gurtin, M.E. and Struthers, A. (1990) Multiphase thermomechanics with interfacial structure. 3. Evolving phase boundaries in the presence of bulk deformation. *Arch. Rational Mech. Anal.* **112**, 97-160.

109. Gurtin, M.E. (1993) The dynamics of solid-solid phase transformations. 1. Coherent interfaces. *Arch. Rational Mech. Anal.* **123**, 305-335.

110. Hadley, D.W. and Ward, I.M. (1965) Non-linear creep and recovery behaviour of polypropylene fibers. *J. Mech. Phys. Solids.* **13**, 397-411.

111. Hammack, J.L. and Segur, H. (1974) The Korteweg–de Vries equation and water waves. 2: Comparison with experiments. *J. Fluid Mech.* **65**, 289-314.

112. Hart-Smith, L.J. (1966) Elasticity parameters for finite deformations of rubber-like materials. *Z. Angew. Math. Phys.* **17**, 608-625.

113. Haughton, D.M. (1987) Inflation of thick-walled compressible elastic spherical shell. *IMA J. Appl. Math.* **39**, 259-272.

114. Hausler, K. and Sayir, M.B. (1995) Nonlinear viscoelastic response of carbon black reinforced rubber derived from moderately large deformations in torsion. *J. Mech. Phys. Solids.* **43**(2), 295-318.

115. Hearn, E.W., Werner, D.J., and Doney, D.A. (1986) Film-induced stress model. *J. Electrochem. Soc.* **133**, 1749-1751.

116. Hill, R. (1968) On constitutive inequalities for simple materials. 1. *J. Mech. Phys. Solids.* **16**, 229-242.

117. Hirota, R. (1971) Exact solution of the Korteweg–de Vries equation for multiple collisions of solitons. *Phys. Rev. Lett.* **27**, 1192-1194.

118. Horgan, C.O. and Knowles, J.K. (1981) The effect of nonlinearity on a principle of Saint-Venant type. *J. Elasticity.* **11**, 271-291.

119. Horgan, C.O. and Payne, L.E. (1990) On Saint-Venant's principle in finite anti-plane shear: an energy approach. *Arch. Rational Mech. Anal.* **109**, 107-137.

120. Hrusa, W.J. and Nohel, J.A. (1985) The Cauchy problem in one-dimensional nonlinear viscoelasticity. *J. Diff. Eqns.* **59**, 388-412.

121. Huilgol, R.R. (1979) Viscoelastic fluid theories based on the left Cauchy-Green tensor history. *Rheol. Acta.* **18**, 451-455.

122. Huppler, J.D., Ashare, E., and Holmes, L.A. (1967) Rheological properties of three solutions. 1. Non-Newtonian viscosity, normal stresses, and complex viscosity. *Trans. Soc. Rheol.* **11**, 159-179.

123. Huppler, J.D., MacDonald, I.F., Ashare, E., Spriggs, T.W., Bird, R.D., and Holmes, L.A. (1967) Rheological properties of three solutions. 2. Relaxation and growth of shear and normal stresses. *Trans. Soc. Rheol.* **11**, 181-204.

124. Hutchinson, W.D., Becker, G.W., and Landel, R.F. (1965) Determination of

the stored energy function for rubber-like materials. *Bull. 4th Meeting Interagency Chem. Rocket Propulsion Group–Working Group Mech. Behavior.* CPIA Publ. 94U. **1**, 141-152.

125. Isihara, A., Hashitsume, N., and Tatibana, M. (1951) Statistical theory of rubber-like elasticity. 4. Two-dimensional stretching. *J. Chem. Phys.* **19**, 1508-1519.

126. James, R.D. (1981) Finite deformations by mechanical twinning. *Arch. Rational Mech. Anal.* **77**, 143-176.

127. Janssen, L.P.B.M. and Janssen–van Rosmalen, R. (1978) An analysis of flow induced formation of long fibers. *Rheol. Acta.* **17**, 578-588.

128. John, F. (1960) Plane strain problems for a perfectly elastic material of harmonic type. *Comm. Pure Appl. Math.* **13**, 239-296.

129. John, F. (1964) Remarks on the nonlinear theory of elasticity. *Seminari Ist. Naz. Alta Matem. 1962/63*, 474-482.

130. Kaschta, J. and Schwarzl, F.R. (1994) Calculation of discrete retardation spectra from creep data. 1. Method. *Rheol. Acta.* **33**, 517-529. 2. Analysis of measured creep curves. *Rheol. Acta.* **33**, 530-541.

131. Kaye, A. and Kennett, A.J. (1974) Constrained elastic recovery of a polymeric liquid after various shear flow histories. *Rheol. Acta.* **13**, 916-923.

132. Knauss, W.G. and Kenner, V.H. (1980) On the hygrothermomechanical characterization of polyvinil acetate. *J. Appl. Phys.* **51**, 5131-5136.

133. Knowles, J.K. (1977) A note on anti-plane shear for compressible materials in finite elastostatics. *J. Austral. Math. Soc.* **B20**, 1-7.

134. Knowles, J.K. and Sternberg, E. (1981) Anti-plane shear fields with discontinuous deformation gradient near the tip of a crack in finite elastostatics. *J. Elasticity.* **11**, 129-164.

135. Kochetkov, V.A. and Maksimov, R.D. (1990) Thermal deformation of unidirectional hybrid composites. 2. *Mech. Composite Mater.* **25**, 690-699.

136. Lai, W.M., Rubin, D., and Krempl, E. (1993) *Introduction to Continuum Mechanics.* Pergamon Press, Oxford.

137. La Mantia, F.P. (1977) Non linear viscoelasticity of polymeric liquids interpreted by means of a stress dependence of free volume. *Rheol. Acta.* **16**, 302-308.

138. La Mantia, F.P. and Titomanlio, G. (1979) Testing of a constitutive equation with free volume dependent relaxation spectrum. *Rheol. Acta.* **18**, 469-477.

139. La Mantia, F.P., Titomanlio, G., and Acierno, D. (1980) The viscoelastic behavior of nylon 6/lithium halides mixtures. *Rheol. Acta.* **19**, 88-93.

140. Landau, L.D. and Lifshitz, E.M. (1969) *Statistical Physics.* Pergamon Press, Oxford.

141. Laun, H.M. (1978) Description of the non-linear shear behavior of a low density polyethylene melt by means of an experimentally determined strain

dependent memory function. *Rheol. Acta.* **17**, 1-15.

142. Lax, P.D. (1964) Development of singularities of solutions of nonlinear hyperbolic partial differential equations. *J. Math. Phys.* **5**, 611-613.

143. Lee, A. and McKenna, G.B. (1990) The physical ageing response of an epoxy glass subjected to large stresses. *Polymer.* **31**, 423-430.

144. Lee, A. and McKenna, G.B. (1990) Viscoelastic response of epoxy glasses subjected to different thermal treatments. *Polym. Eng. Sci.* **30**, 431-435.

145. Leonov, A.I. (1976) Nonequilibrium thermodynamics and rheology of viscoelastic polymer media. *Rheol. Acta.* **15**, 85-98.

146. Leonov, A.I., Lipkina, E.H., Paskhin, E.D., and Prokunin, A.N. (1976) Theoretical and experimental investigation of shearing in elastic polymer liquids. *Rheol. Acta.* **15**, 223-230.

147. Levinson, M. and Burgess, J.W. (1971) A comparison of some simple constitutive relations for slightly compressible rubberlike materials. *Int. J. Mech. Sci.* **13**, 563-572.

148. Lodge, A.S. (1964) *Elastic Liquids.* Academic Press, New-York.

149. Lodge, A.S. and Meissner, J. (1972) On the use of instantaneous strains, superposed on shear and elongation flows of polymeric liquids, to test the Gaussian network hypothesis and to estimate the segment concentration and its variation during flow. *Rheol. Acta.* **11**, 351-352.

150. Lodge, A.S. (1989) Elastic recovery and polymer-polymer interactions. *Rheol. Acta.* **28**, 351-362.

151. Lurie, A.I. (1990) *Non-Linear Theory of Elasticity.* North-Holland, Amsterdam.

152. MacCamy, R.C. (1977) A model for one-dimensional, nonlinear viscoelasticity. *Quart. Appl. Math.* **35**, 21-33.

153. Maccarrone, S. and Tiu, C. (1988) Rheological properties of service weathered road bitumens. *Rheol. Acta.* **27**, 311-319.

154. MacDonald, I.F. (1976) On the admissibility of rate-dependent viscoelastic models. *Rheol. Acta.* **15**, 223-230.

155. Malkin, A.Ya. (1995) Non-linearity in rheology – an essay of classification. *Rheol. Acta.* **34**, 27-39.

156. Marrucci, G. and de Cindio, B. (1980) The stress relaxation of molten PMMA at large deformations and its theoretical interpretation. *Rheol. Acta.* **19**, 68-75.

157. Mathur, P., Apelian, D., and Lawley, A. (1989) Analysis of the spray deposition process. *Acta Metall.* **37**, 429-443.

158. McGuirt, C.W. and Lianis, G. (1970) Constitutive equations for viscoelastic solids under finite uniaxial and biaxial deformations. *Trans. Soc. Rheol.* **14**, 117-134.

159. McHerron, D.C. and Wilkes, G.L. (1993) Apparent reversal of physical aging

421

in amorphous glassy polymers by electron beam irradiation. *Polymer.* **34**, 915-924.

160. McKenna, G.B. and Kovacs, A.J. (1984) Physical aging of poly(methyl methacrylate) in the nonlinear range: torque and normal force measurements. *Polym Eng. Sci.* **24**, 1138-1141.

161. Miura, R.M. (1968) Korteweg–de Vries equation and generalizations 1. A remarkable explicit nonlinear transformation. *J. Math. Phys.* **9**, 1202-1204.

162. Mooney, M. (1940) A theory of large elastic deformations. *J. Appl. Phys.* **11**, 582-592.

163. Morman, K.N. (1986) The generalized strain measure with application to nonhomogeneous deformations in rubber-like solids. *Trans. ASME. J. Appl. Mech.* **53**, 726-728.

164. Morman, K.N. (1988) An adaptation of finite linear viscoelasticity theory for rubber-like viscoelasticity by use of a generalized strain measure. *Rheol. Acta.* **27**, 3-14.

165. Murnaghan, F.D. (1951) *Finite Deformation of an Elastic Solid.* Wiley, New-York.

166. Neville, A.M., Digler, W.H., and Brooks, J.J. (1983) *Creep of Plane and Structural Concrete.* Construction Press, London.

167. Novozhilov, V.V. (1953) *Foundations of the Nonlinear Theory of Elasticity.* Graylock Press, Rochester.

168. Novozhilov, V.V. (1959) *The Theory of Thin Shells.* P. Noordhoff, Groningen.

169. Ogden, R.W. (1972) Large deformation isotropic elasticity: on the correlation of theory and experiment for compressible rubberlike solids. *Proc. Roy. Soc. London.* **A328**, 567-583.

170. Ogden, R.W. (1984) *Non-Linear Elastic Deformations.* Ellis Horwood, Chichester.

171. Oldham, K.B. and Spanier, J. (1974) *The Fractional Calculus.* Academic Press, New-York.

172. Oldroyd, J.G. (1950) On the formulation of rheological equations of state. *Proc. Roy. Soc. London.* **A200**, 523-541.

173. Oldroyd, J.G. (1958) Non-Newtonian effects in steady motion of some idealized elastico-viscous liquids. *Proc. Roy. Soc. London.* **A245**, 278-297.

174. Onogi, S., Masuda, T., and Matsumoto, T. (1970) Nonlinear behavior of viscoelastic materials. 1. Disperse systems of polytirene solution and carbon black. *Trans. Soc. Rheol.* **14**, 275-294.

175. Papo, A. (1988) Rheological models for gypsum plaster pastes. *Rheol. Acta.* **27**, 320-325.

176. Pearson, G. and Middleman, S. (1978) Elongation flow behavior of viscoelastic liquids: modelling bubble dynamics with viscoelastic constitutive rela-

tions. *Rheol. Acta.* **17**, 500-510.

177. Pipkin, A.C. and Rivlin, R.S. (1961) Small deformations superposed on large deformations in materials with fading memory. *Arch. Rational Mech. Anal.* **8**, 297-310.

178. Pipkin, A.C. (1972) *Lectures on Viscoelasticity Theory.* Springer-Verlag, New-York.

179. Plazek, D.J., Ngai, K.L., and Rendell, R.W. (1984) An application of a unified relaxation model to the aging of polystyrene below its glass temperature. *Polym. Eng. Sci.* **24**, 1111-1116.

180. Podio-Guidugli, P. (1987) The Piola–Kirchhoff stress may depend linearly of the deformation gradient. *J. Elasticity.* **17**, 183-187.

181. Polignone, D.A. and Horgan, C.O. (1991) Pure torsion of a compressible nonlinearly elastic circular cylinder. *Quart. Appl. Math.* **49**, 591-607.

182. Powers, J.M. and Caddell, R.M. (1972) The macroscopic volume changes of selected polymers subjected to uniform tensile deformation. *Polym. Eng. Sci.* **12**, 432-436.

183. Prokopovich, I.E. (1963) *Effect of Long-Time Processes on Stresses and Strains in a Structure.* Gosstrojizdat, Moscow (in Russian).

184. Pruss, J. (1993) *Evolutionary Integral Equations and Applications.* Birkhauser, Basel.

185. Rabotnov, Ju.N. (1969) *Creep Problems in Structural Members.* North-Holland, Amsterdam.

186. Rajagopal, K.R. and Wineman, A.S. (1987) New universal relations for nonlinear isotropic elastic materials. *J. Elasticity.* **17**, 75-83.

187. Reinhardt, W.D. and Dubey, R.N. (1995) Eulerian strain-rate as a rate of logarithmic strain. *Mech. Research Comm.* **22**, 165-170.

188. Renardy, M. (1982) Some remarks on the propagation and non-propagation of discontinuities in linearly viscoelastic liquids. *Rheol. Acta.* **21**, 251-254.

189. Renardy, M., Hrusa, W.J., and Nohel, J.A. (1987) *Mathematical Problems in Viscoelasticity.* Longman, New-York.

190. Renardy, M. (1988) Coercive estimates and existence of solutions for a model of one-dimensional viscoelasticity with a non-integrable memory function. *J. Integral Eqns. Appl.* **1**, 7-16.

191. Rivlin, R.S. (1949) A note on the torsion of an incompressible highly-elastic cylinder. *Proc. Cambridge Phil. Soc.* **45**, 485-487.

192. Rivlin, R.S. and Saunders, D.W. (1951) Large elastic deformations of isotropic materials. 7. Experiments on the deformation of rubber. *Phyl. Trans. Roy. Soc. London.* **243**, 251-288.

193. Rivlin, R.S. and Ericksen, J.L. (1955) Stress-deformation relations for isotropic materials. *J. Rational Mech. Anal.* **4**, 323-425.

194. Rivlin, R.S. (1956) Large elastic deformations. In *Rheology. Theory and*

Applications, ed. Eirich, F.R. Academic Press, New-York.

195. Robert, F. and Sherman, P. (1988) The influence of surface friction on the calculation of stress relaxation parameters for prestressed cheese. *Rheol. Acta.* **27**, 212-215.

196. Roitburd, A.L. (1974) Theory of formation of heterophase structure at phase transitions in solid state. *Usp. Fiz. Nauk.* **113**, 69-104 (in Russian).

197. Ross, A.D. (1958) Creep in concrete under variable stress. *J. Amer. Concr. Inst.* **29**, 739-758.

198. Roylance, D. and Wang, S.S. (1978) Penetration mechanics of textile structures: influence of non-linear viscoelastic relaxation. *Polym. Eng. Sci.* **18**, 1068-1072.

199. Rzhanitsyn, A.R. (1968) *Creep Theory.* Stroijzdat, Moscow (in Russian).

200. Scanlan, J.C. and Janzen, J. (1992) Comparison of linear viscoelastic constitutive equations for non-uniform polymer melts. *Rheol. Acta.* **31**, 183-193.

201. Schwarzl, F.R. (1990) *Polymermechanik.* Springer-Verlag, Berlin.

202. Seth, B.R. (1935) Finite strain in elastic problems. *Phil. Trans. Roy. Soc. London.* **234**, 231-264.

203. Seth, B.R. (1964) Generalized strain measure with applications to physical problems. In *Second-Order Effects in Elasticity, Plasticity and Fluid Dynamics.* eds. Reiner, M. and Abir, D. Pergamon Press, Oxford, 162-172.

204. Sewell, M.J. (1967) On configuration-dependent loading. *Arch. Rational Mech. Anal.* **23**, 327-351.

205. Signorini, A. (1943) Transformazioni termoelastiche finite. Memoria 1. *Annali Mat. Pura Appl.* **22**, 33-143.

206. Simpson, H.C. and Spector, S.J. (1984) On barrelling instabilities in finite elasticity. *Quart. Appl. Math.* **42**, 99-111.

207. Skrzypek, J.J. (1993) *Plasticity and Creep: Theory, Examples, and Problems.* CRC Press, Boca Raton.

208. Slaughter, W.S. and Fleck, N.A. (1993) Viscoelastic microbuckling of fiber composites. *Trans. ASME. J. Appl. Mech.* **60**, 802-806.

209. Smart, J. and Williams, J.G. (1972) A comparison of single-integral non-linear viscoelasticity theories. *J. Mech. Phys. Solids.* **20**, 313-324.

210. Smith, T.L. (1973) Physical properties of polymers – an introductory discussion. *Polym. Eng. Sci.* **13**, 161-175.

211. Sneddon, J.N. and Berry, D.S. (1958) The classical theory of elasticity. *Handbuch der Physik.* **6**, Springer-Verlag, Berlin.

212. Souchet, R. (1993) Concerning the polar decomposition of the deformation gradient. *Int. J. Engng. Sci.* **31**, 1499-1506.

213. Soussou, J.E., Moavenzadeh, F., and Gradowszyk, M.H. (1970) Application of Prony series to linear viscoelasticity. *Trans. Soc. Rheol.* **14**, 573-584.

214. Spencer, A.J.M. (1980) *Continuum Mechanics.* Longman, London.

215. Spriggs, T.W., Huppler, J.D., and Bird, R.B. (1966) An experimental appraisal of viscoelastic models. *Trans. Soc. Rheol.* **10**, 191-213.

216. Stafford, R.O. (1969) On mathematical forms for the material functions in nonlinear viscoelasticity. *J. Mech. Phys. Solids.* **17**, 339-358.

217. Stephenson, R.A. (1980) On the uniqueness of the square-root of a symmetric, positive-definite tensor. *J. Elasticity.* **10**, 213-214.

218. Struik, L.C.E. (1978) *Physical Aging of Amorphous Polymers and Other Materials.* Elsevier, Amsterdam.

219. Struik, L.C.E. (1980) Physical aging of margarine. *Rheol. Acta.* **19**, 111-115.

220. Struik, L.C.E. (1987) The mechanical and physical ageing of semicrystalline polymers: 1. *Polymer.* **28**, 1521-1533. 2. *Polymer.* **28**, 1534-1542.

221. Szczepinski, W. (ed.) (1990) *Experimental Methods in Mechanics of Solids.* Elsevier, Amsterdam.

222. Tanner, R.I. (1968) Comparative studies of some simple viscoelastic theories. *Trans. Soc. Rheol.* **12**, 155-182.

223. Tissakht, M. and Ahmed, A.M. (1995) Tensile stress-strain characteristics of the human meniscal material. *J. Biomech.* **28**, 411-422.

224. Titomanlio, G., Spadaro, G., and La Mantia, F.P. (1980) Stress relaxation of a polyisobutilene under large strains. *Rheol. Acta.* **19**, 477-481.

225. Treloar, L.R.G. (1958) *The Physics of Rubber Elasticity.* Oxford Univ. Press, Oxford.

226. Truesdell. C. and Noll, W. (1965) The nonlinear field theories of mechanics. *Handbuch der Physik.* **3/3**, Springer-Verlag, Berlin.

227. Truesdell, C. (1975) *A First Course in Rational Continuum Mechanics.* Academic Press, New-York.

228. Truesdell, C. (1978) Some challenges offered to analysis by rational thermomechanics. In *Contemporary Developments in Continuum Mechanics and Partial Differential Equations.* eds. de La Penha, G.M. and Medeiros, L.A.J. North-Holland, Amsterdam, 495-603.

229. Tsai, C.T. and Dillon, O.W. (1987) Thermal viscoplastic buckling during the growth of silicon ribbon. *Int. J. Solids Struct.* **23**, 387-402.

230. Tschoegl, N.W. (1989) *The Phenomenological Theory of Linear Viscoelastic Behavior. An Introduction.* Springer-Verlag, Berlin.

231. Tsien, H.S. (1950) A generalization of Alfrey's theorem for viscoelastic media. *Quart. Appl. Math.* **8**, 104-107.

232. Valanis, K.C. and Landel, R.F. (1967) The strain energy function of a hyperelastic material in terms of the extension ratios. *J. Appl. Phys.* **38**, 2997-3002.

233. Valent, T. (1988) *Boundary Value Problems of Finite Elasticity.* Springer-Verlag, Berlin.

234. van Dam, J.C. and Mischgofsky, F.H. (1982) Rheology and stircasting of

(organic) alloys. *Rheol. Acta.* **21**, 445-448.

235. Varga, O.H. (1966) *Stress-Strain Behavior of Elastic Materials.* Wiley, New-York.

236. Vinogradov, G.V., Isayev, A.I., Zolotarev, V.A., and Verebskaya, E.A. (1977) Rheological properties of road bitumens. *Rheol. Acta.* **16**, 266-281.

237. Vinogradov, G.V. and Malkin, A.Ya. (1980) *Rheology of Polymers. Viscoelasticity and Flow of Polymers.* Mir, Moscow.

238. Vorp, D.A., Rajagopal, K.R., Smolinski, P.J., and Borovetz, H.S. (1995) Identification of elastic properties of homogeneous, orthotropic vascular segments in distension. *J. Biomech.* **28**, 501-512.

239. Wagner, M.H. (1976) Analysis of time-dependent non-linear stress-growth data for shear and elongation flow of a low-density branched polyethylene melt. *Rheol. Acta.* **15**, 136-142.

240. Wagner, M.H. (1977) Prediction of primary normal stress difference from shear viscosity data using a single integral constitutive equation. *Rheol. Acta.* **16**, 43-50.

241. Wagner, M.H., Raible, T., and Meissner, J. (1979) Tensile stress overshoot in uniaxial extension of a LDPE melt. *Rheol. Acta.* **18**, 427-428.

242. Ward, A.F.H. and Jenkis, G.M. (1958) Normal thrust in dynamic torsion for rubberlike materials. *Rheol. Acta.* **1**, 110-114.

243. Ward, I.M. and Wolfe, J.M. (1966) The non-linear mechanical behaviour of polypropylene fibers under complex loading programmes. *J. Mech. Phys. Solids.* **14**, 131-140.

244. Ward, I.M. (1971) *Mechanical Properties of Solid Polymers.* Wiley–Interscience, London.

245. White, J.L. and Metzner, A.B. (1963) Development of constitutive equations for polymeric melts and solutions. *J. Appl. Polym. Sci.* **7**, 1867-1889.

246. Whitham, G.B. (1974) *Linear and Nonlinear Waves.* Wiley-Interscience, New-York.

247. Wineman, A. and Gandhi, M. (1984) On local and global universal relations in elasticity. *J. Elasticity.* **14**, 97-102.

248. Wing, G., Rasricha, A., Tuttle, M., and Kumar, V. (1995) Time dependent response of polycarbonate and microcellular polycarbonate. *Polym. Eng. Sci.* **35**, 673-679.

249. Winter, H.H. (1978) On network models of molten polymers: loss of junctions due to stretching of material planes. *Rheol. Acta.* **17**, 589-594.

250. Yamamoto, M. (1956) The visco-elastic properties of network structure. 1. General formalism. *J. Phys. Soc. Japan.* **11**, 413-421.

251. Yamamoto, M. (1971) An integral constitutive equation and rate-dependent relaxation spectrum. *Trans. Soc. Rheol.* **15**, 783-788.

252. Zallen, R. (1983) *Physics of Amorphous Solids.* Wiley, New-York.

253. Zapas, L.J. and Craft, T. (1965) Correlation of large longitudinal deformations with different strain histories. *J. Research Nat. Bur. Stand.* **A69**, 541-546.

254. Zdunek, A.B. (1992) Determination of material response functions for prestrained rubbers. *Rheol. Acta.* **31**, 575-591.

255. Zhaoxia, L. (1994) Effective creep Poisson's ratio for damaged concrete. *Int. J. Fracture.* **66**, 189-196.

256. Zhong-Heng, G. and Solecki, R. (1963) Free and forced finite amplitude oscillations of an elastic thick-walled hollow sphere made of an incompressible material. *Arch. Mech. Stosow.* **15**, 427-433.

Index